Virtual Existentialism

Stefano Gualeni • Daniel Vella

Virtual Existentialism

Meaning and Subjectivity in Virtual Worlds

palgrave
macmillan

Stefano Gualeni
Institute of Digital Games
University of Malta
Msida, Malta

Daniel Vella
Institute of Digital Games
University of Malta
Msida, Malta

ISBN 978-3-030-38477-7 ISBN 978-3-030-38478-4 (eBook)
https://doi.org/10.1007/978-3-030-38478-4

This Palgrave Pivot imprint is published by the registered company Springer Nature Switzerland AG.
The registered company address is: Gewerbestrasse 11, 6330 Cham, Switzerland

Foreword: The Computer Gameworld as an Existential Commitment

"I don't like it here. I don't know what's going on. We're both stumbling around together in this unformed world whose rules and objectives are largely unknown, seemingly undecipherable or even possibly nonexistent, always on the verge of being killed by forces we don't understand!"—so laments the character Ted Pikul in David Cronenberg's 1999 thriller *eXistenZ*, in a scene where the events in the hyperrealistic virtual reality game called *eXistenZ* have turned sour and Pikul is contemplating a way out. The movie *eXistenZ* is all about the blurring of the border between the game and reality. This blurring is greatly aided by conventions of cinema: thanks to the ambiguity afforded by the structure of the narrative, and the camera's third-person perspective onto beings in the world, the viewer shares Pikul's anxiety as they are both sometimes unable to tell with any degree of certainty whether events in a given scene are happening inside the actual world or within the gameworld of *eXistenZ.*

The borders between gameworlds and the actual world are also blurred in real life. Those seeking to escape the burden of everyday life can quickly find themselves under the weight of existence also in a computer game, even without the use of game interfaces grown from fertilized amphibian eggs stuffed with synthetic DNA like those conjured up by Cronenberg. Gameworlds can be peculiar places. In them, we find aspects we are familiar with due to actual world experiences like embodiment anchoring us in a location, gravity giving weight to things, and time passing in a linear fashion. We are, however, equally likely to find aspects that are strange and alien to our actual-worldly being, like omnipresent existence and multi-directional time. In this light, it would perhaps be unwise to expect to develop subjectivities in gameworlds in ways that are similar to how we develop them in the actual world.

In the past two decades of academic computer game studies, talk on gameworlds has often referred to the gameworld as an environment or space, a container for an activity. From this perspective, questions pertaining to both the form of the space and its appearance are relevant: how large is the gameworld, how is it structured, and what does it look like? Perhaps due to the affinity of first-wave ludology to game design research, this notion of gameworld is very similar to game designers' pragmatic and utilitarian considerations of gameworlds as arenas consisting of 3D models, scripts, and textures, designed to facilitate the activity of gameplay. If a story of any kind plays a central role in giving meaning to gameplay activities in the particular game, the gameworld in question may be considered a fictional one. Here, the scholar might ask: how do aspects of the game's fictional world correspond with gameplay activities? How do these elements support each other? The primary material for those interested in the gameworld as a gameplay arena or as a fictional universe is the game itself, for example as a software package or a 'text' of some kind.

If the game under scrutiny is a multiplayer online game, the notion acquires a new flavour: the gameworld becomes a site of social activity, and a host of topics and questions becomes relevant. Culture in the gameworld emerges, warranting not only anthropological but also ethical, political, financial, and legal interest. What do people do in the gameworld and why? How are their actions in the gameworld related to what they do outside the game? How do they interact? What kind of ethical relations with each other they enter into? What kind of ideology prevails in the gameworld? Furthermore, if events in the gameworld involve financial interests, economics and legal scholars also express interest, curious about how money moves in this gameworld, and how its movements are regulated. For this perspective, the primary material for analysis is to be found beyond the confines of the software package and its runtime behaviour: the gameworld as a social world is larger than the game itself.

Common to the two perspectives mentioned above is that they consider the gameworld as either a technological-material and/or social 'thing', something 'out there' waiting to be studied. This might appear somewhat precarious to the phenomenologically inclined: to study a world without studying the *being* in the world. Like the psychologist in Jean-Paul Sartre's *Sketch for a Theory of Emotions* (1962), who was forbidden to consider his patients as humans like himself, these perspectives can only indirectly address questions pertaining to subjectivity.

In his account of fiction and embodiment in single-player computer games, Rune Klevjer (2006, 110–111) hints at the possibility for a third kind of perspective on computer gameworlds, a perspective which could perhaps lead us closer to subjectivity: "[s]imulated environments of computer games are 'worlds' not only because they can trigger our imaginations, or because they constitute a rule-based and self-contained 'magic circle' of meaningful activity, or because they may be sensorially immersive, but also, and more importantly, because they are world-like in terms of our mode of interacting with them".

The textually oriented games scholar found the game world in the software package and its run-time behaviour. The anthropologist's gameworld was located in the assemblage formed by the game and its players. Through a phenomenological lens, we can locate the gameworld in the player's experience, where it appears as an 'existential commitment' (Leino 2013, 9). The relevant questions here are: what kind of commitment is this? What is it like to be in a gameworld? Simultaneously facilitating and resisting the player's projects, the game makes its player responsible for the freedom she enjoys as a player. The player faces a risk of failure and a consequent expulsion from the world, and should she want to remain a player, she must be willing to deal with the game's resistance. The player is being subjected to a 'gameplay condition' (Leino 2009).

Similarly, Sebastian Möring (2016) argues that "games are essentially existential structures whose gameplay requires care to ensure its ongoing existence". Our responsibility for our freedom is what keeps in-game interactions meaningful and interesting: without the fear of losing this freedom, our experience of gameworld would quickly oscillate towards boredom (Möring 2014). This dynamic of freedom and responsibility is what turns the game environment, the container, the social realm, into a gameworld.

The nature of computer games as existential structures is what Hubert Dreyfus (2009) missed in his existential critique of the virtual world *Second Life*. Reviewing *Second Life* in terms of its potential for self-discovery and identity exploration, Dreyfus (2009, 102) suggests that "[…] while the safe experimentation of *Second Life* is easy and can give you superficial satisfactions as in a synthetic Mardi Gras, only a bold experiment with the real possibility of having to deal with the consequences of failure could help you discover what is really possible and worthwhile for you".

Commenting on Edward Castronova's (2005, 276) visions of exodus to synthetic worlds—a topic also lucidly addressed by Stefano Gualeni and Daniel Vella in this book—Dreyfus (2009, 93) suggests that *Second Life* does not afford a "sense of wonder and the sacred", as it cannot give us the experience of "being in the grip of mysterious powers that have authority over you. That sort of power is expressed in the traditional myths but is necessarily lacking in the programmed gods and goblins we willfully invent and can completely command and understand" (ibid.). While Dreyfus' observations serve to highlight the specificity of computer gameworlds as a subset of virtual worlds, the assumption that virtual beings suffer from an inherent 'essential poverty' (Sartre 2010, 9) seems off the mark, given that users and players can indeed be surprised by what happens in these worlds (Leino 2010, 229–231), as Gualeni and Vella also note in this book. Most importantly, Dreyfus' failure to find existential significance in *Second Life* by comparing it to the first life is also testimony to Gualeni and Vella's very insightful observation in this book, that the value of virtual worlds to self-discovery and experimentation is not to be found by simply holding up virtual worlds as alternatives to actual world, but rather considering them as new existential domains which are intertwined and overlapping with (and depend on) our being in the actual world.

While I have so far focused explicitly on computer gameworlds, Gualeni and Vella do not restrict the argument of this book to computer gameworlds only. Instead, they talk about virtual worlds and posit computer gameworlds as a subset of virtual worlds. Initially this might seem like a throwback to the time when computer games were not considered worthy of academic study, and all kinds of euphemisms from 'multimedia entertainment' to 'virtual worlds' had to be invented. However, a closer look reveals that the authors' chosen perspective is timely and relevant, since both 'gameness' and 'virtuality' are undergoing transformations. The paradigmatic ideas of games and play as the core ingredients in the software packages we are accustomed to calling 'computer games' need rethinking. For example, a genre of open-ended art games has emerged, consisting of games which present aesthetically interesting worlds for players to explore but lack any gameplay condition (Leino 2009, 2010, 2012) whatsoever. Meanwhile, commercialized forms of computer gaming, known as 'e-sports', risk turning play from autotelic to paratelic (Apter 1992) and thus likening it to work and other 'serious' activities. We have also seen game elements, under the guise of 'gamification', entering places that were previously off-limits. Virtuality, in the sense of both world and reality, has become mundane. Since computer games have gone mainstream, people of all ages visit gameworlds on their smartphones regularly while making no fuss about it. Shopping malls have arcades with virtual reality technology, making use of content and applications that blend both ludic and non-ludic elements and place their users in both active and passive positions. Similar equipment is available for home use, allowing virtual worlds to take their place in one's actual living room. In these circumstances, paying fundamentalist exclusive allegiance to game studies, philosophy of computer games, or existential ludology would have done no good to the framing of the authors' project. It is quite sensible that Gualeni and Vella chose to not restrict the scope of the argument and instead straddled the traditions of philosophy of technology, virtual worlds research, computer game studies, and philosophy of computer games.

Virtual Existentialism: Meaning and Subjectivity in Virtual Worlds manifests intellectual curiosity and dynamism, as the authors engage, in addition to existential thought, with a wide range of perspectives, including those of dramatherapy, religious studies, and live-action role-playing (LARP) studies, and draw also from thinkers who might not at the first glance come across as existentialists, like Helmuth Plessner and Peter W. Zapffe. *Virtual Existentialism* is faithful to the technological specificity and mediality of computer games and virtual worlds and carries no unnecessary theoretical or conceptual baggage.

Game studies will benefit from the account of being-in-the-virtual-world carefully built by Gualeni and Vella in *Virtual Existentialism*, not the least because it allows us to get over the simplifying assumptions of the gameworld as a container and steers our attention to the intentional unity between the playing subject and the gameworld. Those involved in philosophy of computer games will find the vocabulary newly established by Gualeni and Vella, such as 'virtual subjectivity', 'attitude of virtuality', and 'virtual project' useful directly out of the box. For game scholars, there are also plenty of new connections that shed light on questions of play and virtuality. For example, while Eugen Fink is often mentioned when theorizing play, we have seldom seen his work being applied onto virtual worlds and gameworlds. Likewise, while the discussion on en-roling and de-roling may be familiar to those in LARP studies, showing how these notions work if applied to virtual worlds is fresh to many readers. Those in philosophy of technology will find that *Virtual Existentialism* provides an accurate take on contemporary debates on worldness, subjectivity, and experience in philosophy of computer games and shows how the significance of these debates resonates well beyond computer games.

Virtual Existentialism is a timely and highly original account of the positive, negative, and ambiguous existential effects of virtual world experiences and the built-in potential for dealing with virtual worlds that can be leveraged on the grounding of our actual existence.

School of Creative Media
City University of Hong Kong,
Kowloon, Hong Kong

Olli Tapio Leino

References

Apter, M. J. "A Structural-Phenomenology of Play". In M. Apter & J. Kerr (eds.): *Adult Play. A Reversal Theory Approach.* Amsterdam (The Netherlands): Swets & Zeitlinger B. V., 16–18.

Castronova, E. 2005. *Synthetic Worlds. The Business and Culture of Online Games.* Chicago (IL): The University of Chicago Press.

Dreyfus, H. 2009. *On the Internet. (2nd Edition).* New York (NY): Routledge.

Klevjer, R. 2006. *What is the Avatar? Fiction and Embodiment in Avatar-Based Singleplayer Computer Games.* Doctoral dissertation, University of Bergen (Norway).

Leino, O. T. 2009. "On the Logic of Emotions in Play." Proceedings of ISAGA 2009 Conference. Singapore: Society of Simulation and Gaming of Singapore.

Leino, O. T. 2010. Emotions in Play – On the constitution of emotion in solitary computer game play. Doctoral dissertation, IT University of Copenhagen (Denmark).

Leino, O. T. 2012. "Untangling Gameplay: An Account of Experience, Activity and Materiality Within Computer Game Play." In J.R. Sageng, H.J. Fossheim and T. Mandt Larsen (eds.), *The Philosophy of Computer Games.* Dordrecht (The Netherlands): Springer, 57–75.

Leino, O. T. 2013. "From Game Spaces to Playable Worlds". Proceedings of Philosophy of Computer Games 2013 Conference in Bergen (Norway).

Möring, S. 2014. "Freedom in Games – Between Fear and Boredom". Proceedings of Philosophy of Computer Games 2014 Conference in Istanbul (Turkey).

Möring, S. 2016. "The Care-Structure in Computer Games and Computer Game Interpretations". Proceedings of GamesLit – International Conference in Games and Literary Theory in Krakow (Poland).

Sartre, J. P. 1962 [1939]. *Sketch for a Theory of the Emotions.* London (UK): Methuen & Co.

Sartre, J. P. 2010 [1940]. *The Imaginary.* J. Webber (trans.). London (UK): Routledge.

Contents

List of Figures

Introduction and Glossary of Terms

How do experiences in (and of) virtual environments affect the ways in which individual human beings understand and attribute meaning to their own existence? In this book, we adopt a variety of scholarly perspectives in the combined attempt to understand and answer that question.

For the most part, as the title of this volume indicates, our efforts are inspired by the philosophical current of existentialism, a predominantly European twentieth-century approach to the human individual's being, acting, feeling, and pursuit of meaning in the world. Themes such as those of freedom, care, projectuality, duty, and finitude are recurrent in various declinations of existentialism. Generally speaking, and differently from many philosophical currents that preceded it, existentialism does not understand the human subject as a centre of experience that can be conceptually separated from the world (or worlds) that they inhabit. Instead, it approaches the human condition as one that is constituted and constantly shaped by the worlds which human beings find themselves 'thrown into'. Importantly, from an existentialist perspective, our existence as beings-in-the-world precedes our attempts to imbue that existence with meaning.

This book's drive is twofold: It uses existential philosophy as a frame through which to understand and interpret the significance of virtual environments in the context of our existence. At the same time, it considers how our capacity to be in (and towards) these technologically mediated domains might lead to new understandings of the concerns of existential philosophy.

Given this book's investment in technologically mediated experiences, we feel it is necessary to say a few words about the meaning we attribute to technology. We understand 'technology' as a conglomeration of artefacts and forms of knowledge that are employed in the pursuit of something. We find this purpose-oriented interpretation of our relationships with technologies (both as creators and as users) to be useful as a working definition and as a quick reference. In its brevity, however, it is reductive to the point of being misleading: the glossary definition we provide at the end of this introduction does not account, among other aspects, for how technologies are also always involved in wider and less predictable transformative processes. These unanticipated effects foster changes in human beings both at the individual level and as a society that are less obvious and usually slower than those for which that particular technology was originally intended. We could, for example, effortlessly grasp how mobile phones would facilitate communication among people. Before their ubiquitous presence in our daily routine, however, it was arguably less intuitive to envisage how using mobile phones would not only, for example, change the ways in which we interact with our friends, but also our very understanding of what friendship is. A more nuanced account of our relationships with technologies is captured in our glossary definition of 'multistability' as a defining quality of every technological artefacts.

The ambiguous, transformative, and often unforeseen long-term effects of technologies are not recent discoveries: they have been recorded and theorized by philosophers and cultural historians for centuries. Discussions over how our technical extensions affect the way people think and act are, in fact, as old as (written) philosophy. In his *Phaedrus* and *The Seventh Letter*, Plato famously expresses his concerns about how writing (a freshly introduced alpha-technology) would irrevocably transform ancient Greek society (Plato 1995; Levison et al. 1968).

In the present day and age, the ways in which various technologies contribute to reframing and altering social values and processes remain focal interests of academic fields such as the philosophy of technology and science and technology studies (see Verbeek 2005; Feenberg 2005; Bijker et al. 2012). Accordingly, the majority of existing academic work on virtual worlds concentrates its attention on how those technologies in particular affect us cognitively and psychologically, and on ways to harness those effects for purposes that are considered beneficial and socially acceptable (such as education, entertainment, and well-being). This focus on the transformative effects of experiences of virtual worlds at the socio-cultural scale typically ignores what those experiences can mean for the individual. The personal, existential significance of traversing, manipulating, and even creating virtual environments has been, to this day, largely ignored by academia. There are, of course, remarkable exceptions to this claim, for instance in the scholarly fields of cyberpsychology and existential ludology, both of which will be discussed in this book.

By drawing on the philosophical traditions of existentialism and—to a lesser extent—phenomenology, our book aspires to address the academic blind-spot outlined in the previous paragraph. In this pursuit, *Virtual Existentialism* articulates several perspectives from which virtual worlds can be understood as existentially (and even evolutionarily) relevant. To borrow the words of Dreyfus and Spinosa, we aim to show that "by means of our equipment and coordinated practices, we human beings open coherent, distinct contexts or worlds in which we perceive, feel, act, and think" (Dreyfus and Spinosa 2003, 339). More specifically, we claim that, in virtual worlds, human beings can reflect on their values and beliefs, take on new subjectivities, explore previously unexperienced ways of being, and take reflective stances towards their existence and their subjectivity in the actual world.

Given our focus on existentialism, it is worth pointing out two caveats regarding the fundamental assumptions of this philosophical outlook. Firstly, it is important to observe that, in its traditional forms, existential philosophy rests upon a unitary idea of the individual human subject. Secondly, the existential perspective is one that foregrounds the concerns of the individual consciousness, relegating social matters to being, at best, inconvenient distractions and, at worst, active threats to the freedom of one's private interiority. To give an example, writing about Jean-Paul Sartre's *Being and Nothingness*, one of the major works of the existentialist canon, Iris Murdoch describes it as articulating "the psychology of the lonely individual", presenting an "egocentric and nonsocial world" in which the highest virtue is that of "a life devoted to purely private ends" (1953, 51–52). Needless to say, such a perspective not only sidelines the crucial, communal aspects of human life, but also entrenches an evident privilege: the freedom of existential self-determination is only available to those with the right material and socio-cultural conditions. For the purposes of this book, it is pertinent to note that this privilege is reflected in the conditions of access to virtual world experience. Virtual worlds and virtual subjectivities are only available to the small percentage of the global population who have access to the necessary hardware, have acquired the required technological literacy to operate such hardware, and have the leisure time to engage in such experiences.

With regard to the first caveat, by focusing on how virtual world experience complicates the idea of the subject, *Virtual Existentialism* offers a fresh perspective on a number of long-standing questions in the field. Concerning the second observation, instead, we want to emphasize that the individual, existential dimension of our relationship with virtual worlds is not meant in an exclusive sense. We are not arguing, in other words, that experiences and transformations that affect individual human beings can be separated from (or understood apart from) their participation in society. Nor can these experiences be understood outside the context of those individuals' belonging to the human species. In that respect, we want to direct the attention of the reader onto the biological and anthropological ways to approach human existence that are discussed, in particular, in Chaps. 3 and 4.

In pursuing an existential understanding of virtual worlds, we initially introduce and elaborate upon the idea of 'virtual subjectivity' in Chap. 1, and then by examining the mechanisms by which the user is able to transition into and out of a virtual subjectivity in Chap. 2. Chapters 3, 4, 5, and 6 extend and apply the foundational existential notions presented in earlier chapters by building upon the work of four philosophers: respectively, Helmuth Plessner, Peter W. Zapffe, Jean-Paul Sartre, and Eugen Fink. A component of the work of each thinker is repurposed, in their respective chapters, as conceptual tools to reveal how virtual worlds are already understood and employed in ways that are constitutive to how we understand our existence and attribute meaning to it. Our appropriation of the notions and frameworks of those four thinkers also casts light on virtual environments as technologies that can be useful in reframing philosophical questions concerning subjectivity, self-determination, and finitude.

Next to the philosophical tradition of existentialism and phenomenology, the academic field of digital game studies constitutes an important source of material and inspiration for this book. Given the kinship with game studies, and the fact that digital games are frequently invoked as examples in this book, it might be important to clarify in this introduction that we do not intend to portray gameworlds as exceptional instances of virtual worlds. We do not consider them as being qualitatively different from experiences that emerge from engaging with other computer applications that give rise to world-like experiences, such as training simulations or interactive reconstructions of archaeological sites. We believe, however, that the variety and the social diffusion of digital games make them particularly accessible for our readers and particularly convenient (for us and, we hope, for our readers) to analyse and reference.[1]

What follows is the customary list of chapters together with a summary of the respective contents:

Chapter 1: Virtual Subjectivities and the Existential Significance of Virtual Worlds
This chapter introduces the notion of 'virtual subjectivity', which is defined as the subjective sense of 'self' that relates to one's being-in-the-virtual-world. In particular, we propose an understanding of virtual subjectivity as standing in a nested relation to the individual's subjectivity in the actual world and argue that it is this relation that allows virtual world experience to gain existential significance. The foundational notions of virtual subjectivity and virtual projectuality (the kind of subject that the individual aspires to be) pave the way for understanding the transformative, self-transformative, and therapeutic possibilities disclosed by virtual worlds in the chapters that follow.

Chapter 2: En-roling and De-roling in Virtual Worlds
In light of the idea of virtual subjectivity discussed in the previous chapter, Chap. 4 focuses on two aspects of our being in virtual environments: 'en-roling' and 'de-roling'. The first refers to those practices, experiences, and activities by which the virtual world user is led to form an understanding of—and to internalize—a particular subjectivity in relation to the virtual world. The second indicates techniques through which subjects 'disrobe' themselves of their role in the virtual environment. This chapter draws comparisons between engaging with virtual worlds and a range of other cultural practices—including dramatherapy and psychodrama, liminal rites, and live-action role-playing—and the mechanisms of transition between subjectivities used in these practices.

Chapter 3: Helmuth Plessner and Virtual Worlds as Existential Complements
Chapter 3 introduces the second part of this book—four chapters, each of which engage with the thought of a different philosophical figure, applying their ideas to our considerations of the virtual. This chapter frames human existence as inherently lacking a reliable grounding on which to anchor values and meaning. Extending the work of Helmuth Plessner, we

focus on technologies in general—and virtual worlds in particular—as existential tools through which human beings strive to come to terms with the uncertainty and incompleteness that characterize their existence.

Chapter 4: Peter W. Zapffe and the Virtual Tragic
Norwegian philosopher and mountaineer Peter W. Zapffe understands human life as inherently meaningless. In relation to that meaninglessness, he identifies four ways in which human beings typically protect themselves from the existential panic that accompanies the awareness of that meaninglessness—isolation, anchoring, distraction, and sublimation. In this chapter, we use Zapffe's four categories to examine virtual environments as technologies for repressing existential panic.

Chapter 5: Jean-Paul Sartre and Escaping from Being-in-the-World
This chapter approaches our experience of virtual worlds through Jean-Paul Sartre's phenomenology of the imagination. It looks at Sartre's argument that the existential significance of the imagination lies in its giving us the capacity to surpass our being-in-the-world. On this basis, the chapter draws a line between virtual subjectivity and how Sartre's frames the concept of imaginary consciousness, considering how our relationships with virtual environments both reflect the phenomenological structure Sartre attributes to the imagination and puts it into question.

Chapter 6: Eugen Fink and Existential Play
The particular way in which German phenomenologist Eugen Fink understands 'play' is deeply involved with fundamental existential concerns such as the question of freedom. Relying on the existential significance Fink grants to play—specifically, the capacity play grants us to explore unactualized dimensions of our being—this chapter offers a perspective on the relationships between one's actual self and one's roles in virtual environments.

Chapter 7: Conclusions
This concluding chapter articulates a summary of the central themes and the main claims of the book. While doing so, it invites readers to focus their attention on the 'fluidifying sway' of virtual environments, that is to say on the specific ways how virtual world experiences reveal—and grant interactive access to—multiple and possible courses of action that are

latent in any given situation. By experientially disclosing several alternatives to a certain *status quo* through repeated interactions, do-overs, and time reversals, virtual worlds inherently encourage their users to embrace 'what could be' over 'what is'. This chapter approaches our relationships with virtual environments in ways that are not solely concerned with their socially desirable uses and effects, but also identifies aspects of those experiences that constitute a menace to our well-being and survival.

As a consequence of having largely grounded our insights in existential philosophy, many of the technical terms that are used in this book are either directly borrowed from existentialism or can be understood as extensions and elaborations of notions originally developed within that philosophical tradition. This might pose some initial difficulties for readers who are not trained in philosophy. In light of this potential hurdle, we have decided to provide (below) a quick and accessible glossary of key terms that will frequently recur in *Virtual Existentialism*.

In the glossary, the meanings of words such as 'existence', 'freedom', 'situation', and 'subjectivity' are outlined in ways that are either directly derived from, or closely inspired by, existential literature. Existentialism is, however, not the only lens through which we decided to observe the effects of our interactive experiences with (and within) virtual environments. Some of the technical terms used in the book are, instead, understood in accordance with philosophical currents like phenomenology and post-phenomenology (as is the case for words like 'multistability' or 'world'), while others will testify a closer affinity with perspectives offered in the scholarly field of virtual worlds research (which is particularly evident in the case of terms like 'virtuality' or 'actuality') and digital game studies.

This introduction would not be complete if we did not acknowledge some very special individuals for their contribution in this publication. These people assisted us in a number of ways that exceed simple academic criticism, text editing, logistical assistance, or emotional support. Our gratitude goes to Olli Tapio Leino, Johnathan Harrington, and Jonne Arjoranta.

For having crafted beautiful and meaningful digital gameworlds, and having allowed us to use pictures of those worlds in this book, we want to thank Pippin Barr, Paolo Pedercini (La Molleindustria), Subset Games, and CD Projekt RED.

We want to thank the Institute of Digital Games at the University of Malta for generously providing us with the resources and the time to pursue these kinds of humanistic projects in relation to digital media and game culture. We are also grateful to Lucy Batrouney, Ivan Callus, Stefano Caselli, Jos De Mul, Riccardo Fassone, Brendan Keogh, Melinda Sue Mangion, Benjamin Nicoll, Jasper Schellekens, Daniel O'Shiel, and Evan Torner for their help in making this book possible. Last but certainly not least, we want to thank Justyna Janik and Nele van de Mosselaer for sharing this load and this adventure with us.

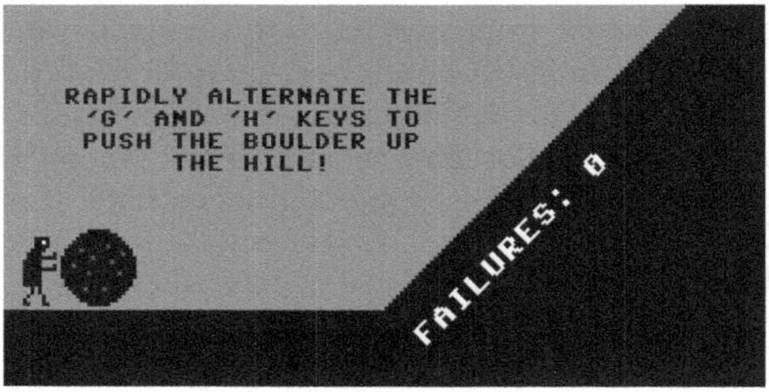

A screenshot of Pippin Barr's 2011 digital game *Let's Play: Ancient Greek Punishment*

Glossary of Terms

Actuality: The adjective 'actual' is used to indicate that some things or events are currently the case in the world that we natively inhabit as biological creatures. The indexical specification 'actual' (e.g. in the case of the 'actual world' or one's 'actual body') clarifies that we are not referring to worlds or bodies that exist virtually (see the definition for 'virtuality').

Existence: We adopt the idea—articulated in Martin Heidegger's *Being and Time* (1962)—that to exist is to be-in-a-world as one concrete being among other beings in (and for) that particular world. This definition holds regardless of whether we are referring to the actual world or to a virtual world. In *Existentialism is a Humanism*, Jean-Paul Sartre contends that existence precedes essence (2007): the simple fact of being-in-the-world is the foundational and primary condition of human being. It is only through finding ourselves in the world, and becoming conscious of our own being-in-the-world, that we can attribute (or fail to attribute) meaning to it.

Existential meaning: With 'existential meaning' we indicate the (inevitably subjective) significance attributed to the set of things, activities, people, and values in a world that a subject cares about and that contribute to make one's existence in that world preferable to non-existence.

Freedom: In the context of this book, we do not understand 'freedom' as state of absence of (or liberation from) limitations and interdictions, as was the case of the modern interpretation of the notion of 'autonomy'. Instead, we use it to indicate the possibilities for self-fashioning and self-determination that emerge in relation to the limitations and interdictions that characterize being in a certain world.

Game: In presenting what we understand by 'game', we follow Ludwig Wittgenstein's argument according to which any comprehensive definition of 'game' is bound to fail at capturing the heterogeneity of the family of practices to which the term has been applied (2009, 37, see also Calleja and Aarseth 2015). In the context of this book, we use the word 'game' to indicate a software package that discloses a virtual environment intended to be engaged with primarily for entertainment purposes. As we are concentrating our attention on virtual environments, the word 'game' functions in this book as a synonym for commonly accepted terms such as 'videogame', 'digital game', or 'computer game'.

Multistability: This term indicates the inherent possibility of every technology to be repurposed and used in unanticipated ways. The quality of being multistable is what makes it possible for a technology to acquire new meanings, functions, and effects within a social context. Our interactions with technologies are thus recognized as not solely determined by the original developers of a technology, but in an ambiguous and dynamic relation with their users. The term 'multistability' was first introduced by Don Ihde in his 1990 book *Technology and the Lifeworld: From Garden to Earth*.

Role: A 'role' is here understood as a specific and recognizable configuration of an individual's being-in-the-world. A role is something we 'take on', and, as such, there is always a distance between the individual's self-perception (what they see as their actual self) and the role. At the same time, since one is oneself when playing the role, the activity of adopting and exploring a particular mode of being-in-the-world that pertains to a particular role might have resonances with, or prove transformative of, one's self-conception. As Robert J. Landy writes, taking on a role involves "being oneself and not oneself at the same time" (1993, 48).

Situation: The term 'situation' is used in the way it is presented in Jean-Paul Sartre's *Being and Nothingness* (1966). It refers to the total, contingent, and undeniable set of facts one finds oneself 'thrown into'—one's socio-technical environment, cultural background, ethnicity, and so on. A situation is equally the product of the facts of the world and of the individual subject's freedom to make of those facts what they will (ibid., 627). Put differently, a situation is a world, understood as the ground for a subject's existence.

Subjectivity: A subjectivity is a consciousness that occupies a particular standpoint towards a certain experiential domain. From its specific perspective, a subjectivity frames that domain as a world (see the definition for 'world'). In other words, a subject participates in a world and conscious of it and of their particular standpoint within it. In relation to the notion of 'situation' presented in this glossary, a subjectivity is a consciousness for whom a world—actual or virtual—is taken up as the ground for their existential project.

Technology: A technology is understood as a conglomeration of artefacts and forms of knowledge that are employed in the pursuit of something. This extremely broad definition includes the specific economic interests, societal norms, and values that influenced the development and the uses of a particular technology (see De Mul 2010, 39). It is crucial to clarify here that the comprehensive and pursuit-oriented understanding of technology we adopt does not exclude the fact that technologies also disclose new social meanings and functions in ways that are not originally intended or anticipated by their creators (see the definition for 'multistability'). In this sense, we argue that technology cannot be understood in a merely instrumental fashion, but needs to be embraced as a characteristic way of how human beings are in the world, a way that is created by humans and—in turn—is constitutive of humans.[2]

Virtuality: The notion of virtuality has been approached and interpreted in a number of different ways in several socio-technical contexts.[3] In this book, we utilize the adjective 'virtual' as referring to something that can be interactively experienced within computer-generated environments (see the definition for 'virtual environment').

Virtual environment: A virtual environment is a particular kind of digital artefact (see 'virtuality'). Virtual environments like the operating theatre of a surgical simulation or a text editor are defined by their possibility to be experientially engaged by its users as worlds (see 'world'). What we are trying to highlight, here, is that an environment is a particular kind of object, whereas a world is a particular kind of relationship. To clarify with an example: a level in a digital game is a virtual environment, and it is our interacting with it that makes it emerge in our experience as a virtual world.

Virtual world: In accordance with the definition of 'world' provided in this glossary, we understand a 'virtual world' as the (relatively) perceptually stable interactive experience that is disclosed by a computer-generated environment (see the definition for 'virtuality'). This interpretation allows us to establish a distinction between virtual world experiences and those of dreams and hallucinations (Gualeni 2015, 22). Virtual worlds can be recognized as worlds precisely because they can be accessed, experienced, and returned to at will and can be experienced in ways that are persistently intelligible in their mechanical and aesthetic aspects (ibid.).

World: In the philosophical tradition of phenomenology, the term 'world' generally indicates two interrelated things. First, a 'world' is a set composed of beings that are understood together with all their properties and mutual relationships. More specifically, a 'world' describes that set as experienced by one of the beings involved in it. To be identified as a world, those properties and mutual relationships need to be experienced in ways that are persistently perceivable and behaviourally consistent for the being in question (Gualeni 2015, 6). Relatedly, in its second meaning, a 'world' indicates the horizon (or ground) against which every object is experienced and understood.

Notes

1. It might be worth mentioning that when academically referencing digital games, we decided to align with the guidelines proposed by Gualeni, Fassone, and Linderoth in their 2019 paper 'How to Reference Digital Games'. Their recommendations are mindful of the specific and specifically unstable qualities of the medium in question and address various bibliographic tensions and lacks that characterize the ways in which digital games are currently referenced in academia.
2. As outlined, technology is understood in this book as something (material or immaterial) of which human subjects are both the creators and the products of. With this approach, however, we do not mean to claim that technology should be exclusively understood as a human affair. The idea that only human beings can create, use, or be affected by technologies would be a ridiculous one to entertain, as there are several examples of animal species capable of producing purpose-oriented artefacts, and as it is obvious that human technologies have effects that are not exclusive to the lives of human animals. What we intend to clarify with this endnote is that the anthropocentrism of *Virtual Existentialism* is not an ideological belief of ours, but a functional decision for this scholarly project. For better or worse, our book has a limited scope and is deliberately meant as a specifically humanistic kind of inquiry into human existence, human technologies, and the mutual relationships thereof.
3. The adjective 'virtual' was originally used to encapsulate the idea of 'potentiality'. More specifically, *Virtualis* is a late-medieval neologism the existence of which became necessary when Aristotle's concept of δύναμις (*dynamis*: potentiality, power) had to be translated into Latin (van Binsbergen 1997, 9). The concept of 'potentiality' lying at the etymological foundation of 'virtual' provides the background for understanding why, at least in one of its interpretations, it is used to indicate the latency of certain possibilities inherent in a specific artefact, combination of artefacts, or state of things (Gualeni 2015, 54–55). 'Virtual' as an adjective can also be attributed to things that we can consider practically actual, but not formally so. It is from that standpoint that Philip Brey highlighted the fact that among the pre-digital connotations of 'virtual' we can also list 'imaginary', 'make-believe', and 'fake'. In this sense, 'virtual' is still used as an antonym for words like 'real', 'actual', and 'physical' (Brey 2008). A widely adopted connotation of the adjective 'virtual' was presented by Pierre Lévy, for whom 'virtual' is not to be understood in opposition to 'actual' in the sense that something 'is currently the case', but to 'actual' as having the quality of existing in the world humans are native to (Lévy 1998, 14). This last understanding of 'virtual' is consistent with a digitalist approach to virtual reality such as the one proposed by David J. Chalmers (2017) and resonates with the one adopted by this book.

References

Bijker, W. E., Hughes, T. P., Pinch, T. and Douglas, D. G. 2012. *The Social Construction of Technological Systems: New Directions in the Sociology and History of Technology*. Cambridge (MA): The MIT Press.

van Binsbergen, W. M. J. 1997. *Virtuality as a Key Concept in the Study of Globalisation*. Den Haag (The Netherlands): WOTRO.

Brey, P. 2008. "Virtual Reality and Computer Simulation." In *Handbook of Information and Computer Ethics*, ed. Kenneth Einar Himma and Herman T. Tavani, 361–384. Hoboken (NJ): John Wiley & Sons.

Calleja, G. & Aarseth, E. 2015. "The Word 'Game': Ontology of an Indefinable Object." Proceedings of the 2015 Foundations of Digital Games Conference. Pacific Grove (CA), June 22–25, 2015.

Chalmers, D. J. 2017. "The Virtual and the Real." In *Disputatio*, *9* (46), 309–352.

De Mul, J. 2010 [2002]. *Cyberspace Odyssey: Towards a Virtual Ontology and Anthropology*. Cambridge (MA): Cambridge Scholars Publishing.

Dreyfus, H. L. & Spinosa, C. 2003. "Further reflections on Heidegger, technology, and the everyday." In *Bullettin of Science, Tecnology and Society*, *23* (5), 339–349.

Feenberg, A. 2005. "Critical Theory of Technology: An Overview." In *Tailoring Biotechnologies*, *1*, 47–64.

Gualeni, S. 2015. *Virtual Worlds as Philosophical Tools*. Basingstoke (UK): Palgrave Macmillan.

Gualeni, S.; Fassone, R.; Linderoth, J. 2019. "How to Reference a Digital Game." Proceedings of the 2019 DiGRA International Conference. Kyoto (Japan), August 6–10.

Heidegger, M. 1962 [1927]. *Being and Time*. Trans. John Macquarrie and Edward Robinson. San Francisco (CA): Harper & Row Publishers Inc.

Ihde, D. 1990. *Technology and the Lifeworld: From Garden to Earth*. The Indiana series in the Philosophy of Technology. Bloomington (IN): Indiana University Press.

Landy, R. J. 1993. *Persona and Performance: The Meaning of Role in Drama, Therapy, and Everyday Life*. London (UK): Jessica Kingsley Publishers.

Levison, M.; Morton, A. Q.; Winspear, A. D. 1968. "The Seventh Letter of Plato." In *Mind*, *77* (307), 309–325.

Lévy, Pierre. 1998. *Qu'est que le virtuel?* Paris (France): La Découverte.

Murdoch, I. 1953. *Sartre: Romantic Rationalist*. Cambridge (UK): Bowes & Bowes.

Plato, 1995. *Phaedrus*. Translated, with Introduction and Notes, by Nehamans, A. and Woodruff, P.). Indianapolis (IN): Hackett Publishing Company.

Sartre, J. P. 1966 [1943]. *Being and Nothingness*. Trans. Hazel E. Barnes. New York (NY): Washington Square Press.

Sartre, J. P. 2007. *Existentialism is a Humanism*. Trans. Carol Macomber. New Haven (CT): Yale University Press.

Verbeek, P. P. 2005. *What Things Do: Philosophical Reflections on Technology, Agency, and Design*. University Park (PA): Penn State University Press.

Wittgenstein, L. 2009 [1953]. *Philosophical Investigations*. Trans. G.E.M. Anscombe, P.M.S. Hacker & J. Schulte. Chichester (UK): Wiley-Blackwell.

CHAPTER 1

Virtual Subjectivities and the Existential Significance of Virtual Worlds

Abstract Drawing upon the existential philosophy of Jean-Paul Sartre and upon work on 'existential ludology' in game studies, this chapter introduces the notion of 'virtual subjectivity', to refer to the subjective sense of 'self' that relates to one's being-in-the-virtual-world. It articulates a projectual understanding of subjectivity (according to which we strive towards the kind of being we wish to be) and proposes the term 'virtual project' for the existential projects we take on in the virtual world. It proposes an understanding of virtual subjectivity as standing in a nested relation to the individual's subjectivity in the actual world, arguing that this relation allows virtual world experience to gain existential significance. This paves the way for understanding the transformative, self-transformative, and therapeutic possibilities disclosed by virtual worlds.

Keywords Subjectivity • Virtual worlds • Sartre • Existentialism • Projectuality • Game studies

Within the tradition of existentialism, the notion of the *project* is of central importance. Philosophers such as Martin Heidegger and Jean-Paul Sartre consider projectuality to be one of the defining aspects of how human beings are in the world. In their work, the idea of the project is a necessary component of how individual subjects shape and attribute meaning to their existence. In Sartre's perspective in particular, one's existence is

S. Gualeni, D. Vella, *Virtual Existentialism*,
https://doi.org/10.1007/978-3-030-38478-4_1

characterized by one's orientation towards the achievement of an overarching existential project (Sartre 1966, 717–722).

In very broad terms, an existential project could be defined as the aspiration to be in a particular way—to be a certain kind of subject. It is through the lens of one's projectual disposition that things and events encountered in a world become meaningful for the individual: they can be recognized as obstacles to the fulfilment of the project, as tools and opportunities that can be leveraged towards the achievement of the project itself or parts of it, and so on (ibid.). In this chapter, we intend to show that the notion of the project is a fruitful lens through which we can observe the way we inhabit virtual environments, and articulate an understanding of 'virtual subjectivity'.

From our standpoint, understanding the existential relevance of a projectual attitude in virtual worlds fundamentally consists in answering two key questions. The first concerns the projectual structure of the individual's being in a virtual context: *how does an existential project give shape to the individual's being in the virtual world?* This question will be tackled in Sects. 1.3 and 1.4. The second question focuses, instead, on the existential significance of virtual worlds in general. More specifically, we will explore the relation between the individual's being in virtual worlds and the existential project that shapes one's life as a whole: *in what way(s) does the practice of taking on a virtual project figure in the overarching project of an individual's being?* Further elaborations on this question and tentative answers to it will be provided in Sects. 1.4 and 1.5.

1.1 The Notion of the 'Existential Project'

In existential philosophy in general, and particularly in the work of Sartre (1966), the project is considered to be fundamental for one's being-in-the-world. For Sartre, projectuality is an organizing principle that contributes to the formation and transformation of one's subjectivity. Approaches that are common in existential philosophy postulate that, as individual human beings, our existence is initially determined by what Heidegger termed a 'thrownness' (*Geworfenheit*) into the world (2008, 174). As a subject, one is always 'thrown' into a world, in the sense that one finds oneself characterized by qualities, capabilities, and conditions that one initially has no control over. Examples of one's thrownness are one's place and date of birth, one's gender, the socio-technical contexts one finds oneself in, one's possible congenital defects, etc. In other words, we can understand one as always being 'thrown into the world' in a certain way, characterized and bounded by a contingent set of facts—a *facticity*—on the basis of which (and against which) one understands and develops one's existence. This facticity constitutes the basis for what Sartre terms our "existential situation" (1966, 127).

Sartre establishes a distinction between one's situated existence and a more essential aspect of one's selfhood (ibid., 147). One can take stock of the contingent facts of one's existence—for example, that one lives in a certain country, that one stands in a network of relations to the members of one's family, and so on. Those facts notwithstanding, according to Sartre, one is always conscious that one's existence is not fully determined by them. One's selfhood, he writes, transcends the contingent facts that are true about oneself in one's present situation. One might presently be employed as a university professor, but this does not define who one is completely: one is always free to make something else of oneself. To think otherwise (to convince oneself that one simply *is* a university professor) would be to live in what Sartre terms "bad faith": refusing or ignoring one's existential freedom, and one's possibilities for self-determining and self-transforming (ibid., 87).

Sartre's radical approach claims that one could go as far as bracketing every external, contingent fact about oneself, leaving one existentially groundless and in a state of indeterminacy. The themes of groundlessness and indeterminacy are also central to the work of another philosopher, Helmuth Plessner, whose anthropological work in relation to technology will be the focus of Chap. 3. For both Sartre and Plessner, in the face of the impossibility to find any stable grounding for one's individual existence—that is, any reliable ways to anchor oneself to a specific set of values or sense of the self—the individual develops an existential need to make something out of oneself: to become, through free self-determination, a particular kind of being. For this reason, we can understand human existence as shaped, by definition, by an overarching project that Sartre calls an "original project" (ibid., 717). He describes the existential drive discussed earlier in this paragraph as a "project of being" (ibid., 722). This project consists in how one *projects* oneself beyond the current conditions of one's existence towards the subject that one wishes to be. The term used by Heidegger to indicate this projectual disposition (*Entworfenheit*) also captures the quality of being always 'thrown ahead' of one's present situation (Heidegger 1962, 184, 185).

Of course, if a projectual structure is inherent to the human subject, then one also needs to understand the human being as *always* projectual. What that means is that the human being cannot ever complete the existential project, as one's existential setup and one's drive to self-determine and surpass oneself are better understood as an unfinishable project. Even if a particular project were to be completed, the existential situation would only constitute newly established contingent facts for one to project oneself beyond.

1.2 Subjectivities in Digital Gameworlds

As we already mentioned, according to existential philosophy by and large, in our everyday existence we find ourselves thrown into a certain existential situation, and it is in relation to that existential situation that one forms the project to be a certain kind of subject. We argue, in a similar way, that when interacting with a digital environment, the users engage with a virtual situation. This perspective implies two theoretical premises:

- the first is that digital environments are understood by the users as an existential situation, and, hence, is engaged as a 'world', and
- the second is that, in order to experience a digital environment as a world in which one can plan, act, and pursue a project, the users must be situated as subjects in relation to the artificial world in question.

If we take this perspective, then the emergence of a virtual world into one's consciousness must be recognized as being cognitively, psychologically, and existentially dependent upon the users being 'thrown ahead' or 'projected beyond' their actual existential situation. In other words, users invest themselves in subjectivities that take that virtual world as their 'situation' (Vella and Gualeni 2019).

To date, the field within which most theoretical investigation has been performed into the question of subjectivity and existentialism in digital environments is that of digital game studies. Most explicitly, Matthew Thomas Payne proposes 'existential ludology' as a method for theorizing player experience in digital gameworlds. His suggestion is that "if existential phenomenology asks about the possibilities and meaningfulness of human action in the lived world, then existential ludology [...] asks similar questions about meaningful play in the virtual world" (2008, 622). For Payne, this means developing a method of game analysis that "works by cataloguing numerous game-play experiences to forward meaningful statements about how particular games evidence recurrent and stable experiential structures" (ibid., 624).

By this understanding, existential ludology rests upon the assumption that our being in digital gameworlds functions according to experiential structures that are analogous to those by which we experience the actual world as biological creatures. In this, Payne is not alone: a great deal of work in game studies has tackled the phenomenology of being in digital gameworlds. In much of this work, one's subjectivity in the gameworld(s) has largely been understood as pivoting on the player's embodiment in the figure of the avatar (Taylor 2002; Grodal 2003; Klevjer 2006, 2012; Bayliss 2007a, b; Gregersen and Grodal 2009; Gee 2008).[1] Just as tied to the concept of embodiment is Ulf Wilhelmsson's notion of the 'game Ego', a "bodily based function that enacts a point of being within the game environment through a tactile motor/kinaesthetic link" (2008, 61). Finally, the idea that the avatar upholds a concrete standpoint for the player within the gameworld has also been formalized by Gordon Calleja with the concept of "incorporation" (2011, 169).[2] In our assessment, and as already outlined in the introduction to this book, the theoretical developments that we have most explicitly and thoroughly sought to reach an understanding of the existential dimension of in-game subjectivity are the idea of the 'gameplay condition' (Leino 2010), that of 'ludic subjectivity' (Vella 2015), and that of the 'gameplay situation' (Kania 2017). As such, we consider it worth providing a brief overview of these concepts in order to anchor our understanding of the existential character of the player's in-game subjectivity.

1. **The 'gameplay condition':** Olli Tapio Leino defines the 'gameplay condition' (2009, 12, 2010, 101) as the condition of responsibility for one's in-game freedom of action resulting from the game's material upholding of the consequences of the player's choices. Drawing on the notion of facticity that we have discussed above, Leino understands a game as an 'extended facticity' for the player (2010, 220). Subjectively adopting a standpoint in relation to this facticity, the players subject themselves to the gameplay condition.

2. **Ludic subjectivity:** The idea of a subjective existence in relation to a game's goals and affordances is further developed in Daniel Vella's theory of ludic subjectivity, "the subjective 'I-in-the-gameworld' that the player develops while engaging with the game" (2015, 22). This subjective standpoint towards the gameworld is structured through a number of interlinked mechanisms, all related to the player's embodiment in the playable figure: the establishment of a spatial 'here' for the player and the related establishment of visual and auditory standpoints, the adoption of the playable figure's capabilities for (and limitations on) action in the gameworld, the orientation of these capabilities towards set goals, and the playable figure's vulnerability to being acted upon by the gameworld (ibid., 266–289). All of these factors contribute to shaping the players' comportment in the gameworld, and, hence, the shape of their ludic subjectivities. The ludic subject, then, is not a pre-existing character that the player finds ready-made and simply steps into (though it can be, and often is, tied to a scripted diegetic character). Rather, the ludic subject refers to the playing individuals' sense of their transposed, subjective existence within the gameworld. As such, the ludic subject is composed of the set of player's subjective experiences of engaging with the gameworld from the standpoint of the ludic subject-position, and is only brought into being by the player's playing.
3. **The gameplay situation:** Marta Matylda Kania articulates the interrelated notions of the 'self-avatar' and the 'gameplay situation' to conceptualize the player's existential 'situatedness' in the gameworld. In her work, she defines the self-avatar as "an emergent being situated within the gameworld, consisting of the player's existence and intentional acts, and the features of the avatar" (2017, 7). In her work, she addresses the gameplay situation as the "perceptual position of the self-avatar towards the gameworld" (ibid., 61) and proposes an understanding of "the self-avatar as a subjective position that enables cognition of the particular gameworld from the point of view of the gameplay situation" (ibid., 57).

In closely related ways, the concepts of the gameplay condition, ludic subjectivity, and the gameplay situation formulate an understanding of the player's subjective existence within the gameworld. All advance positions according to which the digital game environment is understood as a specific kind of facticity that players adopt as the existential situation for a subjectivity that they develop within the virtual world. Though these theoretical insights were developed in relation to player experience in (and of) digital gameworlds, we argue that they can be fruitfully applied beyond the disciplinary boundaries of digital game studies. In particular, the conceptual model that allows for the recognition of a 'ludic subjectivity' could be used, with very little modification, to capture the experiential and existential aspects of virtual worlds in general. For that, we propose to use the term 'virtual subjectivity' (see Vella and Gualeni 2019).

1.3 Projectuality in Virtual Worlds

In the field of game studies, an orientation towards goals has long been recognized as a defining characteristic of games. This emphasis on teleology can be recognized in a number of definitions of 'game', whether in the specific form of the delineation of a game objective or through the notion of a preferred or disequilibrial outcome (Avedon and Sutton-Smith 1971, 7; Suits 1990, 34; Costikyan 2002, 11–14; Juul 2005, 36). The existential implications of this goal-orientation—and its congruence to the notion of the project—is also evident in Hans-Georg Gadamer's philosophical approach to play. Gadamer argues that "every game presents the man who plays with a task" (1989, 107), and to adopt the role of the player is to adopt the task as one's own. However, and in analogy with Sartre's notion of the original project, the existential significance of the project does not reside in completing it. Instead, the players' adoption of this task as their own (and orienting their being towards the achievement of the task) shapes their being into a particular "comportment" (ibid.).

Gadamer's insight resonates with the three existentially-motivated approaches to digital game subjectivity outlined in the previous section. In Vella's model of ludic subjectivity, ***goal-orientation*** is highlighted as one of the primary structures determining the player's subjective 'situatedness' towards the gameworld (2015, 283–286). As he writes, "the setting of goals towards which [the player's] efforts are directed [...] makes the gameworld appear to the player in the light of these goals" (ibid., 284). Likewise, Kania's notion of the gameplay situation is intimately tied to the pursuit of a goal—the gameplay situation is "experienced as purposeful from the internal perspective [that is, internal to the gameworld] with relation to its goal" (2017, 62).

Drawing explicitly on the Sartrean notion of the project in theorizing the player's subjective engagement with digital gameworlds, Leino argues that digital games "simultaneously facilitate and resist a particular (kind of) project, which makes the particular (kind of) project stand out among all possible (kinds of) projects" (2010, 135). From this perspective, the facticity of the game's digital materiality is what renders a particular project possible, but it is also—at the same time—what resists the achievement of that project, thus demanding active effort on the part of the player to overcome obstacles and oppositions. Leino further argues that the project or goal itself cannot be located purely in the game's digital materiality and cannot be simply limited by the game's conceptual and material boundaries. Instead, it is a function of the player's subjective engagement with the facticity of the game—it "*transcends* the game artefact" and becomes a moment of freedom and self-determination (ibid., 146).

This relationship between game goals and playing as an act of self-determination becomes particularly apparent when we begin to account for the fact that digital games will, to varying degrees, allow the player the freedom to engage in actions and practices that do not follow a direct vector to the game's stated goals: for example, exploring the gameworld for its own sake in an open-world game such as *The Witcher 3: Wild Hunt* (CD Projekt RED 2015; see Fig. 1.1) or jumping off the highest mountain in *Just Cause 2* (Avalanche Studios 2010) in order to use the parachute to glide the furthest possible distance (Vella 2015, 286). Pursuing the explicit goals set by the digital game is, thus, only one of the possibilities of being that the player can pursue within the game's virtual world. Needless to say, this capacity for users to freely choose their own virtual project is even more evident in virtual worlds—such as *Proteus* (Key and Kanaga 2013) and *Second Life* (Linden Lab 2003)—that do not present them with explicitly stated goals towards which to orient their virtual subjectivity.

What these approaches to virtual subjectivity in digital gameworlds share is the idea that one's in-game subjectivity takes its shape in the light of the goals one pursues in the gameworld. To reiterate, though, it is not the achievement of the project that matters. On the contrary, its status as an existential project lies in the fact that, as Gadamer writes about the task of play, the goal is only a pretext that serves to shape the player's comportment in the virtual world into a particular form, and, as such, to constitute a particular project of being in the virtual world.

Fig. 1.1 Exploring the world of *The Witcher 3: Wild Hunt* (CD Projekt S.A. All rights reserved. *The Witcher* games are based on the prose of Andrzej Sapkowski)

1.4 Actual Subjectivity and Virtual Subjectivity

Up to this point, we have argued that, during our engagement with virtual environments, we step into a virtual subjectivity that, in phenomenological terms, organizes that digital environment into a virtual world that serves as our existential situation. Furthermore, we have observed that, just as subjectivity takes its shape through the orientation towards an existential project, virtual subjectivity is projectually determined through its orientation towards the user's goals in the digital environment.

However, an approach that merely considers an actual subjectivity in the actual world on the one hand, and an entirely separate virtual subjectivity in the virtual world on the other, would be inadequate and potentially misleading. Virtual subjectivities are fundamentally rooted in actual subjectivities, and there is a definite hierarchy between the two. Virtual worlds are experientially and existentially subordinate to the actual world, and virtual experiences can be considered a subset of actual experiences. By this understanding, the existential structures that we can develop and establish in virtual worlds are ultimately meaningful and valuable only insofar as they accrue meaning and value within the projectual structures of the individual's existence in the actual world.

To begin with, there is a causal, mechanical connection between the actual world and the virtual worlds that are technically created within it. From this perspective, virtual environments are recognized as technological artefacts that are designed, accessed, and interactively experienced on the basis of our *actual* capabilities to build them, perceive them, and meaningfully operate within them. Virtual environments, then, exist *within* the actual world, and this is reflected on an experiential dimension. In our lived experience, the actual world is the base-world we leave behind, and to which we return upon pausing a digital game, upon taking off our virtual reality headset, or upon completing our training session in a flight-simulator. After all, as Elizabeth Grosz argues, virtuality is "a relative or differential concept whose status as virtual requires an actual to which its virtuality can be marked as such" (Grosz 2001, 75). If the virtual is—by definition—the non-actual, it can only be virtual by virtue of its negation of an actual domain. This is what makes a virtual world a "second-order reality" (Poster 2013, 117).

Our previous observations resonate with the thought of the American philosopher Robert Nozick, who, in his famous thought experiment called 'The Experience Machine', asks his readers to imagine a distant technological future in which virtual world experiences are indistinguishable from their actual counterparts. Nozick's thought experiment presents a hypothetical scenario in which human beings could opt to be permanently plugged into experience machines. These machines can disclose virtual worlds that are not only thoroughly realistic in their aesthetic and interactive granularity, but also guarantee a perfectly happy (virtual) existence to whoever decides to plug into them (Nozick 1974). Nozick argues that the awareness of the artificiality of that kind of existence, regardless of how pleasurable, would be enough to deter us from abandoning the one we are native to (ibid.). Nozick's reflections on the experience machine testify to the fact that he considers virtual world experiences to be inferior to actual ones in terms of their experiential depth and their existential significance. At the same time, Nozick's thought experiment shows that virtual and actual subjectivity could completely overlap: if one is ignorant of the fact that one is in a virtual environment, one's 'virtual subjectivity' would be indistinguishable from an actual subjectivity, and there would no longer be any possibility to separate one's actual project from one's virtual project. In any case, the hierarchy between the actual world and the one simulated in his experience machine is so ingrained in Nozick's thought that he goes as far as considering a permanent immersion into a virtual world to be "a kind of suicide" (ibid., 43).

The hypothetical scenario described in 'The Experience Machine' bears, however, little resemblance to how we actually experience virtual worlds in the present day and age. First of all, Nozick presupposes being completely absorbed in a single virtual world, a relationship so exclusive that it requires that its users sever all experiential and existential ties with the actual world. Human beings' actual experiences within virtual worlds are clearly neither exclusive nor an all-or-nothing kind of deal: people can enter and exit a virtual world at will, and—when stepping into a virtual subjective standpoint—they do not entirely leave behin exclusive nor an all-or-nothing kind of deal: people can enter and exit a virtual world at will, and—when stepping into a virtual subjective standpoint—they do not entirely leave behind the actual world, nor are they completely severed from their actual self. In other words, differently from the virtual worlds portrayed in Nozick's thought experiment, the virtual worlds that we can actually access cannot be understood as radical phenomenological and existential alternatives to the actual world. When immersed in virtual worlds, people do not completely stop being themselves any more than they stop being themselves when they take the wheel of a car or sit down for a meal. Where the digital environment allows, the individuals can bring the concerns of their own current, actual project into a virtual world, shaping their virtual subjectivity along the lines of their actual project of being. For instance, one can choose to adopt a vegan lifestyle in digital games where the line of least resistance suggested by the game itself consists in fighting and hunting beasts and consuming animal products (Westerlaken 2017).

We believe that the idea that virtual worlds are always experienced on the backdrop of an actual existential project is useful in opposing what we consider misconceptions about virtual experiences. Virtual worlds are, for instance, often criticized—for a number of different rhetorical purposes—as technological artefacts meant for deceit and compulsion. This is particularly obvious if we turn, once again, to digital games. Casual and social-network games, for example, are notorious for their deliberately exploitative design, leveraging behavioural techniques and social pressure to manipulate players. Evidently, these kinds of interactions can lead to the conclusion that virtual worlds are not suitable contexts for pursuing freedom and authenticity in the ways proposed by existential philosophy.

Along the same lines, several scholars have argued against the possibility of achieving individual freedom or utopian thinking through those devices, with their arguments insisting on the digital (and, as such, inevitably deterministic) constitution of virtual worlds (see, in this regard, Galloway 2006 and Silcox 2019).

The perspectives offered in this book do not deny that some digital games do indeed manipulate and exploit their players, but also offer a wider existential perspective on how virtual pursuits relate to one's actual existential projects. More specifically, we argue that virtual worlds, when treated as virtual possibilities of being, can have a momentous influence on culture, and are existentially relevant. In other words, instead of focusing on them as independent existential domains that require one's complete psychological and cognitive involvement, they can be used as technically-aided means for temporarily adopting new perspectives, to experiment and reflect on one's possibilities and on the meaning thereof. In virtual worlds, understood in this way, aspects of one's actual life and one's relationships with power can be replicated, opposed, and even subverted. They are contexts where we can express ourselves or take up a different project, and where serious reflection overlaps and interweaves with playfulness. It is in this sense that the boundary between the actual and the virtual can be recognized as being porous in both directions. We have already seen that the fundamental values of one's actual subjectivity can have a role in determining one's virtual subjectivities. The reverse movement is also possible: even when a virtual subjectivity is perceived as being shaped by practices, goals, and dispositions that would have no place in the individual's actual subjectivity, it is still possible for some of its aspects to be carried over into other spheres of one individual's life, and potentially having transformative effects (Vella and Gualeni 2019).

In light of the arguments presented, our current relationships with virtual worlds (and our subjective existence within them) can perhaps be better understood as *nested* existential situations. With respect to the ludic subjectivities of digital gameworlds, this is a point that has been observed in the argument for a 'double perspectival structure' of engagement with digital games. According to this structure, the player simultaneously inhabits a subjective standpoint *internal* to the gameworld (the ludic, or virtual, subjectivity) and her own subjective standpoint as an individual *external* to the gameworld. Crucially, the internal perspective (and, hence, the virtual subjectivity) is always experienced through the external perspective (Vella 2015, 55–71). Another way to articulate the same thought would be to say that experientially engaging with a virtual world must always also be recognized as an actual world experience. If we are ready to accept this fundamental standpoint on virtual experiences, then we also need to be open towards recognizing a transformative potential and a more general existential relevance in our interactions with and within virtual worlds. This notion will be the focus of the following section of this chapter.

1.5 Practising Existentialism in Virtual Worlds

Having advanced the perspective that a virtual subjectivity is subordinate to the individual's actual subjectivity, and thereby figures as a practice within the individual's actual existential project, we are able to proceed to the second question posed at the beginning of this chapter: in what way(s) does the practice of taking on a virtual (or in-game) project figure in the overarching project of the playing individual's actual being as a whole?

It is important to highlight, in articulating our answer to that question, that there are several functional reasons as to why one might want to take part in a virtual world. For example, one might be willing to use training simulations with the aim of practicing specific skills in scenarios that would be impractical, unethical, costly, or even impossible to recreate in the actual world (Vella and Gualeni 2019, 125; Silcox 2019, 75). Leaving those functional motivations aside, we decided to explore why one might be willing to dedicate time and resources to derivative worlds such as those of fantasy digital games, which are only loosely related with the practicalities of our existence in the actual world. Why would one psychologically invest into them? Why would one be motivated to pursue experiences that are understood as having less existential significance compared to actual ones, if at all?

A facile answer, which is explored with more philosophical depth in Sect. 4.1 (in the 'Distraction' sub-section), consists in understanding virtual environments as technologies for distraction, as objects that we approach and interact with with the intention of keeping ourselves busy, momentarily diverting our attention from our actual existence and its discontents. Although we consider this to be a valid response to our question, we argue that it would be an answer that is both obvious and incomplete. On top of the functional uses of virtual environments (designed for training and education) and those meant to distract us (built for escapism and entertainment), we deem it important to focus on their capabilities for experientially disclosing routes to (virtual) existential self-realization. In other words, we believe that virtual environments are particularly existentially appealing because they allow one to re-construct and (re)fashion one's virtual self in ways that are significantly more accessible and less resistant to change than its actual counterparts (Gualeni 2015, 128; also see Chap. 5 and Sect. 7.1). In that respect, it is worth noting that the thematic and marketing decisions behind commercial digital games often promise the fulfilment of fantasies and drives that are illegal or impossible to attain in the world that human beings share as biological organisms (ibid., 128). By this understanding, virtual worlds become playgrounds in which we can experiment with possibilities we are, for whatever reason, unable or unwilling to actualize in our everyday existence—this is the idea behind Mark Poster's observation that the function of this "second-order reality" is that it lets us "play with or practice upon the first order" (2013, 117).

Earlier in this same chapter, we identified existential freedom as an 'openness' to making something of oneself that is not limited or deterministically framed by one's present condition. In this regard—and tying the discussion back to our engagement with digital game studies—Sartre's own treatment of the notion of play as an existential disposition can prove crucial to understanding the existential significance of virtual subjectivity. Sartre defines play as being 'the least possessive attitude' one can have in acting (1966, 740–741). Of all attitudes that the individual can harbour towards the world, it is the least motivated by the goal of *having* the object towards which it is directed. Nor is this acting for its own sake—doing for the sake of doing. Instead, Sartre argues:

> The desire to do is here reduced to a certain desire to be. The act is not its own goal for itself; neither does its explicit end represent its goal and its profound meaning; but the function of the act is to make manifest and to present to *itself* the absolute freedom which is the very being of the person. (ibid., 742)

We act in playing, but this form of 'doing' is only a way to explore our own being. As such, "the desire to play is fundamentally the desire to be" (ibid.). For Sartre, then, the activity of play represents an affirmation of the individual's capacity for self-determination, thanks to its status as "an activity of which man is the first origin, for which man himself sets the rules, and which has no consequences except according to the rules posited" (ibid., 741). As such, play is understood as the impulse to put into action a version of oneself—a way of being—that is chosen by one's own free will. This idea, present both in Gadamer's work and in Sartre's, is developed even further by the German philosopher Eugen Fink, who concerned himself with play as the theme of some of his major work.

Some of the most interesting relationships between Fink's work and the ways in which we can act and act upon ourselves in virtual worlds will be the focus of a later chapter of this book (Chap. 6). In that chapter, and as we have already observed in the case of Sartre's approach to play, actions that take place in a make-believe, non-serious context are presented as having actual existential significance as ways to actively explore possibilities of being that lie beyond the individual's already actualized self.

The focus of this book, however, is not on play as a possible context for existential freedom, but on human existence understood and practiced in (and towards) virtual environments. With the intention of exploring virtual worlds as having an existential meaning, we are trying to understand if virtual worlds could be a context where one can practice existential freedom and self-actualization.

The notion that one can pursue existential meaning in virtual worlds can be counterintuitive, if not downright paradoxical. This is especially evident, for example, when engaging with training simulations and digital games, as those virtual environments are specifically designed with certain affordances and objectives in mind: they materialize particular, and particularly limited, ways of understanding and guiding progress, playfulness, meaning, selfhood, etc. One might, thus, object that the existence of specific behaviours and aspirations to which the player must adhere (as long as one wishes to retain her status as a player at all) drains the situation of the freedom that is necessary to pursue an existential project as understood by Sartre. The Norwegian game scholar Espen Aarseth's idea of the 'implied player' of a game indicates "a role made for the player by the game" that can be understood as "a set of expectations that the player must fulfil" (2007, 132). As Feng Zhu points out (2015), this makes the idea of the implied player a normative one—this is a pre-established prospect of the kind of being the player *should* be in the game.

To frame this question in the context of existentialism, we must keep in mind that one of the foundational precepts of that philosophical tradition consists in the denial of the existence of God as a guarantor of meaning.

This fundamental belief makes individuals, and not the transcendent notion of a god, responsible for giving form and meaning to their existence. In other words, it is the rejection of a religious world view that endows the individual with existential freedom. This being the case, how could people ever be able to pursue self-fashioning and self-actualization in artificial environments that are built to constrain their actions and aspirations to pre-determined boundaries, criteria, and ideologies?

As Leino (2019) puts it, existentialism emphasizes ideas like thrownness and indeterminacy, but those conceptual lenses cannot apply to the users' experiences in virtual environments because:

- being in those environments does not correspond to being 'thrown' into worlds, since the users freely and voluntarily decide whether to engage with virtual environments, and because
- the users' existences in virtual worlds cannot be considered formless and indeterminate, as criteria for existential meaning and self-realization in virtual worlds are always already established by its creators in the forms of designed affordances and the goals towards which they are oriented. This might lead us to conclude that—to return to Sartre's term—when playing according to the games' rules and expected comportments, players are acting in *bad faith*. To act in *bad faith* is, according to Sartre, to deny oneself the freedom of self-determination. What that means in the context of interactively taking part in a virtual world is that the players subject themselves to modes of being that are given to them without questioning them, and without considering other possibilities to be and behave in that world. Kania reached the same conclusion (2017, 100), arguing that, even if the phenomenological situation of digital game play allows players to take a reflective distance towards their own gameplay situation, their being-in-the-virtual-world remains inevitably determined by the situation as long as they wish to keep playing the game. (ibid.)

A similar criticism could be raised in relation to our experience of death in virtual worlds. One might object that established existential approaches to finitude and death cannot be applied to virtual subjectivities because death in virtual worlds, generally speaking, does not feature the irrevocability and the absurdity that, instead, characterize actual, biological death.

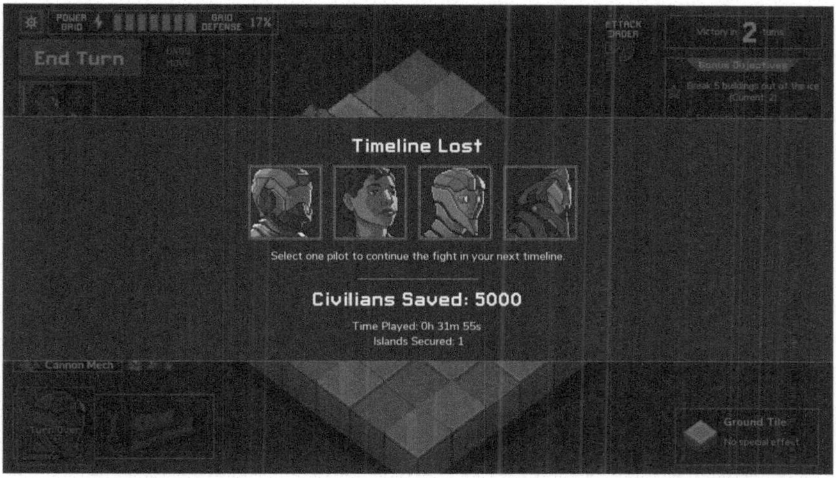

Fig. 1.2 The irrevocability of the death of most of the player's squad members in *Into the Breach*

A knee-jerk response to that objection would be to point out that virtual environments where conclusions and failures are permanent states do exist. We could mention examples of virtual environments designed to be accessed only once, or gameworlds that feature 'permadeath' mechanics.[3] In those instances, aspects connected to failure and death have degrees of irrevocability that range from permanently losing characters in one's team (for instance in Subset Games' 2018 *Into the Breach*; see Fig. 1.2) to not ever being able to access the virtual environment in question again (as is the case with the perma-permadeath condition that characterizes RobotLovesKitty's 2014 *Upsilon Circuit*). These are instances where the application of a traditional, existential conceptualization of death to understand virtual worlds would be less outlandish (we will further elaborate this idea in the 'de-roling' section of Chap. 2 and in the 'isolation' section of Chap. 4; see Fig. 1.2).

The outlined sceptical position towards the possibility of applying existential notions and frameworks to inhabiting virtual worlds (and extracting existential meaning from them) seem to be well-motivated within the framework of twentieth-century existentialism, and particularly so in relation to the outlined notion of 'existential freedom'. In the last section of this chapter, we will argue against these positions and show that the goal-orientation and the conceptual limitations that characterize virtual worlds are not categorically incompatible with the pursuit of individual self-determination and self-realization. We want to argue that the implied designers of virtual worlds (as secular guarantors of meaning) can never eliminate the possibility for the users of those virtual worlds to express themselves and take actively reflective stances towards those experiences (see Van de Mosselaer and Gualeni 2020). To explain how that particular understanding of 'freedom' can emerge in relation to constraints and impositions, we turn to the French philosopher Michel Foucault.

In Foucault's later work (1982, 1988), individual human beings are not considered 'free' when they overcome constraints and interdictions or when they manage to move beyond the controlling reach of power. Rather, they are 'free' when they engage in the critical activity of shaping themselves in relation to their current situation. In Foucault's understanding, then, constraints and limitations are not obstacles to existential freedom, but constitute its necessary preconditions: one can only be free in the Foucauldian sense when one is subject to limitations (Gualeni 2014). This understanding of 'freedom' matured when Foucault was studying classical antiquity. In ancient Greece, he writes, ethics was not practiced with the aspiration of answering the question 'how should I act to be a moral subject?', but rather 'what kind of subject do I want to be?' (Foucault 1988). More simply put, back then ethics was not approached normatively, but existentially: as a practical approach to understanding and negotiating one's possibilities of being.

Seen from a Foucauldian perspective, the affordances that define the possibility horizon of a virtual environment always have the possibility of being addressed critically.[4] To paraphrase Foucault, our interactions with virtual environments have, thus, the potential to be used and understood as 'technologies of the self', that is to say as activities that (when approached critically) encourage and facilitate a number of operations on one's own body and soul, thoughts, conduct, and way of being, so as to transform oneself (Foucault 1988, 18). In a way that is very similar to Sartre's existentialism, the perspective offered by the later Foucault allows us to approach a virtual environment as an existential situation that a (virtual) subject can certainly engage with in *bad faith*, but also in ways that are reflexive and critical.[5] In this vein, Canadian media scholar Felan Parker has considered players' choices to play a game in ways that might diverge from those of the implied player as instances of Foucauldian self-fashioning (2011).

When establishing a critical relationship with a virtual world, then, users give themselves the freedom to resist, reject, circumvent, and repurpose its affordances and ideological positions alike. Michelle Westerlaken's vegan playthrough of *The Legend of Zelda: Breath of the Wild* (Nintendo 2017), to which we have already alluded, is a perfect example. In this case, Westerlaken deliberately chooses not to take advantages of a certain set of affordances the gameworld allows her (hunting animals, eating meat, wearing leather armour, and so on)—a choice through which she is able to give the desired shape to her virtual subjectivity (Westerlaken 2017).

Of course, the initial choice to engage with one virtual world and not another is, itself, already a significant existential decision and has the potential to constitute an act of self-fashioning. One might choose, for example, not to play *America's Army: Proving Grounds* (United States Army 2013), on the assumption that the game will privilege a project of being that adheres to militaristic and nationalistic ideologies that diverge from one's own values. Conversely, having begun to engage with a virtual world, the user might also, at some point, simply decide to disengage from it wholesale. To completely terminate a certain subjectivity is a decision that is evidently laden with existential significance. In line with what Albert Camus wrote at the beginning of his 1955 book *The Myth of Sisyphus*, we consider one's existential decision to abandon a virtual subjectivity (a kind of 'virtual suicide') to be a philosophical question of central relevance with regard to virtual worlds.

Notes

1. It is important to clarify that not every digital game—and not every virtual world—provides the user with a 'virtual body' in the form of a single avatar. Different forms of subject-positioning exist in virtual environments—from situations where the user is given multiple avatars, to ones—such as *Sim City* (Maxis 1989)—where the user's subject-position is not tied to any 'body' (Vella 2016). All of these represent different phenomenological structures of being-in-the-virtual-world. We argue, however, that it is the embodied situation of the single avatar that most closely aligns with the phenomenology of our being-in-the-world as embodied beings, and, as such, it is not surprising that so many virtual worlds situate the user in this way.
2. The idea of a smooth transferral of consciousness from one's actual embodied standpoint into a virtual body within a virtual environment needs nuancing. In order to engage with a virtual environment, the user, in almost every case, needs to act bodily (Nørgård 2011)—performing complex patterns of

button presses on a game controller, walking around a room while wearing a virtual reality (VR) headset, and so on. These performances require not only the user's actual bodily abilities, but also the 'embodied literacies' (Keogh 2018, 78) they have acquired through experience with virtual environments and the technological artefacts through which they are accessed—a point we will consider in greater detail in Chap. 2. For this reason, the phenomenology of virtual embodiment is always more complex than can be accounted for by any singular understanding of 'body', playing out as it does across actual and physical bodies (ibid.). However, for our purposes in this book, we shall bracket these concerns, focusing upon the virtual body of the avatar as the seat of the user's virtual subjectivity.

3. 'Permadeath' is a specific quality of certain tabletop and digital games that prevents the players from continuing their game experience after the virtual death of their *self-avatar* (or of a virtual pet). The design decisions involved in permadeath are explicitly meant to counter the fact that virtual death is rendered trivial and deprived of its irrevocability by popular mechanics such as those allowing characters' resurrection at nearby checkpoints ('respawning') or the meta-ludic possibility to load and restore a saved game state to circumvent a state of failure. It might be important to point out that a 'permadeath condition' does not necessarily need to be mechanically enforced by the game's affordances, but can also be a quality of the game experience that is self-imposed by players. In those cases, the 'permadeath condition' is a way for players to recuperate a degree of existential significance in a virtual world that instead encourages respawnings and doovers.
4. Foucault famously defined 'critique' as the act in which one gives oneself "the right to question truth on its effects on power and question power on its discourses on truth" (Foucault 2007, 47).
5. On the basis of the work of Foucault, in the final passages of this chapter we characterized a critical thought to be a matter of individual disposition. By that we mean that a critical attitude towards virtual worlds can only be actualized when the users actively choose to approach the latter in a thoughtful and self-reflexive fashion. We want to clarify, however, that the emergence of this attitude is not a purely subjective responsibility, but can be invited by particular features of the experience in question. As we argued elsewhere, all forms of technology invite forms of engagement that ask to interactively repurpose, extend, and occasionally unsettle one's cognitive and imaginative faculties (Gualeni 2015, 2016). Virtual worlds that explicitly seek to stimulate a self-reflexive, critical attitude can emerge, for example, in relation with digital games such as *Necessary Evil* (Gualeni et al. 2013), *The Talos Principle* (Croteam 2014), and *The Beginner's Guide* (Wreden 2015).

References

Aarseth, E. 2007. "I Fought the Law: Transgressive Play and the Implied Player." Proceedings of the 2007 DiGRA International Conference on Situated Play. Tokyo (Japan), September 24–28, 2007.

Avalanche Studios. 2010. *Just Cause 2* [Sony PlayStation 3]. Digital game directed by Nedfors, M. and published by Eidos Interactive.

Avedon, E. M. & Sutton-Smith, B. 1971. *The Study of Games.* New York (NY): John Wiley & Sons.

Bayliss, P. 2007a. "Beings-in-the-Gameworld: Characters, Avatars and Players." Proceedings of the Fourth Australasian Conference on Interactive Entertainment. Melbourne (Australia), December 3–5, 2007.

Bayliss, P. 2007b. "Notes towards a Sense of Embodied Gameplay." Proceedings of the 2007 DiGRA International Conference on Situated Play. Tokyo (Japan), September 24–28, 2007.

Calleja, G. 2011. *In-Game: From Immersion to Incorporation.* Cambridge (MA): The MIT Press.

Camus, A. 1955. *The Myth of Sisyphus and Other Stories.* New York (NY): Vintage Books.

CD Projekt RED. 2015. *The Witcher 3: Wild Hunt* [Sony PlayStation 4]. Digital game directed by Tomaszkiewicz, K., Kanik, M. and Stępień, S., published by CD Projekt.

Costikyan, G. 2002. "I Have No Words and I Must Design." Proceedings of the Computer Games and Digital Cultures Conference, ed. Frans Mäyrä, 9–33. Tampere (Finland): Tampere University Press.

Croteam. 2014. *The Talos Principle* [PlayStation 4]. Digital game directed by Hunski, D.; Ladavac, A.; Tomičić, D., published by Devolver Digital.

Foucault, Michel. 1982. "The Subject and Power." In *Michel Foucault: Beyond Structuralism and Hermeneutics.* 2nd ed. Dreyfus, H. L and Rabinow, P. (eds.). Chicago (IL): University of Chicago Press.

Foucault, M. 1988. *Technologies of the Self: A Seminar with Michel Foucault.* Martin, L. H., Gutman, H., & Hutton P. H. (eds.). London (UK): Tavistock Publications.

Foucault, M. 2007 [1997]. "What is Critique?" In Hochroth, L. (trans.). *The Politics of Truth.* Los Angeles (CA): Semiotext(e), 41–81.

Gadamer, H. G. 1989 [1960]. *Truth and Method.* London (UK): Sheed & Ward.

Galloway, A. R. 2006. "Warcraft and Utopia". In *Ctheory. 2* (16).

Gee, J. P. 2008 "Video Games and Embodiment." In *Games and Culture, 3*, 253–263.

Gregersen, A. & Grodal, T. 2009. "Embodiment and Interface." In *The Video Game Theory Reader 2*, ed. Bernard Perron and Mark J. P. Wolf, 65–82. London (UK): Routledge.

Grodal, T. 2003. "Stories for Eye, Ear and Muscles: Video Games, Media, and Embodied Experiences." In Wolf, M. J. P. & Perron, B. (eds.). *The Video Game Theory Reader*. London (UK): Routledge, 129–155.

Grosz, E. 2001. *Architecture from the Outside: Essays on Real and Virtual Space*. Cambridge, MA: The MIT Press.

Gualeni, S. 2014. "Freer Than We Think: Game Design as a Liberation Practice." Proceedings of the 2014 Philosophy of Computer Games Conference. Istanbul (Turkey), November 13–16, 2014.

Gualeni, S. 2015. *Virtual Worlds as Philosophical Tools*. Basingstoke (UK): Palgrave Macmillan.

Gualeni, S. 2016. "Self-reflexive videogames: observations and corollaries on virtual worlds as philosophical artifacts." *G|A|M|E - The Italian Journal of Game Studies*, *1* (5).

Gualeni, S.; Dini, D.; Gomez Maureira, M. 2013. *Necessary Evil* [Windows]. Digital game developed by Gualeni, S.; Dini, D.; Gomez Maureira, M. Available online at: http://evil.gua-le-ni.com

Heidegger, M. 1962 [1927]. *Being and Time*, trans. John Macquarrie and Edward Robinson. San Francisco (CA): Harper & Row Publishers Inc.

Heidegger, M. 2008. *Basic Writings*. New York (NY): Harper Perennial – Modern Thought.

Juul, J. 2005. *Half-Real: Video Games between Real Rules and Fictional Worlds*. Cambridge (MA): The MIT Press.

Kania, M. M. 2017. *Perspectives of the Avatar: Sketching the Existential Aesthetics of Digital Games*. Wroclaw (Poland): University of Lower Silesia Press.

Keogh, B. 2018. *A Play of Bodies*. Cambridge (MA): MIT Press.

Key, E. & Kanaga, D. 2013. *Proteus* [Windows]. Digital game developed by Key E. and Kanaga, D., self-published.

Klevjer, R. 2006. *What is the Avatar?: Fiction and Embodiment in Avatar-Based Singleplayer Computer Games*. Doctoral dissertation, University of Bergen (Norway).

Klevjer, R. 2012. "Enter the Avatar: The Phenomenology of Prosthetic Telepresence in Computer Games." In *The Philosophy of Computer Games*, ed. John Richard Sageng, Hallvard J. Fossheim and Tarjei Mandt Larsen, 17–38. Philosophy of Engineering and Technology 7 (2012). Dordrecht (The Netherlands): Springer.

Leino, O. T. 2009. "Understanding Games as Played: Sketch for a First-Person Perspective for Computer Game Analysis." Proceedings of the 2009 Philosophy of Computer Games Conference. Oslo (Norway), August 13–15, 2009.

Leino, O. T. 2010. *Emotions in Play: On the Constitution of Emotion in Solitary Computer Game Play*. Doctoral dissertation, IT University of Copenhagen (Denmark).

Leino, O. T. 2019. "God is a Game Designer – Accelerating 'Existential Ludology'." Proceedings of the 2019 DiGRA International Conference. Kyoto (Japan), August 6–10, 2019.

Linden Lab. 2003. *Second Life* [Windows]. Digital game directed by Rosedale, P., and published by Linden Lab.

Maxis. 1989. *Sim City* [Windows]. Digital game directed by Wright, W., and published by Maxis.

Nintendo. 2017. *The Legend of Zelda: Breath of the Wild* [Nintendo Switch]. Digital game directed by Fujiyabashi, H., and published by Nintendo.

Nørgård, R. T. 2011. "The Joy of Doing: The Corporeal Connection in Player-Avatar Identity." Proceedings of the 2011 Philosophy of Computer Games Conference. Athens (Greece), April 6–9, 2011.

Nozick, R. 1974. *Anarchy, State, and Utopia*. New York (NY): Basic Books, Inc.

Parker, F. 2011. "In the Domain of Optional Rules: Foucault's Aesthetic Self-Fashioning and Expansive Gameplay." Proceedings of the 2011 Philosophy of Computer Games Conference. Athens (Greece), April 6–9, 2011.

Payne, M. T. 2008. "Interpreting Gameplay Through Existential Ludology." In Ferdig, R. E. (ed.) *Handbook of Research on Effective Electronic Gaming in Education*, 621–635. Hershey (PA): Information Science Reference.

Poster, M. 2013 [2001]. *The Information Subject*. Oxford (UK): Routledge.

RobotLovesKitty. 2014. *Upsilon Circuit* [Windows]. Digital game directed by Stolzer, A. and Goble, C. This ephemeral piece of software is no longer accessible.

Sartre, J. P. 1966 [1943]. *Being and Nothingness*. Trans. Hazel E. Barnes. New York (NY): Washington Square Press.

Silcox, M. 2019. *A Defense of Simulated Experience: New Noble Lies*. New York (NY): Routledge.

Subset Games. 2018. *Into the Breach* [Windows]. Digital game directed by Ma, J. and Davis, M., published by Subset Games.

Suits, B. 1990. *The Grasshopper: Games, Life and Utopia* Boston (MA): David R. Godine.

Taylor, T. L. 2002. "Living Digitally: Embodiment in Virtual Worlds." In Ralph Schroeder (ed.), *The Social Life of Avatars: Presence and Interaction in Shared Virtual Environments*, 40–62. London: Springer-Verlag.

United States Army. 2013. *America's Army: Proving Grounds* [Windows]. Digital game directed by Kuykendall, D., and published by the United States Army.

Van de Mosselaer, N. & Gualeni, S. 2020. "The Implied Designer and the Experience of Gameworlds." Proceedings of the 2020 DiGRA international Conference. Tampere, Finland. June 2–6, 2020.

Vella, D. 2015. *The ludic subject and the ludic self: Investigating the 'I-in-the Gameworld'*. Doctoral dissertation, IT University of Copenhagen (Denmark).

Vella, D. 2016. "'Who am 'I' in the Game?': A Typology of Modes of Ludic Subjectivity." Proceedings of the 2016 DiGRA and FDG (joint conference). Dundee (UK), August 1–6, 2016.

Vella, D. & Gualeni, S. 2019. "Virtual Subjectivity: Existence and Projectuality in Virtual Worlds." In *Techne': Research in Philosophy of Technology*, *23* (2).

Westerlaken, M. 2017. "Self-Fashioning in Action: Zelda's Breath of the Wild Vegan Run." Proceedings of the 2017 Philosophy of Computer Games Conference. Krakow (Poland), November 28–December 1, 2017.

Wilhelmsson, U. 2008. "Game Ego Presence in Video and Computer Games." In *Extending Experiences*, ed. Olli Tapio Leino, Hanna Wirman, and Amyris Fernandez, 58–72. Rovaniemi, (Finland): Lapland University Press.

Wreden, D. 2015. *The Beginner's Guide* [PlayStation 4]. Digital game developed by Everything Unlimited Ltd. and published by Everything Unlimited Ltd.

Zhu, F. 2015. "The Implied Player: Between the Structural and the Fragmentary." Proceedings of the 2015 DiGRA International Conference. Lüneburg (Germany), May 14–17, 2015.

CHAPTER 2

En-roling and De-roling in Virtual Worlds

Abstract Though there is a significant body of work on subjectivities in virtual worlds, relatively little has been written on the transitions into and out of these subjectivities. To address this lack, this chapter considers processes of 'en-roling' and 'de-roling' from our being in virtual environments. The first refers to those practices, experiences, and activities by which the user forms an understanding of—and internalizes—a particular subjectivity in relation to the virtual world. The second indicates techniques through which subjects 'disrobe' themselves from their role in the virtual environment. This chapter draws comparisons between engaging with virtual worlds and the mechanisms of transition between subjectivities used in a range of other cultural practices—including dramatherapy and psychodrama, liminal rites, and live-action role-playing.

Keywords En-roling • De-roling • Roles • Subjectivity • Virtual worlds • Game studies

In the previous chapter, we proposed the term 'virtual subjectivity' to refer to the individual's being-in-the-virtual-world. As we have seen, existential philosophy posits that the ways in which we are in the world develop around an existential project. In line with that tradition we presented the argument that virtual subjectivities also develop around a projectual mode of being. In other words, we argued that one's orientation towards goals

S. Gualeni, D. Vella, *Virtual Existentialism*,
https://doi.org/10.1007/978-3-030-38478-4_2

in a virtual world (regardless of their being implicit or explicit, materially set by a designer, or freely embraced by the users) is a determining factor in the formation of one's virtual subjectivity. We also advanced an understanding of virtual subjectivity as standing in a nested hierarchy with respect to one's actual subjectivity. Developing a virtual subjectivity was, accordingly, framed as a practice undertaken from the perspective of one's being-in-the-actual-world.

In this sense, adopting an existential standpoint within a virtual world and developing a virtual subjectivity have obvious affinities with the processes involved in adopting a role in practices such as dramatic theatre, improvisational theatre, and role-playing. We argue, on this basis, that it is precisely in analogy with the mechanisms of role-taking (or en-roling) and de-roling in these practices that we can arrive at a fecund and comprehensive understanding of the transformative and therapeutic possibilities and advantages afforded by adopting virtual subjectivities in virtual worlds. Accordingly, in this chapter we specifically offer an existential understanding of the practices of en-roling and de-roling in the context of the experience of virtual worlds. This constitutes an initial attempt to explore an academic 'blind spot', a lack that we found particularly surprising in the light of the recognition that digital technologies play an increasingly central role in the deconstruction and reconstruction of personal and cultural identity in late and post-modern societies (Glass 1993). Virtual reality (VR) technologies, in particular, allow for frequent and absorbing shifts between multiple and concomitant versions of ourselves. As such, we consider it urgent to examine how en-roling and de-roling practices could structure these transitions and harness the ensuing cognitive and psychological effects.

Before focusing on practices and techniques concerned with en-roling and de-roling from virtual subjectivities, we want to clarify that this chapter takes the evident, and evidently questionable, ideological standpoint that there is indeed something that we can recognize as an 'everyday' existence. An 'everyday' existence indicates an 'ordinary' state of being on the basis of which one can assume another role and to which practices of de-roling structure one's return. In other words, our exploration takes as its starting point the assumption that there is a relatively consistent, and thus identifiable, combination of perceptions, proprioceptions, attitudes, comportments, and emotions that qualify our everyday existence and upholds our individual sense of the 'self'.

The work of Erving Goffman on selfhood, for example, could cast a problematic light on this idea (i.e. that one can discern a stable, unified sense of the everyday 'self'). In his 1956 book *The Presentation of Self in Everyday Life*, Goffman argues that, in everyday life, one adopts different, and potentially contradictory, roles as one moves between the different spheres of one's life—whether this is to project a desired sense of self or even just to maintain face. For Goffman, the boundaries between the various roles we can adopt are blurred and porous, and identity is something that is always performative and in a state of flux (ibid.). While we acknowledge that Goffman's theories and similar perspectives could invalidate our use of the adjectives 'ordinary' and 'basic' in relation to the concepts of 'state of being' and 'personal identity', we still use them in this chapter. Our decision was motivated by two reasons:

- The implicit assumption that there is a basic state of being and an ordinary identity is also embraced by psychodrama and dramatherapy. Both disciplines aim to address and potentially solve the issues of an actual person (or group of people) by isolating and re-framing these issues within a clearly delineated and safe dramatic 'space'. That being the case, questioning whether tackling specific issues helps actual selves or 'performed selves' seems pointless.
- The distinction between an extra-ordinary state of being (with its attendant role-identity) and the assumed persistence of self-and-world in one's 'ordinary' state of being and role becomes particularly important in negotiating the significance of virtual experiences in the virtual environments of digital games, simulations, and VR applications.

For these reasons, with this important *caveat* having been made—and being kept in mind as we go forward—we shall continue to employ the notion of an 'everyday' self that is departed from, and returned to, when the user develops and adopts a virtual subjectivity.

2.1 En-roling and De-roling from Temporary Subjectivities

The idea of taking on a temporary or alternative subjectivity that is consciously perceived as divergent from one's everyday subjectivity is by no means new to the engagement with virtual environments. In fact, in the chapters that follow, we shall in different occasions discuss the idea that virtual subjectivities address an existential need that results from the uniquely self-conscious structure of human being. In this sense, the ability to take on a virtual subjectivity in a virtual environment represents only the most recent technological answer to an aspiration that is inherent to our species.

Accordingly, an understanding of what is entailed in en-roling into, and de-roling from, virtual subjectivities is best considered in the context of the lineage of cultural practices that developed in response to fundamental existential lacks and drives. Our exploration of the possibilities and the psychological implications of en-roling and de-roling from virtual identities and virtual experiences begins by articulating a broader understanding of en-roling and de-roling as a set of procedures and activities aimed at ensuring a less abrupt, more mindful, and more therapeutically efficient transition into, and back from a temporary, extra-ordinary state of being. Here, we use the adjective 'extra-ordinary' in a sense that indicates something that transcends the ordinary, experiences that go beyond one's everyday identity and one's exclusive relationship with the actual world.

In this section, we will specifically focus on the practices of psychodrama and dramatherapy, liminal and post-liminal rituals, threshold spaces in architecture, tabletop and live-action role-playing (LARP), and simulations used in the context of education and training. In each of these cases, we shall trace mechanisms that aid participants in, on the one hand, understanding, internalizing, and transitioning into their temporary role, and, on the other hand, disrobing themselves of this same role and negotiating the return to their everyday subjectivity.

2.1.1 Psychodrama and Dramatherapy

Drama is an extra-ordinary activity by definition: in drama, subjects consciously act *as if* they were someone else or even something else. The dramatic action frames our perception and understanding of events by marking off a 'stage world' and corresponding 'stage identities' that are conceptually separated from the actual world and our actual self within it (Gualeni et al. 2017). Psychodrama and dramatherapy are methods of psychotherapy "in which clients are encouraged to continue and complete their actions through dramatization, role-playing and dramatic self-representation" (Kellermann 1992, 20). According to the original intentions of its inventor (the Romanian-American psychiatrist Jacob L. Moreno), psychodrama needs to rely on specific and repeatable procedures to guide and develop precise therapeutic 'journeys' (1972), while dramatherapy tends to be less structured and spontaneously draws from the full range of dramatic activity as a therapeutic source (Jennings et al. 1994, 129–131).

Functional and therapeutic differences notwithstanding, dramatherapy and psychodrama have several obvious commonalities that are particularly evident in the techniques that both approaches share. In relation to the specific objectives of this section, it is relevant to point out that both methods embrace beginning-of-session activities such as 'warm-ups' or 'en-roling' practices, and end-session activities such as 'cooling down', 'de-roling', and post-session processing. As Robert J. Landy writes, a dramatherapy session is bounded on one side by en-roling and on the other side by de-roling (1993, 252).

Both psychodrama and dramatherapy employ a 'warm-up' phase (Chesner in Jennings et al. 1994) in order to prepare participants for taking on a role. This phase fulfils a number of functions, such as demarcating the boundedness of the space as a means of establishing its safety and building connections between participants. One of its essential functions, however, is that of establishing the role each participant is to play. In most cases, through techniques such as structured physical improvisation and the use of masks, participants are invited to arrive at a role themselves and to experiment with how the role feels when they step into it.

When laying out the 'drama therapy role method', Landy presents the invocation and naming of the role as the first two steps of the method. The invocation of the role refers to "a calling into being of that part of the person" that corresponds to the particular comportment that pertains to a role or character (1993, 47). Once the character is visible on the stage, as it were, the naming process fulfils the important double function of concretizing the role as a distinct, identifiable character and of distinguishing it from the participant's actual self.

At the end of a psychodrama or dramatherapy session, the term 'de-roling' indicates a set of techniques meant to assist an actor in 'disrobing' oneself from a character (Gualeni et al. 2017). In her overview of necessary procedures for good role-playing induction, Krysia M. Yardley-Matwiejczuk (1997) discusses the psychological importance of techniques for structuring and guiding the transition of role-players back to their everyday existences. David Read Johnson and Renée Emunah similarly emphasize the importance of de-roling as a way of dealing with the paradoxical "'continuity of the me and not-me' within drama" (Johnson and Emunah 2009, 76). The aim of de-roling, they argue, is to structure the "shift from one reality, that of imagination, to another one, that of the everyday, for the purpose of reflection" (Johnson and Emunah 2009, 76).

De-roling often takes repetitive and symbolic forms that have several analogies with the formal structures that occur in activities such as religious rituals and games (Cattanach in Jennings et al. 1994, 38). Some of the techniques involved in punctuating and segmenting these transitional moments could be exemplified in activities such as the ritualized removing of masks and scenic costumes or formalized ways of abandoning the space where the therapeutic enactment had been staged (Jennings et al. 1994, 129–131). Other procedures involve dialogue and are meant to encourage and facilitate self-reflection and the verbalization of thoughts and feelings by the subjects of therapy (Gualeni et al. 2017).

While de-roling techniques are considered to be constitutive of therapeutic methods like dramatherapy and psychodrama, its techniques and procedures are not common or commonly used in professional acting and rarely appear in theatre studies literature. The theatre scholar Sally Bailey reports having known many actors—and having read about others—who had been bothered by playing particularly intense roles (Tinkler 2014). According to Bailey, those roles led to alterations in their personality, sometimes leading to depression and heavy drinking, precisely because they did not adequately de-role at the end of the acting sessions. Bailey further observes that "they spend many, many years learning how to get into role but they're never taught techniques about getting out of it" (Tinkler 2014).

In dramatherapy and psychodrama, then, de-roling procedures pursue the following psychological effects:

- They promote the development of a psychological and critical distance from the non-actual scenarios (or worlds) in which the subjects of therapy were acting, thus stimulating self-reflection and consideration for new possibilities of being, both of which are central to the

therapeutic process (Landy 1996, 16–27; Johnson and Emunah 2009, 75–78).
- They encourage and facilitate the verbalization and the sharing of thoughts and feelings (Robbins 1988, 208–217; Jennings et al. 1994).
- They mitigate the potentially traumatic effects of cognitive dissonance that might result from having assumed a new subjectivity, having adopted uncommon ways of perceiving, and having adapted to unfamiliar scenarios (Cattanach in Jennings et al. 1994, 39; Chesner in Jennings et al. 1994, 71–73). This is especially the case in situations where the actors or the subjects of therapy are asked to immerse themselves in emotionally intensive roles (ibid.). The need for deroling in terms of the emphasis on there being a "protective barrier to be crossed between two kinds of reality" (Grainger in Jennings 1992, 175) is especially pressing in cases when the actors or the subjects of dramatherapy are affected by thought-disorders characterized by dissociative conditions such as schizophrenia or multiple personality disorders (see Sect. 2.2.6).

2.1.2 En-roling and De-roling Techniques in Social and Religious Rituals

Rituals mark events and transitions that are transversal in human cultures, including mourning the dead, becoming part of a new social group, or changing social status (e.g. coming of age, marrying, and being recognized as a religious figure). When structuring the passage of individuals from one social situation to another (for instance, from childhood to adulthood) these rites are referred to as liminal: when undergoing such rituals, the individual stands on the threshold between social roles and in a transitory state in terms of his or her personal identity (van Gennep 1977; Turner 1969).

The field of cultural anthropology often characterizes such ritual activities as being subdivided in phases. According to Imber-Black et al. (1988, 8), "[r]ituals are coevolved symbolic acts that include not only the ceremonial aspects of the actual presentation of the ritual, but the process of preparing for it as well." The first phase, at the start of a rite, has been termed the 'separation' phase: this "comprises symbolic behavior signifying the detachment of the individual or group [...] from an earlier fixed point in the social structure" (Turner 1969, 88).

The phase that concludes most rituals is typically the most climactic among these moments of transition: it is in that concluding passage that the subject involved in the ritual is re-integrated into society through symbolic actions and passages (Turner 1969, 155, 156; van Gennep 1977, 21). The activities that take place in this transitional phase serve the main purpose of 're-incorporating' the ritual subjects: to mitigate the disorientation and possible cognitive dissonance ensuing from the 'dissolution' of the subjects' identity during their ceremonial transformations. They help re-establish one's sense of 'self' and rationalize the new possibilities and perspectives that the liminal experience disclosed (Robbins 1988, 208, 209).

In analogy with the conclusive phases of rituals, expressive therapy proposes techniques and activities that can assist the subjects of therapy in structuring their transitional experiences and come to terms with personal transformations (ibid.). For the theatre scholar Ann Cattanach, de-roling is analogous with the formal structures and effects of rituals in terms of their symbolic meaning and their repeated use (Cattanach in Jennings et al. 1994, 38). From her perspective, the de-roling techniques employed in psychodrama and dramatherapy could be understood as the ritualization of procedures that facilitate and accompany subjects in their therapeutic process. As already explained, those transitions consist in shifting from an active, dramatic state of being to one rooted in the subjects' everyday lives. Once established, these de-roling rituals are repeated each time a therapy group meets, and they tend not to change. They can involve, among other procedures, the formalized removal of masks and scenic costumes, or can be focused on emphasizing the abandoning of the space where the dramatic action had been staged (Chester in Jennings et al. 1994, 129–131).

2.1.3 En-roling and De-roling Strategies in Architectural Design: Transitional and Liminal Spaces

In various cultural contexts and historical periods, architects and architecture theoreticians have approached the built environment as an expressive and rhetorical medium (Jencks 1980; Eco 1980; Goodman 1992; Whyte 2006). Those perspectives advance the claim that rooms, buildings, complexes, and landscapes can be studied in their capability to convey meaning to the people traversing and inhabiting them—to communicate information and to elicit feelings and emotions that go beyond the mere functionality of artificial spaces. Accordingly, the production of specific experiences with space, light, materials, movement, and the relationships of built environment with historical and cultural contexts have been analysed in terms of their transformative effects both on individual behaviour and on the formation of social identities (Lynch 1984; Foucault 1984).

The physical arrangements and subdivisions of space, their symbolic potential, the specific activities that take place within structured spaces, and the processes involved in transitioning from one space to another are some of the fundamental psychological tools that architects and dramatherapists share in their respective practices of subjectification (Gualeni et al. 2017). As such, the history of architecture and practices connected to architectural design offer several spatial strategies for role-taking and role-abandoning that are relevant to our quest of exploring a wider and more encompassing horizon for the practices of en-roling and de-roling (ibid.). The architecture theorist Christopher Alexander, for example, identifies several recurrent spatial patterns in sacred and ritual spaces that are significant for the purposes of this chapter. Alexander highlights the fact that both our relationship with the divine and the officiating of rites have traditionally been associated with specific architectural features. Among those features are the recurrence of vertical symmetry, the presence of decoration and symbols, the use of height to demarcate sacred buildings, and the existence of gated, transitional pathways from common, everyday spaces to the private, sacred ones where rites and ceremonies are carried out (Alexander et al. 1977; Alexander 1979).

The progressive and symbolic segmentation of such pathways and the deliberate use of height (height of ceilings, of buildings themselves, the elevation of a spatial element above other architectural elements, as well as the act of climbing involved in reaching spaces that are extra-ordinary) serve as spatial design strategies to define and ritualize transitional experiences. These architectural tools also serve the purpose of structuring the process of abandoning one's sense of self. It is typical for these spatial functions to work in the direction of segmenting the path of the devout (or the religious figures presiding the rites) into a number of recognizable stages between a public dimension and a more private one, between a space that is shared and a space that is intimate, between a common state of being and an extra-ordinary one.

2.1.4 Moving into and out of Character in Tabletop and Live-action Role-playing

Perhaps surprisingly, the question of the transition into and of the transition out of the in-game role has not been granted a great degree of importance with respect to tabletop role-playing. In his landmark sociological study of *Dungeons & Dragons* player communities, the American sociologist Gary Alan Fine argues that "the consequences of stepping out of role are relatively light", perhaps because "role flexibility is expected and allowed" (Fine 1983, 283). Fine draws on Goffman to explain the way in which players tend to shift constantly between various interpretative frameworks, moving seamlessly, from one moment to the other, between the roles of social person, player, and character (ibid., 186). Due to this inherent role-fluidity within the activity as a whole, the final transition at the end of the activity might not be perceived to be particularly problematic.

Matters are different with respect to the practice of LARP, where there have been more explicit attempts at tackling the difficulties associated with the transition between domains and roles at the beginning and the end of the role-play. LARP design literature emphasizes the necessity of creating spaces that allow players to get to know their characters and to lay the foundations of their performance. Ari-Pekka Lappi looks at the process of interpreting and understanding one's character that is the necessary preparatory stage prior to engaging in a LARP session (2004). This process is envisioned as a hermeneutic circle. On the basis of a character document—"a short story, an article, a report, a psychological analysis" or any other textual raw material (ibid., 97), the player extracts 'character facts' that constitute an understanding of the character. The player should then be able to formulate that character's 'world view', at which point an understanding of how the character is to be played out would have been internalized. To close the circle, the player then returns to the character document, re-reading it in the light of this understanding of the character: if this brings new insights to light, then the understanding of the character is modified or refined, and so on (ibid., 102).

What is key here is the establishment of—and the player's stepping into—the character's 'world view'. This idea is, in many ways, similar to the notions of ludic and virtual subjectivity we considered in Chap. 1. The goal of such practices, as a whole, is for the player to internalize the character's subjective perspective and position in relation to the diegetic domain in which they are placed and to then be and act within that domain from that subjective standpoint.

To this end, a variety of techniques are adopted by LARP designers to aid players in understanding, and in being able to step into, their characters. To give one example, when discussing their design for the LARP *Mad About the Boy*, Edland et al. (2011) list suggestions for the pre-game workshop, a multi-day event that they see as an integral element of the LARP process. Suggestions for the event include 'physical workshops' on body language, players' self-presentation as their characters, and a 'hot seat' session—"a theatre impro technique where one player at a time is interviewed in character by a couple of other participants". The goal of this workshop was for "a player to deepen the knowledge of the character she will play before the game itself starts" (105–107).

If the question of en-roling, or getting into character, is largely taken for granted as a necessary step in LARP design, de-roling, conversely, is granted a great deal of importance in both practical and academic discussions. More is perceived to be at stake in the transition out of character and back into one's everyday subjectivity. Among the potential risks inherent to this transition, one of the most prominent is the problem of 'bleed', in which the line between the LARP situation and the player's actual life becomes porous, allowing the flow of intense and potentially troubling emotions from one's fictional role to one's ordinary self (Montola 2010; Bowman 2013).

This has been addressed by the practice of debriefing (Atwater 2016). The Norwegian interaction designer Eirik Fatland writes about the importance of debriefing in validating the individual's subjective experience of the LARP and as a means to process the experience in such a way that it can be translated into "lasting memories, reflection and learning" (2013). According to Fatland, debriefing also offers the opportunity to deal with any residual problems with the individual role-play session (ibid.). Similarly, the American journalist Lizzie Stark speaks of the significance of 'debriefing' at the end of a LARP activity, arguing that it "can help players begin to process their emotions about the game, address things that were or could have been problematic about the game or the way it was played this time, and can provide feedback to organizers" (2013). Though she does not use the term 'de-roling', Stark does state that the first step of a debrief must be "a little ritual to help people say goodbye to their characters or get out of character" (ibid.). In a deliberately open-ended list, she suggests that this may be achieved by means of techniques such as asking players to literally disrobe themselves of their character by placing a costume item associated with the character on the ground or by providing a countdown before the return to the actual domain (therefore, we might say, establishing a temporal threshold between worlds).

Building on Fatland's and Stark's suggestions, Sarah Lynne Bowman and Evan Torner specifically mention de-roling as "a form of debriefing and detaching from one's in-game role, to recover from LARPers' adoption of an alternate persona for long periods of time, which requires a different frame of reference" (Bowman and Torner 2016, 163). The de-roling techniques that they refer to as being commonly used in LARPing include "taking off a piece of one's character's costume and placing it in a circle, saying goodbye to the character for a time; thinking of one or more aspects of one's character that one admires and 'taking' it with them; thinking of one or more aspects of one's character one dislikes and wishes to leave behind; speaking about one's character in the third person during war stories or debriefing to emphasize a sense of distance; and making sure to interact with all the people from a larp both in-character and out-of-character to emphasize the distinction" (ibid., 164).

2.1.5 *De-roling from Training Simulations and De-briefing*

In line with what we have observed, the integration and use of role-play and digital simulations in education and training processes also emphasize the psychological and formative need for assisting trainees in their transition back to their everyday identities. Directly reflecting Landy's above-quoted observation regarding the structure of a dramatherapy session, training simulations are described as being composed of a briefing (or, sometimes, prebriefing), the simulated scenario proper, and a de-briefing (Rhodes and Curran 2005). It might be important to mention that the literature concerned with the formative and educational uses of role-play tends not to make any functional distinctions between role-playing situations in actual environments versus digital ones (Gualeni et al. 2017). This, in itself, is indicative, further emphasizing the (at least perceived) existential similarities between the adoption of special subjective states and roles in ritualistic, dramatherapeutic or role-playing practices, and the practice of stepping into a virtual subjectivity in a virtual environment.

In the context of education and training, and in analogy with the practices discussed above, a briefing stage is often considered necessary for establishing the parameters of the scenario. In a review of literature on the use of such practices in nursing simulations, for example, Karin Page-Cutrara notes that prebriefing is generally deemed necessary for communicating the objectives of a particular simulation session, assigning trainees' roles and orienting them towards the scenario (Page-Cutrara 2014, 138). At the end of a simulation session, de-roling techniques offer a structured space for critical reflection and for facilitating the re-appropriation of the trainees' everyday bodies and identities. Pilots who practice with flight simulators, for example, experience a degree of disconnection between how their bodies respond in actual situations and how they react in virtual simulations. It takes them several hours to fully 're-incorporate',[1] which is why several airlines do not allow their pilots to fly within a specified period after they have been training in a digital simulator.

When trainee medical doctors use computer simulations, they are invited to take a role that treats the suffering and the medical conditions of a simulated patient as if they were those of an actual human being. On those premises, Faith Stafford reports anecdotal evidence of trainees being psychologically affected by the death of a virtual patient (Stafford 2005). In addition to the use of humour, which Stafford considers to be an oft-neglected psychodynamic strategy for distancing oneself from a potentially traumatic event or situation, the author emphasizes the paramount importance of the moment of debriefing that takes place after a training session. The educational as well as psychological benefits of de-roling through debriefing ensue from the fact that this is the moment in which "the meaning of the enactment is clarified and the lessons to be learnt are underlined" (ibid.).

2.2 En-roling and De-roling in Virtual Environments

At first glance, the practice of engaging with virtual worlds—whether in the form of screen-based experiences or mediated by VR technologies—has several commonalities with the various cultural practices described above. As with drama, participants in virtual experiences find themselves stepping into experiential domains within which they adopt roles, pre-designed or otherwise, which they perform at different levels of involvement while maintaining varying degrees of awareness of their actual self-identity. Accordingly, one would expect to encounter techniques of en-roling into, and de-roling from, virtual subjectivities that are, broadly speaking, analogous to the ones we have identified at work across a wide range of cultural practices.

To a degree, this expectation is borne out. As we have just noted in the preceding section, digital simulation scenarios in an educational context generally make careful use of pre- and de-briefing practices. Outside such a context, however, identifying en-roling and de-roling practices for participation in virtual environments proves much more difficult. By and large, digital games, artistically oriented VR installations, VR walkthroughs of museums and cultural heritage sites, and similar virtual environments aimed at a general public tend to lack explicitly formalized transitional practices. In the following sections of the chapter, we will consider the limited transitional practices that do exist when moving into and out of virtual subjectivities, before considering the question of whether the relative dearth of such practices can, in itself, provide a valuable insight into the particular existential character of virtual subjectivities.

2.2.1 Transitioning into Virtual Environments

When the user takes their first steps in a virtual world they have not engaged with before, they are likely to have only a very limited knowledge of the subjectivities they are afforded within that world. They might not yet be aware of the capabilities they are granted or the limitations that define their possibility to inhabit that world. As such, they might have no idea of what kinds of projects it is possible for them to take on, let alone which project(s), if any, are suggested, explicitly or less so, by the implied designer.[2] In short, the user might not know what they *can* do or what they *should* do—and, as a result, because these are integral components of such an understanding, they might not know who they are in the virtual world.

As we have already pointed out, though a great deal has been written in game studies about the player's incorporation as an embodied being in digital game worlds, little has been said about how the player 'steps into' a virtual subjectivity and (potentially) into a virtual body. Some efforts have been made in this direction. Janet Murray, for instance, refers to the notion of 'threshold markers' to delimit virtual experiences (1998, 117). These approaches have, however, been infrequent and relatively cursory. It appears that, in general, it has been taken as a given that the player simply 'is' already in this subject-position and can step into or out of it at will with little to no effort. Brendan Keogh notes that the idea of "an effortless transferal of agency into a virtual world to take on a virtual body" has been, by and large, "the certain ground of videogame theory" (2018, 6). Because of this assumption, little academic attention has been paid to the activities, practices, and experiences by which players learn what is expected of them, internalize the required knowledge, acquire the necessary skills, and learn to perceive and act in the virtual world of the game *as* the virtual subjects they are to be.

This en-roling is by no means a trivial process. To begin with, before anything we refer to as 'en-roling' can even begin to occur, a basic level of acquired ability in terms of operating the technology required to access a virtual world is a necessary prerequisite—whether this means knowing how to operate a game controller or a VR headset, how to navigate a menu, and so on. When it comes to digital games, for example, Keogh writes about the 'embodied literacies' that are required of players before they can achieve a sense of embodied presence in a digital game's virtual environment and that are often the result of long years of experience (ibid., 78).

In addition to this technical knowledge, as an additional requirement, the user must also internalize, even if only on an implicit level, the idea of what a virtual world is, what the 'virtual' entails in distinction from the actual, and so on. In other words, the user needs to adopt what we might call 'an attitude of virtuality'. This attitude entails the internalization of the knowledge that in virtual environments, at least in most cases, failure is reversible and does not lead to permanent losses, injury, or death, and that the choices they make and the actions they take will not have any direct consequences upon the actual world, and so on. It is on the basis of these prerequisites that an individual is able to step into a virtual subjectivity in the first place (also see endnote 2). However, the process of en-roling into a particular virtual subjectivity still needs to take place. En-roling can happen in different ways, as part of different contextual relations, and in different temporal sequences. In order to make these processes easier to explore, we decided to group them conceptually into three categories: *external*, *threshold*, and *internal* processes. Below, we shall describe each of these categories in turn.

Before continuing, we want to clarify that we consider these categories useful for analytical purposes, as we venture to demonstrate, as they offer neatly separated conceptual domains. However, it is important to remind the reader that, in practice, they are intermingled and mutually constitutive, which makes them hard to actually pigeonhole with the same degree of sharpness with which we are discussing them here. On the basis of what was discussed until this point, we propose to divide processes of en-roling as follows:

1. **External en-roling processes** are spatially and conceptually removed from the specific experience of a virtual world, but still significantly influence how a subject 'en-roles' in that world. External processes can, for example, function as propaedeutic steps towards en-roling in a virtual world. In general, these practices belong to the wider socio-techno-cultural context of engaging with virtual worlds.
2. **Threshold en-roling processes** happen, instead in the liminal moments (and in the liminal spaces) that mark the transition between being engaged in the actual world and being absorbed by a virtual world. In this section, we will discuss apparatuses such as manuals and other documentation, software menus, introductory videos, game box art, the artwork on arcade cabinets, trailers, and tie-in comics.
3. **Internal en-roling processes** are internal to a virtual world. They include events and activities that are part of the direct experience of an interactive digital world's possibilities, affordances, and conventions.

Among all virtual worlds, we decided to focus on those of computer games, and *Fallout 4* (Bethesda Game Studios 2015) in particular. Despite evident technical and experiential divergences with the worlds of computer games, we consider that the processes of en-roling emerging—for example—from the experience of virtual training simulations and interactive explorations of virtual reproductions of world heritage sites to be substantially similar, both in terms of the technologies involved and in terms of their phenomenological and existential character. The most salient differences between the two have to do with the social situation that surrounds the experiences and the largely self-contained quality of our experiences with the virtual worlds of computer games, that is to say that the latter are not generally expected to provide practical and conceptual aspects of subjectivity that can be directly applicable to the actual world.

2.2.2 *External En-roling Processes*

Before a user begins to engage with a particular virtual world experience, they will generally foster a set of expectations regarding what the experience will be like—including what kind of virtual subjectivity they are likely to be positioned in. In this section, we briefly consider four factors that might engender such expectations—the game's genre(s), its franchise, its developers, and its adherence to a transmedia universe. This list is not intended to be exhaustive, but illustrative. While there are certainly other factors that might build the player's expectations of a game, the four factors we consider here demonstrate the variety of ways in which the en-roling process can be set in motion well before the player begins playing the game.

Genre Expectations

Games—and virtual environments—have been organized into genres, which define some of the expectations related to our experiences with those objects. The drawing-up of genre labels and boundaries is not necessarily a clean or clear-cut process (Wolf 2002; Apperley 2006)—at a minimum, digital games can be categorized into genres both based on iconography and on interactivity (Wolf 2002). *Fallout 4* evidently belongs within the iconographic tradition of the post-apocalyptic narrative, alongside films like *Mad Max: Fury Road* (Miller 2015) and novels like *The Road* (McCarthy 2006). At the same time, in 'interactive genre' terms, it is an open-world action role-playing game (RPG). Admittedly, such generic categories tend to shift in response to new examples that question established distinctions. Nevertheless, they shape the players' expectations concerning the game's world and their role within it.

For example, during character creation in *Fallout 4*, the player is asked to divide points into their S.P.E.C.I.A.L. statistics—numbers that determine the character's strength, perception, endurance, charisma, intelligence, agility, and luck affect the possibilities of action within the game world. This type of character creation is common in computer RPGs, and while the system includes information about what the different statistics mean, understanding the full implications of those choices involves familiarity with such systems in general and the mechanics of *Fallout* games in particular.

Franchise Expectations

If a game is an instalment in a franchise, familiarity with earlier games in the same franchise sets a variety of expectations about the game in question. *Fallout 4* is, of course, part of the *Fallout* franchise, which consisted of six previously released games. Though there are evident differences between the various games in the franchise, there are a number of distinctive family resemblances: the diegetic domain of the Wasteland, the lore regarding the nuclear conflict that created it, the Vaults and the factions vying for power, recurring elements of iconography (items, enemies, characters, etc.), and mechanical elements like the S.P.E.C.I.A.L. system. Associated with all of this are recurring elements in terms of who the player is, and what they do, in the Wasteland. In each of the games, the player has to travel through the Wasteland on a variety of quests, negotiating their navigation of the terrain in the light of their vulnerability to environmental hazards like radiation and the threat of being assaulted by beasts or people from hostile factions, engaging in tactically oriented combat, and collecting weapons, armour, and other useful items. In most of the games, the protagonist is an individual who has just emerged from one of the Vaults in which survivors sheltered during the war, and, as such, is new to the Wasteland—anticipating the necessity of learning new skills through a levelling system. Taken together, all those aspects facilitate the potential players' piecing together of what kind of role they will be invited to adopt in *Fallout 4*.

Developer Expectations

The idea that the works of a given artist will share a recognizable stylistic and thematic fingerprint, even in all their possible disparity ('auteur theory', see Sarris 1999), is not as critically prevalent in relation to games designers as it is for film directors. Nonetheless, certain individual designers and development studios came to be associated with recognizable iconographic tropes or gameplay styles.

Apart from being a 'post-apocalyptic game', an 'open-world action RPG', and a '*Fallout* game', *Fallout 4* is also a 'Bethesda game'. Bethesda have become known for establishing—largely through the recent iterations of the *Elder Scrolls* franchise of fantasy RPGs—a particular formula for the open-world action RPG, to which *Fallout 3* largely adhered. As such, the player's horizon of expectation for *Fallout 4* would be shaped not only by its connection to *Fallout*, but also to *The Elder Scrolls V: Skyrim*

(Bethesda Game Studios 2011)—the Bethesda game that immediately precedes it and with which it shares the proprietary *Creation* engine.

Inevitably, this will also shape the player's expectations regarding her in-game subjectivity. A single example will suffice to illustrate this point. In the first *Fallout*, which was developed by Interplay Productions in 1997, the Vault Dweller, the game's protagonist, could take on a variety of side-quests and explore the Wasteland as she saw fit. However, there was a hard time limit to the game: if the Vault Dweller did not complete the quest and retrieve the Water Chip in 150 in-game days, the Vault would run out of drinking water, everyone would die, and the game would be over. Conversely, in Bethesda open-world games, the player is free to ignore the main story quest indefinitely, completing side-quests or wandering at leisure, allowing for a more empowered subjectivity in which the importance of choosing what to do is considerably lessened (since it is conceivable for a player to eventually do virtually everything it is possible to do in the gameworld). It is likely that a player familiar with Bethesda's earlier games will expect a similar temporal freedom to luxuriate in the gameworld's possibilities and would be surprised to encounter a time limit that enforces hard choices regarding what to experience and not experience within a given playthrough.

2.2.3 Threshold En-roling Processes

Before entering the virtual environment in question, the user is likely to encounter a range of texts, images, or other materials related to the kind of experience one will be actively involved in. This is particularly the case for digital games, which exist within a saturated media ecology in which every new release comes with a concerted, multimedia marketing push. These can include official trailers released by the game's publishers, previews, or reviews in the gaming press, the game's box art (in the case of a physical release) or its online-shop splash-screen (for a digital release), *Let's Play* videos, arcade cabinet decorations, feelies (Karhulahti 2012), and stories written in game manuals—a common practice during a time when technical limitations made including that information within a game more difficult. All of these materials—which stand in a paratextual relationship to the game when considered in their capacity to inform and frame the play experience even before it starts (Genette 1997; Švelch 2017)—constitute the game's 'primordial speech' (Arsenault and Perron 2009, 118).

Fig. 2.1 A frame from the *Fallout 4* launch trailer showing the game's protagonist in the Wasteland

Bethesda released three trailers in the run-up to *Fallout 4*'s launch. The first, released on June 3, 2015, introduces the player to the gameworld—intercutting scenes of the fictional town of Sanctuary Hills, Massachusetts, before and after the nuclear war that creates the Wasteland. In the trailer's final moments, however, we glimpse the game's protagonist, holding a rifle and striding dramatically down a road towards a ruined city (Fig 2.1). This confirms what generic conventions would lead a player to expect of the protagonist: the stoic male wanderer in a lawless wilderness. The trailer's closing images suggest that this narrative trope will be translated, in gameplay terms, into combat mechanics based on shooting and spatial exploration and traversal of the gameworld—again, conforming to the expectations established by the game's alignment with the open-world action-RPG genre in general, with Bethesda's earlier games in particular, and with the *Fallout* franchise even more specifically.

The launch trailer, then, exemplifies the ways in which a range of materials—the reveal of the box art, subsequent trailers, previews on specialized press, and interviews with the developers—give the players a glimpse into the virtual world that awaits them. As Dominic Arsenault and Bernard Perron point out, this also includes an idea of the subjective existence the player will have in that world:

> By reading the description on the box or some reviews beforehand, the gamer comes to know what kind of content to expect, and to identify the particular genre(s) it will fall under [...] Since the cover of *Gears of War* (Epic Games 2006) shows a massive soldier holding an oversized gun with a chainsaw, and the narrative is about saving humanity from extinction and waging war against a horde of monstrous aliens, it is not difficult to imagine the sort of actions the gamer will be undertaking. (Arsenault and Perron 2009, 118)

2.2.4 Internal En-roling Processes

Before the player has begun to play the game, a specific 'knowledge schema' might already have been presented to them regarding the virtual subjectivity they will be stepping into (Arsenault and Perron 2009, 114). This, however, only constitutes the frame through which they approach the gameworld when they first pick up the controller. At that point, their expectations regarding their virtual subjectivity are either going to be confirmed and fleshed out, amended, or subverted. In any case, in order to play the game, the player has to internalize a virtual subjectivity within its world, and digital games offer a range of mechanisms for aiding players in making this transition.

Recognizing the Other

After an introductory cutscene, the first thing the new player of *Fallout 4* sees is their avatar's face reflected in a bathroom mirror (Fig. 2.2). We could observe, to start with, that this is a moment in which the relation of the player to their own in-game subjectivity is manifested in both its subjective and objective dimensions (Vella 2015, 228–231). The player has stepped into the avatar's subjective perspective, looking out through his or her eyes upon the gameworld—and yet, at the same time, they also find themselves looking upon their own virtual face as the object of their perception. In Jean-Paul Sartre's terms, this is "an indication of the subject to himself" (1966, 123). When we think about ourselves—making ourselves an object in our own consciousness—we are framing some representation of ourselves *to* ourselves. Necessarily, this leads to a split within the self—the 'I' that is seeing is not the same as the 'I' that is seen. The former is a subjective consciousness, the latter an objectified self-representation *of* and *for* that consciousness. Sartre names these, respectively, the subject and the self.

Fig. 2.2 Coming face to face with one's virtual subjectivity in *Fallout 4*

Of course, the face the player is seeing in the in-game mirror is not their own. Even should they make use of the game's character customization options to create a character that looks as much like themselves as possible, it is still a self-representation that they are distanced from. A paradox therefore emerges that requires some form of reconciliation—the face the player sees in this virtual mirror is not their own, and yet they must take it on *as* their own. This is the paradox at the heart of role-taking in general, which can be summed up as "being oneself and not oneself at the same time" (Landy 1993, 48).

Motivations and Goals

Once the initial identification has been achieved, and the player has incorporated this virtual subject as themselves, *Fallout 4* gives the player cues about the expectations being placed upon them by the game. Immediately, the game—in the voices of the non-player characters the player encounters—begins issuing specific instructions. The player-character's wife or husband (depending on the gender one has chosen to play) asks them to check if breakfast is ready; meanwhile, it is not long before the sound of a baby crying in the nursery and the robot-housemaid informing the player that the baby needs some paternal/maternal affection lead the player to take on the task of caring for the baby.

In this way, the player's virtual subjectivity is addressed as that of a spouse and a parent. These highly conventional roles come with ready-made, socially established sets of expected behaviours—we all know how a spouse and a parent *should* act towards their partner and their child. In drawing upon such instantly comprehensible roles, then, *Fallout 4* provides the player with a ready-made set of role expectations to step into—involving, presumably, the duty to protect and care for the child, to support the partner, and so on.

Of course, to live with the idea that we are bound by the expectations that shape our socially determined role is, for Sartre, to live in 'bad faith' by denying our own freedom (1966, 87). The player has some degree of freedom to shirk these expectations and to pursue a different project in the gameworld instead. For instance, a player might decide to complete all of the game's optional side quests and ignore the main quest. The player is also free to abandon the 'implied player' position by playing in a way that consciously goes against the (perceived) intentions of the implied designer (see endnote 2). For example, a player might make it their subversive mission to kill every non-player character in the game. These transgressive impulses notwithstanding, it remains the case that the player is situated against a more or less limited horizon of possibility. In these respects, Marta Matylda Kania has argued that bad faith is inherent to digital game play: as long as we wish to continue playing, we submit ourselves to the expectations that the game places upon our in-game subjectivity (2017, 99–106).

Returning to *Fallout 4*, these expectations come into play at the end of the introductory sequence, in a cutscene where the player is witness to their in-game character's partner being murdered and their infant child being kidnapped. This sets up the player-character's motivation for the main quest as being a dual one, resting on well-worn and immediately familiar narrative tropes: rescuing their child and revenge for their murdered partner. All other goals in the game—completing sub-quests, collecting helpful items, and so on—can then be understood as sub-goals, subordinated to these overarching motivations (Lankoski 2011, 297). The player's adoption of these goals as their own—that is, as their project in the virtual world—constitutes one of the most significant steps in the en-roling process.

Settling into the Instrumental Complex
At this point, control over the player-character is given back to the player. To draw on Petri Lankoski's terms, the opening sequences of the game have established both 'empathic engagement' and 'goal-related engagement' with the player-character (ibid.). In doing so, the game has structured the player's move through what Lankoski, following Murray Smith (1995), describes as the three stages of engagement with a character: *recognition* of the character, *alignment* with the perspective of the character, and *allegiance* with the character (ibid.).

Nonetheless, there is still something lacking for the player to be able to fully inhabit their virtual subjectivity. As we have highlighted already in Sect. 1.2, to a considerable extent one's virtual subjectivity is shaped by the capabilities and limitations one is granted within the virtual world and by the resulting possibilities for one to act within that world. Again, this observation has an existential basis. For Sartre, the projectual structure of human existence, which we discussed at length in the previous chapter, is inherently tied to one's capacities for acting: our projects are "projects of action" (1966, 425). I pursue my projects by acting upon them. To put it differently, I "project myself towards my possibilities" through acting (ibid., 446)—in the will to act, our consciousness surpasses its current condition in the direction of a new state of affairs it wishes to bring about.

The projectual possibilities we perceive in our being are, thus, determined by our capacities for action. Using a striking image, Sartre writes that "the world as the correlate of the possibilities which I *am* appears [...] as the enormous skeletal outline of all my possible actions" (ibid.). From my actual standpoint, a vast network of action-possibilities spreads outwards, each constituting a path I might choose to follow and a new state of affairs I might bring about.

Sartre uses the term 'instrumental complex' (ibid., 428) to refer to this intricate network of action-possibilities radiating outwards from one's current situation. To embark on a project, he suggests, is to plot a path through this space of potentialities and act to make it actual—which means that it is "only within an instrumental-complex that is already established" that projects can be undertaken and that the things we encounter in the world can take their meaning as affordances or obstacles towards that project (ibid., 620). In order for one to step into a virtual subject-position, then, one has to internalize the instrumental complex that shapes it: one needs to know what one's possibilities for action are, before one can proceed to act. In contemporary digital games, which often feature dauntingly extensive instrumental complexes, this is not a straightforward process. It requires experimentation on the part of the player, as they engage in 'real-time hermeneutics' (Aarseth 2003; Arjoranta 2011) to make sense of the gameworld and their place within it. In a similar vein, Arsenault and Perron have discussed the heuristic cycle of player input and feedback from the gameworld as the process by which players build an understanding of how they are to act in a game (2009).

Of course, this experimentation is facilitated by the virtual nature of the domain, which, in most cases, means our actions are reversible (Gualeni 2015, 124)—we can always reload a saved game and try something different if an experiment goes awry. To establish an even greater facility of experimentation at the start of a game, however, it is typical, even of games with an open world topography, to initially position the player off in a small, gated area in which they need to complete a number of low-level tasks before they are allowed to venture into the wider world. *Fallout 4* adheres to this structure: for the first hour or so of play, the player is confined within Vault 111, a small, easily navigable and relatively safe area where they are not as systematically and consistently under threat as they are out in the Wasteland (Fig. 2.3).[3]

In order to render the complex cognitive task of internalizing the instrumental complex more manageable, Peter Howell et al. (2014, 2)

Fig. 2.3 Finding one's feet in the safety of Vault 111 in *Fallout 4*

write that "games present players with a simple 'starting state' to which other game mechanics are incrementally added as the player discovers them or the game introduces them", terming this process 'incremental accretive learning' (ibid.).

There are two primary elements to this process, which *Fallout 4* explicitly employs during the player's stay in Vault 111. The first is the repetition of tasks with increasing difficulty as a means of testing out the player's grasp of a newly granted ability in the gameworld. When the player picks up the first melee weapon they find in the game—a security baton—they are led into a straightforward fight against a single mutant cockroach, the weakest enemy in the game. Should the player survive and progress, having established that they can at least perform the basics of melee fighting, the game then pits them against three mutant cockroaches in the next room. The second element is the gradual introduction of new elements to the player's agential repertoire. It is not until the player has gone through several melee combat encounters of gradually increasing difficulty that they come across a ranged weapon: in effect, the player is being given time to incorporate one ability into their instrumental complex before being asked to add another ability to their repertoire. It is only once the player has absorbed—and been tested on—the basic set of affordances that they are finally allowed to leave the initial gated area (see note 3) and venture out into the wider virtual environment as the virtual subject they have begun to learn to be.

2.2.5 Transitioning Out of Virtual Environments

If, as we have seen in the previous section, there is a paucity of academic work on players' transition *into* their in-game roles and subjectivities, the lack of critical attention to the ways in which players move *out* of these roles and subjectivities is, if anything, even more marked. This is hardly surprising: not only are such practices undertheorized, but they are underrepresented in the first place. Where, as we have pointed out, games make use of an array of techniques and features to aid the player in en-roling into their virtual subjectivity, the transition back to actual subjectivity tends to be far more abrupt and unceremonious. Having said that, it is possible to identify a number of techniques that constitute, at a minimum, token attempts at structuring ways to transition out of a virtual subjectivity. In order, we shall consider the ways in which contemporary games treat the moment of player-character death, scores and achievements, and narrative closure.

Death of the Player-character

In most games, the death of the player-character or avatar rarely constitutes a matter of any finality or of any serious and lasting consequence (Frasca 2000; Mukherjee 2009; Kirkpatrick 2011, 179–186). Yet, it remains the case that the moment of dying often represents a temporary suspension of one's virtual subjectivity and a resulting, albeit brief, distancing from one's in-game role. In *Deus Ex* (Ion Storm 2000), for instance, the instant of death is marked by a sudden departure from the game's first-person perspective. Where the player ordinarily sees the gameworld through the eyes of the player-character, JC Denton, the occurrence of in-game death throws the player out of Denton's head. During death sequences, the player's perspective on the game world shifts to a third-person frontal mid-shot of the protagonist's corpse, slowly spiralling to a more distant shot as the screen fades to white.

Even when there is no such clean break in the continuity of the player's sensory perspective upon the virtual world, the loss of avatarial control already throws the player out of the avatar as the 'body' of their virtual subjectivity. Instead, the avatar comes to be viewed objectively (Vella 2014), as a 'thing' from which the player's consciousness has dissociated. When the little boy one controls in the platform game *Limbo* (Playdead 2010), for example, falls victim to any of the plethora of threats awaiting the player in the game's threatening world, the player, having lost control, is forced to witness the macabre mangling and dismemberment of the figure that, a moment earlier, they had controlled and identified as their in-game self. At that point, the game no longer responds to any input from the player. Speaking of the similarly gruesome death sequences in *Dead Space 2* (Visceral Games 2011), the Italian game scholar Riccardo Fassone argues that the moment of death marks "the morphing of the player into a spectator" (2017, 63).

Death marks, thus, a temporary relinquishing of one's in-game virtual subjectivity, affording the possibility for a moment of reflection before this same virtual subjectivity is restored. By being cast into the spectator mode, the player is forced—to return to the idea of the 'double perspective' of engagement with virtual worlds (Sect. 1.4)—to step out of the internal perspective of their virtual subjectivity, and, instead, to evaluate this same virtual subjectivity, and their choices within it, from an external perspective.

Scores and Achievements

In the classic arcade game format, the end result of a play session is a numerical high score which the player is invited to autograph using (typically) a three-character signature. It is easy to read this signing-off on the outcome of the play session as a way of taking ownership of one's in-game achievements—for instance, by taking pride if one's score is displayed at the top of the high-score chart.

What the score constitutes, essentially, is a quantitative evaluation of one's performance in their in-game role—did one perform this role well or poorly? As such, apart from the possibility we have just highlighted of taking claim of one's own performance of virtual subjectivity, this kind of measurable result also allows one to take stock of one's own performance. The empire-building strategy game *Civilization II* (MicroProse 1996) went further in linking its final numerical score to a judgment on the player's virtual subjectivity within its world. Based upon the score, the player is granted a title befitting the nature of their rule. If one played particularly badly, for example, the game's final screen shows that history shall remember one, for example as "Stefano the Worthless", while, in the best possible case, posterity would know one as, say, "Daniel the Magnificent".

Though there are a number of contemporary games that still employ such a high-score format, a more common contemporary technique is the awarding of trophies, achievements, or badges as a recognition of the player's performance. For our current purposes, achievements or trophies are interesting for at least two reasons.

First, as we have suggested in the previous chapter, they can motivate an alternative projectual structure to a game's stated (or implied but nevertheless evident) main goals or objectives. Certain achievements might be tied to optional sub-goals (Montola et al. 2009) or encourage in-game behaviours that are not conductive to the completion of the game objectives, and that, indeed, might be inimical to them. As such, the gaining of a particular achievement might represent the trace of a conscious choice on the part of the player—it might, in short, say something about the kind of virtual subjectivity they choose to play out, representing a more qualitative form of judgment or summing-up than a numerical score might. In *Fallout 4*, for example, a player might get the 'Lovable' achievement if they attain the maximum relationship level with a companion; alternatively, they might be awarded the 'Masshole' achievement if they kill 300 people or more. Both of these achievements constitute a summative description of the kind of virtual subject the player chose to enact in the game.

Second, achievements and trophies represent, in a sense, something that is *brought back* across the threshold. Achievements are "stored outside the scope of individual game sessions" (Björk 2015), permanently added to the player's collection of achievements on their account (for instance, on *Steam* or on the *Playstation Network*), and visible as a component of their profile (Jakobsson 2011). Because of this, they can become tangible symbols of a particular engagement with a virtual subjectivity that are reincorporated into one's actual subjectivity—insofar as one's collection of achievements becomes a significant element in the construction of one's identity.

Narrative Closure

'Games of progression' (Juul 2002) are games that are projectually structured around a predetermined quest or series of quests that usually present a more-or-less linear "scripted narrative" (Calleja 2011, 120). Those games typically reach an end point where the player has achieved everything the game has asked them to do, and the story has reached its conclusion.

In such cases, despite the difficulties in specifying closure in relation to an open-ended, configurable experience that still promises further unexplored possibilities, games can still employ the conventions of traditional storytelling, in which "closure is used as a means of harmonizing a story by tying its threads in one coherent knot and resuming stasis" (Fassone 2017, 43).

Significantly, this often happens in the form of a cutscene, in which the player's divestment from their character is marked by a loss of control over the character. Writing about an empirical study of players of the first-person shooter *Crysis* (Crytek 2007), Kristine Jørgensen notes that, when players hear their character speak, unprompted, in a cutscene, "the avatar gives the impression of suddenly turning from being completely controlled by the player into being an individual and autonomous being with a will of his own" (2009, 3)—a significant loss of association with a role they had previously invested themselves in (also see Kirkland 2009).

To return to the example we used in the previous section, upon completing the main quest in *Fallout 4*—whichever of the four factions the player chooses to side with, and, resultantly, which of the four endings they achieve—the player is treated to a cutscene. This is already a significant departure in a game that makes scant use of cutscenes, and that, at every point between the intro sequence and the ending, gives the player constant control over their character's actions and dialogue. In the cutscene—which varies only in small details depending on which of the endings is actualized—the player-character speaks, in monologue, of their life in the new world order which, partly as a result of the player's actions, has reached some form of stability. In a shot that fully embraces cliché, the player-character is seen from behind, walking off into the sunset, and into their new world—away from the player.

2.2.6 Concluding Remarks on De-roling from Virtual Subjectivities

Though the techniques outlined in the previous section represent the basic demarcation of a threshold (distancing the player from their character and providing some form of closure, whether in a narrative sense or in the sense of a tangible or quantifiable outcome) it is plain to see that they do not constitute fully developed and sustained practices of de-roling. Given the fact that practices of de-roling are, by and large, only cursorily present in digital games, it was not surprising to observe that the field of game studies has little to say on the matter. However, we consider this, in itself, to be a potentially revealing observation. It shows, in fact, that the lack of systematic implementations of de-roling practices in digital games is not felt as a significant lack. This invites the question: what does this say about the existential character of stepping into a virtual subjectivity in a digital game?

Part of the explanation could be that there is a significant degree of recognition that the player's investment in a virtual role can never be complete and totalizing. Katie Salen and Eric Zimmermann have similarly argued that the players' engagement with their player-characters "is not a simple matter of direct identification", but operates according to precisely such a "double-consciousness of play", with the result that "the player takes on the role of the figure in relation to the gameworld while remaining aware of her own existence as a player manipulating a game object" (Salen and Zimmermann 2004, 453). Vella has attributed this phenomenon to an inherent "double perspectival structure of ludic engagement" (Vella 2015, 55). He argues that, while playing, players shift from moment to moment between inhabiting the experiential position of the ludic subject(s) in the gameworld and that of themselves as players outside the gameworld. Games often put players in the condition of having to switch between multiple roles or ludic subjectivities. In the opening scene of *Fahrenheit* (Quantic Dream 2005), for example, the player controls a character who has committed a murder in a restaurant bathroom while apparently under a trance and is tasked with concealing the evidence and making a getaway. In scene that follows, the player switches to controlling a detective investigating the same crime scene.

Vella argues that these aspects of flexibility and instability that emerge when adopting ludic subjectivities further destabilize the players' investment and identification in a single, totalizing in-game role (Vella 2016). This observation would align the adoption of virtual subjectivities with Fine's earlier remark concerning the frame-switching that frequently occurs during table-top role-playing sessions. However, our review of the existing literature on LARP suggests that, at least for certain categories of play activity, the players' transition into and, even more so, out of the in-game role, might not be such an inconsequential matter. Moreover, developments in VR technology might fundamentally alter the character of users' experiences of digital game worlds, likely rendering the shift into and out of a subjective standpoint within the gameworld less trivial.

Recent research on altered perceptions, automatic mental processes, and behaviours induced by virtual immersion identify the following as factors that are significantly associated with those phenomena: (i) the player having pre-existing medical or psychological conditions, (ii) the duration of a virtual play session, and (iii) the player focusing on experiential dimensions of role-play such as exploration, customization, and escapism rather than on the game's aspects relating to socialization (Ortiz De Gortari 2016, 11). To these, we might tentatively add a number of considerations that are likely to affect the person's degree of psychological and existential investment in the in-game role. Among them, we emphasize the granularity of our sensory engagement with the virtual environment, the ease or difficulty of stepping 'out of character' during role-play, and the time and energy spent preparing for a specific role. On the basis of these observations, we hypothesize that the greater the personal investment in the role, the more necessary de-roling practices become in order to avoid discomfort, confusion, and potential psychological damage.

As already mentioned in the beginning of this chapter, the lack of specific academic literature on de-roling, and specifically on de-roling from virtual experiences, is surprising, especially in light of the increasingly central role played by such experiences in the deconstruction and reconstruction of personal and cultural identity (Glass 1993; De Mul 2010, 125). In terms of their contribution to personal as well as social transformations, virtual experiences are often embraced as a particularly significant (and also particularly conspicuous) application of digital technologies. In the past two decades, interactive virtual worlds have started to be understood and studied as the testing grounds where new forms of subjectivity are experimented with and as the contexts where a new humanism has already started to arise (Gualeni 2015, 12; Vella 2015).

Dutch philosopher Jos De Mul maintains that a typical trait of the experience of virtual worlds is the blurring of the 'dividing line' between the real and virtual identities of people transitioning in and out of them. Citing Kenneth Gergen, De Mul emphasizes the remarkable growth in the number of dissociative disorders in parts of the world where the use of digital technology underwent a significant increase (Gergen 1991, 73–74 in De Mul 2010, 189). Motivated by similar intuitions, Frederick Aardema et al. observe an increase in dissociative events (depersonalization and derealization) as a result of the experience of artificial worlds through VR technologies (Aardema et al. 2010).

We deem it important to clarify that, to an extent, dissociative conditions are occurrences that are frequent in our everyday existence as human beings and that not all forms of dissociations correspond to states of distress or mental illness. Common situations like those of dreaming, daydreaming, religious or sexual ecstasy, or being absorbed in a book, a film, or a digital game, all involve some degrees of psychological dissociation. The same could be said about chemically altered conditions such as being intoxicated with alcohol or other psychotropic agents. Given the wide variety of contexts and degrees in which dissociations can manifest themselves, the term 'dissociation' ends up referring to a wide spectrum of nonclinical as well as clinical phenomena. In order to bypass confusion concerning the uses of the term, James M. Glass (1993) proposes to distinguish between pathological and non-pathological forms of dissociation, based on two key characteristics.

1. In a pathological scenario, dissociations are often completely outside of the control of the individual and are often triggered by external conditions rather than personal decisions.
2. In their non-pathological expression, dissociative states are less psychologically intense than the pathological ones. Being less psychologically totalizing, a non-pathological dissociation event allows the subjects experiencing it to maintain a persistent and coherent sense of their actual selves. It is precisely that degree of control and self-awareness that allows the subjects of a non-pathological event to recognize and experience a dissociation as a dissociation.

As previously observed, the research fields of game studies, human-computer interaction, and virtual world research invest a great deal of academic effort in concepts that have close relationships with states of sensory and/or psychological detachment from the actual world. The frequent use of constructs like immersion, presence, and incorporation are the most evident examples of such a focus. These notions are similar to concepts that are used in psychology to examine dissociative conditions and even appear to be related: according to Aardema et al., a higher degree of immersion or presence in a virtual world corresponds to a greater level of detachment from the actual world (Aardema et al. 2010, 1). Aardema et al. also advance the perspective according to which individuals who are reportedly likely to experience pathological kinds of dissociations will have more difficulty adapting to discontinuities in perceptual environments. On this hypothetical premise, they show that perceptual discontinuities induced by temporarily being absorbed by a virtual environment increase dissociative symptoms among those with a predisposition towards them (Aardema et al. 2010, 2).

Faced with the growing diffusion and commercial success of VR technologies as a particular, and particularly immersive, way of experiencing virtual worlds, we wonder whether these technologies pose additional dangers to technologically developed societies in terms of their dissociative effects. Our questions and concerns are further stimulated by observing how the access to extra-ordinary states of being (in this case, in the form of interactive, artificial worlds) are rendered progressively more experientially granular and accessible. Will the existentially enticing, accessible, and phenomenologically rich worlds disclosed by VR technologies cause even more intense dissociative states in individuals who might be prone to experience pathological kinds of dissociations? And is there a risk that they might trigger psychological and perceptual dissociations of the pathological kind in those individuals who did not originally demonstrate a predisposition to have them?

If we are ready to accept that, like in the case of drama, the existential meaning and the experiential effects of virtual environments emerge from lived, bodily experiences, then the intensity of the experience of virtual worlds can motivate our questions and worries. In other words, the perspective we are proposing here is that VR technology, which we understand as quantitatively exasperating some of the traits and experiential effects of our relationships with digital technology applications, might also aggravate their pathologically dissociative ones. In that scenario, a better understanding of techniques and effects of en-roling and (especially) de-roling will significantly contribute to the mental well-being of the next generations and become a factor in the ways in which they will lead their progressively more fragmented, multiplicitous existences.

Notes

1. This is due to the fact that the experience of mass and acceleration in the virtual world upheld by the digital simulation (and often enriched by mechanical devices) is inconsistent with how the actual bodies of the pilots adapt and respond to them in the actual world. During the period that follows their computer simulation training, which might last up to several hours, they often experience difficulties with balance (De Mul 2010, 203).
2. In a publication with Nele Van de Mosselaer, we define that implied designer as the conceptualization of the designer which the player constructs based on their experience of a virtual environment, to which the player ascribes all those intentions which they think lie at the basis of the creation of this environment, and which guides the way players interpret and act within it. It is relevant to observe, here, that in virtual environments, the idea of the implied designer not only guides the interpretation of what it is like to be in that situation, but also the ways in which the users position themselves (and orient their goals) within them. This means that the intentions the user infer to be the creative force behind a virtual environment contribute to determining how users give form to their projects and identities within it.
3. Extending Ewan Kirkland's view concerning particular kinds of spaces in horror digital games (2009), we propose to those call these safe spaces, which are characteristic of the initial sections of a wide array of role-playing games, 'natal spaces'. By referring to the small, safe, and typically dimly lit

rooms at the beginning of the game as natal spaces, we are drawing an implicit analogy between being born as an infant and taking up a new virtual (or, in this case, ludic) subjectivity. Players' avatars typically awaken in these antechambers alone and occasionally immersed in the game's equivalent of amniotic fluid (see Caselli 2019). In natal spaces, the players can familiarize with some of the game's affordances and controls. As Caselli also notes, it is no coincidence that these spaces often host in-game tutorials or sections characterized by simplified gameplay (ibid.). In addition to offering a safe space for instructive, exploratory experimentation on the part of the player, another characteristic of such spaces is their often highly rigid structure, in which the player is funnelled along a pre-set path to finally be released in the wider virtual environment through a corridor: the metaphorical correspondent of the birth canal.

References

Aardema, F., O'Konnor, K., Côté, S., & Taillon, A. 2010. Virtual reality induces dissociation and lowers sense of presence in objective reality. *Cyberpsychology, behaviour, and social Networking, 13* (4), 429–435.

Aarseth, E. 2003. "Playing Research: Methodological Approaches to Game Analysis." DiGRA'03 – Proceedings of the Digital Games Research Association Conference 2003: Level Up, November 4–6, Utrecht (The Netherlands).

Alexander, C. 1979. *The Timeless Way of Building*. New York (NY): Oxford University Press.

Alexander, C.; Ishikawa, S.; Silverstein, M. 1977. *A Pattern Language: Towns, buildings, Construction*. New York (NY): Oxford University Press.

Apperley, T. 2006. "Genre and Game Studies: Towards a Critical Approach to Video Game Genres." In *Simulation & Gaming, 37* (1), 6–23.

Arjoranta, J. 2011. "Do We Need a Real-Time Hermeneutics? Structures of Meaning in Games." Proceedings of the 2011 DiGRA International Conference: Think Design Play. Hilversum (The Netherlands): September 14–17.

Arsenault, D., & Perron, B. 2009. "In the Frame of the Magic Cycle: The Circle(s) of Gameplay." In Perron, B. & Wolf, M. J. P. (eds.). *The Video Game Theory Reader 2*. London (UK): Routledge, 109–133.

Atwater, B. 2016. "We need to talk: A literature review of debrief". In *International Journal of Role-Playing, 6*, 7–11.

Bethesda Game Studios. 2011. *The Elder Scrolls V: Skyrim* [Windows]. Digital game directed by Howard, T., and published by Bethesda Softworks.

Bethesda Game Studios. 2015. *Fallout 4* [Playstation 4]. Digital game directed by Howard, T., and published by Bethesda Softworks.

Björk, S. 2015. "Achievements." *Game Design Patterns Wiki*. Retrieved from http://virt10.itu.chalmers.se/index.php/Achievements

Bowman, S. L. 2013, April 18. (Nordiclarptalks). Bleed: How emotions affect role-playing experiences [Video file]. Retrieved from https://nordiclarptalks.org/bleed-how-emotions-affect-role-playing

Bowman, S. L., and Torner, E. 2016. "Post-LARP depression." In Trammell, A., Torner, E. and Waldron E. L. (eds.). *Analog Game Studies: Volume 1*, 161–166. Pittsburgh (PA): Carnegie Mellon – ETC Press.

Calleja, G. 2011. *In-game: from immersion to incorporation*. Cambridge (MA): The MIT Press.

Caselli, S. 2019. "Thrown into the world. Transformative aesthetics of avatars' in-game awakenings." Proceedings of the Philosophy of Computer Games Conference. St. Petersburg (Russia), October 20–24, 2019.

Crytek. 2007. *Crysis* [Windows]. Digital game directed by Yerli, C., and published by Electronic Arts.

De Mul, J. 2010 [2002]. *Cyberspace Odyssey: Towards a Virtual Ontology and Anthropology*. Cambridge (MA): Cambridge Scholars Publishing.

Eco, U. 1980. "Function and Sign: The Semiotics of Architecture." In Broadbent, Bunt, R. & Jencks C. (eds.). *Signs, Symbols and Architecture*. Bath (UK): Pitman Press, 11–69.

Edland, T. K., Lindahl, T. L., & Rauum, M. 2011. "Mad About the Boy." In Andresen, L., Nielsen, C.B., Carbonelli, L., Heebøll-Christensen, J., & Oscilowski, M. (eds.) 2011. *Do LARP: Documentary Writings from KP2011*. Copenhagen (Denmark): Rollespilsakademiet, 92–107.

Epic Games. 2006. *Gears of War* [Xbox 360]. Digital game directed by Bleszinski, C., and published by Microsoft Game Studios.

Fassone, R. 2017. *Every Game is an Island: Endings and Extremities in Video Games*. London (UK): Bloomsbury Academic.

Fatland, E. 2013, July 23rd. "Debriefing intense Larps 101" [Blog post]. Retrieved from http://larpwright.efatland.com/?p=384

Fine, G. A. 1983. *Shared fantasy: Role-playing games as social worlds*. Chicago (IL): University of Chicago Press.

Foucault, M. 1984. "Of Other Spaces, heteropias" (Trans. By Miskowiec, J.). In *Architecture, mouvement, continuité*, *5*, 46–49.

Frasca, G. 2000. "Ephemeral games: Is it barbaric to design videogames after Auschwitz". In *Cybertext yearbook*, *2*, 172–180.

Genette, G. 1997. *Paratexts: Thresholds of Interpretation*. Trans. J. A. Lewin. Cambridge (UK): Cambridge University Press.

van Gennep, A. 1977 [1960]. *The Rites of Passage*. London (UK): Routledge & Kegan Paul.

Gergen, K. 1991. *The saturated self: Dilemmas of identity in contemporary society*. New York (NY): Basic Books.

Glass, J. M. 1993. *Shattered selves: Multiple personality in a postmodern world*. London (UK): Cornell University Press.

Goffman, E. 1956. *The Presentation of Self in Everyday Life*. New York (NY): Anchor House.

Goodman, N. 1992. "How buildings mean." In Alperson, P. (ed.). *The Philosophy of the Visual Arts*, 368–376. New York (NY): Oxford University Press.

Gualeni, S. 2015. *Virtual Worlds as philosophical tools*. Basingstoke (UK): Palgrave Macmillan.

Gualeni, S.; Vella, D.; Harrington, J. 2017. "De-Roling from Experiences and Identities in Virtual Worlds." In *Journal of Virtual World Research*, *10* (2).

Howell, P., Stevens, E., & Eyles, M. 2014. "Disrupting the Player's Schematised Knowledge of Game Components." Proceedings of the 2014 DiGRA International Conference. Snowbird (UT), August 3–6.

Imber-Black, E., Roberts, J., & Whiting, R. 1988. *Rituals in families and family therapy*. New York (NY): W.W. Norton.

Interplay Productions. 1997. *Fallout* [Windows]. Digital game directed by Urquhart, F., and published by Interplay Productions.

Ion Storm. 2000. *Deus Ex* [Windows]. Digital game directed by Spector, W., and published by Eidos Interactive.

Jakobsson, M. 2011. "The Achievement Machine: Understanding Xbox 360 Achievements in Gaming Practices." In *Game Studies*, *11* (1).

Jencks, C. 1980. "The Architectural Sign." In Broadbent, Bunt, R. & Jencks, C. (eds.), *Signs, Symbols and Architecture*. Bath (UK): Pitman Press, 72–118.

Jennings, S. 1992. *Dramatherapy: Theory and practice 2*. New York (NY): Tavistock/Routledge.

Jennings, S., Cattanach, A., Mitchell, S., Chesner, A., Meldrum, B. 1994. *The Handbook of Dramatherapy*. London (UK): Routledge.

Johnson, D. R. & Emunah, R. 2009. *Current Approaches in Drama Therapy*. Springfield (IL): Charles C. Thomas Publications Ltd.

Jørgensen, K. 2009. "'I'm Overburdened! An Empirical Study of the Player, the Avatar, and the Gameworld." Proceedings of the DiGRA 2009 International Conference. London (UK): September 1–4.

Juul, J. 2002. "The Open and the Closed: Games of Emergence and Games of Progression." In Mäyrä, F. (ed.). *Computer Games and Digital Culture Proceedings*. Tampere (Finland): Tampere University Press, 323–329.

Kania, M. M. 2017. *Perspectives of the Avatar: Sketching the Existential Aesthetics of Digital Games*. Wrocław (Poland): University of Lower Silesia Press.

Karhulahti, V. M. 2012. "Feelies: The Lost Art of Immersing the Narrative." DiGRA Nordic '12: Proceedings of 2012 International DiGRA Nordic Conference. Tampere (Finland). 6–8 June 2012.

Kellermann, P. F. 1992. *Focus on psychodrama: Therapeutic aspects of psychodrama*. London (UK): Jessica Kingsley Publishers.

Keogh, B. 2018. *A Play of Bodies: How we Perceive Videogames*. Cambridge (MA): MIT Press.

Kirkland, E. 2009. "Horror Videogames and the Uncanny, Breaking New Ground: Innovation in Games, Play, Practice, and Theory". Proceedings of the DiGRA 2009 International Conference. West London (UK), September 1–4, 2009.

Kirkpatrick, G. 2011. *Aesthetic Theory and the Video Game.* Manchester (UK): Manchester University Press.

Landy, R. J. 1993. *Persona and Performance: The Meaning of Role in Drama, Therapy, and Everyday Life.* London (UK): Jessica Kingsley Publishers.

Landy, R. J. 1996. *Essays in drama therapy: The double life.* London (UK): Jessica Kingsley Publishers.

Lankoski, P. 2011. "Player Character Engagement in Computer Games." In *Games and Culture*, *6* (4), 291–311.

Lappi, A. P. 2004. "The Character Interpretation: The Process Before the Immersion and the Game." In Montola, M. & Stenros, J. (eds.). *Beyond Role and Play: Tools, Toys and Theory for Harnessing the Imagination.* Helsinki (Finland): Ropecon, 97–106.

Lynch, K. 1984. *Good city form.* Cambridge (MA): The MIT Press.

McCarthy, C. 2006. *The Road.* New York (NY): Alfred A. Knopf.

MicroProse. 1996. *Civilization II* [Windows]. Digital game designed by Reynolds, B., Kaufman, D., & Briggs, J., and published by MicroProse.

Mitchell, D., Miller, G., & Voeten, P.J. (Producers) & Miller, G. (Director). 2015. *Mad Max: Fury Road* [Motion picture]. Australia and USA: Warner Bros. Pictures.

Montola, M. 2010. "The positive negative experience in extreme Role-playing". Proceedings of the 2010 DiGRA Nordic Conference: Experiencing Games: Games, Play, and Players.

Montola, M., Nummenmaa, T., Lucero, A., Boberg, M., & Korhonen, H. 2009. "Applying Game Achievement Systems to Enhance User Experience in a Photo Sharing Service." Proceedings of the 13th International MindTrek Conference. Tampere (Finland), September 30–October 2.

Moreno, J. L. 1972 [1946]. *Psychodrama, first volume.* New York (NY): Beacon House.

Mukherjee, S. 2009. "'Remembering How You Died': Memory, Death and Temporality in Videogames." Proceedings of the DiGRA 2009 International Conference. London (UK): September 1–4.

Murray, J.H. 1998. *Hamlet on the Holodeck.* Cambridge, MA: The MIT Press.

Ortiz De Gortari, A. B. 2016. "The Game transfer phenomena framework: Investigating altered perceptions, automatic mental processes and behaviors induced by virtual immersion". In *Annual Review of Cybertherapy and Telemedicine*, *14*, 9–12.

Page-Cutrara, K. 2014. "Use of prebriefing in nursing simulation: A literature review." *Journal of Nursing Education*, *53* (3), 136–141.

Playdead. 2010. *Limbo* [Windows]. Digital game directed by Jensen, A., and published by Playdead.

Quantic Dream. 2005. *Fahrenheit* [Playstation 2]. Digital game directed by Cage, D., and published by Atari.

Rhodes, M. & Curran, C. 2005. "Use of the human patient simulator to teach clinical judgment skills in a baccalaureate nursing programme." In *CIN: Computer Informatics Nursing*, *23*, 256–264.

Robbins, A. 1988. Therapeutic presence: Bridging expression and form. London (UK): Jessica Kingsley Publishers.

Salen, K., & Zimmermann, E. 2004. *Rules of Play*. Cambridge (MA): MIT Press.

Sarris, A. 1999 [1962]. "Notes on the Auteur Theory in 1962." In Braudy, L., & Cohen, M. *Film Theory and Criticism*. New York (NY): Oxford University Press, 515–518.

Sartre, J. P. 1966 [1943]. *Being and Nothingness*. Trans. Barnes, H. E. New York (NY): Washington Square Press.

Smith, M. 1995. *Engaging Characters: Fiction, Emotion, and the Cinema*. New York (NY): Oxford University Press.

Stafford, F. 2005. "The significance of de-roling and debriefing in training medical students using simulation to train medical students." In *Medical education*, *39* (11), 1083–1085.

Stark, L. 2013, December 1st. "How to run a post-Larp debrief" [Blog post]. Retrieved from http://leavingmundania.com/2013/12/01/run-post-larp-debrief

Švelch, J. 2017. *Paratexts to Non-Linear Media Texts: Paratextuality in Video Game Culture*. Doctoral dissertation. Charles University in Prague (Czechia).

Tinkler, D. 2014, June the 6th. "Research project looks at how de-roling may help actors shed intense roles" [Blog post]. Retrieved from https://phys.org/news/2014-06-de-roling-actors-intense-roles.html

Turner, V. W. 1969. *The Ritual Process*. Harmondsworth (UK): Penguin Books.

Van de Mosselaer, N. & Gualeni, S. 2020. "The Implied Designer and the Experience of Gameworlds." Proceedings of the 2020 DiGRA international Conference. Tampere, Finland. June 2–6, 2020.

Vella, D. 2014. "Player and Figure: An Analysis of a Scene in *Kentucky Route Zero*." Proceedings of the Nordic DiGRA 2014 Conference. Visby (Sweden), May 29–30.

Vella, D. 2015. *The ludic subject and the ludic self: Investigating the 'I-in-the Gameworld'*. Doctoral dissertation. IT University of Copenhagen (Denmark).

Vella, D. 2016. "Who am 'I' in the game? A typology of modes of ludic subjectivity". Proceedings of the 2016 DiGRA and FDG (joint conference).

Visceral Games. 2011. *Dead Space 2* [Playstation 3]. Digital game directed by Bagwell W., and published by Electronic Arts.

Whyte, W. 2006. "How do buildings mean? Some issues of interpretation in the history of architecture." In *History and Theory*, *45* (2), 153–177.

Wolf, M. J. P. 2002. *The Medium of the Video Game*. Austin (TX): University of Texas Press.

Yardley-Matwiejczuk, K. M. 1997. *Role-play: Theory and practice*. London (UK): Sage Publications.

CHAPTER 3

Helmuth Plessner and Virtual Worlds as Existential Complements

Abstract Extending and re-digesting Helmuth Plessner's work in philosophical anthropology, this chapter frames human existence as inherently lacking a reliable grounding on which to anchor values and meaning. In relation to this understanding of human existence, it focuses on technologies in general—and virtual worlds in particular—as existential tools through which human beings strive to come to terms with the uncertainty and incompleteness that characterize their existence.

Keywords Plessner • Philosophical anthropology • Virtual worlds • Technology

3.1 Virtual Worlds and Existentialism: Four Dialogues (Chaps. 3, 4, 5, and 6)

The two preceding chapters focused on the existential structures of virtual subjectivities and the processes by which we en-role into and de-role from—such subjectivities. The following chapters will continue the discussion by highlighting some of the existential structure of virtual subjectivities and the processes by which we en-role into, and de-role from, such subjectivities, Chaps. 3, 4, 5, and 6 aspire to highlight some of the ambiguous ways in which the practice of adopting virtual subjectivities in virtual environments can be understood as being existentially relevant. In other words, though our perspective so far has largely been an internal one—looking at

S. Gualeni, D. Vella, *Virtual Existentialism*,
https://doi.org/10.1007/978-3-030-38478-4_3

existential meaning and structures within our virtual existences—in Chaps. 3, 4, 5, and 6 we shift to an external perspective and consider the ways in which virtual subjectivities might be existentially meaningful in the larger context of our actual lives. In pursuing this objective, the following chapters focus on specific notions articulated by four philosophers whose work deal, more or less directly, with existential themes: Helmuth Plessner focus on specific notions articulated by four philosophers: Helmuth Plessner (the present chapter), Peter W. Zapffe (Chap. 4), Jean-Paul Sartre (Chap. 5), and Eugen Fink (Chap. 6). Parts of the work of each thinker are discussed and repurposed as conceptual tools to reveal how virtual worlds are already understood and employed in ways that are existentially significant. Our appropriation of these thinkers' notions and frameworks also serves the purpose of characterizing virtual environments as technologies that can be useful in reframing philosophical questions concerning subjectivity, self-determination, and finitude.

As already mentioned, the four philosophers we decided to engage with are not all directly associated with existential philosophy. They were chosen because of the ways in which their work lends itself to an existential perspective on technologically mediated experience and, at the same time, liaises with other perspectives, such as the evolutionary and the anthropological. The general premise that these four chapters have in common is their reliance upon an understanding of being human as an ongoing process, an open question that cannot be satisfactorily answered or even be completely understood without taking into consideration our relationship with our technological creations. Starting from this common ground, each chapter discusses and contextualizes the ways in which we create, approach, and inhabit virtual environments from a different perspective and with a specific set of cultural objectives in mind.

3.2 Helmuth Plessner's Anthropological Laws

For a long time, the work of the German philosopher and sociologist Helmuth Plessner (1892–1985) remained almost unknown to the English-speaking world. Plessner's 1928 book *Die Stufen des Organischen und der Mensch* (*Levels of the Organic Life and the Human*) is often considered his most relevant contribution to philosophy, and was released in English for the first time only in 2019. The original German release of *Levels of the Organic Life and the Human* did not receive as much attention as we feel it deserved, obscured both by the long shadow cast by Martin Heidegger's *Being and Time* (which was published in German just a few months before

Plessner's book) and by the works of Plessner's own mentor Max Scheler, a central figure in the then-blossoming field of philosophical anthropology.

In the past fifteen years, however, Plessner's ideas have progressively gained academic momentum, and his texts have started to be translated and published in languages other than German. To a degree, the possibility to apply his anthropological perspectives to media theory and, more specifically, to the study of information and communication technologies have contributed to the growing academic interest in his ideas and perspectives. Recent academic publications by Jos De Mul (2010, 2013), Stefano Gualeni (2015), and Katharina Block (2019) among others testify to the many fruitful ways in which Plessner's work has been, and continues to be, repurposed and further developed to understand the existential effects and the evolutionary significance of our relationships with technology (and virtual environments in particular).

Strictly speaking, and in relation to the explicit goals of this book, Plessner cannot be considered an existentialist philosopher. We nonetheless feel that the ways in which he discusses life (both in biological and existential terms) offers insights that can complement and enrich the existential tradition. The same academic aspirations motivated the inclusion of the existential perspectives of the Norwegian philosopher Peter Wessel Zapffe in Chap. 4. The uncommon background of both authors (respectively medicine and zoology) as well as their relative obscurity served as additional incentives to include their work in this discussion on the existential dimensions of virtual technologies. With that purpose in mind, we will now discuss and extend Plessner's understanding of technology as one of the cultural contexts in which human beings try to come to terms with their inherent (existential) brokenness. In the focused pursuit of that specific aspect of his philosophical anthropology, we chose not to address one of the most oft-discussed aspects of *Levels of the Organic Life and the Human*: Plessner's theory of positionality (also known as the theory of boundary-realization). This decision was not taken on the basis of our considering his theory of positionality irrelevant for the topic at hand. Reading Plessner's work, in fact, the opposite appears to be evidently true.[1] However, for the sake of readability and succinctness, we opted to introduce Plessner's philosophy of technology without explaining how it emerges from his theory of positionality. In other words, we decided to side-step themes and terms related to boundary-realization and, instead, focus directly on his understanding of the human condition as one of uncertainty and indeterminacy (Plessner 1982, Vol. IV, 321).

In *Levels of the Organic Life and the Human*, Plessner describes humans as beings that are characterized by indeterminacy and incompleteness. Not only are they aware of their own mortality, but human beings—through their capability for self-reflection—also recognize that the meaning and values they attribute to their existence are transient and arbitrary. The unbearable uncertainty that defines the human condition is, according to Plessner, a consequence of the fact that all possible ways in which humans can try and establish a stable and reliable grounding for their existence (be it scientific, religious, philosophical, artistic, or a combination thereof) are always themselves products of mankind. In this sense, he understood human beings as both the producers and the products of their culture (also see endnote 1 in the Introduction section).

In the concluding chapter of his *magnum opus*, Plessner does not simply recognize humans as creatures who are inherently indeterminate and incomplete, but also as beings that are aware of their mortality and 'groundlessness'. On those premises, *Levels of the Organic Life and the Human* identifies the fundamental existential task of human beings as that of 'becoming something', that of making something out of themselves (Plessner 1982, Vol. IV, 385). Through the process of self-making, human beings provide themselves with the meaning and balance that they constitutively lack (ibid.). To complete this overview of Plessner's fundamental anthropological positions, it might help to clarify that, in his work, humans are understood as creatures who both aspire to reach a state of wholeness and satisfaction and cherish dreams of a 'homeland', a reliable ground (both in the form of a stable set of values and a defined sense of the self) in which to 'root' their existence (Plessner 2006; De Mul 2010, 204–205).

In the articulation of these positions, Plessner radicalizes the philosophical theme of the human as a lacking being, a theme that is obviously not unique to his thought. The idea of humankind as a species that is better understood through its lacks and needs, rather than by its instincts and abilities, is in fact relatively common in Western thought. As we have already seen, in Chap. 2, the core of Jean-Paul Sartre's existential understanding of the human being is the 'nothingness' that defines this being, which necessitates an interminable projectual drive towards self-determination. In this sense, both Sartre and Plessner stand in a distinct philosophical tradition. Famously portraying the human being as "something to be surpassed" in *Thus Spoke Zarathustra*, Friedrich Nietzsche also presents our existence as an ultimately endless task: a challenge in which mankind is constantly asked to overcome itself (prologue, part 3). In a comparable fashion, Arnold Gehlen defines man as an inherently "deficient being" (*Mängelwesen*) (Gehlen 1988, 10). More recently, Bernard

Stiegler labelled this state of native lack "the default of being" (Stiegler 1998).

An aspect that is, instead, specific to Plessner's understanding of the human as a 'lacking being' is the way in which, in his work, human life relates to technologies and technological advancements.[2] In *Levels of the Organic Life and the Human*, technology is discussed as one of the ways in which human beings aspire to complement and overcome their inherent incompleteness. This is the reason why, towards the end of his book, Plessner talks about humans as beings that are 'artificial by nature' (Plessner 2006, 334). 'Natural artificiality' is the name of the first of his three fundamental anthropological laws, general and basic invariants that—Plessner argues—need to be always kept into consideration when studying the human being. To keep this text focused and brief, we will concentrate our attention exclusively on the first and the third anthropological laws, thus by-passing the second one—that of 'mediated immediacy'—that would require explaining Plessner's positionality theory.

As outlined in the previous paragraph, the first anthropological law identifies the artificial dimension of culture as a native one for human beings, as their 'second nature', or a new 'fatherland' (*Heimat*), where they can pursue balance and rootedness (Plessner 1982, Vol. IV, 391–395). Both Plessner's first and third anthropological laws present humans as beings that cannot be isolated or thought apart from their cultural contexts and artificial extensions and its technological extensions. According to him, this has been the case since the very moment in which *Homo habilis* manufactured the first stone tools. Ever since, technology has always been directed at overcoming our lacks and our finitude (Plessner 2006, 334). Through culture and technology, according to Plessner, human beings express (and act upon) their desire to bridge the distance that separates them from the world, from their fellow humans, and from themselves (De Mul 2013, 19). It is in the sense outlined in the previous passages that we can be justified in claiming that human beings "have always been cyborgs": beings that are composed of both organic and technological elements (see Haraway 1991).

Up to this point, Plessner's analysis indicates that the artificial grounding provided by a cultural 'second nature' (or, perhaps more aptly, 'second natures') can allow human beings to overcome their indeterminacy and inherent groundlessness. However, Plessner is also adamant when he clarifies that this 'overcoming' cannot happen once and for all: existence cannot be 'solved' or 'fixed'. On the contrary, it is caught in an endless process that can never reach a stable conclusion. What follows from this

observation is that, in order to establish their (temporary) balance, human beings constantly need to aspire to become something new, to surpass their own deeds and achievements (Plessner 1982, Vol. IV, 395). This paints a rather tragic anthropological scenario: even when aided and transformed by cultures and technologies, the human being is doomed to remain an open question, one that has no conclusive answer and that we need to keep asking ourselves over and over again in an open-ended historical process.[3] The congruence between Plessner's thought on this point and Sartre's understanding of the endless process of self-determination through the existential project, which we presented in Chap. 1, should be obvious.

In Plessner's view, human aspirations to achieve a blissful state of completeness and to find a safe grounding are, thus, bound to remain unfulfilled fantasies. This strenuous and ultimately hopeless existential aspiration is captured in Plessner's third anthropological law, which attributes a 'utopian character' to human beings (which is also sometimes translated as 'utopian standpoint'). What this third law posits is, quite simply, that even if the aspirations to "safety, reconciliation with fate, understanding reality, a native soil" cannot be conclusively fulfilled, humans are nonetheless existentially bound to pursue them (Plessner 1982, Vol. V, 420). De Mul observes, in that respect, that even though technology could be said to have taken over the utopian role of religion in a society such as our own, the law of the 'utopian standpoint' remains valid (De Mul 2013, 20). The utopian character of the human being also expresses itself in what Plessner calls our distinctive "focus on the *irrealis*" (Plessner 2006, 334). This defining anthropological trait makes our desires and imaginations gravitate towards what is not (or not yet) the case, rather than on the present situation. This characteristic quality of being driven to ideas and possibilities that exceed one's mundane engagement with the actual world was considered by Johan Huizinga as having a fundamental role in social processes (1950).

In linguistics, the *modus irrealis* accounts for the 'can', the 'could', and the 'might' through which we negotiate our relationship with the actual world. In that sense, the 'focus on the *irrealis*' is not only a necessary component of how the human being acts as a political being, but is also expressed in the inherent need (and pleasure) of thinking in terms of 'what if' and 'it would be desirable if'. It is also the basic drive motivating activities such as daydreaming, being absorbed by fiction, playing, and even designing digital games. Similarly, and in a way that closely resonates with

Plessner's third anthropological law, Mark B. N. Hansen identifies this need as "that capacity, so fundamental to human existence, to be in excess of one's actual state" (Hansen 2003, 51). We find it particularly interesting, in relation to the larger objectives of this book, that Hansen labels that existential capacity 'virtuality' (ibid.).

To conclude this section, and in analogy with the existential role of technology discussed before, this 'subjunctive mindset' must also be recognized as offering the individual a temporary shelter from the seriousness and the boundaries of one's own identity, one's own physical capabilities, and one's own existential situation more generally speaking. From this particular angle, the 'focus on the *irrealis*' can also be understood as a coping mechanism that can protect us from aspects (or periods) of our existence that are particularly unpleasant or troublesome from a psychological point of view. In this sense, the 'focus on the *irrealis*' can be associated and compared with the existential techniques of 'distraction' discussed by Peter Wessel Zapffe (see Chap. 4).

3.3 Virtual Environments as Existential Technologies

Having developed his work before the social diffusion of computers, Plessner could not comment on digital culture or directly apply his theoretical insights to the analysis and criticism of computer applications like social media or virtual environments. Scholarly perspectives on virtual technologies and digital mediation that have close affinities with the anthropological positions discussed until now are, nowadays, relatively common. American media scholar Jay David Bolter, for example, argues that human subjectivity has been dynamically augmented by artificial means from the very onset of civilization and is now natively integrated with computers as the dominant form of mediation in Western cultures (Bolter and Grusin 2000, 41–44). Though he never explicitly mentions Plessner, Bolter's mentions Plessner, Bolter's perspective is evidently indebted to philosophy of technology and philosophical anthropology. Standing on the shoulders of other scholars who have either directly applied and extended Plessner's work in the age of digital media, or whose work draws parallel with Plessner's thought, we will further elaborate on his first and third anthropological laws—respectively, the human beings' 'natural artificiality' and their 'utopian character'.

As we have argued earlier in this book, there are a number of ways in which encounters with virtual environments can stimulate and guide changes in the structure of human experience and subjectivity. If, on a conceptual level, digital technologies encourage modes of thinking that are no longer exclusively focused on actuality, on a psychological level, they can be understood as granting experiential access to artificial, supplementary bodies and subjectivities that exist simultaneously with our actual ones.[4] To put this understanding in historical perspective, at the beginning of the last century, the futurist writer Filippo Tommaso Marinetti predicted that the "futurist transhuman" would be "multiplied by the machine". This new being was to be a hybrid creature, "with replaceable parts" and "multiple and simultaneous awarenesses" (De Mul 2010, 32). Similarly, Sherry Turkle writes that virtual worlds "encourage us to think of ourselves as fluid, emergent, decentralized, multiplicitous, flexible, and ever in progress" (Turkle 2007, 263–264).

In relation to these extensions of our bodies, and as already discussed in Chap. 2, it is relevant to observe that the growth in the number of dissociative mental illnesses appears to have a direct correlation with the increased social diffusion of digital technologies (Gergen 1991, 73–74; Aardema et al. 2010). In a private conversation, De Mul also noted that—from his perspective—the dissociation that characterizes our experiential engagement with virtual environments does not seem to be favouring an overcoming of the fragmentation and incompleteness that are inherent to human existence. On the contrary, he added, with digital technologies humans are effectively multiplying the number of ways in which they are removed from themselves.

Both in their alluring effects and in their accompanying hazards and dissatisfactions, virtual environments can thus be understood as technologies capable of fragmenting and of extending our possibilities for agency as well as our sense of personal identity. Through the lenses of Plessner's first and third anthropological laws, we could recognize virtual environments both as new manifestations of the 'utopian standpoint' and, at the same time, as situations that can experientially become part of one's actual lifeworld (as familiar and intuitive contexts for experience and existence, or as second, artificial natures). Of course, strictly speaking, virtual worlds are—by definition—distinct from one's actual lifeworld. They invite interactions that do not take place in the context that is shared by biological organisms, and yet—as already explained in Chap. 1—experiences in virtual worlds also, inevitably, constitute a subset of actual experiences.[5] This, we think, is the key evolutionary advantage offered by virtual technologies: their offering several virtual *irrealis* in ways that are accessible through one's actual cognitive and perceptual tools. They disclose new worlds and provide new bodies that we can grow familiar with. It is through this process of familiarization that digital technologies achieve their 'fluidifying' effects over various aspects of how we think (and especially concerning our imagination and our sense of possibility, see Chap. 7).

Discussing the cultural impact of digital technology from an analogous perspective, Vilém Flusser enthusiastically announced that we are beginning

> to liberate ourselves from the tyranny of an alleged reality. The slavish attitude with which we [...] approach objective reality in order to master it, has to give in to a new attitude, in which we intervene in the fields of possibilities in- and outside us, in order to intentionally realize some of these possibilities. [...] We are facing a second birth of mankind, a second *Homo erectus*. And this *Homo erectus*, who plays with chance in order to intentionally transform it into necessity, may be called *Homo ludens*. (Flusser 1992, 25)

Notes

1. Should the reader be interested in Plessner's understanding of the notion of boundary and in his positionality theory, our advice is to read Chaps. 5, 6, and 7 of *Levels of the Organic Life and the Human*. For quicker access to those ideas, Jos De Mul clearly summarized and updated Plessner's positionality theory in his 2010 book *Cyberspace Odyssey* (see De Mul 2010, 193–206).
2. To be sure, even Plessner's understanding of technology as one of the cultural practices through which human beings can make up for their shortcomings and lacks is not a new idea. It is, in fact, a notion that is as old as the philosophical branch of the philosophy of technology. Towards the end of the nineteenth century, the German philosopher Ernst Kapp was among the first to propose that all forms of technology could be ultimately understood as artificial extensions of certain capabilities of the human organism. Kapp laid the foundations of the philosophy of technology by advancing the idea of 'organ projections': the notion that all technologies are conscious (or even unconscious) actualizations of original functions of human organs (Kapp 1877). As already discussed elsewhere, we could even trace this way of approaching technological artefacts all the way back to the origin of

(written) Western philosophy to Plato's *Phaedrus* and *Seventh Letter* (see Gualeni 2015, 10, 101, 102).

3. On the basis of these passages of Plessner's, Martino Boccignone observes that the human being is thus not only portrayed as 'the apostate of nature' (as Plessner explicitly did, emphasizing the inherent artificiality of the human constitution), but also as 'the apostate of culture'. What Boccignone means is that, in their endless struggle to overcome themselves, human beings are the apostate of every particular culture that has previously been established in history as a grounding for human existence (Boccignone in De Mul 2013, 184).
4. The necessity for the qualifier 'supplementary' when talking about virtual extensions of our subjectivity emerges in relation to the fact that the bodies afforded by virtual technology cannot (or at least cannot yet) fully override and replace the actual ones. Nick Yee presents an analogous observation in his 2014 book *The Proteus Paradox*. "When we encounter new media and technological devices", he writes, "we fall back on the existing rules and norms we know. [...] And when we enter virtual worlds, this mental baggage hitches a ride with us. We react to digital bodies the way we react to physical bodies" (Yee 2014, 54).
5. For the Finnish scholar Erkki Huhtamo, the dynamism in the relationship between humans and technology in the digital era is particularly evident in the fact that the design of digital interfaces no longer seems to pursue integration—a symbolic association with similar functions in the actual world—or transparency—the quality of being perceptually as close as possible to a non-mediated experience (see Huhtamo 1995, 171). More and more often, according to him, digital applications inform their users and grant them affordances in ways that rely digital canons and traditions instead of referring to actual entities or functions (ibid.). Huhtamo's perspective on the relationship between the virtual and the actual could, however, not be applicable to technologies like virtual reality or augmented reality. Those applications tend, in fact, to be valued for their perceptual realism and their being intuitively accessible on the basis of one's knowledge and familiarity with the actual world.

References

Aardema, F., O'Konnor, K., Côté, S., & Taillon, A. 2010. "Virtual reality induces dissociation and lowers sense of presence in objective reality." In *Cyberpsychology, behaviour, and social Networking*, *13* (4), 429–435.

Block, K. 2019. "Machina Sapiens: Digital Posthumanism from the Perspective of Plessner's Logic of Levels". In *Human Studies*, *42* (1), 83–100.

Boccignone, M. E. 2013. "The Duty of Personal Identity: Authenticity and Irony." In De Mul, J. (ed.). *ARTIFICIAL BY NATURE: Plessner's Philosophical Anthropology: Perspectives and Prospects.* Amsterdam (The Netherlands): University of Amsterdam Press, 177–194.

Bolter, J. D. & Grusin, R. 2000 [1998]. *Remediation: Understanding New Media.* Cambridge (MA): The MIT Press.

De Mul, J. 2010 [2002]. *Cyberspace Odyssey: Towards a Virtual Ontology and Anthropology.* Cambridge (MA): Cambridge Scholars Publishing.

De Mul, J. (ed.). 2013. *Artificial by Nature: Plessner's Philosophical Anthropology: Perspectives and Prospects.* Amsterdam (The Netherlands): University of Amsterdam Press.

Flusser, V. 1992. "Das Ende der Tyrannei." In *Arch + Zeitschrift für Architektur und Stadtebau* (March 1992), 20–25.

Gehlen, A. 1988 [1940]. *Man: His Nature and Place in the World.* Trans. Clare McMillan and Karl Pillemer. New York (NY): Columbia University Press.

Gergen, K. 1991. *The saturated self: Dilemmas of identity in contemporary society.* New York (NY): Basic Books.

Gualeni, S. 2015. *Virtual Worlds as Philosophical Tools.* Basingstoke (UK): Palgrave Macmillan.

Haraway, D. J. 1991. "A Cyborg Manifesto: Science, Technology, and Socialist-Feminism in the Late Twentieth Century." In Haraway, D. J. 1991. *Simians, Cyborgs and Women: The Reinvention of Nature.* New York (NY): Routledge. 149–181.

Hansen, M. B. N. 2003. *New Philosophy for New Media.* Cambridge (MA): The MIT Press.

Huhtamo, E. 1995. "Encapsulated bodies in motion: Simulation and the quest for Total Immersion." In Penny, Simon (ed.). *Critical Issues in Electronic Media.* Albany (NY): State University of New York Press. 159–186.

Huizinga, J. 1950. *Homo Ludens.* Boston (MA): Beacon Press.

Kapp, E. 1877. *Grundlinien einer Philosophie der Technik.* Braunschweig (Germany): George Westermann ed. Available online at http://vlp.mpiwgberlin.mpg.de/library/data/lit39532?

Plessner, H. 1982. *H. Plessner, Gesammelte Schriften*, ed. Günter Dux, Odo Marquard, and Elisabeth Ströker. Frankfurt am Main (Germany): Suhrkamp.

Plessner, H. 2006 [1928]. *I gradi dell'organico e l'uomo. Introduzione all'antropologia filosofica.* Turin (Italy): Bollati Boringhieri.

Stiegler, B. 1998. *Technics and Time 1: The Fault of Epimetheus.* Stanford (CA): Stanford University Press.

Turkle, S. 2007 [2005]. *Life on the Screen.* New York (NY): Touchstone.

Yee, N. 2014. *The Proteus Paradox – How Online Games and Virtual Worlds Change Us—And How They Don't.* New Haven (CT): Yale University Press.

CHAPTER 4

Peter W. Zapffe and the Virtual Tragic

Abstract Norwegian philosopher and mountaineer Peter W. Zapffe understands human life as inherently meaningless. In relation to that meaninglessness, he identifies four ways in which human beings typically protect themselves from the existential panic that accompanies the awareness of that meaninglessness—isolation, anchoring, distraction, and sublimation. This chapter considers the extent to which our engagement with virtual environments can be understood in terms of each of Zapffe's four categories, thereby building an examination of virtual environments as technologies to repress existential panic.

Keywords Zapffe • Tragic • Isolation • Anchoring • Distraction • Sublimation • Virtual worlds

Peter Wessel Zapffe (1899–1990) was a Norwegian thinker, lawyer, and mountaineer whose philosophical work can be positioned across the currents of existentialism and biosophy.[1] The few works of his that are translated into English present biological life as an absurd, cosmic squander of resources. In pursuing an understanding of the relevance and the transformative effects of our relationships with virtual environments, we were particularly intrigued by Zapffe's take on human existence. In his work, Zapffe presents humans as beings who aspire to meaning and purpose in a world that is irredeemably meaningless. A recurring theme in his writings

S. Gualeni, D. Vella, *Virtual Existentialism*,
https://doi.org/10.1007/978-3-030-38478-4_4

is the paradoxical idea that "man has longings and spiritual demands that reality cannot fulfil".[2] This perspective is central to Zapffe's notion of 'the tragic' and of his consideration of the human being as a 'tragic animal'. Interestingly, for Zapffe, the reasons why human existence can be considered tragic are not rooted in humanity being weak, petty, or particularly prone to suffering, but rather in its being too capable for its own good. In a way that might clash with the nominal use of 'tragedy', Zapffe does not associate the term with notions like violence, misery, or scarcity, but with the idea of excess.

His perspectives on this topic, originally presented in the essay 'The Last Messiah' (1933) and further developed in his treatise 'On the Tragic' (1941), resonate with an ancient Greek understanding of tragedy. It is comparable, for example, to the one embraced by Sophocles' *Antigone*. In the 'Ode to Man' (the first choral song of the *Antigone*, lines 340–380), Sophocles characterizes humans as inherently tragic beings precisely because of their exceptional capabilities: their quick intellect and their being resourceful beyond imagination. These qualities, he continues, allow humanity to understand and face all kinds of situations, to adapt, thrive, and subjugate animals and rivers to its needs. Relentlessly, humans direct their remarkable skills "now to destruction/now again to greatness" (Sophocles 1984, lines 408, 409). From Sophocles' perspective, humanity is thus awesome in both senses of the word: at once wondrous and dreadful (see De Mul 2009).

Similarly, in Zapffe's 1933 essay 'The Last Messiah' the human being is regarded as "a biological paradox, an abomination, an absurdity, an exaggeration of disastrous nature" (Zapffe 2004, part II).[3] Zapffe explicitly presents humanity as a mistake: the tragic result of nature having overshot its intended target: "[a] species had been armed too heavily", thereby becoming "a menace to its own well-being" (ibid.). In which sense does he consider the human being a mistake and a threat to itself? Zapffe highlights what he considers a crucial, existential difference between human beings and beasts: whereas the suffering of beasts is self-confined (i.e. limited to the individual experience of each beast), human beings can grasp the interpersonal, cosmic nature of the meaningless and endlessly repeating nightmare that is existence. In other words, and differently from other animals, the human being is aware that struggle and suffering are ubiquitous, eternal, and purposeless. As a consequence of this awareness, humanity finds itself in a constant state of existential panic.

Why, then, has humanity not long ago gone extinct "during great epidemics of madness?" (Zapffe 2004, part III). Why do "only a fairly minor number of individuals perish because they fail to endure the train of living—because cognition gives them more than they can carry?" (ibid.). Zapffe's answer to these questions betrays a debt to the work of Sigmund Freud, arguing that most people simply learn to protect themselves by artificially suppressing their awareness of the absurdity of existence. In order to continue existing, Zapffe maintains, human beings constantly perform a "repression of its damaging surplus of consciousness" (ibid.).

Seen from his perspective, then, states of depression or fits of madness are not symptoms of a sick mind. Instead, they are indications that the mind's protective mechanisms are failing. In 'The Last Messiah', Zapffe groups these protective mechanisms into four macro-categories—*isolation, anchoring, distraction*, and *sublimation*—which can occur in every possible combination and often overlap with one another.

4.1 Panic-repressing Mechanisms in Virtual Environments

As outlined in the previous section, according to Zapffe, a cosmic panic characterizes life as experienced by human beings. The threat of existential despair is such a constant, overwhelming presence that the attempt to stave it off lies at the core of almost every human activity or pursuit.

From Zapffe's perspective, then, it would not be unfitting—in fact, it would even be a foregone conclusion—to consider virtual environments as technologies that participate in controlling and mitigating our awareness of the meaninglessness of existence. This is possible, as we explained in Chap. 1, precisely because virtual subjectivity is a nested component of our actual subjectivity, and the two influence one another. As such, in virtual worlds, those mechanisms aim to protect us from the awareness of both the meaninglessness of our existence and the meaninglessness of our virtual existence. With that purpose in mind, the creators of virtual environments typically design stimuli and motivations to keep us involved and psychologically invested in both worlds. Accordingly, we will address each of the four protective mechanisms identified by Zapffe with a particular focus on how our relationships with virtual environments could be understood as participating in each.

4.1.1 Isolation

With the term 'isolation', Zapffe indicates the ways in which we dismiss disturbing and destructive thoughts and feelings from our consciousness. In 'The Last Messiah', to exemplify the idea of isolation, Zapffe presents the cases of physicians and medical students who typically protect themselves from the tragic and disgusting aspects of their profession by adopting detached and technical stances towards their patients. This attitude is remarkably similar to what Michel Foucault indicated as the 'medical gaze' in his *The Birth of the Clinic* (2002). Another example of the technique of isolation permeating our everyday life can be recognized in the fact that 'tact' (i.e. not confronting people with untimely reminders of sex, bodily functions, decay, and death) is considered a highly desirable social trait.

In terms of our existential relationship with virtual environments and digital games, the idea of isolation can be recognized to be obviously always at work. To play a digital game is, by definition, to willingly operate within a limited possibility horizon[4] and to develop existential projects that are shaped by its affordances and by its narratives (Gualeni 2019; Vella and Gualeni 2019). The re-framing (and even the practical removal) of the disgusting and tragic dimensions of existence is, thus, one of the defining qualities of our experiences with games in general. For example, one's death is commonly presented in digital games as a rather trivial inconvenience, a nuisance rather than a tragic event (Mukherjee 2009; Kirkpatrick 2011). With very rare exceptions,[5] our avatars do not age in virtual environments—instead, we are presented with a fantasy of constant self-improvement, in which our in-game selves can accumulate experiences indefinitely, growing ever stronger and more capable without the shadow of the inevitability of decline, senility, and mortality. Similarly, toilets and the use of toilets are, in the great majority of cases, tactfully removed from virtual environments.

In other words, the virtual worlds of digital games, by and large, present us with 'sanitized' versions of the world, from which potentially upsetting reminders of the banality and futility of human existence have been excised.

For the sake of clarity, we want to highlight a conceptual difference between the functioning of the repressing mechanisms belonging to the 'isolation' category and the ones that can be grouped under the 'distraction' label. Whereas the 'distraction' techniques aim at diverting the subject's attention from situations that might cause troubling thoughts and feelings, the 'isolation' ones attempt to re-frame those situations as something that can be considered not tragic, not disturbing, or not personally affecting us (and thus negligible). What follows is that if we decided to analyse digital games as tools for 'distraction', we would likely focus on their capability to capture and redirect their players' attention. If, on the other hand, we were to approach digital games as technologies for 'isolation', we would concentrate on how their virtual environments invite and foster troubling ideas, or deliberately exclude those themes and situations.

4.1.2 Anchoring

In 'The Last Messiah', Zapffe describes societies as systems bound together by basic cultural ideas and collective values that are, for the most part, inherited from previous generations. He calls these ideas and values 'anchorings', as they can be understood as the fixed foundations which one's individual existence builds upon. Though this tends to be an unconsciously performed operation, the ways in which we adopt anchorings can also happen in a conscious manner, for example, when someone explicitly sets goals for oneself or when deliberately adopting a certain rule of conduct (e.g. a professional deontology).

Anchorings provide us with a sense of meaning and stability, protecting us from unsettling existential realizations concerning the meaninglessness of existence. According to Zapffe, we love our anchorings for the safety they offer, but at the same time we resent them for the ways in which they limit our freedom (Zapffe 2004, part III).

Broadly speaking, there are two ways in which we can understand anchoring to be at work in our engagement with virtual environments. Firstly, in trying to make the purposes and objectives of virtual activities and interactions intuitively accessible, digital games and training simulations often reference and mimic actual anchorings. Some of the values and ideas that guide and constrain our beliefs and behaviours in the actual world are to some degree reproduced in virtual environments. For example, the notion that every single human life is invaluable implicitly underpins the success criteria of surgical training simulations and flight simulators. Another example of actual anchoring being used in virtual environments and activities is evident in the narrative setup of a variety of digital games, where the freedom and the safety of the (fictional) people that our avatars care for are presented as values worth fighting for (and also worth playing for).

In contrast to our existence in the actual world, our accessing of virtual environments (and virtual subjectivities) allows us to exist in a world in which meaning and value are unambiguously established, objectified in computer code, and measured according to quantifiable metrics. In particular, commercial digital games tend to explicitly present unambiguous and resource-oriented world-views for their players. The clear and quantifiable objectives of those world-views are designed to foster feelings of meaningful progress and to alleviate existential anxiety in the players (Gualeni 2015, 76–78, 128). Moreover, although activities within digital games are often autotelic and self-contained, they can achieve additional existential relevance as social performances, for example when these experiences feature social media components. In those cases, digital games can effectively become contexts in which players can attain various kinds of external validation for their in-game behaviour and self-representation. The communities forming in and around digital games also constitute an influential component of social validation to the already existentially alluring possibilities offered by those virtual environments—for example, by flaunting the trophies or achievements that one has obtained (see Chap. 2).

The possibility to achieve a sense of self-realization in virtual environments does not, however, only emerge from pre-designed mechanisms meant to motivate users with clear reward mechanisms. Their existential effects are, in other words, not limited to validating users for executing a certain set of actions or for taking ethical decisions of a specific kind. Virtual environments disclose opportunities to projectually re-construct and aesthetically fashion one's (virtual) self in a number of ways, some of which are intended by the designers of the virtual environment in question, while some others are independently devised and adopted by the users. For instance, players can achieve a sense of self-realization and existential anchoring in pursuing goals and activities that are materially inscribed in *Super Mario Bros.* (Nintendo Creative Department 1985) by defeating the evil Bowser and rescuing his captive, Princess Peach. The games mechanically and narratively invite precisely that course of action. On top of that, the game also grants points for killing enemies, and offers rewards for collecting the coins scattered throughout Mushroom Kingdom.

However, when interactively engaging with the virtual environment of *Super Mario Bros.*, we are not constitutively bound to mindlessly pursue those goals. Instead, we can treat the game as tool to reflect on our beliefs and our conduct. With that critical attitude, we can decide to play *Super Mario Bros.* in accordance with deliberate, personal projects (see Gualeni 2014; Westerlaken 2017). An example of this alternative existential approach would be to carry out the decision to completely avoid collecting any coins while doing so. Acting in this way can be considered a practice of self-fashioning when voluntarily taken as a way to keep oneself aware of how *Super Mario Bros.* is ideologically rooted in instrumental rationality (i.e. that the in-game goals invite and reward activities such as the accumulation of resources and the optimization of tasks).

Generally speaking, existential anchorings in virtual worlds are particularly attractive for players, as the kinds of self-fashioning and self-realization that they offer are not only quicker and less inertial than their actual counterparts, but are also not bound by several of the limitations that characterize everyday existence. After all, our virtual projects do not need to be serious, permanent, determined by biological and cultural factors, or tied to the physical behaviours of the actual world.

4.1.3 Distraction

In the previous sub-section of this book chapter, we argued that virtual environments can disclose opportunities to pursue virtual projects of different kinds, and to existentially anchor ourselves in them. Having previously discussed techniques for isolation, it should be evident that not all the relationships we can establish with virtual environments are necessarily meaningful in terms of existential self-realization. Similar to isolation techniques, those for 'distraction' are existentially relevant not in their contributing to the construction of our 'self', but in their capability for momentarily cutting us off from meaninglessness and despair.

In 'The Last Messiah', Zapffe describes 'distraction' as a mode of protection by which one stunts one's awareness by "constantly enthralling it with impressions. This is typical even in childhood; without distraction, the child is also insufferable to itself" (Zapffe 2004, part III). For Zapffe, the re-direction of the subject's attention from situations that might cause troubling thoughts and feelings is particularly prominent in the lifestyle of high society.

Distraction mechanisms play an obvious and obviously central role in our relationship with virtual environments and with digital games in particular. We often and explicitly approach digital games (especially in the 'casual' sector of the videogames industry) for distraction purposes, busying ourselves with relatively simple and tantalizing virtual objectives and activities that are typically brief and meant to provide constant stimulation.

Indeed, such distraction could even be understood as the other side of the coin of the anchoring function of engaging with games: we distract ourselves from the lack of anchorings for our actual existence by taking on virtual 're-anchorings'.

For the sake of completeness, we want to point out that the dual effect of both distracting players and offering them possibilities for self-reflection and self-realization through anchoring is not exclusive to digital games in the 'casual' sector. Adventure and exploration games, and especially so when characterized by role-playing elements, are typically developed, promoted, sold, and consumed with the explicit objective of interactively disclosing being someone or something else: to develop that version of oneself in extraordinary ways, often under the promise of spectacular successes. Those games offer their players forms of escapism that are not merely distracting, but that clearly have 're-anchoring' existential effects.

4.1.4 Sublimation

Zapffe considers 'sublimation' the rarest of the four survival mechanisms. The characteristic setting it apart from the other three techniques is the fact that sublimation does not aim to repress cosmic panic, but rather to transform it into something perceived as meaningful. According to Zapffe "[t]hrough stylistic or artistic gifts can the very pain of living at times be converted into valuable experiences" (Zapffe 2004, part III). In other words, sublimation is also a generative mechanism and not simply a protective one. Zapffe exemplifies his idea of sublimation with the very activity of his own writing of 'The Last Messiah'. Through his text, his existential anxiety stops being a private and sterile struggle, but becomes something that can be shared with others, something that can inspire and uplift.

Arguably, it is no great leap to consider a game like *Every Day the Same Dream* (Molleindustria 2009, see Fig. 4.1) to be a similar attempt as the artistic, expressive kind of sublimation that Zapffe is discussing. The game depicts a lived experience—that of the alienated office worker in a late capitalist Western society—that might well give rise to existential panic, thanks to its repetition of an unfulfilling routine that appears to hold no meaning or potential for self-development. However, the very act of engaging with such an existence through rendering it in an expressive form already suggests the possibility of sublimating this existential panic. Of course, this suggestion presupposes both an inherent existential value

Fig. 4.1 A screenshot of Molleindustria's 2009 digital game *Every Day the Same Dream*

to creative practices (*à la* Nietzsche) and an understanding of digital games as constituting a distinct, ludic form of artistic expression, and as giving rise to a distinct kind of aesthetic experience (Vella 2016; Nguyen 2019).

The notion that digital games can carry various forms of meaning is firmly established in game studies through notions such as that of 'procedural rhetoric' (Bogost 2007). So is the idea that games can influence the player's existence, as in the idea of 'transformational play' (Barab et al. 2010)—or even that of the game designer through the creative design process itself (Gualeni 2014; Gualeni and Gomez Maureira 2018). However, the chiefly instrumental perspective of these approaches—which frame digital games in terms of their capacity to impart particular skills or change specific beliefs, attitudes, and behaviours—diverges from Zapffe's existential drive. Closer to the mark, arguably, is the idea we have already identified, in Chap. 1 of this book, as being central to existential ludology: all games echo the general existential structures of our being-in-the-world. By this understanding, when creating games, the designers frame their thoughts, feelings, and beliefs in a ludic form, and, when playing games, players are invited to bring before their consciousness the awareness of the existential structures of our being.

4.2 A Menace to Our Own Well-being

In this chapter, we have tried to offer an original perspective on how virtual environments could be understood as fundamentally contributing to our individual existence. The originality of our proposition chiefly consists in the fact that we did not articulate ways for understanding digital interactions and virtual worlds as novel routes to self-understanding and self-realization. What we have tried to do, instead, was to look at experiences and interactions in the virtual as activities that can assist us in coping with the lack of meaning that inherently characterizes human existence from the perspective of existential philosophy.

In the pursuit of this goal, we adopted Zapffe's categorization and used it as a conceptual lens to look at virtual environments and digital games. Our work revealed that there are many ways in which virtual interactions can be, and in several ways already are, involved in protecting ourselves from acknowledging the inherent absurdity of our existence. At the same time, Zapffe's work can help us realize that our repressive mechanisms and the technologies used to facilitate those techniques are not simply survival tools, but are always also "a menace to our own well-being" (Zapffe 2004, part II). What we are arguing here is that, while offering existential mechanisms that can promote our survival and well-being, our interactions with virtual environments can also be understood as contributing to making our species even less fit for life than it already is (also see Chap. 7).

Notes

1. In his 1941 treatise 'On the Tragic', Zapffe described biosophy as the analysis of human social life based on the philosophy of existence and biological facts. According to Zapffe, this narrow definition should not constitute an obstacle to widen its range to encompass all systematic thinking on biological issues.
2. This quote is taken from the 1990 documentary *The Philosopher Peter Wessel Zapffe in His 90th Year*, Original Films AS, Tromsø, Norway.
3. Zapffe further clarifies that, in front of the human being "[...] all things chain together in causes and effects, and everything he wants to grasp dissolves before his testing thought" (Zapffe 2004, part II). We invite the reader to compare this passage to a similar one in Sophocles' *Antigone*, where human beings are presented as having taught themselves "speech and thought, quick as the wind/and the mood and mind for law that rules the city [...] Man, the master, ingenious past all measures" (Sophocles 1984, lines 395, 396, 406).
4. The term 'possibility horizon' references the Ancient Greek origin of the term 'horizon', ὅρος (*oros*), which denotes a frontier, a boundary, a spatial limit. On this etymological foundation, 'horizon' is used in this book chapter to indicate the spatial boundaries that delimit the possibility space within which the player operates when engaging with a virtual environment.

5. Apart from some experimental, independent games such as *Is it Time?* (Fraina 2010), the only counter-example we are aware of in a major commercial game is the role-playing game *Fable* (Lionhead Studios 2004). Even in *Fable*, however, disturbing thoughts are very much isolated from gameworlds: the player-character's age is capped at 65, and aging has only a cosmetic effect with no impact upon the player-character's physical or mental abilities. It is also striking that the game's sequels stepped back from even this sanitized version of aging, with the feature being completely absent by the time of *Fable III* (Lionhead Studios 2013).

References

Barab, S. A., Gresalfi, M., & Ingram-Goble, A. 2010. "Transformational play: Using games to position person, content, and context." In *Educational Researcher, 39* (7), 525–536.

Bogost, I. 2007. *Persuasive games: The Expressive Power of Videogames.* Cambridge (MA): The MIT Press.

De Mul, J. 2009. "Awesome technologies." In *Art and Social Change. International Yearbook of Aesthetics, 13*, 120–139.

Foucault, M. 2002 [1963]. *The Birth of the Clinic.* London (UK): Routledge.

Fraina, J. 2010. *Is it time?* [Windows]. Digital game developed by Fraina, J., available online at: https://jaimefraina.com/2010/01/is-it-time/

Gualeni, S. 2014. "Freer Than We Think: Game Design as a Liberation Practice." Proceedings of the 2014 Philosophy of Computer Games Conference. Istanbul (Turkey), November 13–16, 2014.

Gualeni, S. 2015. *Virtual Worlds as Philosophical Tools.* Basingstoke (UK): Palgrave Macmillan.

Gualeni, S. 2019. "Virtual World-Weariness: On Delaying the Experiential Erosion of Digital Environments." In Gerber, A. and Goetz, U. (eds.) The Architectonics of Game Spaces: The Spatial Logic of the Virtual and its Meaning for the Real, 153–165. Bielefeld (Germany): Transcript.

Gualeni, S. & Gomez Maureira, M. 2018. "Self-Transformative effects of designing videogames and the challenge of capturing them quantitatively: a case study." Proceedings of the 2018 Foundation of Digital Games Conference. Malmö (Sweden), August 7–10, 2018.

Kirkpatrick, G. 2011. *Aesthetic Theory and the Videogame.* Manchester (UK): Manchester University Press.

Lionhead Studios. 2004. *Fable* [Xbox]. Digital Game directed by Molyneux, P.; Carter, D.; Huskins, B.; Webley, M., and published by Xbox Game Studios.

Lionhead Studios. 2013. *Fable III* [Xbox360]. Digital Game directed by Molyneux, P. and Atkins, J., and published by Xbox Game Studios.

Molleindustria. 2009. *Every Day the Same Dream* [Windows]. Digital Game developed by Paolo Pedercini and available online at https://molleindustria.org/everydaythesamedream/everydaythesamedream.html

Mukherjee, S. 2009. "Remembering How You Died: Memory, Death and Temporality in Videogames." Proceedings of the 2009 DiGRA International Conference. West London (UK), September 1–4, 2009.

Nguyen, C. T. 2019. "Games and the Art of Agency." In *Philosophical Review*, *128* (4), 423–462.

Nintendo Creative Department. 1985. *Super Mario Bros.* [Nintendo Entertainment System]. Digital Game published by Nintendo.

Sophocles. 1984. *The Three Theban Plays – Antigone, Oedipus King, Oedipus at Colonus.* Translated by Fages, R. New York (NY): Penguin Classics.

Vella, D. 2016. "The Ludic Muse: The Form of Games as Art." In *Countertext*, *2* (1), 66–84.

Vella, D. & Gualeni, S. 2019. "Virtual Subjectivity: Existence and Projectuality in Virtual Worlds." In *Techne': Research in Philosophy of Technology*, *23* (2).

Westerlaken, M. 2017. "Self-fashioning in action: Zelda's Breath of the Wild Vegan Run." Proceedings of the 2017 Philosophy of Computer Games Conference. Krakow (Poland), November 29–December 1, 2017.

Zapffe, P. W. 1996 [1941]. *Om det Tragiske* [On the Tragic]. Oslo (Norway): Pax forlag.

Zapffe, P. W. 2004 [1993]. "The Last Messiah" (English translation by Tangenes G. R.). *Philosophy Now*, *45*.

CHAPTER 5

Jean-Paul Sartre and Escaping from Being-in-the-World

Abstract This chapter approaches our experience of virtual worlds through Jean-Paul Sartre's phenomenology of the imagination. It looks at Sartre's argument that the existential significance of the imagination lies in its giving us the capacity to surpass our being-in-the-world. On this basis, the chapter draws a line between virtual subjectivity and how Sartre's frames the concept of imaginary consciousness, considering how our relationships with virtual environments both reflect the phenomenological structure Sartre attributes to the imagination and puts it into question.

Keywords Sartre • Imagination • Imaginary • Virtual worlds • Phenomenology

Jean-Paul Sartre never conceived of virtual worlds in the digitally mediated form we now have at our disposal. However, an intriguing passage in his 1940 book *The Imaginary* would seem to open up the possibility of adopting subjecitivit(ies) towards multiple worlds. The passage we are referring to begins by noting that "the very structure of transcendental consciousness implies that this consciousness is constitutive of *a* world" (2010, 179). Here, Sartre is merely restating one of the foundational tenets of the phenomenological understanding of consciousness upon which his brand of

S. Gualeni, D. Vella, *Virtual Existentialism*,
https://doi.org/10.1007/978-3-030-38478-4_5

existentialism is based: namely, that all consciousness is consciousness *of* something. The passage also implies that each *something* is perceived as part of the continuous, unified set of experiences of the world.

What is of particular interest for us is that Sartre then goes on to claim that a phenomenological analysis of consciousness "will not teach us that it must be constitutive of *one* such world, which is to say precisely the one where we are, with its earth, its animals, its people, and the history of its people" (ibid.). What is suggested here is, intriguingly, a claim that the capacity to direct itself towards worlds beyond the actual is inherent to human consciousness.

Though Sartre does not systematically follow up on this suggestion, a consideration of his development of a phenomenology of the imagination—within which this passage is situated—can lead us to an enriched understanding of the existential significance of our experience of virtual environments. In order to arrive at such an understanding, we shall need to pay attention to the ways in which the virtual both reflects and diverges from Sartre's understanding of the imagination.

5.1 Imagination as an Escape from the World

Sartre argues that, in order to organize the entire set of things it perceives in the form of the unified totality of a world, the individual subjective consciousness has to 'stand back' from that totality (2010, 184). As he puts it in *Being and Nothingness*, "it is from the point of view of beyond the world that being is organized into the world" (Sartre 1966, 51). Consciousness of the world, then, is at the same time 'nihilation' of the world—by this, Sartre means the establishment of non-equivalence between the world and the perceiving subject that is distanced from it in the very act of constituting it in perception. Because of this distance, consciousness is free from the world and can turn away from it. However, in turning away from the world, consciousness must turn towards *something*—after all, consciousness cannot be consciousness of nothing. This is where the imagination comes in. In a term Sartre borrows from Heidegger, he frames the imagination as the 'surpassing' of the world by consciousness, and is inherent to human consciousness: "all apprehension of the real as world implies a hidden surpassing towards the imaginary" (2010, 188).[1]

The existential significance that Sartre grants to the imagination becomes more fully apparent when he writes that productions of the imagination "offer us an escape from all the constraints of the *world*, they seem to be presented as a negation of the condition of *being in the world*, as an anti-world" (ibid., 136). We have already touched upon the importance Sartre grants to the idea of play in Sect. 1.5 of this book. In his book on Sartre, magic, and the imaginary, Daniel O'Shiel argues that play and the imaginary are inextricably linked in Sartre's thought, as they both stand in opposition to the ways we commonly relate to the actual world, and also in opposition to what Sartre calls the 'spirit of seriousness' (the attitude in which we renounce our freedom by fitting ourselves to the facts and demands of the world) (O'Shiel 2019, 118). In Sartre's conception, the imagination lets us escape from the existential situation that the world constitutes for us, with all the constraints and responsibilities it places upon us. It frees us from the inevitability of being-in-the-world and lets us be *anti-worldly*. To put it in simpler terms, it allows our consciousness to transcend the world. In that sense, the conceptual link between the way Sartre understood imagination and the theoretical perspectives on virtuality presented in this book are evident, and becomes even more so when we consider Elizabeth Grosz's claim that "the virtual is the space of [...] the unrealized, which at every moment loads the presence of the present with supplementarity" (2001, 77).

To return now to the largely unfulfilled promise of the passage quoted above—can we understand virtual environments as the fulfilment of a promise of an escape from the world that does not simply negate the world, but that provides consciousness with an alternative world to inhabit—a being-in-a-virtual-world?

5.2 Another World

The idea that the production of images—through painting, narrative, film, or any other medium—can be equated with the creation of a "secondary world" (Tolkien 1964) or "heterocosm" (Baumgarten 1954; Doležel 1998) is a long-established one. Oliver Grau maps out a tradition of what he terms "virtual art", going as far back as classical wall frescoes such as the Great Frieze in the Villa dei Misteri at Pompeii (2003, 25) which is invested in the creation of hermetic "image worlds" (ibid., 3).

Sartre, however, rejects the possibility of an "irreal world". For him, the productions of the imagination—even those which we reach through a physical *analogon*, such as a painting of Charles VIII which brings to our mind an image of the long-dead king (2010, 183)[2]—do not meet the constitutive conditions of worldliness: "the very idea of world implies for its objects the following double condition: they must be strictly individuated; they must be in balance with an environment [...] no irreal object fulfils this double condition" (ibid., 132).

Let us examine more closely the reasons Sartre gives for concluding that irreal objects—objects given to our consciousness by the imagination—fail to meet these two criteria. First, they are marked by an "essential poverty" (ibid., 9). An irreal object possesses precisely—and only—those qualities with which I imagine it. If I perceive an object—say, a red cube—I do so from a particular angle and at a certain distance. I am aware that my knowledge of it is not total: I can only see some of its sides, or I am not close enough to discern its texture. Further examination of the object might reveal surprising details—like the fact that one of the sides that were initially invisible from my viewpoint is in fact blue. When we imagine an object, instead, it is impossible for us to be surprised by it in the same way, as the irreal object has no existence outside my consciousness' grasp of it. It appears to me at once, and I am immediately aware of every detail with which it is endowed—for my consciousness has put those details there. If I attempt to find more detail in the image—for example, if, when describing a forest scene I have imagined, my friend asks me to describe the sounds one can hear there—it is not the case that I am discovering details in the scene which were there all along but which I had simply not turned my attention to. I would be furnishing these details through further acts of the imagination. The irreal object has no existence outside my consciousness' grasp of it.

Second, the irreal objects of the imaginary are "given as a perpetual 'elsewhere'" (ibid., 136)—there is "no simultaneity" (ibid., 131) either between these objects and the actual objects of my perception or between one irreal object and another. An irreal object is given to consciousness as "isolated [...,] it acts on nothing and nothing acts on it" (ibid., 135). As such, for Sartre, irreal objects lack both the independent, individuated existence and the interrelatedness to form part of a world—hence his labelling them as anti-worldly.

Here we come face-to-face with marked differences between our consciousness of a virtual object—that is, an object in a virtual environment—and Sartre's irreal object. Virtual objects are 'digital objects' (Hui 2016) that are intersubjectively verifiable—their individuated existence, and the constituent qualities that define them, are upheld by their computational materiality. Moreover, they exist in relation to the other virtual objects occupying the same plane of existence, potentially standing in relationships of mutual affect and causality. In that sense, at least, virtual objects appear to meet the two criteria for worldliness that Sartre stipulates. Yet, as we have already argued, virtual objects are defined in contradistinction to the actual—as such, they retain their status as irreal.

One might argue that the digital object is simply the *analogon* on the basis of which we can image the irreal object it represents. Just as a painting of Charles VIII directs our consciousness towards the long-dead king, so a particular digital object in a virtual environment might bring an image of a unicorn or a spaceship to our consciousness.

Accepting the validity of this denotative relation of representation between the underlying computational materiality and the virtual environment that appears to the user would equate virtual environments—in their capacity to make us conscious of things that are not really there—with the paintings, novels, or plays that Sartre uses as examples. Still, we cannot stop here. In direct contrast to these instances of representation in more traditional media, our experience of virtual objects seems to align more closely with our perception of real objects. As with the example of the red cube given above, I can move around a virtual object and see it from different viewpoints, adding to my knowledge of it with each new point of view. I can pick it up and position it elsewhere in the virtual environment. Depending on the qualities of the object and of the virtual subjectivity I am granted, I can throw it, consume it, transform it, and so on. This is what Espen Aarseth means when he argues that, unlike the entities represented in a novel, which are only given to us through a finite set of signs, the entities we encounter in a virtual environment "allow us to test their limits, comprehend causalities, establish strategies, and effect changes, in ways clearly denied us by fictions, but quite like in reality" (2006, 35).

This also means that our conduct towards virtual objects is different from the ways we generally deal with irreal objects. Sartre writes that, since irreal objects are produced by my imagination, "none of these objects claim a conduct, an action of me [...] they are pure passivity, they wait" (125). For this same reason, "the imaginary actions that I plan have no consequences but those that I want to give them" (ibid., 146). It is impossible for me to imagine performing an action with an outcome I cannot imagine.

In virtual environments, the situation is different. Virtual objects exist outside our individual consciousness—they continue to exist even if we are not aware of them or thinking about them and can affect us in ways we might not have anticipated. A monster in a digital game can bring our existence in the virtual environment to an end even if we did not notice its approach. Within digital game studies, Olli Tapio Leino coined the term 'gameplay condition', which we have already discussed in Chap. 1, to refer to the fact that the computational materiality underpinning a game's virtual environment upholds, and holds the player responsible for, the results of their choices and actions in that environment (2009, 12). Drawing on the work of Karen Barad, Justyna Janik similarly emphasizes the material quality of the encounter between a digital game and its player, highlighting the agential quality of the game as an equal partner in the process of play, as able to act upon and affect the player as the player is upon it (2018). The point regarding the materiality of our encounter with the digital object can be extended beyond digital games to encompass our engagement with virtual environments in general.

5.3 Being Towards Another World

As outlined in the previous section, the objects we encounter in a virtual environment have an independent, individuated existence in a domain within which they exist in a network of relations, thereby appearing to meet Sartre's criteria for worldliness. However, in order for any domain to be a world for our consciousness, our being must take it as its ground: in other words, it is only a world if we find ourselves *in* the world. In that sense, Grau talks about a "constitution of presence" which establishes "the quality of apparently being present in the images" as one of the characteristics of virtual art (2003, 14). The capacity to engender a feeling of being-there that has been theorized through notions of 'immersion', 'presence', and 'telepresence' has, in fact, often been considered a defining characteristic of virtual reality (see Steuer 1992).

This seems to be an even more decisive divergence of the virtual from Sartre's treatment of the imagination, in which such a consciousness of presence *to* and *alongside* the irreal is inadmissible. And yet, there are passages in *The Imaginary* that hint at the ways in which we *can*, in a more indirect way, be present towards the objects of our own imaginary:

> The object as imagined is an irreality [...] I cannot touch it, change its place: or rather I can indeed do so, but on the condition that I do it in an irreal way, renouncing being served by my own hands, resorting to phantom hands that will deliver irreal blows to this face: to act on these irreal objects, I must duplicate myself, *irrealize myself*. (2010, 125)

In other words: in order to be able to act upon the objects I imagine, I need to project an imaginary version of myself onto the stage of the imagination. As he later says, "a consciousness cannot 'be-in' an imaginary world, unless it is itself an imaginary consciousness" (ibid., 170).[3]

We have already discussed at length the idea that we are positioned as virtual subjects within virtual environments—and that it is our adoption of these virtual subjectivities that allows us to experience these digitally mediated environments as worlds *for us*. We propose that the enactment of a virtual subjectivity as a nested subjectivity in relation to or primary, everyday being (see Sect. 1.4) can be understood as precisely such an adoption of an "imaginary consciousness"—in the sense of a consciousness that exists upon, and in relation to, a realm that is not the actual world.

If we adopt this understanding, then, it would appear that we can corroborate the perspective we have already argued for in preceding chapters—namely, that virtual environments can be understood as alternative, secondary worlds towards which our consciousness can surpass the actual world—a being-in-another-world through which we can escape our being-in-the-world. This would constitute an extension of Sartre's ideas on the imagination, according to which the virtual would represent a kind of liminal category—an irreal domain that, in all the significant ways we have considered, mirrors the phenomenology of our perception of the real world. The final question that remains to be asked is: understood thus, what is the existential significance of the virtual?

There are two interpretative possibilities that we could consider here. The first consists in understanding the virtual as a new horizon of possibility for the imaginary, a more dynamic, material, interactive, and intersubjective one than the ones we could access prior to the social diffusion of digital technologies. This new horizon grants consciousness the possibility to depart from the actual world and freely choose to orient towards another, 'irreal' one, if only momentarily. The second, more negative, avenue by which we could approach the virtual from a Sartrean perspective interprets virtual worlds as secondary focuses towards which to direct our perception. According to this perspective, virtual environments only disclose a new world of facts for us to imaginatively surpass. They allow us to escape our being-in-the-world only to be immediately made subject, once again, to the contingencies of another world.

Can these two perspectives be reconciled? It might be useful to return to the idea of the 'double perspective' we adopt when engaging with virtual worlds, which we already discussed in Sect. 1.4 (also see Vella 2015, 55–71). From the standpoint of our virtual subjectivity, the virtual environment constitutes a real (albeit not actual) existential situation (albeit not actual) existential situation, a being-in-the-virtual-world that makes demands of us and curtails our freedom through its contingency—so much so that we find ourselves adopting a new spirit of seriousness within that world (Kania 2015). However, its secondary nature in relation to our actual subjectivity also implies that, when we consciously return to the standpoint of our actual subjectivity—for example, at the end of a virtual world experience, or when, during one's engagement with a virtual world, one stops to reflect on it—the virtual world, and our virtual subjectivity within it, loses its sense of actuality.

The double perspectival structure of our engagement with virtual environments lets us *step back* from our own virtual subjectivity—and from that aspect of our own being which we have invested and brought into presentation in the performance of that virtual subjectivity.

This "standing back from the object contemplated" (Sartre 2010, 193) is, for Sartre, the signal of the adoption of an aesthetic attitude, in which "the aesthetic object is constituted and apprehended by an imaging consciousness that posits it as irreal" (ibid., 191). Looking at a painting, for example, we stand back and adopt the 'imaging attitude', consciously surpassing the physical object before us (functioning as an *analogon*) towards the image our consciousness makes of it. In the same way, our virtual subjectivity—our perceptions of the virtual world, the actions we take as we strive towards our projects in that world, the emotions we feel as those projects succeed or fail—becomes an *analogon* for a different, imaginary version of ourselves, one that escapes the determination of the actual world.

Notes

1. Heidegger writes that "man, as existing transcendence abounding in and surpassing toward possibilities, is a *creature of distance*" (1969, 131). What Sartre picks up on is precisely the idea of "the primordial distance he established towards all being in his transcendence" (ibid.).
2. Sartre uses the term *analogon* to refer to the object—generally a material object of perception, but potentially also a concept or other mental object—which serves as the analogical material for an irreal object. In this example, the painting is the *analogon* which allows us to imagine Charles VIII. The *analogon* is itself surpassed in image-consciousness—we no longer see the painting as an object before our perception, we see the king it allows us to imagine.

3. This is one of the points in *The Imaginary* where, contrary to his own claim that there are no such things as irreal or imaginary worlds, Sartre not only explicitly uses the term, but does in fact appear to be thinking of a domain of irreal existence.

References

Aarseth, E. 2006. "Doors and Perceptions: Fiction vs Simulation in Games." *Intermedialities*, *9*, 35–44.

Baumgarten, A. G. 1954 [1735]. *Reflections on Poetry*. Trans. Aschenbrenner, K. & Holther, W. B. Berkeley (CA): University of California Press.

Doležel, L. 1998. *Heterocosmica*. Baltimore (MD): Johns Hopkins University Press.

Grau, O. 2003. *Virtual Art: From Illusion to Immersion*. Cambridge (MA): The MIT Press.

Grosz, E. 2001. *Architecture from the Outside: Essays on Real and Virtual Space*. Cambridge (MA): The MIT Press.

Heidegger, M. 1969. *The Essence of Reasons*. Trans. Terrence Malick. Evanston (IL): Northwestern University Press.

Hui, Y. 2016. *On the Existence of Digital Objects*. Minneapolis (MN): University of Minnesota Press.

Janik, J. 2018. "Game/r-Play/er-Bio-Object: Exploring Posthuman Values in Videogame Research." Proceedings of the 12th International Conference in the Philosophy of Computer Games, Copenhagen (Denmark), August 13–14.

Kania, M. M. 2015. "Spirit of Seriousness and Bad Faith: On the Meaning of In-Game Life." Proceedings of the 9th International Conference in the Philosophy of Computer Games, Berlin (Germany), October 14–17.

Leino, O. T. 2009. "Understanding Games as Played: Sketch for a First-Person Perspective for Computer Game Analysis." Proceedings of the 4th International Conference in the Philosophy of Computer Games, Bergen (Norway), August 13–15.

O'Shiel, D. 2019. *Sartre and Magic: Being, Emotion and Philosophy*. London (UK): Bloomsbury Academic.

Sartre, J. P. 1966 [1943]. *Being and Nothingness*. Trans. Hazel E. Barnes. New York (NY): Washington Square Press.

Sartre, J. P. 2010 [1940]. *The Imaginary: A Phenomenological Psychology of the Imagination*. Trans. Jonathan Webber. London (UK): Routledge.

Steuer, J. 1992. "Defining Virtual Reality: Dimensions Determining Telepresence." *Journal of Communication*, *42* (4), 73–93.

Tolkien, J. R. R. 1964. *Tree and Leaf*. London (UK): George Allen & Unwin.

Vella, D. 2015. *The ludic subject and the ludic self: Investigating the 'I-in-the Gameworld'*. Doctoral dissertation, IT University of Copenhagen (Denmark).

CHAPTER 6

Eugen Fink and Existential Play

Abstract The particular way in which German phenomenologist Eugen Fink understands 'play' is deeply involved with fundamental existential concerns such as the question of freedom. This chapter discusses the existential significance Fink grants to play—considering, specifically, the capacity play grants us to explore unactualized dimensions of our being through the taking-on of a role in the playworld and presenting in some detail Fink's treatment of the significance of the mask. This understanding of play is then used to offer a perspective on the relationships between one's actual self and one's roles in virtual environments.

Keywords Fink • Play • Role • Virtual worlds • Mask

Eugen Fink is not, strictly speaking, an existential philosopher. His work, however, has a great degree of overlap with existentialist thought. What is particularly relevant for this book is his understanding of the notion of 'play', which he articulated in his 1960 work *Play as Symbol of the World*, and in a series of papers, lectures, and notes written between 1957 and 1973.

For Fink, the idea of play is deeply involved with fundamental existential concerns such as the question of freedom and its relation to one's situation. This is apparent in his consideration of play as an attempt at tackling one of philosophy's oldest questions: "[h]ow are natural causality and

S. Gualeni, D. Vella, *Virtual Existentialism*,
https://doi.org/10.1007/978-3-030-38478-4_6

freedom compatible with each other?" (Fink 2015b, 80). This is a question that has a special importance within existentialism. Another way of phrasing it could be: how is it possible to consider human being as free and self-determining when we know that, in fact, the shape of one's life is to a great degree determined by the facts of one's situation?

In an existential fashion—and with evident influence from Jean-Paul Sartre-Fink considers echo of Jean-Paul Sartre (1966)—Fink considers human being to consist, at base, of an infinite freedom of self-determination within an almost unbounded horizon of possibility. When we are born, we are *nothing*: as infants, we have not actualized any of our potential and, since existence precedes essence, the specifics of our being are not yet determined. Precisely because of this indeterminacy, in the domain of potentiality we are *everything*. As a child, one can aspire to be an astronaut, a veterinarian, a K-pop celebrity, an Olympic swimmer, *and* prime minister. At least hypothetically, all of these are potential ways by which one can be in the world.

As one grows up, though, choices—both those we make for ourselves, and those that are made for us—gradually narrow this possibility space. A person can develop oneself towards one possibility, or a small set thereof. One might, for example, study biology and learn the skills involved in handling and relating to animals, which might eventually lead to the actualization of the possibility of being a veterinarian. However, in choosing to do so, one would have chosen, at least implicitly, not to undertake the demanding daily training regime that would have actualized one's potential Olympian's body. In the process of actualizing the potential for being a veterinarian, the possibility of becoming an Olympic swimmer fades from one's horizon of possibility and so do the possibilities of being an astronaut, a K-pop celebrity, and prime minister. Moreover, choices are sometimes made for us by the situations in which we find ourselves. One might live in a country where studying to be a veterinarian requires one to pay tuition fees that one is unable, due to one's social background, to afford, or one might have to abandon one's studies to care for a sick relative. In such cases, one's being is partly determined—and actualized—by conditions that are external to one's own conscious will.

In Fink's words,

> [t]he path of life, so to speak, is determined by an uncanny, accompanying contraction of our possibilities. Every activity that we carry out makes us

> more determinate and at the same time less possible [...] The child is indeterminately everything, the old man is determinately little—we are born as many and die as one. (2015b, 89–90)

Fink conceives of human life as a move from undetermined possibility to determined actuality. It is only through such a determination that the individual can gain an actual existence in the world—without such a determination, without actualizing any possibility, one would remain unexpressed potential: nothing. At the same time, each successive determination—each choice made, each experience lived—narrows the available horizon of possibilities for the individual, and, thereby, reduces the extent of one's freedom.[1]

6.1 The Playworld and Existential Freedom

How can play come to the rescue? To begin with, it is important to note that Fink grants play a clear existential relevance. When compared to all the other kinds of activity we engage in, play stands apart—indeed, ***standing apart*** is, in a way, the defining character of play as an existential phenomenon:

> Play is a fundamental phenomenon of existence, just as original and independent as death, love, work and mastery, but it is *not* directed, as with the other fundamental phenomena, by a collective striving after the final purpose. It stands *over and against them*, as it were, in order to assimilate them into itself by portraying them. We play seriousness, play genuineness, play actuality, we play work and battle, play love and death. (2015a, 21)

All the aspects of our existence are geared towards the achievement of some purpose: we work in order to be able to afford our basic needs of food and shelter, we strive to progress in our career so as to gain wealth and status, we try to find a partner in order to start a family and have children who will remember us after we die, and so on. Play, however, stands apart from this teleological disposition: when we play, we have no purpose in mind outside the act of playing itself.[2] Because of this, play "entices us into an 'aesthetic' bearing toward life and thus into an under-determination of the reality of the human being", through "in-sincere, non-serious activity [...] an 'acting as if'" (2015b, 86).

There is a distinct echo here of Sartre's characterization of play as the stepping-out of "the spirit of seriousness", which we briefly discussed in Chap. 5. Fink follows Sartre in depicting play as the taking of action in a non-serious attitude—a characterization that also aligns with Johan Huizinga's argument that one of the defining qualities of play is precisely its being perceived as "not serious" (1950). However, Fink goes much further in articulating the implications of acting in play. Central to his understanding of play is the concept of the 'playworld'. The playworld is anchored in an encounter with concrete material things; in its first moment, "playing is always a confrontation with beings" (2015a, 24), taking place in the player's encounter with the materiality of the *plaything* as an actual thing in the actual world. Upon the material foundation of the plaything, however, *something* else is brought into view. Fink writes that "play is creative bringing-forth [...] the product is the playworld" (ibid., 28). The playworld, then, is essentially a representational relationship of appearance to matter: it is "an objectively present appearance, which rests on simply actual things and overlays them in an entirely unique way" (2015b, 92). In this sense, for Fink, the playworld is "a sphere of appearance" (2015a, 28).

In Fink's example, a girl playing with a doll produces a playworld in which the doll is a child and the girl, taking on a role with the playworld, is the child's mother (ibid., 24). The child is represented by an actual doll, but constitutes a "non-actual" representation superimposed upon the plaything, an 'as-if' that is "not nothing and yet is nothing actual" (ibid., 25). Even more crucially—at least for our purposes—is Fink's observation regarding what happens to the playing girl. She is also, he argues, taken up into the playworld: not as herself, but as the role she adopts in that world:

> In every kind of play in which the role-character of the players predominates and one moves about with a fictive conception of oneself, a 'non-actual comportment' is portrayed in an actual comportment, and such self-rapture in the realm of an 'appearance' is felt with excitement and pleasure. (2015b, 91)

The taking-up of a role in the playworld—a role that is, by definition, distinct from the player's actual, determinate self—leads to the appearance of the player, as a different actualized possibility of herself, and, hence, of her freedom to be someone different to the self she has actualized. The young girl 'plays at' being a mother, taking on this role in the playworld and adopting the comportment it requires towards things in the playworld: in this case, the comportment is likely to be structured as a relation of care and responsibility towards the doll-child, in which the girl-mother might perform actions such as feeding the child-doll, brushing its hair, rocking it in her arms, and so on.[3]

6.2 Play and Virtuality

Before considering the existential significance of this understanding of play, we consider it necessary to bring the idea of the virtual into the discussion. Though Fink clearly did not have digitally mediated virtual worlds in mind, there is an evident congruence between his idea of the 'non-actual' as the mode of existence characteristic of the playworld and the notion of the virtual. This congruence goes beyond the observation that, by at least one definition, the virtual is precisely 'non-actual'—that is, having some form of existence, "not nothing" in Fink's words, and yet standing in contradistinction to the actual.

Digital technologies disclose virtual situations in which digital objects take on the role of playthings, in Fink's specific sense of the term. The question of fictionality in virtual environments (and digital games in particular) is a contentious and theoretically fraught one (see Aarseth 2006; Tavinor 2012; Meskin and Robson 2012; Jørgensen 2013; Van de Mosselaer 2018). Without venturing into this debate, however, we can make the relatively uncontroversial assertion that, in many—if not most—digital games, actual digital objects represent non-actual referents in the same way that Fink's actual, material doll represents a non-actual child. By this understanding, a virtual world (as we have defined it in our glossary section) is a digitally mediated and digitally upheld playworld.

Moreover, as we have already seen in Chap. 1, the academic field of game studies has long been aware of the player's adoption of a 'role' in relation to the world of the game. Just as the girl playing with the doll takes on the role of the mother in the playworld, the player of a digital game takes on a particular 'role' towards the things in the gameworld, determined in great part by the affordances they are granted in the game, the goals towards which they are directed, and so on (Vella 2015; Kania 2017).

This assertion claims a kind of continuity between the activity of taking on a virtual subjectivity in a virtual world and other cultural practices that Fink would classify as 'play', in which a non-actual domain is represented through materially actual stand-ins and in which individuals take on a role that differs from their actual selves. This aligns with the way this theme was treated in Chap. 2 of this book, where we proposed an understanding of virtual world experience in relation to practices like dramatherapy and psychodrama, role-playing, and religious rituals. More importantly for our current purposes, this recognition of the fit between engagement with virtual worlds and Fink's notion of playworlds means that, by looking at the existential significance Fink grants to play, we can develop a perspective upon the existential significance of the virtual.

6.3 Playing with Selves

The existential significance of play is perhaps most apparent in Fink's discussion of the mask. He writes that "the mask is the enchanted item whose possession transforms the human being" (2015b, 151). Behind the magical vocabulary used by Fink—which derives from the couching of this discussion within a discussion on ritualistic and cultic play—this is an inherently existential argument. This becomes more apparent when he notes that the donning of the mask represents "the passage to a multifariousness of existence" (ibid.). By means of the mask, "one can be everything again" (ibid.): the presence of the single, specific, individuated being that one actually is, is set off against the vast reservoir of non-actualized possibilities of one's being. What the mask reveals is the possibility of being something or someone *other than* one's actualized being.

By picking up a microphone in a karaoke bar, jumping onto a table and pretending to be a K-pop celebrity (assuming that one is not, in actuality, a K-pop celebrity), one is performing an appearance of oneself that diverges from one's everyday being, thereby exploring a non-actualized possibility of one's being. One might be a reserved, composed individual in one's everyday existence. This might be the case, for example, out of insecurity or out of an investment in maintaining a certain degree of propriety. When taking up this role, one is able to perform and explore a much more extraverted comportment and to see what it feels like to be the centre of attention, to perform for the gaze of others, and to sound one's voice aloud. This, for Fink, is experienced joyfully, since it allows one to see in oneself aspects of one's being that do not normally see the light. By his argument, this extraverted version of oneself is no less a part of oneself than the mild-mannered, reticent self that one would usually express in everyday life: it just so happens the conditions one happens to live in allow for some aspects of one's being to come to the surface, while others remain non-actualized. The joy comes, for Fink, in the realization that one already contains, in potential, a K-pop celebrity, just waiting for the right conditions in order to rise to the surface of actuality—and, moreover, that this is only one of the near-infinite potentialities of being one contains. Play allows us to recognize that we contain multitudes.[4]

The donning of a virtual subjectivity could easily be understood as such an exercise. Besides, the idea that virtual worlds provide us with a venue in which we can become 'someone else' is, as David J. Gunkel observes (2018, 39), commonplace in advertising and popular discourses. Scott Rettberg's description of the appeal of massively multiplayer online role-playing games like *World of Warcraft* (Blizzard Entertainment 2003), for example, is typical in situating such games directly in relation to this understanding of play:

> [w]hile we all find ourselves living lives that we have in part determined by our choices and in part been thrown into by virtue of being born into them, in *World of Warcraft* and in other MMORPGs we have the opportunity to wipe the slate clean, to start again and choose new lives in a new world. (2008, 23)

Within this discourse, virtual worlds are understood as experiential domains where we can adopt a new self and perhaps even create an 'ideal self' (as an alternative for an actual self that one is, for whatever reason, dissatisfied with, see Bessière et al. 2007, 531). As is the case with Fink's notion of play, though, this is not just about the ability to take on one specific other identity. Rather, it is about the fact that virtuality destabilizes the idea of a single self in a single world, and might allow—at least in theory—for a more fluid and multifaceted understanding of selfhood, thus blurring the distinction between the actualized self and its potentialities.

In this vein, the distribution of the player's self across any number of virtual identities establishes a 'multiple personality order' within which different avatars allow for the expression of different sides of the self (Gilbert et al. 2011). The formative conceptual assumptions underpinning this idea can be traced back to Sherry Turkle's extensive research into online identities. Turkle argues that the dispersal of subjectivity across multiple computer-mediated domains creates "a distributed self that exists in many worlds and plays many roles at the same time" (1997, 74).

Though Turkle is not concerned with the idea of play, the echo of Fink's concepts is clear: for Turkle, the possibility for a multiplication of roles, and hence of identities, follows on from the multiplication of worlds.

6.4 Playing and the Doubling of the Self

The existential promise that Fink identifies in play—and that we have considered in the virtual as a particular technological manifestation of play—needs to be qualified. This fluidity of the self is not absolute, and one's actual self continues to occupy a central position. Fink acknowledges the fact that, when we take on a role in the playworld, we do not stop being who we actually are: "the human being who masks himself indeed remains in the form that nature has given to him; he cannot escape from it" (ibid., 150). One retains a consciousness of who one really is while one takes on a role in the playworld. The role one adopts in play, then, is perceived from the standpoint of (and in relation to) one's actual being, leading to a kind of double consciousness. As he writes, "by means of the mask [...] the human being is for the very first time transposed into the possibility of *appearing* ambiguously and polysemously" (ibid.).

This appearing is a *self*-appearing: there is the played-self, and the playing-self, to whom the played-self is an appearance. Fink talks about a "self-rapture in the realm of 'appearance'" which is "felt with excitement and pleasure" (ibid., 91). This perspective fits in well with the idea—which we have articulated in Chap. 1 of this book—of a hierarchical relation between actual and virtual subjectivities, according to which any virtual subjectivities one adopts stand in a nested relation to one's actual subjectivity. In the relation between the actual self and one's role in play, we can see an encapsulation of the individual human being's capacity to be in excess of their actual existence. Play grants the player a "double existence", for she is, in every instance of play, both player and role, and each functions as a perspective on the other. "This doubling", Fink writes, "belongs to the essence of playing" (2015a, 25).

In the final analysis, Fink offers something of a definition of play and its function in relation to human being. He writes that

> In human play an ecstase of existence toward the world takes place. In play the human being 'transcends' himself, surmounts the determinations with which he has surrounded himself and within which he has 'actualized' him-

> self, makes the irrevocable decisions of his freedom revocable [...] leaps free from himself, and plunges from every fixed situation into the possibilities that stream forth in the primordial ground of life. (2015b, 206–207)

For Fink, then, play is the practice that answers to the eccentric world-position of human being. It addresses an impulse to *be otherwise*, to attain—if only in the form of an 'appearance'—a sense of exploring possibilities of being beyond the one shape that our choices and our determined condition have moulded us into. Moreover, play allows us to explore these possibilities in a 'revocable' way, to use Fink's term—that is, in a way which does not lead to a further determination of our being, and, hence, to new limitations upon our future possibilities of being. In this sense, we might say that there is an inherently playful quality to the virtual, especially—to return to a point we have already touched upon in Chap. 1—when we keep in mind Mark B. N. Hansen's claim that the virtual addresses "that capacity, so fundamental to human existence, to be in excess of one's actual state" (2003, 51).

Understanding virtual environments as facilitators of Finkian play, then, means understanding them to grant us the opportunity of being otherwise, facilitating the ecstatic structure of human being—'ecstatic' being used here in the technical, phenomenological sense, to refer to the decentred nature of our being, which always projects itself beyond our current situation. The relation of our virtual subjectivities to our actual subjectivity reflects this ecstatic structure: we step into our virtual self while continuing to stand with one foot in our actual self. The virtual self, then, represents a departure and an experiment, "in the realm of an appearance", with our being: it lets us appear to be otherwise to ourselves.

Notes

1. Though, as we have already mentioned, Fink's treatment of the idea of freedom bears similarities to Sartre's, here there is a divergence on the matter between the two philosophers. Sartre is certainly aware of the limitations placed upon one's being by one's situation—nonetheless, his emphasis is on the freedom that an individual always possesses. As we have already seen, he claims that to deny oneself this freedom is to live in 'bad faith' (1966, 87). With a perspective we might perhaps label more pragmatic, Fink emphasizes the concrete limitations that are placed upon our future being by the choices,

events, and conditions that shape us, bringing to the fore a temporal aspect of freedom that Sartre does not develop to the same extent.

2. An acknowledged difficulty in a number of theories of play is that of reconciling this supposed disinterested character of play with the goal-oriented comportment we adopt in most games, in which we certainly adopt. Arguably, a solution can be found in Hans-Georg Gadamer's observation—already remarked upon in Chap. 1—that although "every game presents the player with a task", this task is not valuable *per se*, but only insofar as it serves to structure a particular comportment for the player in striving towards it. C. Thi Nguyen makes a similar point in greater detail when developing the notion of "striving play" (playing to win). "In-game", he argues, "a player is interested in achieving the lusory goal. But when we step back and consider their purpose for taking on the activity of pursuing that lusory goal, we see that striving play is disinterested. I care not if my striving succeeds, but am only in it for the aesthetic qualities of the experience of striving itself" (2019, 16). This is what leads him to characterize striving play as "disinterested interestedness" (ibid.).
3. One could say it is disingenuous to talk about freedom concerning this example of play. Fink highlights an instance of play in which a contemporary reader might struggle to see an expression of the supposed freedom play grants us. Instead, we might be more likely to interpret this play-situation as a mechanism by which the girl is conditioned into a specific gender stereotype and trained for the maternal role she will be expected to adopt when she is older. This is undeniably the case, and it speaks to the potential ways in which play can be appropriated by institutions and take part in systems of social control that Fink did not account for. However, for the purposes of Fink's argument, the point remains—the girl is still departing from the actual role she inhabits at that point in her life and exploring a different role that, at that point, is for her entirely non-actual.
4. To link this back to the discussion in Chap. 2—specifically to the observation that virtual worlds stand in a long tradition of cultural practices—it is interesting to note the extent to which, without specifically drawing upon Fink's work, dramatherapy has operated upon similar assumptions regarding the link between the activity of play and the exploration of the possibilities of one's self. Robert Landy writes, for example, that "the primary rule of the game is to play, to enter into an 'as if' context where I am both who I am and who I am not" (1993, 104). Jessica Williams Ciemnyjewski and Tamar Brown similarly link dramatherapy to the experimentation with "unfamiliar [roles] through play, through the whole notion of make-believe" (2010, 15). More explicitly aligning the practice with a Finkian understanding of play, Alida Gersie writes that dramatherapy is invested in "the exploration of alternatives which is inherent to play" (46).

References

Aarseth, E. 2006. "Doors and Perceptions: Fiction vs. Simulation in Games." In *Intermedialities, 9*, 35–44.

Bessière, K., Fleming Seay, A., & Kiesler, S. 2007. "The Ideal Elf: Explorations of Identity in World of Warcraft." In *CyberPsychology and Behavior, 10* (4), 530–535.

Blizzard Entertainment. (2003). *World of Warcraft* [Microsoft Windows]. Digital game directed by Rob Pardo, Jeff Kaplan, and Tom Chilton, published by Blizzard Entertainment.

Fink, E. 2015a [1957]. "Oasis of Happiness: Thoughts toward an Ontology of Play." In Moore, I. A., & Turner, C. (trans.), in *Play as Symbol of the World and Other Writings*. Bloomington and Indianapolis (IN): Indiana University Press, 14–31.

Fink, E. 2015b [1960]. *Play as Symbol of the World*. In Moore, I. A., & Turner, C. (trans.) in *Play as Symbol of the World and Other Writings*. Bloomington and Indianapolis (IN): Indiana University Press, 33–215.

Gilbert, R. L., Foss, J. A., Murphy, N. A. 2011. "Multiple Personality Order: Physical and Personal Characteristics of the Self, Primary Avatar and Alt." In Peachey, A., Childs, M. (eds.), *Reinventing Ourselves: Contemporary Concepts of Identity in Virtual Worlds*. London (UK): Springer, 213–234.

Gunkel, D. J. 2018. *Gaming the System: Deconstructing Video Games, Game Studies, and Virtual Worlds*. Bloomington (IN): Indiana University Press.

Hansen, M. B. N. 2003. *New Philosophy for New Media*. Cambridge (MA): The MIT Press.

Huizinga, J. 1950. *Homo Ludens*. Boston (MA): Beacon Press.

Jørgensen, K. 2013. *Gameworld Interfaces*. Cambridge (MA): The MIT Press.

Kania, M. M. 2017. *Perspectives of the Avatar: Sketching the Existential Aesthetics of Digital Games*. Wroclaw (Poland): University of Lower Silesia Press.

Landy, R. J. 1993. *Persona and Performance: The Meaning of Role in Drama, Therapy and Everyday Life*. London (UK) and Bristol (PA): Jessica Kingsley Publishers.

Meskin, A., & Robson, J. 2012. "Fiction and Fictional Worlds in Videogames." In Sageng, J. R., Fossheim, H., Larsen, T. M. (eds) *The Philosophy of Computer Games*. Dordrecht (The Netherlands): Springer, 201–218.

Nguyen, C. T. 2019. "Games and the Art of Agency." In *Philosophical Review*, 128 (4), 432–462.

Rettberg, S. 2008. "Corporate Ideology in World of Warcraft" In Corneliussen, H., & Walker Rettberg, J. (eds.), *Digital Culture, Play, and Identity: A World of Warcraft Reader*. Cambridge (MA): The MIT Press, 19–38.

Sartre, J. P. 1966 [1943]. *Being and Nothingness*. Trans. Hazel E. Barnes. New York (NY): Washington Square Press.

Tavinor, G. 2012. "Videogames and Fictionalism." In Sageng, J. R., Fossheim, H., Larsen, T. M. (eds.) *The Philosophy of Computer Games*. Dordrecht (The Netherlands): Springer, 185–199.

Turkle, S. 1997. "Multiple Subjectivity and Virtual Community at the End of the Freudian Century." In *Sociological Inquiry*, *67*, 72–84.

Van de Mosselaer, N. 2018. "Fictionally Flipping Tetrominoes? Defining the Fictionality of a Player's Actions." In *Journal of the Philosophy of Games*, *1* (1).

Vella, D. 2015. *The Ludic Subject and the Ludic Self: Analyzing the 'I-in-the-Gameworld'*. Doctoral dissertation, IT University of Copenhagen (Denmark).

Williams Ciemnyjewski, J., & Brown, T. 2010. "A Question of Identity and Belonging: The Self Between Familiar and Unfamiliar, Foreign Spaces." In *Dramatherapy*, *32* (1), 14–16.

CHAPTER 7

Conclusions

Abstract This concluding chapter articulates a summary of the central themes and the main claims of the book. While doing so, it invites readers to focus their attention on the 'fluidifying sway' of virtual environments, that is to say on how virtual world experiences reveal—and grant interactive access to—multiple and possible courses of action that are latent in any given situation. By experientially disclosing several alternatives to a certain *status quo* through repeated interactions, do-overs, and time reversals, virtual worlds inherently encourage their users to embrace 'what could be' over 'what is'. This chapter approaches our relationships with virtual environments in ways that are not solely concerned with their socially desirable uses and effects, but also identifies aspects of those experiences that constitute a menace to our well-being and survival.

Keywords Virtual worlds • Virtuality • Existentialism • Possibilities • Fluidifying sway

In this book, we have approached virtual environments as technologies capable of disclosing world-like experiences. In articulating our arguments concerning the existential relevance of these technologies, we largely focused our attention on experiences of virtual worlds that transcend—and often controvert—aspects of human beings' everyday relationships with the actual world.

S. Gualeni, D. Vella, *Virtual Existentialism*,
https://doi.org/10.1007/978-3-030-38478-4_7

Our choice to emphasize the possibilities for virtual environments to grant interactive access to existential situations and possibilities does not imply that we consider 'the virtual' as an insular experiential domain. We have argued that virtual world experiences cannot be understood apart from their relationships and ties with 'the actual'. In Chap. 1 in particular, we made the case that virtual worlds are experientially available to human beings precisely because both their creators and their users are grounded in the actual worlds as bodily beings.

In extreme synthesis, we have recognized virtual worlds as existentially relevant in their particular ways of disclosing (virtual) existential situations from the perspective of a plurality of (virtual) subjectivities and in their allowing us to adopt and pursue virtual existential projects in virtual worlds. From that standpoint, we have observed that, in those situations, the users can engage in practices that encourage and facilitate reflections about (and, potentially, transformations of) their actual selves.

To this fundamental way in which virtual worlds can be understood as existentially significant, we could add others in which virtual experiences complement and extend our capabilities for perceiving, thinking, and imagining. As we discussed elsewhere, digital environments can be effectively employed to assess problematic, hypothetical situations in their possibilities and consequences in ways that resonate with the philosophical use of thought experiments (see Gualeni 2018). Moreover, technologies that disclose virtual worlds can also be used, for example, with the deliberate objective of familiarizing oneself with prosthetic extensions and with different (virtual) possibilities for embodiment (see Gualeni 2014; Ahn et al. 2016; Hook 2019; Arjoranta 2019).

In the face of a great variety of potential applications, in this concluding chapter we will exclusively address what we consider the most general and perhaps most relevant socio-cultural effect of our relationships with virtual environments. We are specifically referring to their encouraging users to orient their thinking towards the 'modus irrealis'. In that mindset, the desires and imaginations of the users tend to gravitate towards what is not (or not yet) the case, rather than on the present situation. This effect was initially discussed in Chap. 3, where virtual worlds were also discussed in terms of their potential to 'fluidify' the ways in which we think about ourselves, our relationships with others, and the existential situations we find ourselves in. It was also a central theme in Chap. 5, where also central to Chap. 5, where we considered the virtual in relation to Sartre's understanding of the 'irreal'.

7.1 Technologies for 'Fluidifying' How We Think

It is important to emphasize, once again, that with *Virtual Existentialism* we did not aspire to use established existential notions and framework to virtual worlds in a way that was literal and literally consistent with the philosophical tradition from whence they originated. Instead, our academic efforts have been aimed specifically at:

- developing new concepts that help us understand virtual worlds as new existential domains (this is the case for notions like 'virtual subjectivity', 'virtual existential project', and 'attitude of virtuality'), and at
- understanding how experiences in virtual worlds emerge from actual experiences, overlap with them, and can encourage certain existential operations on one's actual self that might not be possible—or, at least, might not be as readily available—without VR technologies.

In relation to the second point, and as we anticipated in the previous section, the transformative effect of our relationships with (and within) virtual worlds that we consider to be the most central and encompassing is their 'fluidifying sway'. With 'fluidifying sway' we indicate the specific ways in which interactions with virtual environments reveal—and grant cognitive access to—multiple and often unfamiliar courses of action that are latent in any given situation.[1] By experientially disclosing several alternatives to a certain *status quo* through repeated interactions, do-overs, and time reversals, virtual worlds inherently encourage their users to embrace 'what could be' over 'what is' (Gualeni 2015, 59, 124).[2] Our consideration of virtual worlds in the light of Plessner's recognition of the 'utopian character' of the human being (Chap. 3), of Sartre's understanding of 'the imaginary' as consciousness surpassing the real (Chap. 5) and of Fink's notion of 'play' as the exploration of possibilities in non-actuality (Chap. 6) resonate with such a perspective.

By virtue of this 'fluidifying sway', and as we discussed at length in the last two chapters, technologies that disclose virtual worlds are recognized as facilitators of specific transformative processes in the individual.[3] They can stimulate our imagination and our sense of possibility, allowing us to comport ourselves in ways that diverge from—and that might even be antithetical to—the current configuration of our actual self. Also, they invite us to imagine, and experiment with, different versions of ourselves.

Beyond the individual, existential level that has been our focus in this book, it is also possible to envisage that virtual worlds can—through the active experiencing of new ways of being and alternative possibilities for socio-economic organization—promote attitudes towards social and ethical problems that are both less provincial and less fixated on the present state of affairs. What we mean, more specifically, is that the active experience of new ways of being (and previously unexperienced possibilities for socio-economic organization) could put into question the incontrovertibility of the *status quo* and make alternatives to it more intuitive and less challenging to imagine.[4]

7.2 The Other Side of the Virtual

As already discussed, we consider all technologies as having the quality of being multistable, meaning that the designers of a technology can never fully anticipate and determine the ways in which the users of a certain technology will use it and attribute meaning to it. In other words, we treat the relationships between technologies and societies as always dynamic and open-ended. In that sense, the multistability of technologies that disclose virtual worlds guarantees that the latter do not exclusively produce expected (and presumably socially positive) effects, but also invite the emergence of unanticipated functions and implications.

The Canadian media scholar Marshall McLuhan advances another way of framing this ambiguity in our relationships with virtual technology. In his 1964 book *Understanding Media: The Extensions of Man*, McLuhan portrays technologies not only as enhancements of the human capabilities for thought and action, but also as forms of self-amputation (McLuhan 1994). New technological extensions, he argues, always entail a balance between the increase in acuity of certain cognitive functions and the desensitization of others (ibid.). McLuhan understands technology both as a functional solution to some of our limitations and dissatisfactions and, at the same time, as a source of new problems and inconveniences. Relatedly, in his *Disseminations* (1972), the philosopher Jacques Derrida offers a close reading of Plato's dialogue *Phaedrus*, concentrating his attention on Socrates's use of the ancient Greek word *pharmakon* to describe an alpha technology in particular: writing. In his text, Derrida points out that *pharmakon* is a particularly ambivalent word, as its etymology refers to both the concepts of 'poison' and 'remedy' (Derrida 1981).

Adopting a way of approaching our relationships with technology that is not solely concerned with the socially desirable uses and effects, we can identify aspects of our relationships with virtual worlds that, instead, are configured as "a menace to our own well-being" (see Sect. 4.2). The elements of danger involved in our mutually constitutive relationships with virtual environments can be of various kinds. Virtual worlds research has typically focused its scholarly attention on the ways in which they can affect our psychological health and stability, with particular interest in phenomena like addiction, dissociative disorders, and social isolation (which we briefly discussed in Sect. 2.2.6). However, this is not the limit of the possible dangers associated with virtual world experiences. As an important case in point, an obvious threat is present in the very 'fluidifying sway' that we identified as the most general effect of experiences of virtual environments. An earlier publication of ours identified a potential downside of our becoming more prone to thinking in terms of possibilities and alternative courses of action in the ensuing, general trivialization of experiences and values (see Gualeni 2015, 126–128). In that work, we argued that the more engaged and fragmented into virtual subjectivities we are, the less likely it is that we will psychologically invest in actual situations (ibid.). An conceivable consequence of this effect is a diminished participation in socio-political processes in the actual world.

Other potential threats are made apparent by the participation of the discourses and technologies of the virtual in contemporary ideologies of 'smart power'—that is, power structures that operate not through visible restrictions and oppression, but, instead, by appropriating the logic of freedom. In other words, smart power is a way to exercise power that "does not negate or oppress freedom so much as exploit it" (Han 2017, 15). The idea of freedom—freedom from being subject to the facts and constraints of the world, freedom from the determinations of our own actualized subjectivities—has animated much of the discussion about the virtual in this book. Though the primary drive of our argument has not been political, a remark about the implications this idea might have in this regard is in order.

Writing about digital games, Möring and Leino (2016) have argued that the theories of play which contemporary digital game studies continue to rely on are rooted in a liberal construction of subjectivity. Following the Industrial Revolution and the institutionalization of the working day, the liberal ideology of capitalism contributed to the formation of a split subject, structured according to "the separation between work and spare time, duty and fun" (2016, 148). Play then comes to be understood as freedom *from* work—but, of course, this is a very compartmentalized form of freedom that not only results in a splitting of the self, but also leaves untouched the oppressive conditions of labour under capitalism. Given the close conceptual links between virtuality and play that we have traced in Sartre (Sect. 1.5 and Chap. 5) and Fink (Chap. 6), it bears questioning the extent to which the restatement of a similar idea at the core of the virtual might merely be a repetition of the same ideological move. The idea that the virtual can provide us with spaces in which we can 'be ourselves', unconstrained by the limitations placed upon us by our everyday existence, might, as its consequence, exacerbate our alienation from our actual, situated existence, through an 'escape' into a virtual existence that evades, rather than tackling, the actual conditions which motivated such an escape in the first place.

Nor is the tying of the virtual to existential self-fashioning or its understanding as a Foucauldian technology of the self (Sect. 1.5) free from potential criticism. Echoing the projectual understanding of human being that has been foundational to our approach, philosopher Byung-Chul Han writes that "today, we do not deem ourselves subjugated *subjects*, but rather *projects*: always refashioning and reinventing ourselves" (2017, 1). The inherent danger of this, he argues, is that the purported freedom to be anything very quickly morphs into a compulsion to constantly reproduce ourselves anew. We become our own oppressors, exploiting ourselves in "unlimited self-production" (ibid., 6). In this way, "the neoliberal regime utterly claims the technology of the self for its own purposes: perpetual self-organization—as the exemplary neoliberal technology of the self—represents nothing so much as a highly efficient mode of domination and exploitation" (2017, 28).

Finally, it can be argued that the idea that virtual subjectivities allow us to 'take on' and explore alternative identities rests on a flattening and objectification of the idea of identity. In other words, it requires that the notion of identity can somewhat be separated from the complex social, historical, cultural, and personal conditions within which it is rooted, and that it can be transformed into the homogeneity of a commercial product. Writing about the usage of virtual worlds as venues for putting on gender or ethnic identities that differ from one's own, for example, Chris Kerich argues that such a usage represents "the aestheticization and commodification of identity in digital spheres, advancing a neoliberal ideology that holds identity as an aesthetic commodity to be bought and sold" (Kerich 2019, 1).

7.3 Concluding Remarks

In this book, we have discussed virtual environments as technologies: as extensions and complements that influence how we inhabit and experience worlds, how we relate to one another, and reflect on the meaning of our individual existence. As technologies, virtual environments are part of our lived experience: they disclose virtual experiences in persistent digital worlds that are occasions to reflect on our values and beliefs, put them to the test, and operate changes on ourselves and our conduct. As technologies, virtual environments are also multistable tools, meaning that they can elicit effects and trigger social dynamics that are unpredictable and often threatening for our well-being and our survival as a species.

With the purpose of highlighting some of the ambiguous ways in which technologies are constitutive to how we understand our existence and attribute meaning to it, we have engaged with the philosophical work of Helmuth Plessner, Peter W. Zapffe, Jean-Paul Sartre, and Eugen Fink (respectively in Chaps. 3, 4, 5, and 6). It is important to emphasize once again, as we did in the introduction, that our work in these chapters was not meant to be a slavish and literal application of the theories of four continental thinkers, but consisted in a re-digestion and an extension of their work in the context of our existential relationships with virtual environments. Our appropriation of their notions and frameworks helped us reveal multiple and profound ways in which virtual worlds are existentially relevant. They also further reinforced our belief that an approach to virtual environment as having effects that can be univocally determined and completely designed would be both naïve and misguided.

Inspired by those thinkers, in this book we understood being human as an ongoing process, and, thereby, bound to remain an open question. On those premises, it should be evident that our technological complements can frame and influence the way we understand and lead our existence, but not ultimately define or resolve it. Accordingly, we have not discussed virtual environments as sweeping technological answers to the questions and discontents posed by existence. Instead, we have presented them as technological artefacts that are constitutive of our actual existence both in their allowing us to transcend certain aspects of our everyday relationship with the actual world, and in disclosing previously unexperienced ways in which our incompleteness, our indeterminacy, and our limitations can be experienced and understood (Fig. 7.1).

Fig. 7.1 A screenshot of Pippin Barr's (2011) digital game *Let's Play: Ancient Greek Punishment*

Notes

1. In relation to this point, and inspired by Arthur Schopenhauer's notion of 'transcendental will', the Italian philosopher Dario Compagno recognizes a crucial difference between expressing our will in the actual world and in virtual worlds. The main difference consists, for him, in how we conceptualize the notion of 'choice'. According to Compagno, human beings do not have freedom of choice in the actual world, simply because, in their everyday life, they cannot know in advance the outcome of their choices (Compagno 2008). Compagno argues that the options for action available to human beings in their ordinary, linear understanding of time are based on intuitions and on the rationalization of the accounts of their past experiences. Humans are, thus, ultimately uncertain about the consequences of their actions. For these reasons, we cannot technically understand human actions as expressions of will: they are perhaps better understood as hopeful decisions that have the quality of being irrevocable (ibid.). What Compagno is arguing is that, by grace of virtual environments, for the first time in history we can finally experience making choices. This is clearly due to the fact that virtual environments typically grant their users the opportunity to explore alternative possibilities and outcomes before committing to a certain course of action. Lamentably, Compagno does not further elaborate on this point or on the moral aspects involved in being able to make the kinds of choices discussed in his book chapter. Our hypothesis is that the possibility to take fully informed decisions might orient the users of a virtual world towards adopting a utilitarian moral stance over aprioristic ones founded on duties and/or values.
2. For the sake of rigor, we want to clarify that virtual environments do not categorically disclose multiple possibilities for acting and doing. Not all virtual experiences, in other words, invite their users to actively explore new

perspectives and alternative courses of action. In the case of serious digital games or games used for specific persuasive purposes, for example, social meaning often emerges from the specific ways in which those environments restrict the agency of their users. The impossibility to successfully accomplish tasks in any other way than following the procedures established by the designers is frequently used as a rhetorical tool to make the users experience subjugation and hopelessness—as, for example, when being forced to complete one test chamber after another, with no way out and only one solution, in *Portal* (Valve Software 2007)—or to convince them of the necessary nature of certain courses of action.

3. Virtual environments are relatively commonly used in psychotherapy to help subjects overcome traumas and phobias (see Riva et al. 2016). They are also commonly used to disclose persuasive experiences meant to change one's attitude and beliefs towards the actual world. The latter is, for example, the case of serious games, games for change, and propaganda games (see Bogost 2007).
4. By arguing that virtual worlds inherently feature a utopian impulse, we are taking a position that opposes that outlined by American media scholar Alexander R. Galloway in his 2006 article titled "Warcraft and Utopia". In it, Galloway claims that "[…] virtual worlds are always in some basic way the expression of utopian desire […] but that the very act of creating an immaterial utopian space at the same time inscribes a whole vocabulary of algorithmic coding into the plane of imagination that thereby undoes the play of utopia in the first place" (Galloway 2006, 6).

References

Ahn, S. J., Bostick, J., Ogle, E., Nowak, K. L., McGillicuddy, K. T., & Bailenson, J. N. 2016. "Experiencing nature: Embodying animals in immersive virtual environments increases inclusion of nature in self and involvement with nature." In *Journal of Computer-Mediated Communication*, *21*(6), 399–419.

Arjoranta, J. 2019. "Playing the Nonhuman." In Karkulehto, S., Koistinen A., and Varis E. (eds.) *Reconfiguring Human, Nonhuman and Posthuman in Literature and Culture*. New York (NY): Routledge.

Barr, P. 2011. *Let's Play: Ancient Greek Punishment*. Digital game developed by Barr, P. and available online at http://www.pippinbarr.com/games/letsplay-ancientgreekpunishment/LetsPlayAncientGreekPunishment.html

Bogost, I. 2007. *Persuasive games: The Expressive Power of Videogames*. Cambridge (MA): The MIT Press.

Compagno, D. 2008. "I am Link's Transcendental Will: Freedom from Hyrule to Earth." In Cuddy L. (edited by) *The Legend of Zelda and Philosophy*, ed. Luke Cuddy, 185–202. Chicago (IL): Open Court.

Derrida, J. 1981. "Plato's Pharmacy." In Derrida, J. 1981 [1972]. *Dissemination*. Trans. Barbara Johnson. London (UK): Athlone Press.

Galloway, A. R. 2006. "Warcraft and Utopia." In *Ctheory*, *16* (2).

Gualeni, S. 2014. "Augmented ontologies or how to philosophize with a digital hammer." In *Philosophy & Technology*, *27* (2), 177–199.

Gualeni, S. 2015. *Virtual Worlds as Philosophical Tools*. Basingstoke (UK): Palgrave Macmillan.

Gualeni, S. 2018. "A Philosophy of 'DOING' in the digital." In Romele, A. and Terrone, E. (eds.). *Towards a Philosophy of Digital Media*. Basingstoke (UK): Palgrave Macmillan, 225–255.

Han, B. C. 2017. *Psychopolitics: Neoliberalism and New Technologies of Power*. Trans. Erik Butler. London (UK): Verso.

Hook, A. 2019. "Exploring speculative methods: Building artifacts to investigate interspecies intersubjective subjectivity." In *Alphaville: Journal of Film and Screen Media*, *17*, 146–164.

Kerich, C. 2019. "Polygonal Modeling: The Aestheticization of Identity." Proceedings of the 2019 DiGRA International Conference. Kyoto (Japan), August 6–10.

McLuhan, M. 1994 [1964]. *Understanding Media: The Extensions of Man*. Cambridge (MA): The MIT Press.

Möring, S., Leino, O. T. 2016. "Beyond Games as Political Education: Neoliberalism in the Contemporary Computer Game Form." *Journal of Gaming and Virtual Worlds*, *8* (2), 145–161.

Riva, G., Wiederhold, B. K., Gaggioli, A. 2016. "Being different: The transformative potential of virtual reality." In *Annual Review of Cybertherapy and Telemedicine*, *14*, 3–6.

Valve Software. 2007. *Portal* [Windows]. Digital game directed by Swift, K. and published by Valve Corporation.

Index[1]

A

B

C

D

E

[1] Note: Page numbers followed by 'n' refer to notes.

S. Gualeni, D. Vella, *Virtual Existentialism*,
https://doi.org/10.1007/978-3-030-38478-4

The Right to Silence in Transnational Criminal Proceedings

Fenella M.W. Billing

The Right to Silence in Transnational Criminal Proceedings

Comparative Law Perspectives

Springer

Fenella M.W. Billing
Department of Law
Aarhus University
Aarhus, Denmark

ISBN 978-3-319-82484-0 ISBN 978-3-319-42034-9 (eBook)
DOI 10.1007/978-3-319-42034-9

Printed on acid-free paper

This Springer imprint is published by Springer Nature
The registered company is Springer International Publishing AG Switzerland

This book is dedicated in loving memory to Paul and Leanne.

Preface

This book is about the complex balance between effectiveness of law enforcement, in bringing wrongdoers to justice, and ensuring fairness to a criminal defendant. From my experience as a criminal law practitioner, I was concerned and intrigued by a perception that these two objectives may be mutually exclusive. In particular, the book focuses on the significance of maintaining the balance between limiting and protecting the right to silence and the right against self-incrimination in transnational proceedings. This spotlight came about as the result of an initial comparison I undertook of coercive measures in Denmark and Australia, which revealed that the greatest divergence between the uses of such measures in the two systems was the manner in which the rules regulating the right to silence functioned in administrative investigations. In addition, while talking to criminal justice practitioners involved in transnational cases, I learned of the potential difficulties that may arise in relation to differing approaches to the right to silence, for example, when the investigative and prosecution authorities want to question a suspect abroad and must decide as to which state's rules should apply. By ensuring there is a proper foundation of fair trial rights in the national systems, which accords with minimum standards under international human rights law, the admissibility of evidence across borders is maximised, leading to more effective criminal prosecutions.

The book is largely the result of my research towards a PhD qualification. Being in the somewhat unusual position of having 'grown up' as a common lawyer and finding myself in a foreign setting, I set about the formidable task of immersing myself in the Danish system and trying to come to terms with the many alternative viewpoints on the criminal law that I observed. It is my hope that I have been able to shed some light and understanding on the complexities and uniqueness of trans-national cases, for the benefit of criminal justice practitioners, the judiciary and policy makers alike—across systems and traditions. I have attempted to state the law as of 1 July 2015.

This book would not have been possible without the support and assistance of many others. First and foremost, I would like to thank the Department of Law and staff at the University of Southern Denmark. In particular, I wish to express my gratitude to my PhD supervisors, Professor Thomas Elholm and Associate Professor Birgit Feldtmann, for their ongoing direction and support and for helping me to demystify Danish criminal law. I would also like to thank the Law Faculty at the University of Tasmania, where I was temporarily based as a guest researcher. I defended my PhD in Odense in May 2014 and I thank the members of the assessment committee, Professor John Jackson of the University of Nottingham and Associate Professors Anette Storgaard and Bugge Thorbjørn Daniel of Aarhus University and the University of Southern Denmark, respectively, for a challenging and constructive process. There are many others along the way who expressed their interest in my research and shared their many good ideas, including academics, prosecution lawyers and other practitioners from various corners of the globe. From amongst them, I would particularly like to thank Mr. Bruce Gardner for his dedicated guidance and friendship.

Finally, this research would not have come to fruition without the continued support and help of family and friends, both in Denmark and Australia. Most of all, I give thanks for the love and devotion of my husband, Frank, and our two lovely children, Felix and Oliver—it was tough on all of us.

Aarhus, Denmark
May 2016

Fenella M.W. Billing

Contents

Part I
The Right to Silence in Context

Chapter 1
Introduction

Abstract Is it fair to use international cooperation to obtain evidence of a confession or the suspect's silence in one legal system and use it at trial in another? This is an important question that relates to the problem of maintaining a balance between the effectiveness of law enforcement, on the one hand, and effective defence rights, such as the right to silence, on the other, in transnational criminal proceedings. Procedural differences between legal systems may mean that the international cooperation process disrupts the continuity of law between the investigative and trial phases in a national criminal proceeding that relies on confession evidence or evidence of silence obtained abroad. The differences between legal systems may be more obvious and detrimental when cooperation in the gathering of evidence takes place between countries that are applying different minimum standards of human rights protection, where they come under different human rights frameworks. This chapter introduces these central ideas and concepts and explains how comparative law, primarily based on the functional method, will be used to examine them.

An English Court of Appeal case presents a scenario that illustrates the problem of balancing the right to silence in transnational proceedings, as follows: Two men are offered £300 to take a car from England to the Netherlands. On the return journey they are stopped at a frontier port in Belgium. The men are taken into a customs office while their car is searched. It is found to contain 5 kg of amphetamines and nearly 10,000 LSD tablets, with a total street value of over £100,000. An investigating magistrate instructs the customs officer to interview one of the men through an interpreter and to take him to the court the next morning. During the interview the interpreter keeps a written record of the person's answers. It contains statements made by him to the effect that he has no idea who owns the drugs. Evidence of the interview is later presented at his trial in England and the accused concedes that these statements were lies. On the accused person's behalf an objection is raised in relation to the interview. The interview had been conducted properly and in accordance with Belgian law. However, had the interview been conducted in England, a number of the investigative steps taken would have been in breach of English codes of practice. In particular, no caution had been administered and the

F.M.W. Billing, *The Right to Silence in Transnational Criminal Proceedings*,
DOI 10.1007/978-3-319-42034-9_1

accused had not been advised that he could have a legal representative present during the interview[1].

1.1 A Problem of Balance

Is it fair to use international cooperation to obtain evidence of a confession (or evidence that turns on the suspect's right to silence) in one legal system and use it at trial in another? This was the question that the English Court of Appeal had to decide when the case of *R v Konscol*[2] eventually went on appeal. The Court of Appeal agreed with the trial judge's decision to allow the evidence. The Court found that the procedure was fair and that by using the evidence obtained abroad in an English trial there was still a just balance between the effectiveness of the law enforcement aims involved and the rights of the accused.

This case serves to illustrate the difficulties that may be involved in cross-border transfer of evidence for trial in another system. However, in the 20 years that have followed since Konscol's case was decided in 1993, following the lead of the European Court of Human Rights (ECtHR), there has been a shift in the approach to certain pre-trial due process rights, such as the right to silence and the right against self-incrimination. Today, some of these procedural rights are considered to be so fundamental that evidence obtained in breach of them will be inadmissible. Nevertheless, the same scenario may be envisaged today, even within the EU, as a result of differences in national procedural rules.[3] This may be a particular risk due to a general increased use of international cooperation instruments to gather evidence in transnational criminal cases.

As mentioned in the Konscol case, there has to be a just *balance* between law enforcement interests in bringing wrongdoers to justice and the fair trial rights of the accused, such as the *right to silence*.[4] At the core of the problem is the question of balance between effectiveness of law enforcement and fairness to a suspect or accused. The main assumption of this work is that the overall regulation of this

[1] *R v Konscol* [1993] Crim LR 950.

[2] *R v Konscol* [1993] Crim LR 950. See Chap. 7, Sect. 7.3.1.2 for a full discussion of the case.

[3] In Belgium today, for example, a suspect has the right to remain silent (including the right against self-incrimination) and the right to access a legal representative prior to questioning—though not within the first 24 h of police custody, when questioning may take place. There is no legal obligation on the police to inform a suspect of these rights. This is despite the possibility of legal consequences following from remaining silent. There is a right to be informed that their statements may be used as evidence in court: Cape et al. (2010), pp. 78–80 and 86; European e-justice portal, Rights of defendants in criminal proceedings, Belgium; see also (ECtHR) *Stojkovic v France and Belgium*, Application no. 25303/08, 27 October 2011.

[4] See Thunberg Schunke (2004), p. 37, where the problem of balance between law enforcement interests and the interests of an accused was also one of the perspectives taken in examining international cooperation instruments.

balance generally relies on a *continuity of law* between the investigative and trial phases of a national criminal justice system.

This is a problem that may be viewed from three perspectives: national, transnational and international. Looking at the problem from the *national* perspective, this work seeks to establish that there is continuity between the investigative and trial phases in the law that limits or protects the right to silence in the legal systems of Denmark, England and Wales and Australia. Some national laws are used to limit rights. Other laws create safeguards which enable a suspect or accused to knowingly participate in the criminal justice process, to protect against abuse and restore balance. Sometimes, where rights are limited in the investigative phase of criminal proceedings, safeguards are in place in the trial phase and vice versa. This is the way in which a national system may be built-up in order to regulate the balance between effectiveness and fairness.

Having examined the nature of the balance between effectiveness of law enforcement and fairness to the accused in the national perspective, from a *transnational* perspective, the question is to determine whether it is fair to transfer, for example, confession evidence between different legal systems. Thus, this work looks more specifically at the implications of continuity of national law for transnational criminal cases. Transnational crime is having a growing impact on law enforcement, particularly in regions with open borders, such as the EU. International cooperation is now a standard tool for fighting cross-border crime, particularly organised crime, within the EU and beyond. However, is it always possible in transnational cases to gather confession evidence or evidence of silence under one set of rules and transfer it for use at trial in another system, without disrupting the national balances between limiting and protecting the accused's right to silence and thereby creating unfairness? Problems in relation to jurisdictional 'cross-admissibility' of evidence may arise when evidence of a suspect's confession or silence has been obtained in a manner that is considered to be irregular or unlawful by the trial courts in the requesting state. In transnational cases this may mean that a suspect misses out on rights in the transfer process.[5] This may be due to a perceived inconsistency between the way the evidence has been gathered and the way in which fundamental rights, such as the right to silence, are expressed in the national law of the trial jurisdiction. As a result, the right to silence and the right against self-incrimination may be diluted or undermined.

Within a national legal framework, the circumstances in which evidence is obtained during the investigation may trigger a particular legal response to the manner in which that evidence can be received at trial. Therefore, the circumstances of an investigation may determine the manner in which evidence is received and used at trial. Another question is, therefore, how can we maintain the national balance between effectiveness in obtaining confession evidence or evidence of silence and fairness in protecting the right to silence, when transferring evidence across national borders? Additional measures in transnational proceedings may be required.

Finally, the problem of the fairness of transferring confession evidence (or evidence of silence) across borders may be viewed from a vertical, *international*

---

[5]See also Bantekas and Nash (2003), pp. 367–369.

perspective. Decision making bodies under international and regional human rights frameworks, such as the 1966 International Covenant on Civil and Political Rights (ICCPR) and the 1950 European Convention for the Protection of Human Rights and Fundamental Freedoms (ECHR), set minimum standards for human rights protection. What is the effect of applying the same minimum human rights standards within the frameworks of either the ICCPR or the ECHR in relation to the right to silence on the process of international cooperation? If states are applying the same minimum standards the transfer process may be fairer and, thereby, more trusting and effective. The application of varying standards within different human rights frameworks, different national legislative structures and with differences in the pervading culture of cooperation may create fundamental divergences between systems, in the way in which the national balance between effectiveness and fairness is regulated. Being a part of the same human rights framework may minimise the risk of disrupting national balances and creating unfairness. This work also aims to reveal the nature of the relationship between minimum standards for human rights protection and international cooperation.

1.2 The Right to Silence and the Right Against Self-Incrimination

This work looks at the balance between limiting and protecting the right to silence—a composite fair trial right, which may be taken to include the right against self-incrimination. Along with other defence rights, the right to silence and the right against self-incrimination are considered central to fairness in criminal proceedings—without fair trial processes, other abuses of power cannot be challenged.[6]

The importance of a fair trial to notions of justice can be traced back to early philosophical discussions and the first national rule of law documents, such as the English Magna Carta of 1215 and later in the revolutionary texts of the 1689 English Bill of Rights, the 1789 US Constitution and Declaration of the Rights of Man, and the 1789 French Declaration of the Rights of Man and of the Citizen.[7] Today, the right of an accused person to a fair trial is guaranteed nationally by the constitutions, statute, criminal codes and common law of almost every country in the world.[8] The right to silence, including the right against self-incrimination, has developed in international and national law as an important aspect of the right to a fair trial. Nevertheless, despite its centrality the right to silence is approached in different ways in different national legal systems.

---

[6]Moeckli et al. (2010), p. 304.

[7]See Moeckli et al. (2010), pp. 18–24; Breay and Harrison (2015); and see the texts of the 1689 English Bill of Rights and the 1789 Declaration of the Rights of Man published online by the Yale Law School's Avalon Project, see Yale Law School (2015b) and (2015a), respectively.

[8]Cape et al. (2010), p. 7.

Countries from the civil law jurisdictions, such as Denmark, tend to view issues about the right to silence as forming a part of the right against self-incrimination. For example, the need to caution a suspect about the right not to say anything in response to questioning by the state is categorised as an aspect of the right against self-incrimination; a prohibition that generally applies to documents as well.[9] However, common law jurisdictions tend to focus on the right to silence. For example, a witness's 'privilege' against self-incrimination may have been extended to other types of information, such as documents, due to the procedure of in-court production of documents, where witnesses are subpoenaed to 'give evidence and produce' items to the court. In common law, there is some uncertainty whether a privilege against self-incrimination, in fact, exists in relation to documents—uncertainty which has been compounded by lack of clarity in the jurisprudence of the ECtHR dealing with the scope of the rights.[10] For the reasons explained below, throughout this work, the right against self-incrimination is presented in the context of suspect and accused persons' statements, forming a part of the right to silence.

1.2.1 Protecting the Right to Choose to Speak or to Remain Silence

As part of the right to a fair trial, the right to silence is viewed as a composite right that is made up of a number of more specific rights, which serve to protect the suspect or accused person from abusive coercion and aim to preserve human dignity. The right to silence does so by protecting the suspect or accused's right to choose whether or not to speak to the authorities and enabling him or her to knowingly and willingly participate in his or her own defence.

In particular, the most relevant aim of the protection of the right to silence is to ensure that the suspect or accused is not compelled to make self-incriminating statements when giving an account of the facts, for example, in an oral or written statement in response to police questioning. Thus, by allowing a suspect to *choose whether to speak* to the police (and to provide the evidence upon which he or she may be convicted) or to remain silent, evidence is not gathered in defiance of the *will* of the suspect or accused person through physical or psychological coercion. In addition, by avoiding the gathering of unreliable evidence, these principles contribute to preventing a miscarriage of justice in a criminal case. Along with other fundamental guarantees such as the presumption of innocence and equality of arms in participating in criminal proceedings, these rights contribute to the overall aim of fairness in criminal proceedings.

[9]See §752(1) of the Danish Administration of Justice Act and §10 of the Coercive Measures Act. See further Chap. 3, Sects. 3.5.1 and 3.7.2.

[10]See Emmerson et al. (2012), pp. 616–620.

1.2.2 *The Right to Withhold Self-Incriminating Information as a Part of the Right to Silence*

In this work the right against self-incrimination is seen as part of the right to silence. The right to silence is centred upon the will of the suspect or accused to choose whether to speak or remain silent. Clearly, it covers the right not to say anything at all. However, this overall choice also inherently covers the decision to speak, which in turn involves a decision about what to say and whether or not to incriminate oneself. Therefore, the right against self-incrimination forms a part of the broader right to silence.[11] However, this is not the complete picture. While the right to silence is broader because silence and all types of statements are protected (including statements of innocence and self-incriminating statements), the right against self-incrimination may be broader in terms of the types of information sources that are protected, including statements as well as documents and other real evidence that exists independently of the will of the accused.[12]

One reason why it may be difficult to define the right to silence and the right against self-incrimination and their relationship to each other is that the rights have not had a clear lineal development together, but have evolved in response to various influences at different times. The precise origins of a right against self-incrimination may be traced back to ancient Christian and Talmudic writings.[13] There are also various links to the development of both the rights in common law, starting with the development in medieval canon law of the maxim *nemo teneter prodere seipsum* (meaning that no one should be required to bear witness against themselves) and the accompanying move against interrogation under oath. In addition, the extension in the late 1700s of the witness privilege rules, firstly to witnesses in *criminal* proceedings, and then to the accused in the mid-1800s, also had an effect on the definition of the rights. Moreover, the rights in common law were influenced by the changes in the status of the criminal defendant from the 1600s, as an undefended person who had restricted rights to call witnesses and who

---

[11]See (HL) *R v Director of the Serious Fraud Office, Ex p. Smith* [1993] A.C. 1, 30–31; (ECtHR) *John Murray v. The United Kingdom*, Application no. 18731/91, 8 February 1996, [32]; but see Brøbech (2003), p. 164, in which the author suggests that the right to silence is a part of the right against self-incrimination.

[12]Trechsel and Summers (2006), p. 342; see also Jackson and Summers (2012), p. 249, where it is argued that one of the downfalls of the ECtHR's jurisprudence in this area is that the cases refer to both rights as if they are one and the same; see, for example, the ECtHR's admissibility decision in *H and J v The Netherlands*, Application nos 978/09 and 992/09, 13 November 2014, [69] where the court refers to the primary concern of the right against self-incrimination being to respect the will of an accused person to remain silent yet goes on to explain that it does not extend to preventing the use in criminal proceedings of material obtained under compulsion which exists independently of the accused's will, such as documents and other forms of real evidence. This paragraph would have been more clearly understood had they separated the two rights and found that the right to silence did not protect against such use of real evidence obtained under compulsion.

[13]Jackson and Summers (2012), p. 241.

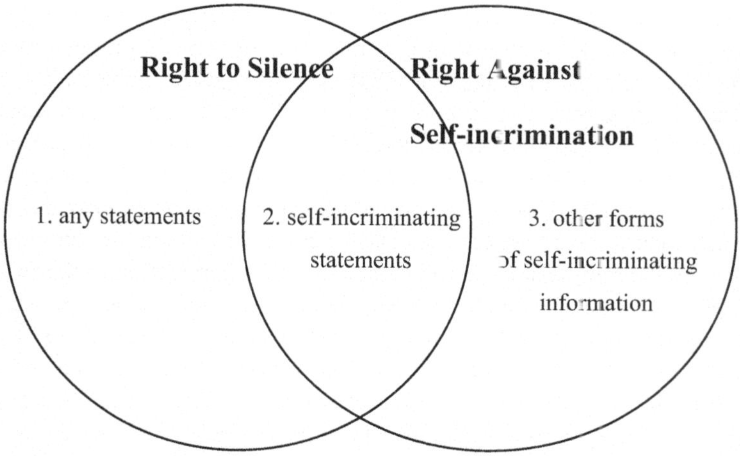

Fig. 1.1 The overlap between the right to silence and the right against self-incrimination

was only allowed to make unsworn statements, to the mid to late 1800s, when the defended accused became a non-compellable witness who was competent to give sworn testimony.[14] On the Continent, the self-incrimination principle has for a long time been a focal point of inquisitorial criminal procedure. This patchwork development of the right to silence and the right against self-incrimination may account for why the rights, when viewed separately and together, are deeply interrelated yet different in certain aspects.[15]

As Trechsel and Summers suggest, perhaps the relationship between the right to silence and the right against self-incrimination is best described as two overlapping circles, as illustrated in Fig. 1.1.[16]

First, the right to silence protects everyone's right not to make a statement or to make any kind of statement about him or herself (including a statement about innocence) and is broader than the right against self-incrimination in this respect. The suspect or accused is commonly cautioned prior to police questioning (or examination), "...you do not have to say anything."[17] Although there may be a civic duty for witnesses to assist in the disruption, prosecution and prevention of crime (and a general requirement to answer a summons to testify in court), even witnesses cannot be forced to speak to the police about private matters relating to

[14]See Helmholtz et al. (1997), pp. 6–8, 84–89, 107, 148–153 and 185–201; Langbein (2003), pp. 107, 178, 254–257, 268, 278–279 and 281; Bentley (1998), pp. 147–149; and see *R v Warickshall* (1783) 1 Leach 263, 263–264 [168 ER 234].

[15]See further Jackson and Summers (2012), pp. 241–243.

[16]However, Trechsel and Summers (2006) refers to the right to silence as protecting acoustic communication rather than statements: 342; see further Jackson and Summers (2012), p. 249.

[17]See the discussion about the *Miranda* warning in Jackson and Summers (2012), p. 244.

themselves.[18] Therefore, the right to silence acknowledges human dignity and everyone's right to keep information about him or herself private, especially when the state seeks their participation in the criminal justice process. The actions of the state in procuring the suspect or accused's participation in criminal proceedings may be the divide between the right to silence (as a fair trial right) and other related rights and freedoms such as privacy and freedom of expression.[19]

Second, within the right to silence, additional protection is provided against making self-incriminating statements. This is the area in which the right to silence *overlaps* with the right against self-incrimination. Here protection of the suspect or accused person—a person who has been engaged by the state in the criminal justice process—against making self-incriminating statements is given extra weight.[20] This means that police questioning should be prevented until certain safeguards are in place. The only effective way of ensuring that a suspect or accused person is guaranteed the right not to incriminate him or herself is by ensuring a right not to say anything at all. Otherwise, a suspect's selective answers would obviously work against him. In addition to respecting the will of the suspect or accused to speak or remain silent, the need for special protection against making self-incrimination statements is also a ground for 'cautioning' the suspect or accused. Witnesses under subpoena who may be compelled to give evidence in a criminal trial under the control of the court are also protected against self-incrimination. Otherwise, it may be a difficult task to bring any witnesses to court in the proof process.

Third, unlike the right to silence, which only deals with statements, the right against self-incrimination may also protects the suspect or accused (and witnesses in court) from being required to produce other sources of self-incriminating information, such as documents, data, fingerprints, DNA or other bodily samples without grounds. For example, subject to the principle of proportionality, the right against self-incrimination may protect against the state compelling the suspect under pain of punishment to come forward with evidence, such as documents, where there are otherwise insufficient grounds to obtain a warrant.[21] Therefore, the right against self-incrimination may be considered the broader of the two rights in this respect.

---

[18]Witnesses are generally compellable to give trial testimony in the controlled atmosphere of the courtroom—otherwise it would be extremely difficult to prove any criminal case beyond reasonable doubt. A witness may be a co-accused when his unwillingness to assist indicates a level of involvement in a crime by being an accessory after the commission. See further Jackson and Summers (2012), p. 249.

[19]But see Jackson and Summers (2012), p. 249.

[20]Jackson and Summers (2012), pp. 275–277.

[21]See, for example, [ECtHR] *Funke v France*, Application no. 10828/84, 25 February 1993; *JB v Switzerland*, 31827/96, 3 May 2001; *Weh v Austria*, Application no. 38544/97, 8 April 2004 and the discussion about this case in Chap. 2, Sect. 2.3.2.4. But see Gans and Palmer (2014), p. 289, citing *Sorby v The Commonwealth* [1983] HCA 10, [8], to say the privilege against self-incrimination under common law and uniform evidence law in Australia only extends to 'testimony'. See further Easton (2014), pp. 194–197, about the suspect's body as a source of evidence.

The distinction between the right to silence and the right against self-incrimination can also be demonstrated by the decisions of the ECtHR. A violation of art. 6 has been found by the ECtHR in two types of cases: first, where *criminal proceedings are pending or anticipated*, and compulsion has been used to obtain potentially *self-incriminating information* from the suspect, including statements and 'real evidence' sources of information such as documents[22]; second, where *incriminating information in the form of statements only*, which has been obtained under compulsion at a time when there are *no criminal proceedings on foot*, is used in subsequent criminal proceedings.[23] In the second situation, the protection afforded by the right to silence and the right against self-incrimination does not appear to apply to the use in subsequent proceedings of real evidence previously obtained under compulsion. In other words, where no criminal charges are anticipated at the time compulsion is used, for example, under compulsory powers in an investigation by an administrative authority, documents or other real evidence sources of information may then be used in subsequent criminal proceedings; where any statements obtained under compulsion may not. Therefore, a suspect or accused is given extra protection against making self-incriminating *statements*.

In international case law, the reason for the distinction in the level of protection afforded has not been made clear.[24] However, the most obvious difference between statements and real evidence is that real evidence is tangible and capable of independent examination, such as forensic testing. In this sense, it is independent of the "...will of the accused." Real evidence is, therefore, perhaps more trustworthy and less open to manipulation by physical or psychological compulsion. Today, subject to human error in the lab and its complicated nature, DNA evidence is particularly probative of a suspect or accused's identity. Fingerprint evidence may be even more probative of an accused person's presence at a crime scene, given its objective reliability and probative value. Real evidence, such as documents and DNA, is often obtained by the police under the court's supervision by means of a warrant system. If not by consent, then it is commonly accepted that fingerprints can be obtained by force, without destroying the reliability of the evidence. On the other hand, due to the unlikelihood that a person will make a false confession,

---

[22][ECtHR] *Weh v Austria*, Application no. 38544/97, 8 April 2004 [42], which refers to *Funke v France* [44]; *Heaney and McGuinness v Ireland* [55–59] and *J.B. v Switzerland* [66–71]. Here, the case of *Weh* does not refer to the defiance of the will of the accused as being an element of the violation of the right against self-incrimination; but see Jackson and Summers (2012), p. 251, including fn 55.

[23][ECtHR] *Weh v Austria*, Application no. 38544/97, 8 April 2004 [43–45 and 50], referring to *inter alia* [GC] *Saunders v The United Kingdom*, Application no. 19187/91, 17 December 1996 [67]; *Allen v. The United Kingdom*, Application no. 76574/01, decision of 10 September 2002; and *Vasileva v Denmark*, Application no. 52792/99, 25 September 2003 [34]

[24]See Emmerson et al. (2012), p. 618; Jackson and Summers (2012), pp. 248–256 and 269–271; and see further [ECtHR GC] *Jalloh v Germany*, Application no. 54810/00, 11 July 2006, where the level of compulsion involved in administering emetics to force the suspect to regurgitate a packet of drugs was considered to be in breach of the prohibition against ill-treatment in art. 3 of the ECHR.

confession evidence is generally highly probative and the community interest in capturing it is high. Therefore the right to silence is central to defining pre-trial due process. these distinctions between statements and real evidence are significant and, therefore, between the right to silence and the right against self-incrimination in broader contexts. It is against this background that three is an important focus on the right to silence, primarily where it overlaps with the right against self-incrimination. In other words, in the present context, in discussing the transfer of confession evidence and evidence of silence across borders, the approach is that the right against self-incrimination is a part of the broader right to silence.

1.2.3 The Right to Silence and the Criminal Justice Process

At certain points, whether during an investigation or at trial, a suspect or accused person will be required by the state to participate in the criminal proceedings being brought against him or her. If what is being asked of the suspect or accused is that he provides an account of the facts, then the right to silence (including the right against self-incrimination) is the procedural right that protects the suspect or accused's choice of whether to speak out or to remain silent.[25] The state may call on a person to make such a choice, for example, when the police openly seek to interrogate or during the course of a trial. Alternatively, by the use of compulsory powers, the state may question a suspect in the absence of the right. Further, the choice of whether to speak or remain silent is also relevant when a suspect decides to talk to a person whom he believes is not an agent of the state, such as a family member, friend or business associate, and where the conversation may be the subject of covert surveillance.

In all the above examples, the purpose of the right to silence is the same—and it is closely connected to the presumption of innocence.[26] By protecting the suspect or accused's choice to speak or to remain silent, the burden of proof will always remain on the prosecution. A suspect or accused person is protected psychologically and physically against coercive and abusive questioning. As well as being protective, the right to silence may also be described as an enabling, participatory fair trial right, which ensures that the suspect or accused person's participation in the proof process of his own prosecution is on an informed, knowing and willing basis.[27] In turn, the reliability of potential suspect evidence is protected, as is the integrity of the criminal justice process as a whole.

[25]See further Jackson and Summers (2012), pp. 277 and following.

[26]See [ECtHR GC] *Saunders v The United Kingdom*, Application no. 19187/91, 17 December 1996 [68].

[27]Jackson and Summers (2012), pp. 241, 276–277 and 285 and following, in relation to defence participation; Jackson (2009), pp. 860–861.

Although the right to silence is a foundational fair trial right and central to any criminal justice system, the legal mechanisms that limit and protect the right to silence, and thereby ensure overall balance between effectiveness and fairness, are significantly different between legal systems. Therefore, it is highly relevant to emphasise the right to silence when considering the fairness of transferring evidence in transnational cases.

1.3 Methodology

This thesis is based on the hypothesis that there is continuity between investigative and trial procedures in the course of criminal proceedings in every legal system. The regulation of national balances between limiting and protecting the right to silence in transnational cases is a particular focus. As with the overarching right to a fair trial, the right to silence and the right against self-incrimination are ongoing rights and exist to protect and enable the suspect or accused person throughout the criminal proceedings. For example, the suspect's choice to speak to the police or to remain silent in the pre-trial phase has consequences for the way in which the evidence unfolds at trial. These consequences may be different depending on which legal system is under consideration. Therefore, in order to assess the fairness of transferring confession evidence or evidence of silence for trial in another system (the transnational perspective) and to examine influence of the international human rights frameworks on the cooperation process (the international perspective), a thorough examination of the right to silence and the right against self-incrimination in each cooperating system is necessary in order to discover the similarities and the differences. Such an examination will give way to an understanding of how the rights are structured in the national system and how they work in a trial in another system. Detailed examination of the rights in question in national law is the foundation for subsequent analyses about the fairness of relying on confession evidence or evidence of silence gathered abroad.

1.3.1 Comparative Analysis Based on the Functional Method

Comparative analysis based on functionality is the primary method by which the laws limiting and protecting the right to silence and their jurisdictional cross-admissibility will be considered. Zweigert and Kötz state that the basic methodological principle in comparative law is that of functionality.[28] Thus, the national laws that function to limit or protect the right to silence are of central focus to the question of the fairness of cross-border transfer of evidence. The types of investigative measures that are considered, which tend to limit the right to remain silent

[28] Zweigert and Kötz (1998), p. 223.

and relieve the prosecution of the burden of proof, include administrative compulsory information-gathering powers; interrogation of suspects; and covert surveillance. The national law that determines how the evidence gathered by these measures is used at trial is also relevant. This includes any national laws that consider the fairness of relying on the evidence obtained, such as rules about admissibility of evidence. Therefore, in line with the comparative legal theory of Zweigert and Kötz,[29] detailed reports of the law about the right to silence in each national jurisdiction under comparison, as well as the international human rights law under the ICCPR and the ECHR that forms the background to the national law and may influence its development, is presented.

From a horizontal, transnational perspective, the reports are the foundation for an analysis of what happens when the individual systems are forced to interact through the process of international cooperation in the gathering of evidence. In addition, the reports are used to examine the vertical, international perspective. It is a common assumption that the process of international cooperation in gathering usable (admissible) evidence is assisted by the *application* of common minimum standards in relation to fair trial rights. Common minimum standards reduce the gaps between national legislation and build trust among the cooperating states. In the EU, for example, the principle of 'mutual trust' in international cooperation in criminal matters is based upon the assumption that Member States are applying the minimum standards set by the ECtHR. In order to test such assumptions and define the role of international human rights instruments in the transnational transfer of evidence, the international reports will establish the minimum standards that are applicable in each national system under consideration. Of significance, the national reports will be used to examine whether these minimum standards are in fact being met.[30] Any residual unfairness in the national system as a result of an inadequacy in using safeguards may carry over to the international cooperation process.

However, as Bell states, any comparative analysis based on the functional method must inherently involve some consideration of the legal setting in which legal problems emerge.[31] No comprehensive comparative analysis using the functional method can completely avoid consideration of the operation of the rules in action in the conceptual setting of the system under examination. The explanation for some of the conclusions in this comparison have been found by looking at the underlying procedural tradition or structure of the systems more broadly.

[29]Zweigert and Kötz (1998), p. 43.

[30]But see Zweigert and Kötz (1998) where it is suggested that each individual system report should be neutral and free from criticism: 43.

[31]Bell (2011), p. 170.

1.3.2 *The Human Rights Frameworks and Systems of Law Under Comparison*

In examining the operation of national law that limits and protects the right to silence, the question of continuity between investigation and trial phases will be considered in relation to the jurisdictions of Denmark, England and Wales and Australia. According to the analysis of Van Hoecke and Warrington, the legal systems of Denmark, England and Australia would fall all within the one general legal family based on Western culture, all having European roots.[32] Common *legal culture* includes shared understandings on a number of paradigmatical elements, which may include: concept of law; theory about legal sources and legal interpretation; methodology of law; theory of argumentation; theory of legitimation of the law; and a common basic ideology.[33] Some of these elements may make what is referred to as legal tradition a part of legal culture.[34] When comparing the legal doctrine (the description and systemisation of the law) of different systems they conclude that, due to anthropological and sociological differences, a strict legal comparison is only possible within a single legal cultural family, such as the Western legal cultural family.[35] In addition, the authors consider the plurality of law and argue that a new, common European legal language is developing due to the "intersystematicity" forced by instruments and institutions of EU and European integration.[36]

Although writing primarily from a private law and civil procedure perspective, Zweigert and Kötz have found that there are four identifiable legal families in Europe: Romanistic, German, Nordic and the Common Law family, with its roots in the law of England.[37] Alternatively, Mattei suggests that in order to move away from a Western-centric classification of systems and achieve a more dynamic approach to legal families, legal systems should be grouped based on patterns of law.[38] He described three groups: first, rule of professional law, in which the legitimacy of law is neither political nor religious, but technical (this group includes most western democratic systems); second, rule of political law, in which the political process and the legal process are not autonomous (such as countries in

---

[32]Other legal cultural families may include Asian, African and Islamic: Van Hoecke and Warrington (1998), pp. 502–508.

[33]Van Hoecke and Warrington (1998), pp. 514–515; at 498 the authors refer to the definition by John Bell of legal culture as "a specific way in which values, practices, and concepts are integrated into the operation of legal institutions and the interpretation of legal text."

[34]On the relationship between legal culture and legal tradition see Glenn (2004), pp. 8–14.

[35]Van Hoecke and Warrington (1998), pp. 523 and following. At 533, the authors suggest that a purely technical comparison is only possible at a third level of comparison between systems within a family that have the same paradigmatical theories within the six areas mentioned above. It is not suggested that this is a purely technical comparison, or pure functionality; although it is primarily based on the functional method.

[36]Van Hoecke and Warrington (1998), pp. 527 and 533.

[37]Zweigert and Kötz (1998), pp. 68–69.

[38]Mattei (1997), pp. 11–12 and 19–44.

development transition, including African and socialist or post-socialist systems); and rule of traditional law, where the separation between law and religion or philosophical tradition has not taken place, such as China and Hindu. However, the result of Mattei's grouping does not add value to the present analysis. Mattei's description of the 'rule of professional law' group sub-divides civil law systems, inspired by either French or German law, from common law systems, such as English and American, and from what he terms as 'mixed' systems, including Scandinavian systems. This subdivision, therefore, is similar to the grouping into legal families by Zweigert and Kötz.[39]

Along with Sweden, Denmark is considered to be a mature legal system from the Nordic legal family, in addition comprising Iceland, Norway and Finland.[40] The Nordic countries have always been politically and culturally closely related. Although Roman law has played a smaller role in the development of Nordic law, than in the systems of the Romanistic or German legal families of Continental Europe, Denmark may nevertheless be categorized as a civil law country, which principally relies on statute as a source of law.[41] Historically, the laws of all the Nordic systems started with the common base of old German law and the process of transferring the law into statute and code[42] began in the twelfth century.[43] Nordic law continued to be influenced by the law of Continental Europe. However, by comparison, its statute law became relatively free of dogma.[44] The Nordic countries have also developed a strong interest in legal cooperation with each other, including in criminal law.[45]

Zweigert and Kötz suggest that the three legal families of Continental Europe (Romanistic, German and Nordic) are closer to each other than to the common law family.[46] This is despite the amount of ideological borrowing that takes place between modern systems of law. The gap between civil law and common law traditions in Europe may be gradually decreasing, at least in part as a result of the unification of European law. However, even today, the dichotomy between civil law and common law forms a basis for comparing Danish law with the law of England and Wales and explains many of the divergences between the different systems under consideration.[47]

---

[39]Although, by grouping them together as he does, Mattei (1997) highlights his view that the contrast between civil law and common law is gradually fading: 12.

[40]Zweigert and Kötz (1998), pp. 41 and 277.

[41]Zweigert and Kötz (1998), pp. 277 and 285.

[42]Statute is the written law made by the legislature; code is legislation that exclusively covers the field in a particular area of law.

[43]Zweigert and Kötz (1998), pp. 278 and 284.

[44]Zweigert and Kötz (1998), pp. 41 and 278–279.

[45]See Zweigert and Kötz (1998), pp. 280–284 and the 1962 Helsinki Treaty (as amended), signed 23 March 1962 and entered into force on 1 July 1962.

[46]Zweigert and Kötz (1998), pp. 68–69 and 73.

[47]See further Zweigert and Kötz (1998), pp. 68–72. About the civil law and the common law traditions, see Glenn (2014), pp. 135–155 and 237–252, 260–267, respectively.

Roman law was traditionally 'scholar's law', which had a theoretical quality and was controlled by academic teaching at universities; whereas common law was originally established and controlled by the practitioners in the King's courts in London and was to be found in extensive law reports rather than text books.[48] Another related distinguishing feature has been the rigid application of the doctrine of *stare decisis*, or binding precedent, in English law. At common law, the courts are bound by the precedent set by courts further up the hierarchy.[49] It was not until 1966 that the most superior courts were no longer bound to follow their own precedent.[50] While today there is a tendency among common law countries to rely on legislation as the primary source of law, the principles regarding the binding or persuasive quality of judicial decisions (and distinguishing precedent on the basis of factual differences in specific cases) are still highly relevant in relation to statutory interpretation. The objective meaning of a statute and the legal principles that apply in a given situation are to be found in judicial decisions.[51] As part of the interpretive process, where considered necessary these decisions may refer to a parliamentary reading speech, for example, as indicative of the intention of the legislature.[52] In addition, the common law courts may find inspiration on the meaning of the law and its application by looking to other jurisdictions for inspiration (particularly within the common law family or in international courts). In continental systems, the courts' endeavour is typically more inclined towards discovering the 'subjective' meaning of law. Therefore, the civil law judge's natural habitat is the national jurisdiction; and there is generally no formal legal rule compelling a court to follow the decisions of higher courts.[53] However, lower courts do usually follow the decisions of superior courts—also in civil law jurisdictions.[54] Furthermore, modern civil law judges are playing an increasing role in the development of the law, as can be seen in Danish law.[55]

---

[48]Zweigert and Kötz (1998), pp. 69 and 258.

[49]Zweigert and Kötz (1998), pp. 259–260.

[50]Zweigert and Kötz (1998), p. 261.

[51]Passing legislation is usually a process of enacting, simplifying and modifying (even internationalising) existing common law principles. On the English objective approach to contractual interpretation, see further Van Hoecke (2004), pp. 179 and 182–185.

[52]Van Hoecke and Warrington (1998), pp. 501–502, suggest that there has been a paradigmatic shift in the consideration by the English courts of *travaux préparatoire*, which has changed the purpose of statutory interpretation from an objective to a subjective process, or at least 'considerably weakened' the English objective approach. However, numerous common law judgments today attempt to discover the meaning of legislation without reference to the legislature's intention, or by referring to it as a means of clarifying the objective meaning.

[53]See further Smits (2004), p. 231; and Zweigert and Kötz (1998), p. 262.

[54]Zweigert and Kötz (1998), p. 262; see also Ladeur (2004), p. 92.

[55]Zweigert and Kötz (1998), pp. 268–269; see further Chap. 3, Sect. 3.9; about the legal thinking and the traditional features of the civil law tradition, see further Smits (2004), pp. 230–234.

Procedurally, civil law and common law systems have developed quite differently.[56] This is in relation to the presentation and evaluation of evidence and the respective functions of the judge, the parties and legal representatives in that process. A civil law judge is usually a 'career judge' who starts his or her professional life as a trainee.[57] In a typical civil trial, the trial judge exercises considerable control over the presentation of evidence. Civil law jurisdictions generally require the trial judge to hear all the relevant evidence in the case and to determine the facts based on the principle of the free evaluation of the evidence.[58] This may include a power to demand that certain evidence be called (even against the parties' intentions).[59] In addition, the principle of free evaluation of evidence may allow the court to consider the weight of unlawfully or unfairly obtained evidence. The civil law judge also typically has a larger role in questioning witnesses.[60]

Even though jury trials in England and other common law jurisdictions today may be limited to serious contested criminal cases, the broader rules of procedure have stemmed from the adversarial method of trial and the use of lay jurors to exclusively decide on the question of guilt. At trial, the legal representatives have the responsibility of examining and cross-examining witnesses. It is very rare for a common law judge to interfere with the process of calling and questioning witnesses. Numerous rules of evidence apply that allow the qualified and experienced judges (who are appointed to be judges after gaining some years of professional experience) to evaluate the admissibility of evidence and, in certain circumstances, to reject evidence before it is presented to the lay jury on the question of guilt. Therefore, it is anticipated that the lay jury will not be distracted by misleading evidence and the integrity of the criminal justice process is thought to be preserved.

A comparison between the right to silence (and the right against self-incrimination) in Danish law and in the law of England and Wales, two mature systems from different legal families, is central to the analysis of what happens to a national balance between obtaining evidence from a suspect in one system and using that evidence at trial in another.[61] Although modern Danish trials are familiar with some of the adversarial features of the common law, Danish legal thinking and the judge's standing and role in relation to the presentation and evaluation of evidence in the course of criminal proceedings indicates that the Danish legal system does not have many of the 'hallmarks' of common law and stills falls to be classified as a system coming from the civil law tradition.[62] Furthermore, Zweigert and Kötz find that the Nordic legal family is closer to the other continental

[56]Zweigert and Kötz (1998), p. 271.

[57]See, for example, Zweigert and Kötz (1998), p. 124.

[58]Zweigert and Kötz (1998), p. 274.

[59]See, for example, § 874(4) of the Danish Administration of Justice Act.

[60]See, for example, § 867 of the Danish Administration of Justice Act.

[61]See further, Zweigert and Kötz (1998), p. 41.

[62]Zweigert and Kötz (1998), pp. 277 and 285.

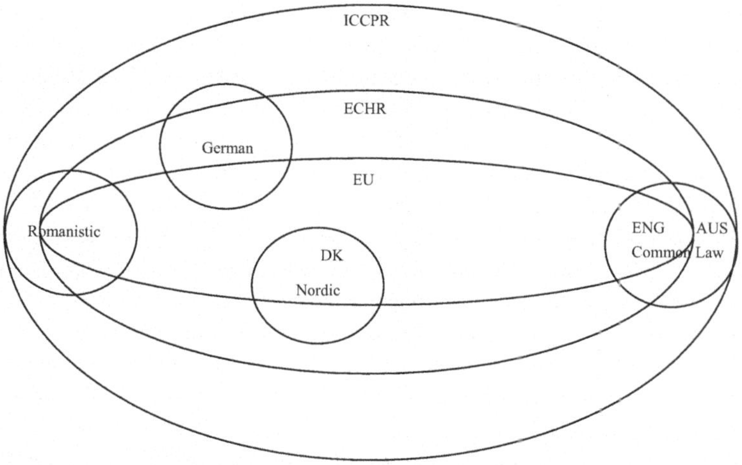

Fig. 1.2 Legal families within the frameworks of both the ICCPR and the ECHR

law families than to common law.[63] Therefore, it is most relevant to compare the operation of the right to silence in Danish criminal proceedings with what happens in criminal proceedings in England and Wales. The contrast between the civil law and common law traditions, described above, may be the basis for significant differences between Danish and English law.

Examining the right to silence in the three jurisdictions outlined above also allows some reflections to be made on the relationship between human rights and international cooperation. The legal divergences created by different legal traditions in the treatment of the right to silence is highlighted by the fact that Denmark and England are both countries that are positioned within the same legal frameworks of UN ICCPR as well as the Council of Europe's ECHR; and, more specifically, they are governed by human rights instruments and actions specific to the EU. To cast light on the effect of being a part of this particular union on the question of balance between effectiveness and fairness in relation to the right to silence, it is relevant to consider the treatment of the right outside the EU and the ECHR. Of all the common law countries outside Europe and the EU, Australia is the country whose law is closest to the law of England.[64] Significantly, Australia's human rights obligations are defined directly by reference to the ICCPR. Therefore, in the context of this comparison, examining the treatment of the right to silence in the law of Australia focuses attention on the significance of the two international human rights frameworks under consideration. The relationship of these three legal systems and their operation under the two human rights frameworks in question is illustrated in Fig. 1.2.

[63] Zweigert and Kötz (1998), pp. 278–279, 282, 284.

[64] Zweigert and Kötz (1998), pp. 221–223.

Legrand argues that in the process of a comparative analysis a common lawyer can never understand a civil law system *like a civil lawyer* and vice versa, due a fundamentally different epistemological mind-set.[65] However, Van Hoecke points out that, despite the shortcomings of comparative law research (based on functionality), it has attained results that are "...clearly beyond pure description".[66] Scholars and other lawyers the world over have been involved in comparative research projects, harmonisation initiatives and drafting uniform laws. In reality, judges in international and national courts have had to deal with diverging legal cultures and uniform laws continuously. Therefore, Van Hoecke suggests that the comparatist ought to consider the concepts and theories of law found in history, sociology or politics, in addition to functionality.[67] International cooperation based on harmonisation as well as acceptance of legal differences between countries with different legal traditions is a reality within the EU; as well as with and between countries outside the EU.[68]

There may well be differences in approach to the law between common law and civil law lawyers (even between lawyers from different systems within those traditions) that influence a comparatist's perception of the law. Nevertheless, by taking all the relevant material into account it should be possible to identify the similarities and differences in the law of each the systems in Denmark, England and Wales and Australia that limit the right to silence and then restore balance and to reflect upon the extent to which the law is influenced by different legal structures and traditions, supranational institutions and human rights frameworks. Taking these factors into consideration, the aim is to draw some conclusions about how these laws will relate to each other in a cross-border context.[69]

1.3.3 Legal Sources and Interpretation

In order to discover how the national balances between limiting and protecting the right to silence function it is necessary to explore all the relevant sources of applicable law. The question of sources of law is approached by reference to national law and the application of international human rights standards. Therefore, there are three issues to consider: national sources of law; international sources of law; and the treatment of international sources of law in the national setting.

---

[65]Legrand (1996), pp. 77–78.

[66]Van Hoecke (2004), p. 172.

[67]Van Hoecke (2004), pp. 189 and 191–192. Zweigert and Kötz (1998) also suggest that the comparative lawyer may have to turn to sociology to find out how far to cast the net in the comparative enquiry: 11.

[68]Van Hoecke (2004), p. 172.

[69]About the functional comparative method see also Samuel (2014), pp. 65–69; Siems (2014), pp. 25–28.

A source of law is whatever shapes the living law in a chosen system.[70] Therefore, statute and code,[71] ministerial orders, case law, legal writing with authoritative weight, standard form documents (such as a Letter of Rights), a practice or custom may all be sources of law.[72] Zweigert and Kötz argue that sources of law in comparative law research should be given the same weight and value as a local lawyer would assign.[73] Overall, in accordance with Zweigert and Kötz, in this work an all-encompassing approach is used, by taking everything into account that is relevant to the functioning of the right to silence in national and transnational cases.[74]

1.3.3.1 National Sources of Law

Until quite recently, in general the question about 'sources of law' was handled rather differently in civil law and common law systems. In civil law systems, the principle source of law has always been statute and code. Common law systems are traditionally based on judge made law (in the absence of legislation); although in recent times, many common law principles in these systems have been 'codified'.[75] Today, while all relevant sources of law may be equally legitimate points of reference,[76] in the vast majority of western democratic legal systems (including common law systems) the tendency is to focus on statute enacted by the legislature as the highly formalised, primary source of law, which undergoes textual interpretation by the courts or the administration.[77] Statute is, therefore, often a starting point in relation to sources of law; and what comes next may be a matter of prioritisation rather than a hierarchy, as such.[78] The exception here in common law systems is that not all of the common law has been codified or abrogated by the legislature.[79]

In the chapters that follow, about the national law in Denmark, England and Wales and Australia, the starting point for the discussions of the applicable law is the national legislation, which sets out the relevant rules about limiting, and protecting the right to silence, including subordinate legislation made through

---

[70]Zweigert and Kötz (1998), p. 35.

[71]Here code is taken to mean comprehensive legislation that is intended to encompass all the law in a single area; whereas statute is legislation in an area of law, where other legislation or judge made law may also contain relevant principles.

[72]Zweigert and Kötz (1998), p. 35.

[73]Zweigert and Kötz (1998), p. 35.

[74]Zweigert and Kötz (1998), p. 71.

[75]Wild and Weinstein (2013), p. 43.

[76]See, for example, Blume (2009), p. 74.

[77]Blume (2009), pp. 65 and 74.

[78]Blume (2009), p. 74.

[79]Wild and Weinstein (2013), p. 43.

delegated power. Judicial decisions are also an important formal source of law. What comes next is a matter of relevance and priority rather than hierarchy. For example, ministerial advice, guidelines and reports may be relevant. Legal texts written by qualified publicists in all three jurisdictions have been relied on to develop an understanding of the valid law in each, but in this work are not generally treated as a source of law *per se*.[80]

Despite the differences, most western democratic legal systems recognise three basic rules of interpretation. Firstly, formal sources of law, such as legislation (acts or parliament and subordinate legislation based on delegated power) and judicial decisions, will usually be more authoritative than other non-binding sources, such as ministerial advice, guidelines and reports (soft law).[81] Second, where there is a conflict, newer law will generally carry more weight than 'old'.[82] Third, specific rules may be given more weight than general rules.[83]

In accordance with Scandinavian legal realism, the valid law is the law that is being enforced by the courts.[84] Therefore, in determining the applicable law about the right to silence in the national legal systems, the way in which the national legislation is interpreted by the national courts is clearly highly relevant in all three systems under consideration. However, a nuanced understanding of the authoritative value of case law in the national systems must be seen in light of how judicial precedent operates in the national setting. In common law systems, despite the process of converting common law principles into legislation, judicial precedent is still an extremely important influence when it comes to the meaning of the law. Statute may be interpreted in case law in light of changing social circumstances.[85] The doctrine of *stare decisis* still applies to the extent that an interpretation of the law by a higher court is binding on a lower court.[86] The decision of a court that is not superior, or *obiter dicta* comments made in passing (compared with the reasons of a decision), are persuasive only.[87] The use of precedent in developing a coherent line of reasoning, interpreting statute by developing legal principles from case to case, in the courts of common law countries is easily discernible, even today. This may account for why so many legal decisions are formally reported in England and Australia; or at least recorded, transcribed and made accessible in written form as an 'unreported decision'. Although there is no equivalent formalised doctrine about

---

[80] Legal literature is described by Blume as the mid-way point between 'hard' law, such as legislation and judicial decisions, and less formalised 'soft' law: Blume (2009), p. 68.

[81] In Denmark, law created by delegated power generally creates obligations for the state and can either be binding or non-binding to the citizen, in the form of circulars or guidelines issued to public authorities such as the police: Blume (2009), pp. 119–122. On subordinate legislation in England, see Wild and Weinstein (2013), p. 34.

[82] See, for example, Blume (2009), p. 94.

[83] These three rules may be referred to as *lex superior, lex posterior* and *lex specialis*, respectively.

[84] See Nielsen (2013), p. 96; Blume (2009), p. 144.

[85] Zweigert and Kötz (1998), p. 18.

[86] Wild and Weinstein (2013), pp. 45–46.

[87] Wild and Weinstein (2013), pp. 45–46.

binding precedent in the Danish system, the decisions of higher courts are considered more authoritative and will usually be followed by the lower courts.[88]

In common law systems such as England and Australia, focus on achieving an objective, literal interpretation of legislation means that sources such as Hansard, a volume which documents the parliamentary debates and second reading speeches, and law reform reports will not be frequently consulted.[89] When reference is made to Hansard it will usually be with a view to finding out the purposive meaning of the text (objective) rather than the actual purpose of the drafters (subjective), for example, by discovering the gap or the problem with the previous law.

In Denmark, however, no conscious choice has been made to pursue an objective rather than a subjective meaning of legislation.[90] Today, although the administration may rely more heavily on the *travaux preparatoires* than the courts, preparatory documents are considered a source of law to be used to find out the purpose of specific legislation in connection with determining its meaning.[91] Arguments for and against the use of *travaux preparatoires* may be mounted, where only the text of the statute binds the citizen. However, some decisions may place great emphasis on them. This is especially true in areas where there is a lack of judicial decisions about the textual meaning or where the legislation is recent.[92]

Comparing case law from different national systems allows a comparatist to develop an understanding of how legal rules work in practice in the national systems.[93] However, depending on the size of the sample, it may also indicate the prevailing legal culture.[94] For example, a collection of national cases (or national cases from a given region, such as the EU) dealing with the admissibility of confession evidence obtained 'irregularly' under the national laws in another system may indicate the willingness on the part of the judiciary to assist cooperation—perhaps even as a priority over the interests of the accused. This practice may be indicative of a broader culture towards cooperation. Both of these results are relevant to the current analysis of what happens to the right to silence in the process of international cooperation in gathering evidence in transnational cases—including transnational cases within the EU and those involving the transfer of evidence outside the framework of the ECHR.

[88]Blume (2009), p. 67; see further, Van Hoecke and Warrington (1998), p. 500.

[89]Wild and Weinstein (2013), pp. 40–41.

[90]Blume (2009), p. 95.

[91]Blume (2009), pp. 66 and 93–96.

[92]Blume (2009), pp. 66 and 94–95; see also Smits (2004), pp. 240–241.

[93]Van Hoecke (2004), p. 169.

[94]But see Van Hoecke (2004), p. 171.

1.3.3.2 International Sources of Law

1.3.3.2.1 General Principles About International Sources of Law

Article 38 of the Statute of the International Court of Justice sets out what are the generally recognised sources of international law. Briefly stated, these include: international conventions (expressly accepted as law by contesting states); international custom; general principles of law; and, as a subsidiary source, judicial decisions and the teachings of the most highly qualified publicists.

International conventions are generally accepted as formal sources of law, where they create rules of general application and are legally binding on the ratifying states.[95] As a formal source of law, a national court can accept the value of a treaty as law, without looking for other evidence. The ICCPR and the ECHR are examples of international treaties that are formal sources of law.[96] Customary law and general principles of law, if established, are also generally applicable and legally binding.[97]

The large variety of material sources of law all provide evidence of a normative consensus among states and other actors that may amount to custom or general principles of law that, when established, are binding and of general application.[98] Certain types of material sources of law, such as decisions of an international court (the ICJ), resolutions of the UN General Assembly or legally binding multilateral treaties (that are not of general application) may be of higher value as evidence of consensus than other material sources, on account of the interactive process which they embody.[99] Other types of material sources of law, also known as 'soft law', include concluding statements from an international conference or similar forum and the writing of highly qualified publicists.[100] The practice of political organs, such as the UN General Assembly and Security Council may even provide evidence of the law, provided that it is apparent that legal matters were sufficiently considered before a particular decision is taken.[101] The value of different material sources of law varies according to the circumstances.[102]

In art. 38 above, judicial decisions have been described as subsidiary sources of law. The article also refers to art. 59 of the ICJ Statute, which provides that there is no doctrine of precedent before the ICJ. This reflects the general principle in

---

[95]See further Crawford (2012), p. 20; Moeckli et al. (2010), p. 105.

[96]See further Chap. 2, Sects. 2.2.1 and 2.3.1.

[97]Crawford (2012), p. 20; Moeckli et al. (2010), p. 105. The distinction between formal and material sources of law is sometimes difficult to establish. For example, treaties and conventions may be a formal source of law and may also be a material source of law, providing evidence of customary international law.

[98]Crawford (2012), pp. 20–21.

[99]Crawford (2012), p. 21.

[100]Crawford (2012), pp. 42–43; Moeckli et al. (2010), pp. 118–122.

[101]Crawford (2012), p. 194; Moeckli et al. (2010), p. 120.

[102]Crawford (2012), p. 24.

international law that a decision of a court is only binding between the parties to the action.[103] Nevertheless, the reality is that judicial decisions contribute greatly to the development of international law by interpreting the treaties and filling out their meaning. In doing so, they may provide evidence of consensus about the law through exposition.[104] International and national courts make frequent cross-references to the way in which an area of law has been developed by another court. International organisations and NGOs may also find it appropriate to follow the line of reasoning of an international or national court.

Although, under the ECHR, the decisions of the ECtHR are not on their face binding on the contracting parties generally, they nevertheless contribute to the development of general principles that allow a fuller understanding of the ECHR (and the relevant minimum standards of protection). This is augmented by the system of individual complaints and the EU apparatus (see further Chap. 2, Sect. 2.3.1, and about national implementation in Denmark and England and Wales in Chaps. 3 and 4, respectively). Unlike EU law with direct effect, the ECHR does not say how implementation of its provisions should be secured.[105]

Also an example of material sources of law, the general comments and decisions of the treaty body known as the Human Rights Committee (HR Committee), are also important in giving meaning to the text of the ICCPR (as discussed in Chap. 2, Sect. 2.2.1).

1.3.3.2.2 The Interpretation of Treaties in International Law

The international law rules about the interpretation of treaties are to be found in art. 31 of the 1969 Vienna Convention on the Law of Treaties. The general rule is that a treaty "...shall be interpreted in good faith in accordance with the ordinary meaning to be given to the terms of the treaty in their context and in the light of its object and purpose."[106] In addition to the text of a treaty, including the preamble and annexes, a court may consider an agreement or instrument made and accepted between the parties, in connection with the conclusion of the treaty.

1.3.3.3 International Sources of Law in the National Setting

International law describes a state's legal obligations under international law—obligations that must be given effect in national law in order to be applied in that setting.[107] Many civil law systems are monist, whereby national and international

---

[103]Moeckli et al. (2010), p. 115; Crawford (2012), p. 37.

[104]Moeckli et al. (2010), pp. 115–117; Crawford (2012), p. 37.

[105]Elo Rytter (2013), p. 50.

[106]Article 31(1) Vienna Convention on the Law of Treaties, opened for signature on 23 May 1969, entered into force 27 January 1980.

[107]See Crawford (2012), pp. 48–50.

law are interlocking parts of one coherent legal order. Therefore, in general, international law can be applied directly in the national system. This applies in relation to treaty and customary international law—some European systems even giving some form of constitutional recognition to customary law.[108] However, there are many variations of the approach between the civil law systems.[109] In European jurisdictions, it is also common for treaties to prevail over domestic law. Therefore, there are careful constitutional controls over the signature and ratification of treaties.[110]

Common law systems are usually strong dualists such that international law and national law are separate legal orders.[111] In the case of a conflict, either national law will apply or the national system decides which rule must prevail. Therefore, implementation of treaties generally requires incorporation into the national law by legislation, for example, in order for individual rights under international law to be enforceable in domestic courts.[112] Provisions of an international treaty may be directly enacted or national law may be employed to give effect to an international treaty.[113] In addition, legislation may be interpreted in light of international treaties, both incorporated and unincorporated.[114] The decisions of international courts and tribunals interpreting a treaty are not binding, although they are usually followed due to a presumption that Parliament does not intend to act in breach of international law obligations.[115] For example, the UK Human Rights Act 1998 incorporates ECHR rights into domestic law. Under s. 2(1) of the Act, national courts must consider relevant decisions of the ECtHR but are not bound to follow them (as referred to in Chap. 4, Sect. 4.2.1). Australia has incorporated ICCPR rights into its domestic law and the national and Australian state courts refer to the General Comments and the decisions of the UN Human Rights Committee.[116] This interpretive presumption also applies to interpretation of the common law.[117]

Denmark is not typical of European continental systems because the traditional starting point is that Danish law is dualist. However, there is not a watertight separation between international and national law in Denmark—international law is used by legal decision makers even when it has not been implemented in Danish

[108]Crawford (2012), p. 88.

[109]Crawford (2012), p. 88.

[110]Crawford (2012), p. 93.

[111]Crawford (2012), p. 63; about monism and dualism in the relations of international law and national law, see Crawford (2012), pp. 48 and following.

[112]Crawford (2012), p. 63.

[113]Crawford (2012), p. 65.

[114]Crawford (2012), p. 65.

[115]Crawford (2012), pp. 65–66.

[116]See, for example, *Iliafi v The Church of Jesus Christ of Latter-Day Saints Australia* [2014] FCAFC 26 (19 March 2014). A search of the Australasian Legal Information Institute online database indicates that many Australian courts have referred to the General Comments or decisions of the United Nations Human Rights Committee.

[117]Crawford (2012), p. 67.

law.[118] In Denmark there are three ways that international law can be implemented: formal notice by state authorities that the existing Danish law is consistent with Denmark's obligations under international law; enacting a law which incorporates Denmark's international obligations into its text; and by 'incorporation', meaning that a Danish law is enacted, stating that an international treaty is now incorporated into Danish law.[119] The ECHR has been implemented by incorporation in 1992 (as discussed in Chap. 3, Sect. 3.2.2). In addition to implementation of international law, there are interpretive rules that give preference an interpretation of Danish law that is consistent with international law, in addition to presuming that the Danish legislature and administrative authorities with particular powers do not intend to breach Denmark's international obligations, whether they have been implemented domestically or not.[120] The Danish courts tend to focus on a more literal and purposive interpretation of legislation, as the central source of law that must be applied.[121] Nevertheless, even in a system such as Denmark care is generally taken to ensure that the minimum standards set by the ECtHR are also applied in Danish law.

From an international perspective, implementation and consistent application of the minimum standards set by international law in relation to fair trial rights reinforces the effectiveness of international treaties at achieving uniformity and the importance of international sources of law, including the development of general principles of law.[122] In addition, from a transnational perspective, the application of uniform standards underpins the international cooperation process by developing a common base of understanding between cooperating states about how foreign nationals or residents and their cases should be dealt with. Today, within the EU, a deeper level of international cooperation in criminal matters is sought, founded upon the common minimum standards that are being particularised by frequent reference to the decisions of the ECtHR. However, the extent to which legislatures and national courts refer directly to the Convention and the decisions of the ECtHR varies between systems, seemingly in accordance with legal and interpretive tradition. In relation to the right to silence and, more generally, the right to a fair trial, the English courts may more readily discuss 'fairness' and the impact of art. 6 ECHR than the Danish courts. English cases that discuss the admissibility of evidence and the discretion to exclude evidence based on unfairness not surprisingly discuss the term 'fairness' at length (as discussed in Chap. 4, Sect. 4.8.2). However, the extent to which the discussion by the English courts about 'fairness' is a result of the influence of the ancestry of the term in common law principles or the influence of the ECHR is not entirely clear. The relationship

---

[118]Blume (2009), p. 212.

[119]Blume (2009), p. 212.

[120]Blume (2009), pp. 212–213.

[121]On Scandinavian legal realism's theory about what is law, see Nielsen (2013), pp. 94–97.

[122]See, Harris et al. (2014), pp. 26–28. See further, Chap. 2, Sects. 2.2.1.2 and 2.3.1.3 about implementation of international minimum standards in Danish, English and Australian law.

between the Supreme Court (UKSC) and the ECtHR has been a matter of some controversy. In some decisions the law Lords have expressed the view that they are bound by the ECtHR, in particular the Grand Chamber, and in other decisions, for example relating to evidence, they have indicated their unwillingness to follow decisions in relation to certain matters that do not appear to fit well in the English system.[123] In Denmark, as discussed in Chap. 3, cases can be decided on issues that might typically be discussed in relation to 'fairness', without reference to the term or a detailed analysis of art. 6.

With increasing 'Europeanisation', the variety of sources of law in systems within the EU has increased.[124] Today, when dealing with issues under EU law, the law is also derived from the institutions of the EU and the Council of Europe, such as the EU parliament, the CJEU and the ECtHR.[125] EU law (derived through the EU democratic process) that has direct effect must be directly taken into consideration by national courts.[126] Further, there is an increasing need to take into account what other national legislators and courts are doing with the EU law, so that there is a consistent approach.[127] Very little interpretive freedom is allowed for when implementing or interpreting EU regulations or directives.[128] National courts, including the lower courts, can refer a case for a preliminary ruling to the CJEU on a question of EU law, circumventing the authority of higher or constitutional courts.[129] Therefore, EU sources of law can have direct impact on the valid national law.

Different national systems may rely on various legal theories about applicable law, sources of law and interpretation, such as positivism, legal realism and institutional theory. However, today, these theories have all been adapted to modern western democracy and by and large lead to identical results.[130] "All the theories presuppose that individual legal rules are part of a legal system and must be interpreted in light of the system."[131] With the plurality of law within the EU, Ruth Nielsen argues that the EU can be seen as one big legal system.[132] The EU legal system is made up of increasingly interlocking elements consisting of EU law, public international law and national law.

Ruth Nielsen points to certain differences between the theories and how well they fit in with the notion of legal plurality within the EU. For example, Nielsen

---

[123]Dennis (2011), pp. 336–338 and 340–341.

[124]On legal pluralism see Smits (2004), p. 234; Bell (2011), pp. 260–261; and Crawford (2012), p. 50.

[125]See Smits (2004), p. 234.

[126]On general principles about EU sources of law see Chap. 2, Sect. 2.3.1.3.1.

[127]Smits (2004), p. 235.

[128]Smits (2004), p. 235.

[129]Dyevre (2013), pp. 69–70.

[130]Nielsen (2013), p. 120.

[131]Nielsen (2013), p. 120.

[132]Nielsen (2013), p. 107.

refers to Tuori's critical legal positivist concept of examining the law at three levels: surface level, being legislation, texts of judgments and standpoints in legal scholarship; a medium level of legal culture, consisting of concepts, principles, legal doctrines and theories that make up the normative, conceptual reservoir at the disposal of the surface level actors; and a deep structure of settled law.[133] Thus, the application of minimum standards which derive from the international sources of law may create uniformity at the surface level of the pluralistic EU system of law, while diversities between national legal systems may still prevail at the sub-surface levels.[134]

Ruth Nielsen also highlights the horizontal dimension of EU law.[135] Therefore, she refers to Scandinavian legal realism to conclude that all the courts in the EU (CJEU and other international courts, such as the ECtHR, and national courts) enforce valid EU law because they share the same doctrine of sources of law.[136] Clear legislation, which implements common minimum standards found in international law, may be applied in all the courts with a greater degree of certainty than law that is based on more vague considerations of cultural traditions.[137] It may be assumed, therefore, that the application of more precisely defined minimum standards of a fair trial will make the use of evidence obtained through international cooperation in the EU more effective. If national law and the national courts in two legal systems within the EU, such as Denmark and England and Wales, are implementing and enforcing the same common minimum standards under the ECHR in relation to the protection of the right to silence then the transfer of confession evidence, for example, should be more effective. If, however, co-operating states do not share the same minimum procedural standards, more difficulties in the use of evidence obtained abroad may be expected.

1.3.3.4 Other Material Sources of Law

To assist in compiling the national reports about the valid law in relation to the right to silence, sources of law such as legislation and conventions, ministerial orders or official guidelines, practice codes and judicial decisions that interpret the law, have been supported by information gathered through informal interviews with police, prosecutors and policy makers from the different systems.

133 Nielsen (2013), p. 87.

134 See Nielsen (2013), pp. 107 and 121.

135 This is contrary to legal theories that interpret sources of law as being structured as a hierarchy, such as Kelsen's pure legal positivist theory: Nielsen (2013), pp. 120–121; about Hans Kelsen's theory, see also 82–85.

136 Nielsen (2013), pp. 120–121.

137 See Nielsen (2013), p. 96.

1.3.4 Scope of the Research

This work examines what happens to the national balance in relation to limiting and protecting the right to silence, when transferring evidence across national borders in transnational cases. It is concerned with the maintenance of procedural due process and fairness in the cross-border setting, which is influenced by structural differences in the law and a lack of continuity. As such, the scope of the examination does not take into account issues that could be of a purely national concern, such as the reliability of evidence due to coercion. The aim is to determine the fairness of the actual *process* of transferring evidence by international cooperation. There are a number of international and regional mutual legal assistance instruments that facilitate gathering evidence abroad. In general, these instruments are based on *locus regit actum*, which means that the evidence is gathered according to the rules of the requested state. However, more recent instruments, such as the 2000 EU Convention on Mutual Assistance in Legal Matters[138] and the European Investigation Order,[139] use the *forum regit actum* so that in general evidence is gathered according to the rules of the requesting state. Although it would appear, as it was intended, that the latter forum would assist with problems related to jurisdictional cross-admissibility, this result does not necessarily follow in all cases. Almost all international cooperation instruments aim to facilitate the widest measure of cooperation possible. In practice, instruments such as the Council of Europe 1959 European Convention on Mutual Assistance in Criminal Matters[140] and the UN Convention Against Transnational Organized Crime 2000[141] may be used, and in many cases are being used, to obtain evidence according to the requirements of the requesting state whenever possible. Therefore, this work has not focused on the different options available under the various international cooperation instruments, but rather on the international cooperation *process* itself.

Central to considering the fairness of the cooperation process in this research, is an examination of evidence obtained from a suspect or accused person contrary to the right to silence, being the choice whether to speak (or make a statement) or to remain silent. Specifically, the types of suspect evidence that are relevant are confession evidence or evidence of the suspect or accused remaining silent, when investigative or prosecutorial agents (or the court) have sought an account of the facts from him or her. Accordingly, the law in each national system may be

---

[138]COUNCIL ACT of 29 May 2000 establishing in accordance with Article 34 of the Treaty on European Union the Convention on Mutual Assistance in Criminal Matters between the Member States of the European Union OJEC 12 July 2000 (2000/C 197/01).

[139]Directive 2014/41/EU of the European Parliament and of the Council of 3 April 2014 regarding the European Investigation Order in criminal matters, OJEU L 130/1, 1 May 2014.

[140]European Convention on Mutual Assistance in Criminal Matters, Council of Europe, 20 April 1959, ETS 030.

[141]United Nations Convention Against Transnational Organized Crime, entered into force 29 September 2003 UNTS, vol. 2225, no. 39574.

categorised in a number of groups that represent the most relevant rules in relation to limiting and protecting the right to silence. The first group includes rules that relate to the right to silence during police questioning of suspects, such as those that deal with cautioning, access to legal representation, recording of interviews and using an undercover agent to covertly interrogate a suspect.[142] The questioning of legal persons, such as a company, is not specifically covered. In general, legal persons are questioned through the responsible officers in much the same way as an ordinary suspect.[143] The second group of rules are those that pertain to the adverse inferences that may be drawn at trial from the silence of the suspect or accused, or from statements that amount to evidence of lies and selective answering during a police interview. The third group, are those that provide for compulsory questioning powers to obtain statements from a suspect or accused. In addition, the use of information gathered under compulsory questioning powers to obtain derivative real evidence will be considered. The fourth group of rules that are of interest relate to self-incriminating statements obtained covertly from a suspect via a coercive measure, but not through police questioning. This includes, for example, suspect statements recorded by a listening device which were made after the suspect had already asserted the right to remain silent.

Accomplice evidence can provide invaluable information in relation to crimes that are otherwise difficult to investigate.[144] Gathering and using accomplice evidence in return for immunity, plea-bargaining or support by the prosecution for obtaining a sentencing reduction is controversial and is regarded as a fair trial issue.[145] Often an accomplice will provide details about a crime that only a perpetrator can know. Therefore, accomplice evidence may be viewed as indirectly undermining a suspect or accused person's real choice about whether to speak or to remain silent. Nevertheless, being a question about reliability and not being suspect evidence or evidence obtained from an accused, national rules relating to accomplice evidence will not be presented in this research.

One of the purposes of the right to silence is to protect a suspect against abusive conduct on the part of interrogators aimed at compelling confessions. The

---

[142] As referred to in Sect. 1.2 above, this research focuses on the right to silence (in particular, where it overlaps with the right against self-incrimination) and therefore the national rules in relation to obtaining or using real evidence as sources of information, such as documents or DNA, will not be covered in detail.

[143] See, for example, Howes (2013), p. 143.

[144] See, for example, in Denmark, Straffelovrådets betænkning nr. 1424/2002; Smith et al. (2008), p. 135; and Rørdam (1996), p. 289; in England see *R v Makanjuola* [1995] 1 W.L.R. 1348 CA, 1351, followed in *John Dawes and Others v The Queen* [2007] EWCA Crim 1165, [78]; see also s. 73, Serious Organised Crime and Police Act 2005 in relation to plea agreements in serious organised crime cases and the case of *R v Dougall* [2010] EWCA Crim 1048, [20–23], where the court concludes that the final sentence in plea agreements should be left to the sentencing court's discretion.

[145] See further [ECtHR] *Cornelis v. Netherlands*, Application no. 994/03, decision as to admissibility, 25 May 2004; and *X v. United Kingdom*, Application no. 7306/75, decision of the Commission, 6 October 1976, 115.

international debate about torture and other forms of ill-treatment is serious and highly relevant. However, this type of unfairness can never arise 'lawfully' and may arise as a problem within national borders, rather than as a direct result of the process of international cooperation.[146] By contrast this research focuses on the problem of irregularly obtained evidence—that is evidence that is obtained lawfully in a foreign legal system but according to investigative rules that are structurally different to those being applied in the trial jurisdiction. Likewise, cases dealing with dishonesty or inducement are not directly relevant and go to the question of reliability. Nevertheless, judicial decisions of national and international courts dealing with torture, ill-treatment or dishonesty may be indirectly relevant where they add to the interpretation of the meaning of 'fairness' in the context of the right to silence.

Often, when the question of fairness and admissibility of evidence is raised in criminal cases, this is in relation to cases of entrapment. In situations of entrapment there is a direct involvement of law enforcement officers in the commission of crime. However, cases of entrapment do not specifically involve violation of the right to silence or the right against self-incrimination. Nevertheless, entrapment cases may also be relevant to discussions about the definition of fairness.

Today in England and Australia, an accused person is competent to give evidence in his own defence but is not a compellable witness. This has the effect that if an accused chooses to give evidence and knowingly and corruptly gives an untruthful account of the facts, he or she can be liable for perjury.[147] In these systems, there may be some pressure on a suspect not only to break the silence when questioned by the police, but also to tell the truth and confess, in light of the fact that they could be charged with perjury if they later give the same false account in sworn testimony. However, the problem of prosecuting an accused for perjury based on his allegedly false evidence of his innocence at a previous trial is best categorised as an issue relating to double-jeopardy rather than to the right to silence. The elements of the offence of perjury will be different to the elements of another substantive

---

[146]On the use of oppression by the police see [ECtHR] *Funke v. France,* Application no. 10828/84, 25 February 1993, [44]; [GC] *Saunders v. The United Kingdom*, Application no. 19187/91, 17 December 1996, [68–69]; [GC] *Jalloh v Germany*, Application no. 54810/00, 11 July 2006, [100]; Human Rights Committee, General Comment 32, [41]; Human Rights Committee, *Paul Kelly v Jamaica*, Communication No 253/1987 (1991), [5.5]; on the inadequacy of legal representation, which is another 'human error' issue not covered here, see Wu (2011), pp. 103–104; and Cape et al. (2010), pp. 38–44; see also [ECtHR] *Imbrioscia v Switzerland*, Application no. 13972/88, 24 November 1993, [41].

[147]s. 314 Crimes Act 1958 (Vic), makes the offence of perjury punishable by a sentence of imprisonment of up to 15 years; in English law a person is liable to be punished for perjury by imprisonment of up to 7 years (or by a fine or both), see ss. 1 and 13–15 Perjury Act 1911 and s. 5 in relation to making a false statement, which is punishable by a fine or imprisonment of up to 2 years, a fine or both; in relation to defining perjury, 'judicial proceedings' for the purposes of giving sworn evidence include proceedings in a court or tribunal that are presided over by someone with the power to hear, receive or examine evidence on oath, and include proceedings before a person who has the power by law to administer an oath and authenticate a statement; see further Starmer et al. (2001), p. 208.

offence and, therefore, there may not be a case of *autrefois convict* or *autrefois acquit*.[148] Nevertheless, an accused does not risk being charged with perjury simply by falsely denying his guilt at his trial.

This reasoning is consistent with the decision of the High Court of Australia in the case of *The Queen v Carroll*, which highlighted such factors as the incontrovertibility of an acquittal.[149] The Court found that a general provision creating the offence of perjury cannot abolish legal principles about double jeopardy.[150] The Court unanimously found that it would be an abuse of process by the prosecution to have an accused tried for perjury on the basis of his denial of guilt and what is essentially the same evidence about the material facts for which he has been previously acquitted or convicted.[151] For example, if the accused gives evidence on a murder trial denying that he killed the victim, he cannot later be tried for perjury on essentially the same evidence that was led in the murder trial. If there is some distinctively new and different evidence then it might be possible.

There may be an indirect connection between perjury and the right to silence. In some countries the right to silence is associated with the freedom of a suspect or accused to say anything in their own defence, including telling lies.[152] However, the connection between a prosecution for perjury and the risk of double-jeopardy is more direct. Therefore, in this work perjury will not be presented as a relevant limitation on the right to silence.

1.3.5 Terminology

1.3.5.1 Effectiveness

Effectiveness in law enforcement may be described as realising the community's interest in bringing a wrongdoer to justice. The public interest in punishing the those who commit crime is clear. The central aim of the criminal trial is to convict the guilty, and acquit the innocent.

In the Danish legal system, as in other civil law jurisdictions, the judge is obliged to get to the truth in a case.[153] This is called the principle of material truth (*den*

[148] See *The Queen v Carroll* [2002] HCA 55, [11–16, 129].

[149] *The Queen v Carroll* [2002] HCA 55, [35–38].

[150] *The Queen v Carroll* [2002] HCA 55, [147].

[151] *The Queen v Carroll* [2002] HCA 55, [45–51, 114–116, 134]. The High Court of Australia referred to the following decisions from other common law jurisdictions: *H M Advocate v Cairns* 1967 SLT 165; *Gridic v The Queen* [1985] 1 SCR 810; *R v Humphreys* [1977] AC 1; *R v Moore* [1999] 3 NZLR 385.

[152] European e-justice portal (Europa), Rights of defendants in criminal proceedings, Belgium.

[153] See, for example, §§ 867 and 874(4) of the Danish Administration of Justice Act in addition to § 880 in relation to the free evaluation of evidence; Smith (2013), p. 18; Smith et al. (2008), pp. 26–27.

materielle sandheds princippet). Similarly, prosecutorial ethics in many criminal justice systems dictate that, as a representative of the community interest, prosecutors should guard against the conviction of the innocent. For example, in *McCullough v The Queen*, the Tasmanian Court of Criminal Appeal stated that:

The Crown has no interest in procuring a conviction. Its only interest is that the right person should be convicted, that the truth should be known, and that justice be done.[154]

Therefore, certain legal and ethical rules connect effectiveness to the truth. Indeed, the balance between effectiveness and fairness may be described as the balance between seeking the truth and fairness.

In relation to transnational cases, prosecutors are using international cooperation instruments as part of the weaponry in the fight against serious transnational crime. International cooperation in criminal matters allows relevant and reliable evidence to be placed before the trial court, even when that evidence was found outside the jurisdiction. Police powers and other legal parameters, including coercive measures, are the weapons of effectiveness that are in place to assist the police and other public authorities in exposing the truth, and thereby convicting criminals. The powers/measures that achieve effectiveness generally limit the safeguards that protect defence rights such as the right to silence.

1.3.5.2 Fairness

Fairness is a vague concept, which is difficult to define. In civil law systems, particularly those traditionally based on inquisitorial legal method, ensuring fairness is largely incorporated into the objectivity of the investigating magistrate and the prosecutor. An inquisitorial system trusts the state's legal professionals to conduct a neutral investigation into the truth.[155] In a Danish trial the balance between effectiveness and fairness may be more closely interpreted as the balance between the desire to come as close as possible to the truth, on the one hand, and the presumption of innocence, on the other.[156] Long before the enactment of the ECHR and other major human rights provisions, the common law developed a right to a fair trial, or the fair trial principle, as it is sometimes known.[157] The common law concept of fairness is 'multi-faceted', involving interests of the accused person and also of the wider community.[158]

---

[154]*McCullough v The Queen* [1982] Tas. R. 43 at 57; see further Lord Denning in *Jones v National Coal Board* [1957] 2 QB 55 at 63, where it is stated that the object of a trial is to ascertain the truth and to do justice.

[155]Wu (2011), p. 88.

[156]Waaben and Langsted (2011), p. 18.

[157]Victorian Law Reform Commission (2009), p. 26.

[158]*McCullough v The Queen* [1982] Tas. R. 43 at 57.

With the growing significance of international human rights law, fairness in both civil law and common law systems has become increasingly connected to the right to a fair trial as a fundamental human right.[159] This is particularly so in the legal systems of Council of Europe countries that have ratified the ECHR and come under the supervision of the ECtHR. The right to a fair trial has become the most relevant way of defining and analysing fairness in the modern criminal trial.

The right to a fair trial encapsulates the safeguards that are necessary to bring about justice, also for the suspect or accused. Using this definition of fairness, in the balance between effectiveness and fairness, the state authority's weapons of effectiveness, such as police powers, are weighed against the safeguards of the individual suspect or defendant. Fairness includes all the procedural rights and minimum guarantees that the right to a fair trial entails. Truth-finding and fairness are the two distinctive aims of criminal justice and the legitimacy of any legal system depends upon its ability to do both.[160]

For balance to be achieved in transnational cases, using international cooperation to gather evidence should not affect the national balance in the requesting or issuing state's trial, by undermining the suspect or accused person's fair trial rights. Ideally, the process of international cooperation does not, of itself, achieve a tactical advantage for the prosecution.[161] A balance between effectiveness and fairness in national and transnational cases means that a suspect or accused's defence rights may be limited and protected in such a manner that the criminal proceedings are fair when viewed as a whole.

1.3.5.3 Adversarial and Accusatorial Trials

The principles of adversarial and accusatorial trial are interrelated and the terms are frequently used interchangeably.[162] The adversarial trial is connected to the early development of the common law criminal trial in England—a development that was the work of defence lawyers, following their introduction into criminal trials in the 1700s.[163] The idea is that a case is contested by two adversaries (opposing parties), the prosecution and the accused in criminal cases, before a neutral court. Incumbent in this is the right of both parties to make legal submissions and to present and challenge evidence through examination and cross-examination of witnesses. The adversarial system is based on the belief that the truth will emerge by placing the

---

[159]See further Wu (2011), p. 86.

[160]Wu (2011), p. 89.

[161]On the development of the concept of fairness in Denmark, England and Australia see Chap. 6, Sect. 6.4.3.

[162]See, for example, Summers (2007), pp. 6–10, where comparisons between accusatorial and inquisitorial methods of trial are presented.

[163]Helmholtz et al. (1997), pp. 87 and 180; see also Langbein (2003), p. 173.

adversaries in opposite corners, both sides having the responsibility of gathering, presenting and testing evidence.[164]

The central principle of an accusatorial trial is that the accuser, historically the victim but now the prosecution, bears the burden of proof throughout the criminal proceedings. This means that an accused does not have to say anything (the right to remain silent) or to present any evidence in his own defence. Along with the presumption of innocence, the accusatorial principle (and the right to remain silent on caution) is a cornerstone of the adversarial trial (in the common law sense).

1.3.5.4 Irregular Evidence

In this project, the phrase 'irregularly obtained evidence' refers to the situation where evidence is gathered lawfully in a foreign legal system in a way that is inconsistent the law of the requesting state.

1.3.5.5 Admissibility and Cross-Admissibility

Admissible evidence is evidence that is relevant, reliable and 'fair' and does not present any other problems in relation to the rules in particular systems, which determine whether or not evidence may be used for conviction.

Cross-admissibility is commonly a term used to describe the situation where evidence that is relevant on one charge is also relevant on another and vice versa, for example, evidence which shows that an accused had used the same criminal system on another occasion. This project has used the term 'jurisdictional cross-admissibility' to refer to the situation where evidence gathered in one legal system is admissible in another.

1.3.5.6 Charging and 'Sigtelse'

The usual practice in English and Australian law is that a suspect is charged when the investigation is complete. At this point the suspect is handed a formal notice of the charges (which also becomes a part of the police 'brief of evidence') and becomes the 'accused'.

The English term 'suspect' means something more than the Danish term, *mistænkt*. By comparison, in Denmark a *mistænkt* who becomes suspected on reasonable grounds is often 'charged' early on in the investigation and becomes the *sigtet*. This means that he or she is the one person who the investigators suspect on reasonable grounds has committed an offence. *Sigtelse* (the process of 'charging')

[164]Wu (2011), p. 87; by contrast, the inquisitorial trial is presided over by an investigating judge, who is also said to be neutral because he does not have a prosecuting role, but has the function of conducting the examination of all the witnesses and the accused: Summers (2007), pp. 31–34.

occurs by an oral communication of the accusation to the particular suspect, which is then noted in the police 'report'. The person does not stop being the *sigtet* until the prosecution files an indictment or the case is dismissed. Upon indictment, the *sigtet* becomes the *tiltalt* (accused). *Tiltalelse* (indictment) is the process by which the prosecuting authority makes a formal accusation against a person, marking the end of the investigative phase and the beginning of the trial phase.[165]

Using common law terminology, the state of being the *sigtet* lies between being a suspect and being an accused. Compared to the formal charging process in a common law jurisdiction, this process is relatively informal. However, in Denmark a distinction is drawn between being *regarded* as *sigtet* and formal *sigtelse*.

The distinction between being the suspect or accused and being the *sigtet* is important to understand in the context of when certain due-process rights accrue in the different systems being examined. In a transnational case, for example, it may be of significance whether the requested state is able to provide the same rights as would normally apply in the requested state at the time of arrest or the first interrogation.

The differences between the processes described above are explained in Table 1.1, Appendix section.

1.4 The Importance of the Right to Silence in Transnational Cases

It might be assumed that pursuing effective law enforcement through international cooperation and promoting adherence to procedural rights are mutually exclusive. Perhaps the international movement to promote respect for human rights stands in the way of effectiveness. Where there is continuity of law between the investigative and the trial phases of each national legal system, the international cooperation process may mean that certain balancing mechanisms are lost when these two phases take place in different legal systems. Alternatively, the process of international cooperation in gathering confession evidence in cross-border criminal cases may be hindered by efforts to ensure the application of common minimum standards in relation to the right to silence. Reluctance to seek evidence from a suspect in another country may be justified, if different standards of protection of the right to silence are being applied.

On the other hand, in the EU for example, the implementation of the ECHR, coupled with the push for effective law enforcement across national borders based on a culture of trust may have led to an over-confidence in the ability of transnational criminal proceedings to remain fair. Promotion of human rights in the EU has encouraged a high level of confidence and trust between EU Member States. However, judges and prosecutors in transnational cases may pass over real issues

[165]Under §§ 834(1), 895 and 896 of the AJA, the prosecution prepares, signs and registers an indictment with the court and the person becomes known as the *tiltalt* ('the indicted').

of fundamental unfairness that result from confessions being received 'lawfully' under different procedural rules, on the basis that the unfamiliar rules of a fellow EU state 'must' be fair because of the state's participation in the Convention. What is considered fair in the context of one national system may not be fair when viewing transnational criminal proceedings as a whole. The high level of confidence in the fairness of other member states' processes is what the EU Commission is driving at with the notion of mutual trust in EU international cooperation. EU states have, after all, had to deal with the rising level of transnational crime that has accompanied the opening of national borders.

This research aims at getting to the bottom of what really happens to the individual's right to silence when confession evidence or evidence of silence is transferred from one legal system to another in the process of international cooperation. The starting point in Chap. 2 is to set out the frameworks of protection of the right to silence under the ICCPR and the ECHR in order to determine the common minimum standards that are being set within these international frameworks as a background to the development of the rights in national law. Implementation of the ECHR in national law varies among the high contracting parties and, therefore, the standards discussed in Chap. 2 may be relevant in the national perspectives contained in Chaps. 3–5. In addition, the human rights frameworks are highly relevant to the international perspective in Chap. 7, which considers the extent of the influence of the frameworks on the international cooperation process. Chapters 3–5 are an in-depth study, on the national plane, of the ways in which the laws in Denmark, England and Wales and Australia regulate the balance between effectiveness and fairness in gathering and using suspect evidence, such as confession evidence or evidence of silence. The results of these examinations form the basis of Chap. 6, which looks at jurisdictional 'cross-admissibility' of confession evidence to identify problem areas in the transnational perspective. From the international perspective, Chap. 7 looks at whether the different international frameworks encourage different investigative, prosecutorial and judicial approaches to balancing the right to silence in transnational proceedings. The chapter also considers whether uniform minimum standard setting under the ICCPR and the ECHR affects the question of balance in the international cooperation process, by establishing a foundation for mutual trust between cooperating states. The final chapter sums up the major conclusions and provides some suggestions for strengthening the maintenance of balance in gathering confession evidence in the international cooperation process.

The 'mosaic of international conventions', with the various forms of mutual legal assistance available and the range of procedures for which assistance can be sought, are exceptionally flexible and capable of facilitating cooperation between different systems.[166] Therefore, attention needs to be paid to ensure that the protection of foundational fair trial rights, such as the right to remain silent, are also preserved in the process of transferring evidence in transnational cases.

---

[166]Van Daele (2010), p. 163.

Appendix

Table 1.1 Comparative status of the suspect/accused person

England and Wales		Denmark	
Title (person)	Action	Title (person)	Action
Suspect	Grounds for suspicion being developed	Mistænkt ('Suspected') Meaning a person who is under suspicion	Grounds for suspicion being developed
Suspect	Reasonable grounds for suspicion established. Investigation is still ongoing. Obtains the following rights: Right to silence and right against self-incrimination; Caution before questioning; Access to legal advice of own choosing and opportunity to access a duty lawyer call centre for all suspects at a police station prior to being questioned.	Sigtet ('Charged') Meaning a person who has been charged, whereby they have been informed by the investigating officers of the accusations against them.	Reasonable grounds for suspicion established. Investigation is still ongoing. Obtains the rights of a *sigtet*, including: Right to silence and right against self-incrimination; Caution before questioning; Right to legal advisor of own choice and to be told of the right to access a publicly appointed lawyer and how to go about it; Right to information about the charges and disclosure.
Suspect	Arrest: Based on reasonable grounds of suspicion and necessity. Investigation still ongoing. Obtains the additional rights pertaining to detention, such as: Information about the grounds and reasons for arrest; Right to silence under caution; Clear right to access legal advice, including free legal advice, where relevant.	Sigtet ('Charged')	Arrest: Based on reasonable grounds of suspicion and necessity. Investigation still ongoing Obtains the additional rights pertaining to detention, such as: Information about the grounds and reasons for arrest; Right to silence under caution; Right to access legal advice, including free

(continued)

Table 1.1 (continued)

England and Wales		Denmark	
			legal advice, where relevant.
Accused	Formally charged: Investigation is complete and there should be a prima facie case to answer. Accused receives written charges. Commencement of the trial phase. Obtains the additional rights pertaining to being charged, including: Caution on charging; Right to legal advisor, including free means tested legal advice in preparation for hearing; Right to information about the charges and disclosure.		
Accused	First court hearing to determine bail/remand. Charge sheets, along with an application to the court, form the basis of a bail/remand hearing.	Sigtet	First court hearing to determine bail/remand An application to the court prepared by the police in conjunction with the public prosecutions service forms the basis of the bail/remand hearing. Obtains additional rights: Legal advice, including free legal advice as relevant.
Accused	Indictment: Indictment prepared by the public prosecutions service and filed with the court.	Tiltalt ('Indicted'): a person who has been accused on indictment	Indictment: Investigation is complete. Indictment prepared by the public prosecutions service and filed with the court. Commencement of the trial phase.

References

Australasian Legal Information Institute (2015) Austlii. http://www.austlii.edu.au/. Accessed 15 Oct 2015

Bantekas I, Nash S (2003) International criminal law, 2nd edn. Cavendish, London

Bell J (2011) Legal research and the distinctiveness of comparative law. In: Van Hoecke M (ed) Methodologies of legal research. Hart, Oxford

Bentley D (1998) English criminal justice in the nineteenth century. The Hambledon Press, London

Blume P (2009) Juridisk Metodelære, 5th edn. Jurist- og Økonomforbundets Forlag, Copenhagen

Breay C, Harrison J (2015) Magna Carta in context. In: Treasures in full, Magna Carta, articles. The British Library. http://www.bl.uk/treasures/magnacarta/themes/index.html. Accessed 4 Jan 2015

Brøbech B (2003) Ulovligt tilvejebragte beviser i straffeprocessen. Jurist- og Økonomforbundets Forlag, Copenhagen

Cape E, Namoradze Z, Smith R, Spronken T (2010) Effective criminal defence in Europe. Intersentia, Antwerpen

Crawford J (2012) Brownlie's principles of public law, 8th edn. OUP, Oxford

Dennis I (2011) The human rights act and the law of criminal evidence: ten years on. Syd Law Rev 33:333–357

Dyevre A (2013) Outline of a legal realistic approach to legal integration. In: Neergaard U, Nielsen R (eds) European legal method. DJØF, Copenhagen

Easton S (2014) Silence and confessions: the suspect as a source of evidence. Palgrave Macmillan, Hampshire

Elo Rytter J (2013) Individets grundlæggende rettigheder. Karnov Group, Copenhagen

Emmerson B, Ashworth A, Macdonald A (eds) (2012) Human rights and criminal justice, 3rd edn. London, Sweet & Maxwell

European e-justice portal (2014) My rights during the enquiries and investigation and before the matter is referred to the court: arrest and questioning by the police. In: Rights of defendants in criminal proceedings, Belgium. https://e-justice.europa.eu/content_rights_of_defendants_in_criminal_proceedings_-169-BE-maximizeMS-en.do?clang=en&idSubpage=2#No1. Accessed 4 Jan 2014

Gans J, Palmer A (2014) Uniform evidence. OUP, Sydney

Glenn HP (2004) Legal cultures and legal traditions. In: Van Hoecke M (ed) Epistemology and methodology of comparative law. Hart, Oxford

Glenn HP (2014) Legal traditions of the world. OUP, Oxford

Harris DJ, O'Boyle M, Bates EP, Buckley CM (eds) (2014) Harris, O'Boyle and Warbrick: law of the European Convention on Human Rights, 3rd edn. OUP, Oxford

Helmholtz RH, Gray CM, Langbein JH, Moglen E, Smith HE, Alschuler AW (1997) The privilege against self-incrimination: its origins and developments. University of Chicago Press, Chicago

Howes T (2013) England. In: Ligeti K (ed) Toward a prosecutor for the European Union, vol 1, A comparative analysis. Hart, Oxford

Jackson JD (2009) Re-conceptualizing the right to silence as an effective fair trial standard. Int Comp Law Q 58(4):835–861

Jackson JD, Summers SJ (2012) The internationalisation of criminal evidence: beyond the common law and civil law traditions. CUP, Cambridge

Ladeur K (2004) Methodology and European law – can methodology change so as to cope with the multiplicity of law? In: Van Hoecke M (ed) Epistemology and methodology of comparative law. Hart, Oxford

Langbein JH (2003) The origins of adversary criminal trial. OUP, Oxford

Legrand P (1996) European legal systems are not converging. Int Comp Law Q 45(1):52–81

Mattei U (1997) Three patterns of law: taxonomy and change in the world's legal systems. Am J Comp Law 45:5–44
Moeckli D, Shah S, Sivakumaran S (2010) International human rights law. OUP, Oxford
Nielsen R (2013) New european legal realism – new problems, new solutions? In: Neergaard U, Nielsen R (eds) European legal method. DJØF, Copenhagen
Rørdam T (1996) Bekæmpelse af den organiserede kriminalitet. Paper presented with Bay Larsen L, Thiesen H at Gruppdiskussion Bekämpandet av den organiserade brottsligheten, 34th Nordiske Juristmøde, Stockholm, 1996
Samuel G (2014) An introduction to comparative law theory and method. Hart, Oxford
Siems M (2014) Comparative law. CUP, Cambridge
Smith E (2013) Straffeprocess, 7th edn. Forlaget Thomson, Copenhagen
Smith E, Jochimsen J, Kistrup M, Poulsen JL (2008) Straffeprocessen, 2nd edn. Forlaget Thomsen, Copenhagen
Smits JM (2004) The Europeanisation of national legal systems: some consequences for legal thinking in civil law countries. In: Van Hoecke M (ed) Epistemology and methodology of comparative law. Hart, Oxford
Starmer K, Strange M, Whitaker Q (2001) Criminal justice, police powers and human rights. Blackstone Press Limited, London
Straffelovrådet (2002) Straffastsættelse og strafferammer I – Almindelig del. Betænkning nr. 1424/2002. http://www.statensnet.dk/betaenkninger/1401-1600/1424-2003-1/1424-2003-1_pdf/printversion_1424-2003-1.pdf. Accessed 15 Oct 2015
Summers S (2007) Fair trials. Hart, Oxford
Thunberg Schunke M (2004) Internationell Rättslig Hjälp I Brottmål Inom EU: Effektivitet v. Rättsäkerhet. Iustus Förlag, Uppsala
Trechsel S, Summers S (2006) Human rights in criminal proceedings. OUP, Oxford
Van Daele D (2010) Mutual assistance between Belgium, France, Germany and the Netherlands. A comparative analysis of possibilities and difficulties. In: Fijnaut C, Ouwerkerk J (eds) The future of police and judicial cooperation in the European Union. Martinus Nijhoff, Leiden
Van Hoecke M (2004) Deep level comparative law. In: Van Hoecke M (ed) Epistemology and methodology of comparative law. Hart, Oxford
Van Hoecke M, Warrington M (1998) Legal cultures, legal paradigms and legal culture: towards a new model for comparative law. Int Comp Law Q 47:495–536
Victorian Law Reform Commission (2009) Jury directions: final report. http://www.lawreform.vic.gov.au/sites/default/files/VLRC_JuryDirections_FinalRep ort.pdf. Accessed 15 Jan 2014
Waaben K, Langsted LB (2011) Strafferettens almindelige del I – Ansvarslæren, 5th edn. Karnov Group, Copenhagen
Wild C, Weinstein S (2013) Smith and Keenan's English law, 17th edn. Pearson Education Ltd., United Kingdom
Wu W (2011) European perspective and standards on interrogational fairness. In: Cools M, De Ruyver B, Easton M, Pauwels L, Ponsaers P, Vander Beken T, Vender Lænen F, Vande Valle F, Verhage A, Vermeulen G, Vynckier G (eds) EU Criminal justice, financial & economic crime: new perspectives. Governance of Security Research Paper Series, vol. 5. Maklu, Antwerp
Yale Law School (2015a) The declaration of the rights of man – 1789. In: The Avalon Project. http://avalon.law.yale.edu/18th_century/rightsof.asp. Accessed on 15 Jan 2015
Yale Law School (2015b) The English Bill of Rights – 1689. In: The Avalon Project. http://avalon.law.yale.edu/17th_century/england.asp. Accessed on 15 Jan 2015
Zweigert K, Kötz H (1998) An introduction to comparative law, 3rd edn. OUP, Oxford

Chapter 2
Development of the Right to Silence in International Human Rights Law

Abstract International frameworks for human rights protection provide an important background to the transnational process of evidence gathering in cross-border cases. Their main functions are setting and monitoring minimum standards of individual protection and providing an avenue of complaint against the state when these standards are not being met. Through these functions, international attention is drawn to the legal development of human rights. There is an argument at least in Denmark that the principles developed under a framework such as the one provided for in the ECHR, due to the nature of the ECtHR's dynamic interpretation, may be applicable in national law through incorporation. In addition, the protective frameworks may create certain assumptions between cooperating states about, for example, the types of procedures that operate to the benefit of suspects, victims or other individuals caught up in the process. Further, depending on the influence of theses frameworks, they may encourage uniformity of minimum procedural standards to such an extent that the overall aim of gathering admissible evidence in transnational cases is optimised. This chapter lays out the legal background of the right to silence in relation to two significant international human rights frameworks, the ICCPR and the ECHR.

2.1 Introduction

In modern history, the right to a fair trial has been central to the international human rights movement that followed the atrocities of the Second World War and culminated in the adoption by the United Nations in 1948 of the Universal Declaration of Human Rights (UDHR). Although the international military tribunals that followed in Nuremburg and Tokyo were criticised in a number of respects, the right to a fair trial was an important subject of discussion. The centrality of the right to a fair trial, the right to silence and the right against self-incrimination is also reflected by the fact that these rights are protected in all major international human rights treaties.[1]

[1]Cape et al. (2010), p. 7; see further articles 14–16 of the International Covenant on Civil and Political Rights 1966 (ICCPR); art. 6 of the European Convention on Human Rights and

F.M.W. Billing, *The Right to Silence in Transnational Criminal Proceedings*,
DOI 10.1007/978-3-319-42034-9_2

A significant number of historical legal documents, of both national and international origin, have contributed to the setting of minimum international standards to ensure fairness in criminal proceedings and provide a relevant background to the way in which procedural rights, such as the right to silence, have developed in national and international contexts.[2]

Two international frameworks, in particular, have had an important influence on the development and enforcement of minimum standards for fair trial rights protection at the national level: the United Nations' political enforcement system under the 1966 International Covenant on Civil and Political Rights (ICCPR)[3]; and the legal enforcement system within the Council of Europe, based upon the 1950 European Convention on Human Rights and Fundamental Freedoms (ECHR).[4] These two frameworks continue to carve out foundational understandings about human rights, such as the right to a fair trial, including the right to silence and the right against self-incrimination. The substance of these rights in international law, as reflected in the decisions of international decision-making bodies, to varying extents inspire and influence the development of minimum standards amongst nations, for example, in relation to the maintenance of balance between effectiveness of law enforcement and fairness to a suspect or accused person. Their sphere of influence includes national and transnational criminal cases inside Europe, outside Europe and inside to outside Europe.[5]

Certain elements of the right to a fair trial are cornerstones in ensuring a balance between effectiveness and fairness throughout the criminal proceedings. For example, a suspect or accused is to be presumed innocent until proven guilty according to law.[6] The Human Rights Committee (HR Committee) has classified this guarantee

---

Fundamental Freedoms 1950 (ECHR); art. 3 and 8–10 of the American Convention of Human Rights 1969 (ACHR); articles 3 and 7, African Charter on Human and People's Rights 1981 (ACHPR); articles 12–13 and 15–19 Arab Charter on Human Rights 1994 (revised 2004); and articles 6–7 of the Commonwealth of Independent States Convention on Human Rights and Fundamental Freedoms (1995).

[2]As examples of the way in which the right to a fair trial is established in international criminal law, see the following articles of the Rome Statute of the International Criminal Court: art. 55(1)(a) in relation to the right to silence and the right against self-incrimination, as well as the right to access a legal advisor prior to and during interrogation; art. 55(2)(b) on the effect of the investigative right to silence, which cannot be used against an accused person at trial; art. 55(2)(c) and 55(2)(d) about the right to legal representation prior to and during interrogation; and articles 66 and 67 in relation to the presumption of innocence and a broad range of other fair trial guarantees.

[3]International Covenant on Civil and Political Rights, opened for signature in New York on 19 December 1966 and entered into force 23 March 1976 (with the exception of Article 41), UNTS 999/171.

[4]Convention for the Protection of Human Rights and Fundamental Freedoms 1950 87 UNTS 103; ETS 5.

[5]Within the EU the Charter of Fundamental Rights and Freedoms is also of continually growing significance. The ICCPR and the ECHR have also influenced the development of defence rights in international criminal cases.

[6]Human Rights Committee, General Comment 30; Art. 6(2) ECHR.

as a peremptory norm of international law that cannot be deviated from by the operation of the law during times of emergency, or in any other circumstances.[7] This presumption has fundamental significance for gathering and using evidence. The principle of objectivity requires investigators and prosecutors to work on the assumption that a suspect may be innocent. Nevertheless, necessary investigatory steps such as fingerprinting, taking DNA samples or searching property do not appear to violate the presumption.[8] Equality of arms may also be said to be central to achieving a fair trial. The HR Committee has described equality of arms as ensuring that both the prosecution and the defence have the same procedural rights, "…unless distinctions are based on law and can be justified on objective and reasonable grounds, not entailing actual disadvantage or other unfairness to the defendant".[9] During criminal proceedings, the right to an effective defence also demands that an accused is accorded the right to an impartial, timely and adversarial trial, including the immediacy principle.[10]

Acting together with the rights described above, under both human rights frameworks, the right to silence and the right against self-incrimination are said to be rights that lie at the heart of a fair trial.[11] All of these rights have 'enabling' and 'protective' aspects that ensure that the individual's interests are established and maintained throughout the criminal justice process.[12] Through the cases and commentary that are generated in each of the frameworks being examined, a contribution is made to identifying minimum standards of protection in relation to these foundational rights that may be applicable to criminal proceedings in the national setting, depending on the form and value of implementation (see Chap. 1, Sect. 1.3.3.3).

It is generally recognised that achieving *balance* between effectiveness and fairness in national prosecutions means that the right to a fair trial is not absolute. Defence rights, such as the right to silence, may be lawfully limited so long as there are safeguards in place that restore balance in the criminal proceedings when viewed as a whole. Limitations are typically placed on a suspect or accused's

---

[7]CCPR, General Comment No. 29: States of emergency (Article 4), CCPR/C/21/Rev.1/Add.11, [11].

[8]Moeckli et al. (2010), p. 324.

[9]Human Rights Committee, General Comment No 32, Art. 14: Right to equality before courts and tribunals and to a fair trial, UN Doc. CCPR/C/GC/32 (2007), [13].

[10]This requires the defence has the right to call, examine and cross-examine witnesses. See further Article 14(3)(e) ICCPR and Human Rights Committee, General Comment No 32, Art. 14: Right to equality before courts and tribunals and to a fair trial, UN Doc. CCPR/C/GC/32 (2007), [39]. There are other issues relevant to an effective defence, such as the right to be present; right of appeal; compensation for miscarriage of justice; *ne bis in idem*; prohibition on retrospective criminal legislation: see further Human Rights Committee; General Comment 32, [47]; and see Kälin and Künzli (2009), pp. 456–460.

[11](ECtHR) *John Murray v The United Kingdom*, Application no. 18731/91, 8 February 1996, [45].

[12]See also Jackson and Summers (2012), p. 276, referring to protective and participatory defence rights and that the right to silence may be classified as both.

right to silence, including the right against self-incrimination, in situations such as police interrogation; the use of administrative compulsory questioning powers; the drawing of adverse inferences of guilt from the decision to remain silent; and using covert surveillance measures to obtain confession evidence. Within the frameworks of the ICCPR and the ECHR, international legal obligations in the form of minimum standards in relation to the right to silence are set that determine how the balance between effectiveness and fairness should operate in these circumstances. However, despite the significance of the right, the scope and content of protection in national law may vary significantly depending on how the rights have developed in national law and the influence of a particular international human rights framework. Therefore, this variation in protection may reflect *inter alia* differences in constitutional law, diversity of legal traditions, the nature of the international frameworks, the substance of the minimum standards under the frameworks and how the international obligations under each framework are implemented and enforced in national law.

This chapter aims to identify the international minimum standards under the human rights frameworks of the ICCPR and the ECHR that form the international background to the right to silence and the right against self-incrimination as they are applicable under national law in criminal proceedings in Denmark, England and Wales, and Australia. These minimum standards contribute to defining the function of the right to silence in the gathering and use of confession evidence or evidence of a suspect or accused person's silence.

The chapter commences by examining the development and nature of the frameworks for protection of fair trial rights under the ICCPR and the ECHR. It discusses how the international obligations may be implemented in the national systems so that they become applicable national law. In addition, the chapter will review the case law and commentary in relation to the right to silence under each framework. By doing so, the precise scope of these rights under international law may be understood, as an indication of the minimum standards that are being sought in the investigation and trial phases in national and transnational criminal proceedings.

2.2 Fair Trial Rights and the Right to Silence Under the ICCPR

2.2.1 The ICCPR Framework

2.2.1.1 The Development of Legally Binding Fair Trial Rights Under the ICCPR

The task of creating international legal obligations in the field of human rights has thrown up complex political and philosophical questions, particularly in the light of the political, economic, social and cultural differences between nations and human rights theories based on relativism.[13] Therefore, instead of a legally binding treaty, on 10 December 1948, the UN General Assembly adopted the Universal Declaration of Human Rights (UDHR). Although the UDHR was not a formal source of law at the time of its adoption, it is now a topic of debate whether in fact the UDHR has gained legal force by becoming part of international customary law. At the 1993 World Conference on Human Rights in Vienna, 171 countries reaffirmed their political commitment to the UDHR.[14] The 'growing consensus' is that most of the rights elaborated in the UDHR, if not all, are legally binding and form part of international customary law.[15] This would include the right to a fair trial in art. 10 and the right of a person charged with a criminal offence to be provided defence rights.

Following the adoption of the UDHR, work began on more detailed provisions and it was decided to divide the UDHR into two sets of rights. Civil and political rights were to be realized immediately; and economic, social and cultural rights, to be achieved progressively. Finally, 10 years after approval by the General Assembly, in 1976 two human rights treaties achieved the requisite number of ratifications for entry into force.[16] Together with the UDHR they are known as the International Bill of Rights. They are considered primary sources of international law in relation to the rights contained in them. They are of general application and are legally binding in international law upon the ratifying states.[17]

Article 14 of the International Covenant on Civil and Political Rights (ICCPR) provides for a right to a fair trial in much greater detail than the UDHR. Under art. 14(1) and (2) a person is guaranteed equality of the law, a fair hearing and a public hearing (where appropriate), judicial impartiality, the provision of a judgment and the presumption of innocence. The right to equality guarantees equal access to and lack of discrimination before courts and tribunals, as well as 'equality of arms'.[18]

Article 14(3) sets out a number of "minimum guarantees" available to anyone in the determination of a criminal charge. These guarantees include being fully informed; access to translation services as required; adequate preparation time and facilities, including access to counsel; being able to defend him or herself in

[13]Moeckli et al. (2010), p. 35.

[14]American Civil Liberties Union (2012); on the importance of the UDHR as a moral constitution for the human rights movement, see Steiner et al (2007), p. 136.

[15]De Schutter (2010), pp. 16 and 50; Steiner et al. (2007), p. 137.

[16]See the International Covenant on Civil and Political Rights, opened for signature in New York on 19 December 1966 and entered into force 23 March 1976 (with the exception of Article 41), UNTS 999/171; and the International Covenant on Economic, Social and Cultural Rights, opened for signature in New York on 19 December 1966 and entered into force 3 January 1976, UNTS 993/3.

[17]The ICCPR and the ECHR could be described as law-making treaties, which not only create reciprocal undertakings (as do 'contractual treaties') but also provide for rights and liabilities: Emmerson et al. (2012), p. 83.

[18]Human Rights Committee, General Comment No 32, art. 14: Right to equality before courts and tribunals and to a fair trial, UN Doc. HR COMMITTEE/C/GC/32 (2007), [13].

person or through counsel; to examine and cross-examine witnesses; and a right "not to be compelled to testify against himself or to confess guilt."[19]

Despite the significance of the ICCPR it has only a moderate rate of ratification.[20] Therefore, the dividing line is clear between international support of human rights law as something countries aspire to (as with the UDHR) compared with commitment to international legally binding minimum standards. The difference may indicate the difficulty the international community has in overcoming the gaps in minimum human rights protection that stem from divergences in legal tradition, culture and institutions. In transnational criminal cases, these are the gaps that may serve to hinder the state's ability to seek admissible evidence abroad. These rights have political and moral force under international law. Amongst the countries that have ratified the ICCPR, the question is to what extent the international legal commitments are actually being applied and enforced in national law (see also Chap. 1, Sect. 1.3.3.3 above; and Chaps. 3–5 in relation to Denmark, England and Wales and Australia, respectively).

2.2.1.2 Implementation and Enforcement of the ICCPR

Ratification of an international treaty is generally an executive action. Therefore, the question of enforcement and how the ICCPR sits alongside national law is determined by national rules and may vary from jurisdiction to jurisdiction, influenced by monism or dualism theories and constitutional law (referred to in Chap. 1, Sect. 1.3.3.3).[21]

In Denmark, the ICCPR has not been incorporated into Danish law (see Chap. 3, Sect. 3.2.4). However, this is not generally of great significance in relation to the right to a fair trial since the incorporation of the ECHR.[22] Though it may be relevant in relation to the occasional legal rule that has developed under the ICCPR in international law, but not by the ECtHR. The requirement for audio-visual recordings of police interviews under the HR Committee's guidelines on the ICCPR is one example. Likewise, in England, the ICCPR has not been directly incorporated into English law (referred to in Chap. 4, Sect. 4.2.4). However, the English courts may

[19]Article 14(3)(g) ICCPR. Article 14(4), (5), (6) and (7) also provide for special rights for juveniles; a right to appeal conviction and sentence; compensation for wrongful imprisonment and the rule against double jeopardy. Article 15 protects against retroactive effect of criminal liability or sentencing practices that are detrimental to a person charged with a criminal offence.

[20]There are 74 signatories and 168 state parties to the ICCPR and 71 signatories and 164 parties to the ICESCR (the rates are even lower for the optional protocol to each treaty): United Nations (2015).

[21]Crawford (2012), pp. 55–57.

[22]See Sect. 3.2.2 below.

occasionally refer to the ICCPR and the HR Committee, illustrating some level of influence in the national setting that goes beyond incorporation.[23]

As discussed in Chap. 5 (Sect. 5.2.1), Australia has incorporated the rights contained in the ICCPR through its national (and state) legislation, such as anti-discrimination and statutory evidence laws. The Human Rights (Parliamentary Scrutiny) Act 2011 (Cth) requires a Statement of Compatibility to accompany all new federal legislation; the Act also provided for the establishment of the Parliamentary Joint Committee on Human Rights. A presumption also prevails that Australian law should be interpreted in accordance with international obligations (as discussed in Chap. 1, Sect. 1.3.3.3 and Chap. 5, Sect. 5.2.1).

If an individual of a signatory state has exhausted national remedies in relation to a breach of an individual right stemming from the ICCPR, they may make an individual complaint (called a communication) to the HR Committee under the first Optional Protocol to the ICCPR.[24] Unlike the United Kingdom, Denmark and Australia have ratified the Protocol in 1972 and 1991, respectively. Even in Denmark and Australia, while the HR Committee has the power to adjudicate in such complaints, the decisions are not binding. However, they arguably carry moral and political weight *vis a vis* the future application of the ICCPR.[25] Inter-state complaints about a right under the ICCPR are also possible under art. 41. Under art. 40 of the ICCPR, the HR Committee receives periodic reports from each of the three countries about every 4 years, which go through a review process before the Committee. The results are put before the General Assembly.[26]

The above procedures provide the evidence upon which the HR Committee base their guidelines about minimum standards in the form of general comments. Again, general comments do not create binding international obligations but form part of the system of political enforcement of human rights at the international level, which can influence the legal development of the rights in the national settings (to varying extents). Upon incorporation into national law, the minimum standards suggested by the HR Committee are to some extent 'persuasive' in the national law.

2.2.1.3 Other International Treaties and Sources of Fair Trial Rights

The significance of the right to a fair trial under international law is evidenced by the fact that a number of other generally applicable international treaties,

[23]See the BAILII online database for case examples referring to either the ICCPR or the HR Committee: British and Irish Legal Information Institute (2015).

[24]Optional Protocol to the International Covenant on Civil and Political Rights, opened for signature on 19 December 1966 and entered into force on 23 March 1976 UNTC 999/171.

[25]There have been communications initiated against Denmark and Australia: see, for example, Office of the High Commissioner (2014). The question whether the ICCPR individual complaints procedure is effective in encouraging reform in the respondent states has not been assessed here.

[26]This reporting procedure is also backed by the Universal Periodic Review process conducted by the HRC under the UN Charter.

declarations and guiding principles, enact certain elements of the right. For example, art. 40 of the Convention on the Rights of the Child (CRC),[27] art. 15 of the Convention against Torture (CAT)[28] and the UN Basic Principles on the Role of Lawyers[29] all incorporate elements of the right to a fair trial. Other treaties may be formal sources of law and all the above are examples of material sources of law that provide evidence of the general acceptance of the principles they set out.

The numerous international fair trial treaty provisions have frequently been the subject of decisions by international and supranational courts, as well as domestic courts—thereby adding to the law coming from the formal sources identified in art. 38 of the Statute of the ICJ. In addition, fair trial rights have frequently been the focus of discussions by international human rights bodies and NGOs. For example, art. 15 CAT has been the subject of General Comments by the Committee Against Torture. This is true to such an extent that the 'fundamental principles of fair trials' are accepted to form part of the body of international customary law.

The HR Committee considers fair trial rights to be peremptory norms of international law, thereby creating legal obligations among all states.[30]

2.2.2 *The Scope of the Right to Silence Under the ICCPR*

2.2.2.1 Commentary of the HR Committee as a Primary Source of Law

As stated at Sect. 2.2.1.1 above, art. 14(1) guarantees *inter alia* a fair trial in the determination of a criminal charge and, specifically, under art. 14(3) an accused person cannot be compelled to make self-incriminating statements. In order to understand the precise scope of state obligations by virtue of these formal and binding international provisions, being the minimum standards that may be applicable in national law, the general comments of the HR Committee, along with the ICCPR itself, may arguably be regarded as a primary source law.[31] Although not a formal source of law, the general comments of the HR Committee are arguably a high value material source of law as they are formed through a consultative process,

[27]Convention on the Rights of the Child, opened for signature in New York on 20 November 1989, entered into force on 2 September 1990, UNTC 1577/3.

[28]Convention Against Torture and Other Cruel, Inhuman or Degrading Treatment or Punishment, opened for signature in New York on 10 December 1984, entered into force on 26 June 1987, UNTC 1465/85.

[29]Basic Principles on the Role of Lawyers agreed at the 8th UN Congress on the Prevention of Crime and the Treatment of Offenders, Havana, Cuba, A/CONF.144/28/Rev.1, 118 (1990).

[30]HR Committee, General Comment No. 29: States of Emergency (article 4), 31-08-2001, CCPR/C/21/Rev.1/Add.11.

[31]Jackson and Summers (2012) have taken a similar approach in considering the way in which evidentiary rules and principles are dealt with under international law, focusing primarily on the HR Committee's general comments and the jurisprudence of the ECtHR, as the applicable minimum standards under the ICCPR and ECHR frameworks, respectively.

which includes the participation of specialised agencies and NGOs, academics and other human rights treaty bodies.[32] In addition, the opinions of the HR Committee about the scope of the obligations assumed by states is carried out on the background of the knowledge obtained through monitoring work via the state reporting system and through adjudicating individual communications (as referred to above at Sect. 2.2.1.2).

2.2.2.2 Police Questioning and the Right to Silence

2.2.2.2.1 Do Fair Trial Rights Apply to Pre-Trial Procedures Under the ICCPR?

In general, once a person has been 'charged', establishing equality of arms involves providing a suspect or accused person with a number of specific minimum guarantees that enable an accused person to commence formulating a full and proper defence. These guarantees include ensuring that an accused fully understands the nature and cause of the accusations against him,[33] and is aware of the evidence in the possession of the prosecution that tends to support or disprove the accusations or charges,[34] including evidence of investigative impropriety.[35] In the criminal law context, these rights are becoming known as the 'right to information'.[36]

However, it is not clear to what extent art. 14 applies to pre-trial procedures. The HR Committee has determined that the right of a person who has been charged with a criminal offence to be informed of the charges applies to persons in detention and those who are not in detention, but does not apply to criminal investigations preceding the laying of charges.[37] After a suspect is formally charged with the commission of a criminal offence, they have a right to be informed of the charge

[32] The HR Committee constitutes representatives of 18 signatory states from different regions. About the work of treaty bodies as a source of law, see further Moeckli et al. (2010), p. 119.

[33] See for example, the right to be informed of a charge in a language which the accused person understands in art. 14(3)(a) ICCPR; see also art. 9 ICCPR with respect to the right to information in order for a detainee to challenge the lawfulness of his detention.

[34] See art. 14(3)(b) ICCPR where the use of the word 'facilities' would include disclosure of documents. An accused should also be tried without undue delay, where a defendant in pre-trial detention may be innocent, or where delay may weaken the defendant's chances of challenging the prosecution evidence: see, for example, art. 14(3)(c) ICCPR. See further Kälin and Künzli (2009), pp. 455–466.

[35] Human Rights Committee, General Comment No 32, Article 14: Right to equality before courts and tribunals and to a fair trial, UN Doc. CCPR/C/GC/32 (2007), [33]; see also Moeckli et al. (2010), p. 325.

[36] Human Rights Committee, General Comment No 34, Article 19: Freedoms of opinion and expression, under "Rights of Access to Information".

[37] Human Rights Committee, General Comment No 32, Art. 14: Right to equality before courts and tribunals and to a fair trial, UN Doc. CCPR/C/GC/32 (2007), [31]. The HR Committee goes on to note that notice for reasons of an arrest is separately protected in art. 9(2) of the ICCPR.

promptly.[38] Therefore, taking a common law approach to charging, for example, one might conclude that the right to a fair trial does not apply in the pre-trial phase; though the opposite is arguable, based on the Danish notion of charging/*sigtelse*.[39]

The HR Committee has found that the right not to be compelled to testify against oneself or to confess guilt (art. 14(3)(g)) must be understood as requiring an absence of any physical or undue psychological pressure "...from the investigating authorities on the accused, with a view to obtaining a confession of guilt".[40] The use of the word "accused" may imply that an interviewee has been formally charged with criminal offences. However, in most legal systems, police interrogation of a suspect usually occurs before formal charges have been laid or the accused has been indicted. Therefore, this statement can only be meaningful if it is implied that the right to silence and the right against self-incrimination arises at least at the first time a suspect is interrogated. This interpretation would be consistent with the HR Committee's clear statements about the non-use of evidence obtained by torture or other ill-treatment prohibited by art. 7 of the ICCPR, which is not reliant on the question of whether a person has been formally charged with a criminal offence.[41] Therefore, as an unclear, inconclusive exception, the right to silence and the right against self-incrimination may apply from the first interrogation. However, the HR Committee has not made any clear recommendations about when the need to caution arises.

2.2.2.2.2 Access to Legal Representation

Achieving equality of arms includes providing access to a legal advisor. Articles 14 (3)(b) and (d) of the ICCPR, as with all the major human rights treaties, guarantee the right of a person who has been charged with a criminal offence to have adequate time and facilities to prepare a defence and communicate with counsel of own choosing; and to defend him or herself in person or through legal assistance of his own choosing. If an accused person cannot afford to be defended by a legal adviser, then there is a state obligation to provide a state-funded lawyer where the interests of justice so require, depending on the circumstances of the case.[42] Moreover, if the interests of justice require a lawyer to represent the 'accused' then counsel should

[38]Human Rights Committee, General Comment No 32, Art. 14: Right to equality before courts and tribunals and to a fair trial, UN Doc. CCPR/C/GC/32 (2007), [31].

[39]For a clarification of this point, see Table 1.1 in Chap. 1, Appendix section.

[40]Human Rights Committee, General Comment No 32, Art. 14: Right to equality before courts and tribunals and to a fair trial, UN Doc. CCPR/C/GC/32 (2007), [41].

[41]Human Rights Committee, General Comment No 32, Art. 14: Right to equality before courts and tribunals and to a fair trial, UN Doc. CCPR/C/GC/32 (2007), [41]; see further Jackson and Summers (2012), pp. 247 and 250.

[42]See further Kälin and Künzli (2009), p. 457.

be assigned, even if the accused wishes to represent him or herself.[43] However, the HR Committee has again been unclear when elaborating on the right to communicate with counsel under art 14(3)(b). The HR Committee has stated that the right may only be effective with the assistance of an interpreter during the pre-trial phase, without specifying at what point in the pre-trial phase this right arises.[44] The UN Basic Principles on the Role of Lawyers recommends that the right to access a lawyer extends to all stages of criminal proceedings, which ".. presumably include interrogations before arrest and charge."[45]

Under ICCPR and other UN initiatives, attention has been drawn to the inherently coercive nature of a police interrogation—in particular the custodial interrogation. A number of bodies have asserted that access to a lawyer for prisoners, and those being interrogated for the first time, is a key safeguard against abusive coercion. Specifically, the HR Committee and the Committee Against Torture have concluded that all persons deprived of their liberty should have a right to promptly receive independent legal assistance.[46] The Special Rapporteur on the Question of Torture has confirmed that all persons held in detention by a law enforcement agency must be allowed the right of access to a legal adviser, including the right to consult a lawyer in private.[47]

2.2.2.2.3 Recording of Police Interviews

The United Nations Special Rapporteur on the Question of Torture has confirmed the view that all interrogations by law-enforcement officers should be recorded, preferably by way of video-recording.[48] Arguably, this adds an additional layer of protection against coercion in the interrogation room. Recording of interviews is a measure relied on by a number of state parties in various circumstances, including England and Australia.

---

[43]See Kälin and Künzli (2009), p. 456. The circumstances where a lawyer must be assigned may include, for example, if the defendant is charged with a serious offence and is unable to represent himself, or where the defendant should be prohibited from cross-examining and thereby intimidating witnesses: HR Committee, General Comment No. 32, [37].

[44]Human Rights Committee, General Comment No 32, Art. 14: Right to equality before courts and tribunals and to a fair trial, UN Doc. CCPR/C/GC/32 (2007), [32]. Under art. 14(3)(d) of the ICCPR, in connection with being tried, a person's right to legal assistance includes the right to have free legal assistance assigned to him in the interests of justice.

[45]Jackson and Summers (2012), p. 277, referring to Basic Principles on the Role of Lawyers agreed at the 8th UN Congress on the Prevention of Crime and the Treatment of Offenders, Havana, Cuba, A/CONF.144/28/Rev.1 (1990).

[46]UN Committee Against Torture, General Comment no. 2, 24 January 2008, CAT/C/GC/2, [13].

[47]United Nations Economic and Social Council (2004), [67].

[48]United Nations Economic and Social Council (2004), [68].

2.2.2.3 A Lack of Minimum Standards About the Right to Silence Beyond Police Questioning

Through its general comments and case adjudication, the HR Committee has limited the protection of the right to silence of suspects and accused persons under the ICCPR by setting minimum standards only in relation to police questioning. Furthermore, the precise scope of these pre-trial standards is somewhat unclear. Significantly, the guidelines have not ventured to comment the application of the right to silence in relation to adverse inferences from silence, compulsory questioning powers or self-incriminating statements obtained by covert surveillance.

2.3 A Fair Trial and the Right to Silence Under the ECHR

2.3.1 The ECHR: A Regional Framework for Human Rights Protection

2.3.1.1 The Development of the Right to a Fair Trial Under the ECHR

In addition to the international response to the atrocities of the Second World War, the regional community in Europe came together to avoid a repeat of what had occurred on European soil by advocating for peace, security and freedom. With this goal in mind, the Council of Europe was formed in 1949, with 10 founding Member States, and now has 47 Member States. The Council adopted the ECHR in 1950 and it entered into force in 1953. This was the first international instrument providing for legal enforcement of human rights. Through two bodies provided for in the ECHR, the European Commission of Human Rights[49] and, since 1959, the European Court of Human Rights (ECtHR), the contribution of the Council of Europe to the interpretation and application of individual civil and political rights has been significant. The ECtHR was established as a permanent court in Strasbourg in 1998. The authority of the Court is derived from the continuing consensus of the member states of the Council of Europe.[50] However, legal enforcement of Convention rights in the ECtHR depends on the policy of national legislatures and judiciaries. Nevertheless, it is arguable that the authority of the ECtHR in the national setting is more persuasive by virtue of this consensus.[51]

[49]The European Commission on Human Rights was abolished in 1998 under Protocol 11.

[50]Wu (2011), p. 91.

[51]See further Chaps. 3 and 4 about the implementation of the ECHR and the value of the decisions of the ECtHR in the national settings of Denmark and England and Wales.

Article 6 of the ECHR protects the right to a fair trial and has been the subject of the majority of complaints brought before the ECtHR.[52] Under art. 6, the ECtHR has developed a doctrinal framework for establishing minimum standards of protection in Europe for criminal suspects and accused persons. The court has striven towards establishing a human rights framework that accords with the various legal traditions of the contracting states.[53]

Article 6(1) guarantees a fair and public hearing by an impartial tribunal in the determination of a criminal charge. The first paragraph encompasses the broader notion of a fair trial, which stands behind any criminal justice system. A complaint under art. 6(1) essentially claims that the proceedings, when viewed as a whole, were unfair. The jurisprudence of the ECtHR deals with any one of a number of constituent elements of a fair trial, which have the capacity to adversely affect the fairness of proceedings. The list of fair trial rights that the ECtHR has been concerned with includes judicial impartiality, the length or proceedings, equality of arms, the right to silence and the right against self-incrimination.

Under art. 6(2), anyone 'charged' with a criminal offence has the right to the presumption of innocence and, under art. 6(3), to a number minimum defence rights that are considered fundamental to achieving fairness in criminal proceedings. These include the right to receive information about the accusations against him and to receive legal assistance of his own choosing, or free legal assistance where necessary in the interests of justice. Although defence rights are considered fundamental features of a fair trial, they are not absolute and can be restricted by the state if the circumstances so justify.[54]

Research suggests that "…the various elements that can be found in the fair trial requirements are very much interwoven."[55] For example, the ECtHR has often stated that the guarantees in paragraphs (2) and (3) are specific aspects of the right to a fair trial; and the general 'fair hearing' guarantee has elements that supplement the rights in paragraphs (2) and (3).[56] When the presumption of innocence or the minimum guarantees are raised under paragraphs (2) or (3), they will normally be considered together with art. 6(1) in the resolution of complaints before the Court.[57] In other words, the processes and the overall results of criminal proceedings must be fair. The Court has frequently reiterated that the fairness of each case must be examined having regard to the development of the proceedings as a whole. However, specific aspects of a case, such as the right to legal assistance prior to a

---

[52]European Court of Human Rights, Public Relations Unit (2012), p. 5.

[53]Wu (2011), p. 85.

[54]See for example (ECtHR) *Pishchalnikov v. Russia*, Application no. 7025/04, 24 September 2009, [66].

[55]Cape et al. (2010), p. 53.

[56]See further Harris et al. (2014), p. 409.

[57]See for example (ECtHR) *Pishchalnikov v. Russia*, Application no. 7025/04, 24 September 2009, [64] citing *Poitrimol v France*, 23 November 1993, [29].

custodial interrogation, may be so decisive as to enable the fairness of the proceedings to be determined at an earlier stage.[58]

Regional human rights treaties, such as the European Convention for the Protection of Human Rights and Fundamental Freedoms (ECHR), are also formal international sources of law, setting minimum standards for the protection of the right to a fair trial to be applied within the signatory states that have incorporated the ECHR into national law.[59]

2.3.1.2 Interpretation of the ECHR by the ECtHR

Article 32 ECHR establishes the jurisdiction of the ECtHR in all matters concerning the interpretation and the application of the ECHR. The ECHR is to be interpreted in accordance with art. 31 of the Vienna Convention. Therefore, the provisions of the ECHR are to be interpreted with the general purpose of the protection of individual human rights and maintenance democratic values.[60] Nevertheless, when interpreting the right to a fair trial in art. 6, the general purpose of the Convention does not necessarily require that the article is absolute in its protection, or that any limitation of defence rights will amount to a violation. Article 6 has been frequently interpreted to require that the individual's right to a fair trial is protected by considering the overall fairness of the proceedings when viewed as a whole.[61] A limitation of an art. 6 right in the pre-trial phase, which is not overcome by balance-restoring safeguards at trial, may result in a violation, particularly if the evidence gathered as a result of the limitation is decisive on the question of guilt.

The emphasis placed on the object and purpose of the Convention leads to the adoption by the ECtHR of a dynamic, or 'evolutive', method of interpretation.[62] Therefore, the ECHR is regarded as a living instrument that must be interpreted in light of the present-day conditions in the (majority of) member states of the Council of Europe, including the application of commonly accepted national policy standards.[63] This has allowed the ECtHR to read certain implicit rights, for example, the right to silence, into the broad terms of art. 6.[64] However, a line is drawn between judicial interpretation and judicial-lawmaking, so that the Convention is not

---

[58](ECtHR) *Pishchalnikov v. Russia*, Application no. 7025/04, 24 September 2009, [64].

[59]See further Art. 6, European Convention for the Protection of Human Rights and Fundamental Freedoms (1950); Article 8, American Convention on Human Rights (1969); Articles 3 and 7, African Charter on Human and People's Rights (1981); Articles 6 and 7, Commonwealth Convention on Human Rights and Fundamental Freedoms (1995) from the Commonwealth of Independent States; Articles 7, 9 and 18, Arab Charter on Human Rights (1994) from the League of Arab States.

[60]See the preamble of the ECHR; see further Harris et al. (2014), p. 7.

[61](ECtHR GC) *Gäfgen v Germany*, Application no. 22978/05, 1 June 2010.

[62]Harris et al. (2014), p. 8.

[63]See Harris et al. (2014), p. 9.

[64]see further Harris et al. (2014), pp. 8–9.

interpreted in such a way as to introduce the protection of rights that clearly go beyond the general intention of the Convention drafters.[65] The Court must also avoid results that are manifestly absurd or unreasonable.[66] In this context, the ECtHR has made very occasional reference to the *travaux preparatoires* of the Convention.[67] On the other hand, in cases where there is a wide range of diverging policies among the contracting states about a particular issue, and therefore no discernible majority practice in Europe, the ECtHR may prefer to decide on the basis of "...clear and uncontested evidence of a continuing international trend".[68] Dynamic interpretation, therefore, leads the court to rely on a large variety of sources of law as evidence of state practice around the world and general principles of law. This may explain, for example, why the ECtHR has considered national case law from Canada and Australia when developing a line of reasoning to define the right to silence and 'fairness' in the context of covert surveillance.[69] In addition to national case law from outside Europe, the ECtHR has considered other human rights treaties and instruments and principles of public international law.[70]

Dynamic interpretation also means that the case law of the ECtHR is, in itself, an extremely important source of law, secondary only to the ECHR. The judgments of the Court, and also the admissibility decisions, are central to the application of the right to silence in the national jurisdictions within the Council of Europe. This is evidenced by the frequent reference made to the Court's jurisprudence in national and EU policy, judicial or doctrinal discussions about effect of art. 6. Council of Europe policy documents, such as resolutions of the Committee of Ministers, are also valuable material sources of law that provide evidence about the development of general principles of law in relation to the ECHR guarantees.

2.3.1.3 Application of the ECHR in National Law

2.3.1.3.1 Implementation of the ECHR in Denmark and England

The ECHR and the jurisprudence of the ECtHR have an enormous impact on the development and interpretation of fair trial rights in the national systems under the Council of Europe.[71] As referred to at Sect. 2.2.1.2 above, the ECHR has been

---

[65] Harris et al. (2014), p. 9.

[66] See further, art. 32 Vienna Convention on the Law of Treaties.

[67] Harris et al. (2014), pp. 19–20.

[68] [ECtHR GC] *Christine Goodwin v The United Kingdom*, Application no. 17488/90, 11 July 2002, [85]. Alternatively, where a clear European consensus is lacking, the ECtHR sometimes interprets a provision in such a way that allows states a wide margin of appreciation. About the margin of appreciation, see Harris et al. (2014), pp. 14–17.

[69] (ECtHR) *Allan v. The United Kingdom*, Application No. 48539/99, 5 November 2002.

[70] Harris et al. (2014), p. 11.

[71] See, for example, Van Hoecke and Warrington (1998), pp. 501 and 527.

incorporated into Danish and English law (also discussed at Sect. 3.2.2 in Chap. 3 and Sect. 4.2.2 in Chap. 4 below). In Denmark, the legislature has passed a law stating that the Convention is now part of Danish law.[72] In England, incorporation of the ECHR is via the UK Human Rights Act (1998). This means that fair trial rights under the ECHR are justiciable in the domestic law of both countries.

2.3.1.3.2 Enforcement of the ECHR Framework's Minimum Standards

Enforcement of convention rights in the ECtHR give these rights a special status and focus throughout the Council of Europe contracting party states. Under art. 56 (3) of the ECHR the contracting parties are obliged to apply the provisions of the Convention in its named territories "...with due regard ... to local requirements." However, following incorporation of the ECHR, the question is what value the jurisprudence of the ECtHR has, as setting minimum standards under the ECHR, in the national setting.

Under art. 34 of the ECHR, individuals who have exhausted national remedies may make a complaint to the ECtHR.[73] However, the ECtHR is neither a constitutional court nor a court of appeal, and does not have the power to strike down national law or to quash a judgment and order a rehearing of proceedings. The court's jurisdiction by intergovernmental agreement is declaratory only, whereby a finding of a violation of the ECHR allows the court to award compensation.[74] Execution of the Court's award is a political enforcement process. In light of the fact that the court is not a democratic institution as such, a political approach to some extent protects state sovereignty.[75] Nevertheless, the decisions of the ECtHR have been highly influential in setting minimum standards, despite a 'democracy deficit'. As Ma explains, "...most states do choose to change national laws that are questioned by the Court."[76]

There are two limbs to enforcement of the minimum standards set by the ECtHR in their decisions. Firstly, under Art 46(1) entitled "Binding force and execution of judgments", all High Contracting Parties must abide by the decisions of the ECtHR "...in any case to which it is a party." If a contracting party state does not abide by the Court's decision, then the matter is referred to the Committee of Ministers under sub-paragraph (3), to decide the measures to be taken. Thus, enforcement of a judgment ultimately ends with political persuasion and peer pressure.[77] Cases such as *Cyprus v Turkey* and *Ilascu v Moldova and Russia*, and England's refusal to implement prisoners' voting judgments highlight the limits of what is essentially an

[72] Elo Rytter (2013), p. 49.

[73] See art. 34 and 35(1), ECHR.

[74] No action can be taken in the national system for a failure to pay the awarded sum.

[75] Wu (2011), p. 91.

[76] Wu (2011), p. 91.

[77] Harris et al. (2014), pp. 198–199.

inter-governmental system of execution.[78] However, Council of Europe states do generally follow through with the awards for compensation and adjust national law to bring it in line with the ECtHR's jurisprudence—within the allowable margin of appreciation.

The second limb of enforcement of the minimum standards of the ECtHR is based on the willingness of high contracting state parties in connection with implementation. States are often conscious of generally complying with the decisions of the ECtHR in order to avoid unnecessary liquidated damages in similar cases. For example, national practices in criminal proceedings, such as the way the authorities conduct an investigation, have altered in response to the universal standards set by the decisions of the ECtHR. In addition, the Committee of Ministers of the Council of Europe are also responsible for publishing rules that are relevant as a secondary material source of law to the question of procedural rights in criminal proceedings in Europe. For example, the Standard Minimum Rules for the Treatment of Prisoners[79] provides for, *inter alia*, legal representation for untried prisoners immediately upon detention (Rule 93).

Nevertheless, it is unlikely that all states apply the minimum standards set by the ECtHR at all times or in the same way. The manner in which states choose to implement/incorporate the ECHR will to a significant extent determine the persuasiveness of the ECHR framework at establishing uniformity throughout the Council of Europe States. Some states are more receptive to accepting the supervision of the ECtHR than others.

Chapters 3 (Sect. 3.2.2) and 4 (Sect. 4.2.1) describe the ways in which Denmark and England and Wales have incorporated the ECHR and the extent to which the ECtHR's jurisprudence is influential.

2.3.1.4 Reinforcement of the ECHR by the EU Legislative Framework

The application of common minimum standards of a fair trial set by the institutional framework of the ECHR are tied into and reinforced by the EU legislative framework and other EU action as will be explained here. Establishing commonly applied minimum standards for the protection of human rights under EU law has been central to developing mutual trust between EU Member States and thereby facilitating a deeper level of international cooperation in the area of EU criminal justice.

2.3.1.4.1 General Principles About EU Sources of Law

The primary source of EU law is the foundational legislation: the Treaty on European Union (TEU) 1992, the Treaty on the Functioning of the EU 1997

[78]Harris et al. (2014), p. 199.

[79]Resolution 73(5) of the Committee of Ministers of the Council of Europe.

(TFEU) and, since the Lisbon Treaty 2009, the Charter of Fundamental Rights of the European Union 2007.[80] These three treaties set out the EU's constitutional, human rights and cooperative framework. In general, where state obligations and individual rights under these treaties are clear and unconditional they can be invoked with direct effect before national courts. Regulations are examples of secondary EU law with direct effect. Directives, such as directives about procedural rights, are also binding secondary acts.[81] However, in general, directives do not have direct effect and must be implemented in national law.[82] If a directive is aimed at protecting human rights, the CJEU may recognise the direct vertical effect (between the individual and the State) of a directive that has not been transposed by the deadline, where the provisions are clear and precise.[83] In similar circumstances, a binding 'decision' has direct vertical effect, but needs to be implemented in national legislation.[84] This indicates a tendency in favour of protecting individual rights.

Other documents may be issued by EU institutions that are not binding, but may nevertheless be material sources of law. This category includes, for example, reports or proposals issued by the European Commission or the European Council. In addition, in Europe, the important role of the European Courts (the CJEU and the ECtHR) in legal interpretation is generally accepted by legislators, scholars and practitioners.[85]

2.3.1.4.2 Fair Trial Rights Prior to the 2009 Lisbon Treaty

Since the 1992 TEU,[86] the 1997 TFEU[87] and the Tampere Conclusions of the European Council meeting in Finland in 1999,[88] there has been a growing concern within the EU about the application of minimum standards for procedural rights of

---

[80] Daniel et al. (2011), pp. 141–144. About the future accession of the EU to the ECHR, see Barnard and Peers (2014), pp. 249–252; Harris et al (2014), pp. 31–33; and Daniel et al. (2011), pp. 156–158; and see [CJEU] Opinion on the draft agreement on the accession of the European Union to the European Convention for the Protection of Human Rights and Fundamental Freedoms 2/13, 18 December 2014.

[81] Daniel et al. (2011), p. 158.

[82] Secondary legislation is adopted by EU institutions on the basis of the EU founding treaties. Regulations are an example of secondary legislation that have direct effect. See further, Daniel et al (2011), pp. 158–161.

[83] See further Daniel et al. (2011), pp. 269–270.

[84] On direct effect of decisions, see further Daniel et al. (2011), pp. 165 and 273; Framework decisions, such as the European Arrest Warrant Framework Decision, enacted prior to the entry into force of the Lisbon Treaty did not have direct effect: Chalmers et al. (2014), p. 300.

[85] Van Hoecke and Warrington (1998), p. 500.

[86] Treaty on European Union 1992 (TEU), entered into force 1 November 1993.

[87] Treaty on the Functioning of the EU 1997 (TFEU), entered into force on 1 December 2009.

[88] Tampere European Council Presidency Conclusions, 15 and 16 December 1999.

suspects and accused persons.[89] Judicial cooperation in the EU, which is based on the principle of mutual recognition of judicial decisions, requires mutual trust between Member States that minimum standards of fundamental and international human rights will be met in the exchange process of cooperation.

In December 2000, the Charter was a political declaration proclaimed by EU Member States at the Nice Intergovernmental Conference. The Charter recognised the scope of the equivalent convention rights as minimum standards for protection among EU states.[90] However, the rights it contained were non-justiciable prior to the Lisbon Treaty entering into force in December 2009. Nevertheless, prior to the Lisbon Treaty, the European Court of Justice had ruled on defence rights in criminal proceedings and referred to the jurisprudence of the ECtHR.[91]

2.3.1.4.3 Fair Trial Rights in EU Law Post-Lisbon

Following the Lisbon Treaty's entry into force on 1 December 2009, legislative decision-making power in relation to EU matters is now shared between the European Parliament and the national governments in the Council.[92] Legislative activity is based on art. 82(2) of the TFEU which allows the European Parliament and the Council to establish minimum rules in relation to the mutual admissibility of evidence between Member States, and individual criminal procedural rights. Taking into account the differences in legal traditions and systems within the EU, establishing minimum standards is said to harness mutual trust between Member States.[93] The purpose is to facilitate police and judicial cooperation in criminal matters, including through mutual recognition of judgments and judicial decisions, which form the basis of key instruments such as the European Arrest Warrant (EAW) and the European Investigation Order (EIO).

On the eve of the Lisbon Treaty coming into force, the Swedish Presidency presented a 'roadmap' for reinforcing the procedural rights of suspects and accused persons. The roadmap refers to the steps involved in fundamental protective measures, including the enactment of a number of directives about minimum standards in relation to basic procedural rights.[94] Directives have already been adopted in relation to access to interpreters and translators,[95] the right to

[89]See further Cape et al. (2010), pp. 8–10.

[90]Article 52(3), Charter of Fundamental Rights of the European Union, OJ 2007 L303/01.

[91]See, for example, (ECJ GC) Case C-105/03 *Pupino* [2005] ECR I-5285, 16 June 2005, [58–60] in relation to the operation of a framework decision on victims' rights in national criminal proceedings.

[92]Though note the optional positions of Ireland and the United Kingdom, and the position of Denmark.

[93]For a detailed analysis of mutual trust in international cooperation within the EU see Chap. 7.

[94]See further Cape et al. (2010), p. 11.

[95]Directive 2010/64/EU of the European Parliament and of the Council of 20 October 2010 on the right to interpretation and translation in criminal proceedings, OJEU L 280/1.

information[96] and about access to a lawyer.[97] Replies to a Green Paper launched in 2011 have also been reported about the use of detention in the EU.[98]

Ongoing legislative action is anticipated regarding the other areas of interest in the roadmap, such as the presumption of innocence and the right to be present at trial. The directive about the right of access to a lawyer and the right to communicate with a family member, employer or consulate official, is to be applied in accordance with the principle of subsidiarity and Member States have until 27 November 2016 to implement it.[99] After this time, the Directive may arguably have direct effect if it is considered to be unconditional, In order to enhance mutual trust between the Member States in the functioning of instruments such as the EAW, the directive builds on the minimum standards being set by the ECtHR and approximates the procedural rules of Member States by requiring that suspects and accused persons have access to a lawyer from the first stage of police questioning in criminal proceedings.[100]

The Court of Justice of the European Union (CJEU; previously the ECJ) has recognised and applied fundamental and international human rights as part of a body of general principals of law, insofar as they have been guaranteed by the ECHR and followed by EU (previously EC) Member States as common constitutional, democratic principles.[101] Since the Lisbon Treaty came into force in December 2009, these rights, as general principles, have been maintained by art. 6(3) of the TEU. The Charter is now a primary source of law and the rights have the status of legally binding rights under art. 6(1) of the TEU. Thus, the right to a fair trial and other related rights in Articles 47 and 48 of the Charter generally apply in relation to EU legislation and national implementing law. It is likely that any new

[96]Directive 2012/13/EU of the European Parliament and of the Council of 22 May 2012 on the right to information in criminal proceedings, OJEU L 142/1.

[97]Directive 2013/48/EU of the European Parliament and of the Council of 22 October 2013 on the right of access to a lawyer in criminal proceedings and in European arrest warrant proceedings, and on the right to have a third party informed upon deprivation of liberty and to communicate with third persons and with consular authorities while deprived of liberty, OJEU L 294/1, 6 November 2013.

[98]European Commission (2011).

[99]Directive 2013/48/EU of the European Parliament and of the Council of 22 October 2013 on the right of access to a lawyer in criminal proceedings and in European arrest warrant proceedings, and on the right to have a third party informed upon deprivation of liberty and to communicate with third persons and with consular authorities while deprived of liberty, OJEU L 294/1, 6 November 2013; after 27 November 2016, the directive may arguably have direct effect, if it is regarded to be unconditional and sufficiently clear and precise: see the (CJEU) Case C-41/74 *Van Duyn v Home Office*, 4 December 1974.

[100]Directive 2013/48/EU of the European Parliament and of the Council of 22 October 2013 on the right of access to a lawyer in criminal proceedings and in European arrest warrant proceedings, and on the right to have a third party informed upon deprivation of liberty and to communicate with third persons and with consular authorities while deprived of liberty, OJEU L 294/1, 6 November 2013; see below Sect. 2.3.2.2.5.

[101]Elo Rytter (2013), p. 31.

judicial cooperation directives that are adopted by the EU, like the EIO directive,[102] will require, at least implicitly, that implementing states respect fundamental and international human rights under the Charter, the Convention and other international treaties. Paragraphs 12 and 39 of the EIO directive refer to the need to respect fundamental defence rights contained in the Charter and by reference to art. 6 TEU. The CJEU now plays a part, alongside the ECtHR, in developing minimum standards in relation to the procedural rights of suspects and accused persons, as they relate to the process of judicial cooperation and the directives that are adopted in accordance with Article 82(2) of the TFEU.

This opens the gate to the enforcement mechanisms under the TFEU, which are different in nature and impact than an *ex post* complaint under the ECHR.[103] In national criminal proceedings it is now possible under art. 267 of the TFEU for a court or tribunal to make a request for a preliminary ruling on a relevant issue.[104] According to art. 258 of the TFEU, the Commission also has the power to make a complaint against a Member State that it considers has failed to carry out the obligations created under the TEU or the TFEU. A finding against a Member State will require it to bring its national legislation in line with EU obligations.[105]

In the future, in accordance with art. 6(2) of the TEU, the EU will become a member of the Council of Europe and therefore the ECHR will be incorporated into EU law and as such create binding minimum standards that are also justiciable in the CJEU.[106] However, accession by the EU to the ECHR is controversial.[107]

2.3.1.4.4 The Right to a Fair Trial in Articles 47 and 48 of the Charter of Fundamental Rights

Chapter VI of the Charter concerning "Justice" appears to recognize all the fundamental principles that are already established in certain national constitutions and the major international conventions. Under Article 47 of the Charter, everyone is entitled to a fair hearing. This explicitly includes the possibility of being "...advised, defended and represented." According to art. 48(1) of the Charter,

---

[102]Directive 2014/41/EU of the European Parliament and of the Council of 3 April 2014 regarding the European Investigation Order in criminal matters, OJEU L 130/1, 1 May 2014: The UK is taking part in the directive, but Denmark and Ireland are not taking part and will not be bound by its provisions.

[103]Cape et al. (2010), p. 13.

[104]Note the procedures for urgent hearings in relation to detainees: Cape et al. (2010), p. 13; see, for example, (CJEU GC) Case C-396/11 *Radu* [2013] and (CJEU GC) and Case C-399/11 *Stefano Melloni v Ministerio Fiscal* [2013].

[105]See further Cape et al. (2010), p. 14.

[106]Elo Rytter (2013), p. 32.

[107][CJEU] Opinion on the draft agreement on the accession of the European Union to the European Convention for the Protection of Human Rights and Fundamental Freedoms 2/13, 18 December 2014.

everyone is to be presumed innocent until proven guilty; and art. 48(2) demands respect for the defence rights of a person who has been charged. The Charter does not list in detail specifically what comprises the defence rights. Article 6(3) ECHR provides significantly more detail about specific defence rights, in particular, taking into account the jurisprudence of the ECHR. Nevertheless, the CJEU, in general, is taking on a greater role in the protection of Charter rights.

Significantly, articles 52(3) and 53 of the Charter guarantee everyone within an EU jurisdiction that the meaning and scope of Charter rights is no less than corresponding rights in other European treaties and international conventions (most importantly, including the ECHR), and in national constitutions of the Member States.[108] As well as the duty in domestic law to comply with the ECHR discussed above, the Member States are obliged under art. 6 of the Treaty on European Union (TEU) to respect fundamental human rights, which includes charter and convention rights, when dealing with EU law.[109] Further, when implementing Union law the national courts are bound to comply with the law in relation to fundamental rights as it is interpreted by the Court of Justice of the European Union (CJEU).[110]

2.3.1.4.5 The EU as a Consumer of and a Buttress for Convention Rights

Today, it is hard to imagine the application of minimum fair trial standards in the framework of the Council of Europe, without the EU perspective. On the one hand, international cooperation between European states would operate in absence of the EU international cooperation instruments and bodies, such as Eurojust, that strive for a deep level of cooperation. Therefore, the demand for definition of minimum standards for human rights protection would be less. On the other hand, there would not be the need for the extra layers of human rights standardisation carved out by actions such as the EU roadmap, which enlarge the impact of Convention rights.

However, there would probably still be efforts to achieve more effective cooperation by virtue of geography and the significant focus on facilitating cooperation envisaged in the Council of Europe criminal justice instruments alone; and European states would still be guided by the decision-making of the ECtHR. In the absence of an EU, states may arguably have been less inclined to follow the minimum standards set by the Court and may have paid less attention to the standards actually being applied by fellow contracting states. From this perspective, the EU legislative framework has certainly added to the need for and the

[108]As examples of constitutional rights within the EU, the Portuguese and the Spanish constitutions protect the presumption of innocence. The Spanish constitution also establishes the right not to be compelled to confess. The Italian legal system recognizes that a citizen is *not considered guilty until final judgment is passed*: D'Amico (2010), pp. 300–301.

[109]Article 6, Treaty on European Union (TEU). See the Consolidated Version of the Treaty on European Union (TEU), OJ C 115/13 May 2008.

[110]Section 3(1) of the European Communities Act 1972.

persuasiveness of the ECHR framework, particularly with the door opening wider to the enforceability of Convention rights in the CJEU.

2.3.2 *The Scope of the Right to Silence Under the ECHR*

2.3.2.1 The Application of Fair Trial Rights to Pre-Trial Proceedings

Under art. 6(1) of the ECHR, the right to a fair and public hearing applies in the determination of a criminal charge.[111] The more specific protections in art. 6(2) and 6(3) are explicitly provided to "everyone charged with a criminal offence". For example, art. 6(3)(a) provides a person who has been charged with the right information about the charges (in a language he or she can understand)[112]; and art. 6(3)(b) ensures adequate time and facilities for the preparation of a defence, which is said to include disclosure.[113]

Nevertheless, in its case-law, the ECtHR has underlined the connectivity between the investigative and trial phases in national criminal proceedings and the concept of evaluating the fairness in the proceedings *as a whole*. The Court has referred to the significance of "...the investigation stage for the preparation of the criminal proceedings, as the evidence obtained during this stage determines the framework in which the offence charged will be considered at the trial."[114] In *Pishchalnikov v Russia*, the ECtHR found that the protections provided "...may be relevant before a case is sent to trial if and so far as the fairness of the trial is likely to be seriously prejudiced by an initial failure to comply with its provisions".[115] This statement clearly acknowledges the importance of investigative proceedings to the overall conduct of the trial. The Court held that the question in each case dealing with a restriction of a defence right in pre-trial proceedings, is whether the

[111]The European Commission on Human Rights expressed the opinion that a person is charged with a criminal offence for the purposes of the ECHR when he or she is substantially affected by the proceedings against them: *Deweer v. Belgium,* Application no. 6903/75, 27 February 1980, [46].

[112]See also art. 6(3)(e) about the right to an interpreter and art. 5(2) ECHR with respect to the right to information in order for a detainee to challenge the lawfulness of his detention.

[113]See further Cape et al. (2010), pp. 32–36; Wu (2011), p. 104. In most European jurisdictions, the authorities do not disclose all the material relating to an alleged offence in the investigation stage. Most states do allow a right to full disclosure of the material evidence to the accused or his lawyer prior to trial. Recently, the ECtHR has been resolving complaints relating to non-disclosure under the right to an adversarial trial, in art. 6(3)(d): Harris et al. (2014), pp. 416–417.

[114](ECtHR) *Salduz v Turkey*, Application no. 36391/02, 27 November 2008, [54]; *Pishchalnikov v Russia,* Application no. 7025/04, 24 September 2009, [69].

[115](ECtHR) *Pishchalnikov v Russia,* Application no. 7025/04, 24 September 2009, [65]; see further (ECtHR) *Imbrioscia v Switzerland,* Application no. 13972/88, 24 November 1993, [36] and (GC) *Öcalan v Turkey*, Application no. 46221/99, 12 May 2005, [131].

restriction was justified and, if so, considering the *proceedings as a whole*, whether the accused has been denied a fair hearing.[116]

Although art. 6 is formulated in terms of the determination of a criminal 'charge', the ECtHR has taken a non-restrictive interpretive approach. The word 'charged' has an autonomous meaning, which is to be understood within the terms of the Convention, rather than solely according to the interpretation under national law.[117] Under the Convention a person is 'charged' when they have been notified by the authorities of an allegation that has been made against them; correspondingly this means that the situation of the person has been significantly affected by an accusation.[118] This is a substantive rather than a formal meaning, which has been used in relation to art. 14 ICCPR, as discussed at Sect. 2.2.2.1 above.[119] Therefore, in accordance with art. 6, the determination of a criminal charge may even commence prior to formal charges being laid. For example, a suspect may be 'charged' if he or she has been arrested for a criminal offence, or has been told officially about a criminal investigation and begins to be affected by it.[120]

2.3.2.2 Police Questioning and the Right to Silence Under Article 6

2.3.2.2.1 Cautioning About the Right to Silence

According to the case law of the ECtHR, guaranteeing an effective defence and the right to information also includes an obligation to make a suspect or accused person fully aware of his fundamental defence rights.[121] Judicial and other authorities are obliged to take positive steps to ensure that art. 6 is complied with, which includes an obligation to advise (caution) suspects and accused persons who have been charged (within the autonomous meaning of the Convention), or are regarded as charged, of their right to silence and the right to legal advice, including legal aid.[122]

In *Zaichenko v Russia,* the Applicant was stopped at a road check and was unable to produce proof that he had purchased the diesel that was found in his car. The road checks were being carried out at the request of a company director because there had previously been cases where workers were pouring out diesel from their service vehicles for their personal use. The Court found that at the time they questioned the

---

[116](ECtHR) *Pishchalnikov v Russia,* Application no. 7025/04, 24 September 2009, [67]; see also (ECtHR) *Imbrioscia v Switzerland*, Application no. 13972/88, 24 November 1993, [38].

[117]Van Dijk et al. (2006), p. 540.

[118](ECtHR) *Deweer v Belgium*, Application no. 6903/75, 27 February 1980, [44]; *Shabelnik v Ukraine*, Application no. 16404/03, 19 February 2009, [57]; see also Van Dijk et al. (2006), p. 540.

[119](ECtHR) *Deweer v Belgium*, Application no. 6903/75, 27 February 1980, [44, 46].

[120]See further Harris et al. (2014), pp. 376–377.

[121]It may not be sufficient that this is done by a letter of rights: see (ECtHR) *Panovits v Cyprus*, Application no. 4268/04, 11 December 2008, [67–68]; see further, Cape et al. (2010), p. 34.

[122]Cape et al. (2010), p. 34. The right to be 'cautioned' first appeared in the US Supreme Court *Miranda* case, in 1966: Trechsel and Summers (2006), p. 352.

applicant at the road check, the police should have suspected the applicant of theft and therefore should have advised him of the right to remain silent and the right against self-incrimination, to ensure that he understood the consequences of his being questioned.[123] The Court concluded that the lack of caution was a breach of pre-trial due process and examined whether this breach had been remedied at trial. In the circumstances of the case, the incriminating statement obtained prior to being cautioned was the decisive evidence for conviction. Therefore, it was decided that art. 6 had been violated.[124]

In *Panovits v Cyprus*, the Court highlighted that the authorities are obliged to take all reasonable steps to ensure that a suspect is fully aware of the nature and seriousness of the investigation and about the importance of his defence rights in the proceedings, including the right to remain silent and the significance of that right during police questioning.[125] The suspect or accused person must be advised adequately about fundamental rights, taking into account the circumstances of the case, including the age, maturity or other vulnerability of the person involved.[126] Significantly, the Court held that a mere oral caution about a suspect's right to remain silent may not be enough, in the circumstances of the case, for the particular suspect to sufficiently comprehend the nature of the right.[127]

Research shows that the practice of providing suspects with information on defence rights varies widely across EU jurisdictions. In the majority of EU jurisdictions, information on defence rights is only provided orally.[128]

Defence rights that accrue when a person becomes suspected of criminal offending, and therefore 'charged', namely the right to silence, the right against self-incrimination and, in custodial situations, the right to legal assistance must also apply to persons being questioned as witnesses whenever they become suspected of criminal offences.[129]

2.3.2.2.2 The Right to Access Legal Representation

Access to a legal advisor is a good example of the enabling and protective features of the overarching principle of equality of arms. A legal adviser is often essential to

[123](ECtHR) *Aleksandr Zaichenko v. Russia*, Application no. 39660/02, 18 February 2010, [52 and 55]. The Court did not actually specify a test to determine the extent of suspicion required when a person is to be regarded as charged.

[124](ECtHR) *Aleksandr Zaichenko v. Russia*, Application no. 39660/02, 18 February 2010, [55–60] It should be noted that a violation of the right against self-incrimination may be found even where the incriminating statement is not used at a trial to found a conviction: (ECtHR) *Marttinen v Finland*, Application no. 19235/03, 21 April 2009, [64].

[125](ECtHR) *Panovits v. Cyprus*, Application no. 4268/04, 11 December 2008, [67–68].

[126](ECtHR) *Panovits v. Cyprus*, Application no. 4268/04, 11 December 2008, [67].

[127](ECtHR), *Panovits v. Cyprus*, Application no. 4268/04, [74].

[128]See Spronken (2010).

[129](ECtHR) *Brusco v France*, Application no. 1466/07, 14 October 2010, [47–50].

enable the accused to participate effectively in the criminal justice process by understanding the charges, accessing the right to information and responding to the allegations appropriately. In addition, a great deal of the ECtHR's jurisprudence in relation to access to legal advice focuses on the significance of the role of a lawyer as a safeguard to protect the accused when his fundamental rights are at risk of being restricted, thereby ensuring that he receives a fair trial.

Article 6(3)(c) ECHR provides that a person charged with a criminal offence has the right to access legal assistance of his own choosing, or to receive legal assistance free, if the interests of justice so require. As referred to at Sect. 2.3.2.1 above, the word 'charged' has a substantive meaning under the ECHR. Therefore, from the time a person has been advised by the authorities of criminal allegations, he or she has the right to receive legal advice, including free legal assistance, where the interests of justice so require. At trial mandatory representation by a lawyer may be justified in the interests of the legitimacy of the tribunal, by ensuring that matters are dealt with expeditiously.[130]

Recent decisions of the ECtHR indicate the Court's current approach: in order to guarantee the right to silence and the right against self-incrimination, access to a lawyer should be given from the first time a suspect is interrogated, as a right under art. 6(1) and art. 6(3)(c). The Court now considers early access to a lawyer is a key procedural safeguard in ensuring that the right to silence and the right against self-incrimination are upheld throughout the course of criminal proceedings.[131]

Decisions such as *Panovits v Cyprus* serve as a reminder of the importance of making suspects, particularly minors, aware of their right to legal representation.[132] Some suspects may not appreciate the consequences of proceeding to be questioned in relation to serious charges without legal assistance. Where a suspect is not made aware of their right to a lawyer, a violation of art. 6 is likely to result if incriminating statements are used at a later trial, even if such statements are not the sole evidence against an accused.[133] It is still not clear whether the right to consult a lawyer at the first interrogation includes the presence of a lawyer in the course of the interrogation itself.

In a leading judgment of *Salduz v Turkey*, the applicant's right of access to a lawyer was restricted during his police custody because he was accused of committing offences that fell to be determined in the jurisdiction of the State Security Court. As a result, he made self-incriminating statements to the police, the public prosecutor and the investigating judge in the absence of prior legal assistance. As referred to at Sect. 2.3.2.1 above, the ECtHR has underlined the importance of the investigation stage for the preparation of criminal proceedings, as it is the evidence

[130]This is the approach of the ECtHR, which has followed the International Criminal Tribunal for the Former Yugoslavia: Cape et al. (2010), pp. 36–38.

[131](ECtHR) *Salduz v. Turkey*, Application no. 36391/02, 27 November 2008, [54]; *Pishchalnikov v Russia*, Application no. 7025/04, 24 September 2009, [69].

[132](ECtHR) *Panovits v Cyprus*, Application no. 4268/04, 11 December 2008; see also (ECtHR) *Adamkiewicz v. Poland*, Application no. 54729/00, 2 March 2010 in relation to the need for a minor to have broad access to defence lawyer prior to pre-trial questioning.

[133](ECtHR) *Panovits v Cyprus*, Application no. 4268/04, 11 December 2008, [76].

that determines the framework in which an offence will be considered at trial.[134] The Court has repeatedly stated that pre-trial failures, such as denial of access to a lawyer prior to interrogation, may affect the overall fairness of the proceedings, even if there has been an opportunity to challenge evidence at trial—particularly in the case of custodial interrogation.[135]

The rationale of the Court in emphasising the significance of early access to a legal adviser has been explained as follows:

> ...an accused often finds himself in a particularly vulnerable position at that stage of the proceedings, the effect of which is amplified by the fact that legislation on criminal procedure tends to become increasingly complex, notably with respect to rules governing the gathering and use of evidence. In most cases, this particular vulnerability can only be properly compensated for by the assistance of a lawyer whose task it is, among other things, to help to ensure respect of the right of an accused not to incriminate himself. This right indeed presupposes that the prosecution in a criminal case seek to prove their case against the accused without resort to evidence obtained through methods of coercion or oppression in defiance of the will of the accused...
>
> Against this background, the Court finds that in order for the right to a fair trial to remain sufficiently "practical and effective" ...as a rule, access to a lawyer should be provided as from the first interrogation of a suspect by the police, unless it is demonstrated in the light of the particular circumstances of each case that there are compelling reasons to restrict this right ...such restriction – whatever it's justification – must not unduly prejudice the rights of the accused under Art. 6. The rights of the defence will in principle be irretrievably prejudiced when incriminating statements made during police interrogation without access to a lawyer are used for a conviction.[136]

Finally, the Court stressed the fundamental importance of access to a lawyer by a detainee who is a minor, while confirming that the above reasoning applied irrespective of the age of the suspect.[137]

Although not an absolute right, from the above statement it is clear that the Court views the assistance of a lawyer from the first interrogation is paramount in two respects: firstly, to protect the suspect from being overwhelmed by the inherently coercive environment of a custodial interrogation and the potentially improper coercive methods of investigators; and second, to ensure that the suspect is able to make fully informed decisions about speaking to the police or remaining silent, despite the complexity of the rules surrounding the interrogation and their potential ramifications at trial.[138] In this connection, the Court highlighted that national laws may attach consequences to the attitude of the accused at the initial stage of police interrogation, which are decisive for the defence case in any subsequent

---

[134](ECtHR) *Salduz v. Turkey*, Application no. 36391/02, 27 November 2008, [54].

[135](ECtHR) *Salduz v. Turkey*, Application no. 36391/02, 27 November 2008, [50], [58] and [62]; see also the Council of Europe Prison Rules 2006.

[136](ECtHR) *Salduz v. Turkey*, Application no. 36391/02, 27 November 2008, [54–55].

[137](ECtHR) *Salduz v. Turkey*, Application no. 36391/02, 27 November 2008, [60–61].

[138]See further Wu (2011), p. 99.

proceedings.[139] Neither assistance at trial nor adversarial proceedings can rectify the adverse effect of these factors on the overall fairness of the proceedings.[140]

This judgment suggests that as a starting point it should not be underestimated how inherently coercive an interrogation is and how easily inroads are made that irretrievably undermine the right to silence and the right against self-incrimination. Therefore, by providing access to a lawyer at the initial stages of a police interrogation, miscarriages of justice are prevented and a suspect is guaranteed the broader protection of art. 6, by establishing equality of arms between the investigating and prosecuting officers and the suspect or accused.[141] In addition, the decision recognises that limitations on fundamental pre-trial rights have ramifications further into the criminal justice process, if measures are not in place to ensure a fair balance in the criminal proceedings as a whole.

The impact of the judgment in *Salduz* has been limited by the ECtHR's decision in *Aleksandr Zaichenko v Russia*. In this case the Court distinguished the case of custodial interrogation from the situation of the police making enquiries at a road check. The Court found that there was no violation of art. 6 because the applicant was not formally under arrest and was not being interrogated in police custody.[142] The applicant should have been suspected of theft and, therefore, should have been regarded as 'charged' for the purposes of cautioning about the right to silence. Nevertheless, although he was not free to leave, the circumstances of the case disclosed no significant curtailment of his freedom that warranted access to legal assistance.[143] Furthermore, questioning in public may often be in the presence of witnesses.[144] It is uncertain how *Salduz* and *Zaichenko* apply to the situation of a suspect attending a police station voluntarily to be formally questioned; or the situation of a suspect who has attended the station for questioning and has requested access to a legal representative, but is still awaiting the lawyer's arrival.[145]

In *John Murray v. the United Kingdom* (discussed at Sect. 2.3.2.3 below), under anti-terrorism legislation, the applicant was denied access to a lawyer for the first 48 hours in police custody, during which time he was interrogated. In the circumstances of the case, the applicant faced the strong possibility at a trial that the jury would infer his guilt if he remained silent when questioned by the police. Therefore, the Court concluded that to deny the applicant access to a lawyer prior to interrogation "...in a situation where the rights of defence may well be irretrievably prejudiced..." is incompatible with the applicant's rights under art. 6, whatever the

---

[139](ECtHR) *Salduz v. Turkey*, Application no. 36391/02, 27 November 2008, [52].

[140](ECtHR) *Salduz v. Turkey*, Application no. 36391/02, 27 November 2008, [58].

[141](ECtHR) *Salduz v. Turkey*, Application no. 36391/02, 27 November 2008, [53]; *Pishchalnikov v Russia*, Application no. 7025/04, 24 September 2009, [68].

[142](ECtHR) *Aleksandr Zaichenko v. Russia*, Application no. 39660/02, 18 February 2010, [47].

[143](ECtHR) *Aleksandr Zaichenko v. Russia*, Application no. 39660/02, 18 February 2010, [48].

[144](ECtHR) *Aleksandr Zaichenko v. Russia*, Application no. 39660/02, 18 February 2010, [47].

[145]See further the discussion of *Pishchalnikov* and waiver of the right to access a lawyer at Sect. 2.3.2.2.3.

justification for such denial.[146] A violation was found in relation to art. 6(1) in conjunction with art. 6(3)(c).[147] The ECtHR highlighted the importance of access to a legal advisor prior to a suspect being interrogated by police, as it acts as a safeguard to ensure that a suspect knows and fully understands the right to silence and the implications of the right for a future trial.

The question of the fairness of what has become known as "safety interviews" was recently raised in the 2014 case of *Ibrahim and Others v The United Kingdom.*[148] The ECtHR found that there were *compelling reasons to justify* the denial of access to a lawyer prior to the questioning, in safety interviews, of three suspects, who denied their involvement in the London bombings on 21 July 2005 (which had followed the attacks of 7 July 2005) and one witness, whom the police continued to interview (as a witness, without arrest or advice about the right to silence and access to a lawyer) after he had started to incriminate himself. In addition, the Court found *no undue prejudice* resulted, whereby the relevant legislation only allowed for denial of access to a lawyer in relation to safety interviews on a case-by-case basis rather than systematically. Furthermore, the Court referred to the following factors in coming to their decision: once there was sufficient information received to ensure public safety, the legislation required the safety interview to stop; there were no allegations of coercion or abusive police conduct; the first three applicants acknowledged their involvement in the bombing at trial (on the basis that the bombs were not meant to explode); the fourth applicant did not retract his statements and there was sufficient opportunity at trial to challenge the statements made; there was a considerable body of other evidence. On 1 June 2015, the case was referred to the Grand Chamber.

The decision in *Murray* was distinguished on the facts, in the case of *Brennan v. the United Kingdom.*[149] In Brennan, the applicant was also denied access to a lawyer under anti-terrorism legislation. However, the applicant made no admissions during the deferral period, and no adverse inferences were to be drawn from the applicant's statement or silence within this timeframe. The applicant's complaint was not that he was denied access to a lawyer prior to making incriminating statements to the police, but rather that he made the statements during an interrogation that was conducted after the deferral period ended but prior to the arrival of his solicitor. Therefore, the Court found that the absence of a lawyer when the statements were made did not infringe art. 6(1) or 6(3)(c) of the Convention.

---

[146](ECtHR) *John Murray v United Kingdom*, Application no. 18731/91, 8 February 1996, [66].

[147](ECtHR) *John Murray v United Kingdom*, Application no. 18731/91, 8 February 1996, [66, 70]; art. 6(3)(c) gives a person who has been charged with a criminal offence, a right to defend themselves through legal assistance; in relation to denial of access to a lawyer, see the more recent decision of ECtHR (GC), *Öcalan v. Turkey*, Application no. 46221/99, 12 May 2005, [131].

[148](ECtHR) *Ibrahim and Others v the United Kingdom*, Application no. 50541/08 and others, 16 December 2014.

[149](ECtHR) *Brennan v. the United Kingdom*, Application no. 39846/98, 16 October 2001.

This decision may be difficult to reconcile with the judgment of the ECtHR in the case of *Pishchalnikov v Russia*.[150] Here the applicant complained that he had been denied access to a lawyer in the first few days of police custody, despite being interrogated during that time. Shortly after his arrest for robbery, the applicant had made an entry in the arrest record, requesting to be assisted by a particular lawyer. The investigator then proceeded to question the applicant, despite the applicant's pending request for legal assistance. The applicant confessed to robbery. The following day the applicant was further interrogated and confessed to a number of criminal offences, including murder, kidnapping, hijacking and unlawful possession of weapons, committed as part of a group.[151] The applicant asserted that he had repeated his request for legal assistance prior to the interrogation.

In relation to the rationale of the right to access a lawyer, the ECtHR emphasised that every person charged with a criminal offence has the right to a practical and effective defence and that being defended by a lawyer, while not absolute, is fundamental to a fair trial.[152] Again, the Court emphasised that the vulnerability of a suspect during the investigative stage is amplified by the increasingly complex legislation on criminal procedure.[153] National laws "...may attach consequences to the attitude of an accused at the initial stages of police interrogation which are decisive for the prospects of the defence in any subsequent criminal proceedings."[154] In the Court's view, the complexities of criminal proceedings are often incomprehensible to laypersons, and the absence of legal assistance in this case meant that the applicant had been unable to make an informed assessment of the consequences of his confession to the outcome at trial.[155] Moreover, the Court pointed out that the police and prosecution authorities, all experts in the field of criminal proceedings, are well-equipped with various, often psychologically coercive, interrogation techniques that are devised to facilitate the receipt of information from a suspect.[156] These difficulties can only be properly compensated for by the assistance of a lawyer whose task it is to ensure respect of the right of a suspect or accused not to incriminate himself.[157]

Similar to protection under the ICCPR and other UN conventions, as referred to above (Sect. 2.2.2), Rule 93 of the Council of Europe's Committee of Ministers' Standard Minimum Rules for the Treatment of Prisoners provides for, *inter alia*, legal representation for untried prisoners immediately upon detention.[158] The

---

[150] (ECtHR) *Pishchalnikov v. Russia*, Application no. 7025/04, 24 September 2009.

[151] (ECtHR) *Pishchalnikov v. Russia*, Application no. 7025/04, 24 September 2009, [8, 72 and 75].

[152] (ECtHR) *Pishchalnikov v. Russia*, Application no. 7025/04, 24 September 2009, [66].

[153] (ECtHR) *Pishchalnikov v. Russia*, Application no. 7025/04, 24 September 2009, [69].

[154] (ECtHR) *Pishchalnikov v Russia*, Application no. 7025/04, 24 September 2009, [65].

[155] (ECtHR) *Pishchalnikov v. Russia*, Application no. 7025/04, 24 September 2009, [84–85].

[156] (ECtHR) *Pishchalnikov v. Russia*, Application no. 7025/04, 24 September 2009, [86].

[157] (ECtHR) *Pishchalnikov v. Russia*, Application no. 7025/04, 24 September 2009, [69].

[158] Resolution 73(5) of the Committee of Ministers of the Council of Europe, Standard Minimum Rules for the Treatment of Prisoners, Rule 93.

atmosphere of a custodial interrogation is pointed to as being particularly coercive. The rules are non-binding but serve as another material source of law.

2.3.2.2.3 Waiver of the Right to Access a Lawyer

Although the right to access a lawyer prior to being interrogated may be fundamental, the ECtHR found that it is not absolute and is capable of being restricted where compelling reasons exist.[159] The Court did not rule out that a suspect or accused may voluntarily waive his right to legal assistance. Due to the fundamental nature of the right, it requires "…the special protection of the knowing and intelligent waiver standard."[160] Therefore, after initially being advised of his rights, a suspect may waive these rights and respond to interrogation.[161] Nevertheless, being advised of fundamental defence rights and then signing a form stating these rights will not necessarily establish a knowing and intelligent waiver.[162] For example, a caution administered by investigating officers informing a suspect of the right to remain silent is a minimum recognition of the right and barely meets the minimum aim of acquainting an accused with the right.[163] If a suspect asks to see a lawyer then he or she should not be subjected to further interrogation by the authorities until counsel has been made available, unless the suspect or accused initiates further communication with the police or prosecution.[164]

If a suspect is advised of his rights and asks to see a lawyer in connection with the interrogation, then a subsequent implicit waiver of that right cannot be established by showing only that he responded to further police-initiated interrogation.[165] If the state claims that a suspect implicitly waived an important right under art. 6, in order to establish a knowing and intelligent relinquishment of a right "it must be shown that he could have reasonably foreseen what the consequences of

---

[159](ECtHR) *Pishchalnikov v. Russia*, Application no. 7025/04, 24 September 2009, [67].

[160](ECtHR) *Pishchalnikov v. Russia*, Application no. 7025/04, 24 September 2009, [78].

[161](ECtHR) *Pishchalnikov v. Russia*, Application no. 7025/04, 24 September 2009, [78].

[162](ECtHR) *Salduz v. Turkey*, Application no. 36391/02, 27 November 2008, [59]; *Pishchalnikov v. Russia*, Application no. 7025/04, 24 September 2009, [79].

[163](ECtHR) *Salduz v. Turkey*, Application no. 36391/02, 27 November 2008, [59]; *Pishchalnikov v. Russia*, Application no. 7025/04, 24 September 2009, [79]; see also *Lazarenko v The Ukraine*, Application no. 22313/04, 28 October 2010 the applicant signed a waiver without being aware of the seriousness of the allegations and the penalty he may face and the ECtHR found a violation of art. 6; but see *Trymbach v The Ukraine*, Application no. 44385/02, 12 January 2012, in which the ECtHR found that voluntarily (and genuinely) signing a document which detailed his procedural rights was compliant with art. 6 because the waiver did not appear to have any bearing on the fairness of the proceedings, where the applicant consistently advanced the same version of events: [63–65].

[164](ECtHR) *Pishchalnikov v. Russia*, Application no. 7025/04, 24 September 2009, [79].

[165](ECtHR) *Pishchalnikov v. Russia*, Application no. 7025/04, 24 September 2009, [79].

his conduct would be."[166] Therefore, in the *Pishchalnikov* case, it could not be ruled out that, where his request for legal assistance had been left without adequate response, the applicant, who had no previous encounters with the police, did not know what was required to stop the interrogation.[167] The Court concluded that if a suspect asks for a lawyer and is without a lawyer during police interrogations, there is less chance that he will be fully informed of his rights and that his rights will be respected.[168] In *Sejdovic v Italy* the Grand Chamber found that for a waiver to be effective, in addition to being unequivocal, it must not run counter to the important public interest in legal representation of accused persons.[169]

The ECtHR's jurisprudence that develops the right to equality of arms in relation to issues such as access the legal advice has often emphasised the importance of the appearance of the fair administration of justice, which echoes the English procedural bias rule that 'justice must be done, as well as be seen to be done'.[170] Equality of arms is essential to the appearance of justice and public confidence in the criminal justice system, including the confidence of the accused person. As Cape et al state:

> An ultimately successful appeal against conviction, secured on the basis that the absence of legal advice meant that a confession made to the police officer was unreliable, which is determined following months, if not years, in pre-trial detention, is likely to leave the accused resentful and distrustful not just of the police, but of the whole criminal justice apparatus.[171]

Cases such as *Salduz* and *Pishchalnikov* have taken a strikingly clear approach to evidence of incriminating statements made by a suspect during a custodial police interrogation, without prior access to a lawyer—these statements may not in principle found a conviction.[172] According to the more recent jurisprudence of the ECtHR, merely cautioning a suspect or accused person does not fully equip them in relation to their defence rights.[173] These decisions are striking when compared to the relatively wide margin of appreciation which has normally been allowed to national courts in cases involving the admissibility of evidence.

2.3.2.2.4 Access to Free Legal Advice Under Article 6 of the ECHR

Article 6(3)(c) of the ECHR guarantees a right to access a lawyer, including free legal representation appointed by the state where the interests of justice so require. Under art. 6(3)(c), if a suspect or accused does not have the financial ability to pay

---

[166] (ECtHR) *Pishchalnikov v. Russia*, Application no. 7025/04, 24 September 2009, [77].

[167] (ECtHR) *Pishchalnikov v. Russia*, Application no. 7025/04, 24 September 2009, [79].

[168] (ECtHR) *Pishchalnikov v. Russia*, Application no. 7025/04, 24 September 2009, [78].

[169] (ECtHR) *Sejdovic v. Italy*, Application no. 56581/00, 1 March 2006, [86].

[170] See further Harris et al. (2014), p. 414.

[171] Cape et al. (2010), p. 4.

[172] Cape et al. (2010), p. 4.

[173] (ECtHR) *Panovits v. Cyprus*, Application no. 4268/04, 11 December 2008, [74]; see also (ECtHR) *Dayanan v. Turkey*, Application no. 7377/03, 13 October 2009, [32].

himself, free legal representation is regarded as necessary in cases involving actual or potential deprivation of liberty and in complex cases.[174]

In the case of *Croissant v Germany*,[175] the ECtHR has examined the right to free legal advice in the context of German law, where a convicted person is obliged to repay the state for the costs of *inter alia* a government appointed defence lawyer.[176] The Court questioned whether a such a scheme complied art. 6(1) and 6(3)(c). The Court did not find a violation of art. 6. However, it concluded that such a system would not be compatible with art. 6 if it adversely affected the fairness of the proceedings.[177] No consideration was specifically given to the issue of the affect of the existence of the repayment system on the decision about whether or not to seek a defence lawyer prior to interview.[178] Similarly, in the admissibility decision of *Mortensen v Denmark*, the Commission found that the fairness of the proceedings had not been adversely affected because the applicant had been provided a proper defence at trial.[179]

2.3.2.2.5 A European Directive on the Right of Access to a Lawyer

On 7 October 2013, as part of the fair trials roadmap, the EU Council adopted a directive on the right of access to a lawyer and the right to communicate upon arrest, for example with family members, employers and support persons.[180] Member States have 3 years to implement the directive in national law. Under the preamble at paragraphs 58 and 59, the United Kingdom, Ireland and Denmark are not bound by the directive or subject to its application by virtue of Protocols 21 and 22 of the TEU and the TFEU. Nevertheless, it is hoped that every state will implement the directive.[181]

---

[174](ECtHR) *Tsonyo Tvsonev v. Bulgaria no. 2*, Application no. 2376/03, 14 January 2010, [40].

[175](ECtHR) *Croissant v Germany*, Application no. 13611/88, 25 September 1992.

[176]The obligation to repay may be based on an assessment of the convicted person's economic status at the time he or she is required to reimburse the state: (ECtHR) *Croissant v Germany*, Application no. 13611/88, 25 September 1992, [34–35].

[177](ECtHR) *Croissant v Germany*, Application no. 13611/88, 25 September 1992, [36]. The court emphasised that the German scheme allowed for the opportunity for debt relief based on the inability of the convicted person to pay: [21, 37–38].

[178]See further, (ECtHR) *Croissant v Germany*, Application no. 13611/88, 25 September 1992, [27–38].

[179](ECommHR) *Mortensen v Denmark*, Application no. 24867/94, 15 May 1996.

[180]European Commission, Directive 2013/48/EU of the European Parliament and of the Council of 22 October 2013 on the right of access to a lawyer in criminal proceedings and in European arrest warrant proceedings, and on the right to have a third party informed upon the deprivation of liberty and to communicate with third persons and with consular authorities while deprived of liberty. OJEU 6 November 2013 L 294/1.

[181]Legal Experts Advisory Panel (LEAP) (2014), [46]; but see [30–31] and [41] where concerns are raised about implementation of the roadmap directives generally and more specifically, *inter alia*, in relation to Denmark, about effective participation of a defence lawyer at police interviews, the legal aid system and an inadequate system of redress in the evaluation of evidence to address breach of access to a lawyer at [49], [52] and [53], respectively.

The directive is based on art. 82(2) of the TFEU which supports actions such as the European Arrest Warrant (EAW) and the European Investigation Order (EIO) that are based on mutual recognition and underpin police and judicial cooperation in criminal matters. The Directive aims to set out minimum requirements at the EU level governing the right of suspects and accused persons to access legal assistance and therefore promotes articles 4, 6, 7, 47 and 48 of the Charter of Fundamental Rights and articles 3, 5, 6 and 8 of the ECHR.[182] In general, the directive builds upon the jurisprudence of the ECtHR as the basis for the minimum standards in relation to these rights.[183] The proposal to the directive stated:

> Without proper implementation of the ECtHR's case-law, Member States are likely to incur substantial costs stemming from liquidated damages awarded by the Court to successful applicants.[184]

Under articles 2 and 3 of the directive, a right of access to a lawyer during the investigation stage applies from the time a suspect or accused is made aware that they are suspected or accused of having committed an offence, irrespective of whether he or she is in custody. It applies in relation to questioning or some other evidence-gathering act which requires or permits the suspect or accused person's presence under national law. The right applies to a witness who becomes a suspect or accused person during questioning or a hearing pursuant to article 2(3).

In addition, the directive applies explicitly in relation to surrender under an EAW from the time of arrest. Under art. 10(4) and (5) the person the subject of surrender proceedings has the right to appoint a lawyer in the issuing state to assist the lawyer in the executing state, a right which is facilitated by the issuing state.

The rights in articles 3 and 10 can be subject to a voluntary and unequivocal waiver in accordance with art. 9, where the suspect or accused has been provided with information on the consequences of a waiver of the right to a lawyer.

[182]European Commission, Directive 2013/48/EU of the European Parliament and of the Council of 22 October 2013 on the right of access to a lawyer in criminal proceedings and in European arrest warrant proceedings, and on the right to have a third party informed upon the deprivation of liberty and to communicate with third persons and with consular authorities while deprived of liberty, OJEU 6 November 2013 L 294/1, [preamble 12].

[183]See the preamble of European Commission, Directive 2013/48/EU of the European Parliament and of the Council of 22 October 2013 on the right of access to a lawyer in criminal proceedings and in European arrest warrant proceedings, and on the right to have a third party informed upon the deprivation of liberty and to communicate with third persons and with consular authorities while deprived of liberty, OJEU 6 November 2013 L 294/1.

[184]European Commission, Proposal for a Directive of the European Parliament and of the Council on the right of access to a lawyer in criminal proceedings and on the right to communicate upon arrest, Brussels, 8 June 2011, COM (2011) 326, 2011/0154 (COD), [13].

2.3.2.2.6 Questioning of a Suspect by a Covert Human Intelligence Source

The ECtHR has further developed the scope of the right against self-incrimination and the right to silence by considering whether the use of undercover agents to elicit self-incriminating information is a form of improper coercion.

In *Allan v. the United Kingdom*,[185] the applicant consistently asserted his right to remain silent in his interviews with the police. Following that, a police informer was placed in his prison cell at the police station, and then in the same prison, with the specific purpose of eliciting information from the applicant about his involvement in the offences with which he had been charged. Some of the conversations between them were recorded. The high point of the applicant's alleged admissions, that he was present at the crime scene, was not part of the covert recording.

Based on the reasoning in previous decisions, the Court confirmed that:

> The right not to incriminate oneself is primarily concerned with respecting the will of an accused person to remain silent and presupposes that the prosecution in a criminal case seeks to prove the case against the accused without resort to evidence obtained through methods of coercion or oppression in defiance of the will of the accused.[186]

In examining whether a procedure has extinguished the 'very essence' of the right against self-incrimination, for the first time the Court in *Allan* clearly formulated a broad three-pronged test. The Court stated that it will examine: the nature and the degree of the compulsion; the existence of any relevant safeguards; and the use to which the incriminating material is put.[187]

In applying this approach to the facts of the case, the Court stated:

> While the right to silence and the privilege against self-incrimination are primarily designed to protect against improper compulsion by the authorities and the obtaining of evidence though methods of coercion or oppression in defiance of the will of the accused, the scope of the right is not confined to cases where duress has been brought to bear on the accused or where the will of the accused has been directly overborne in some way. The right, which the court has previously observed is at the heart of the notion of fair procedure, serves in principle to protect the freedom of a suspected person to choose whether to speak or to remain silent when questioned by the police.[188]

In determining whether the right to silence had been undermined to such an extent as to give rise to a violation of art. 6 in the circumstances of the case, however, the ECtHR considered case law of Canada and Australia.[189] Commencing with the Canadian decision of *R. v. Herbert*, these cases developed a two stage test

[185](ECtHR) *Allan v. The United Kingdom*, Application No. 48539/99, 5 November 2002.

[186](ECtHR) *Allan v. The United Kingdom*, Application No. 48539/99, 5 November 2002, [44].

[187](ECtHR) *Allan v. The United Kingdom*, Application No. 48539/99, 5 November 2002, [44].

[188](ECtHR) *Allan v. The United Kingdom*, Application No. 48539/99, 5 November 2002, [50].

[189]At [30–33] the ECtHR considered the decisions of *R. v. Herbert* [1990] 2 Supreme Court Reports 151; *R. v. Broyles* [1991] 3 Supreme Court Reports 595; *R. v. Liew* [1999] 3 Supreme Court Reports 227; and *R. v. Swaffield and Pavic* [1998] High Court of Australia 1. At [51] the Court also referred to s. 7 of the Canadian Charter of Rights and Freedoms, which guarantees a fair trial.

to be used in the situation of covertly questioning a suspect. Following the reasoning in these cases the right to remain silent and the right against self-incrimination will be infringed if the subterfuge is conducted by a police officer (or a person acting as an agent of the police) and a confession is elicited in unfair derogation of the suspect or accused person's right to exercise a choice whether to speak or to remain silent. A confession will be 'elicited' if the covertly recorded conversation is the functional equivalent of an interrogation.[190] The Canadian and Australian case law considered the fairness to the accused, in the absence of reference to the gravity of the offending.

In analysing the facts, the Court noted that after the applicant had sought to remain silent when questioned by the police, the police coached the informer to conduct his covert conversations with the applicant in a way that can be regarded as the functional equivalent of an interrogation.[191] The applicant's statements were not spontaneous and unprompted, but rather were induced by persistent questioning. The interrogation was not accompanied by the usual safeguards that would attach to a formal police interview, including the attendance of a solicitor and the usual caution.[192] In addition, the Court pointed to the psychological pressures the applicant would have already been subjected to by virtue of his situation, which would have impinged on the voluntariness of the disclosures allegedly made.[193] The Court concluded that the information obtained by the informer "...may be regarded as having been obtained in defiance of the will of the applicant and its use at trial impinged on the applicant's right to silence and the privilege against self-incrimination."[194] Accordingly, the Court found a violation of art. 6(1).[195]

The situation in *Allan* may be contrasted to that of self-incriminating statements being made to police officers in a police cell while the officers attempt to engage a suspect in conversation after he or she has repeatedly asserted the right to remain silent under formal interrogation. Jackson and Summers warn that while this would not infringe the ECtHR's test in *Allan*, because the suspect makes spontaneous ('voluntary') statements, it is nevertheless problematic.[196] They suggest that the test in *Allan* does not go far enough to protect an accused in the oppressive environment

---

[190](ECtHR) *Allan v. The United Kingdom*, Application No. 48539/99, 5 November 2002, [51].

[191](ECtHR) *Allan v. The United Kingdom*, Application No. 48539/99, 5 November 2002, [52].

[192](ECtHR) *Allan v. The United Kingdom*, Application No. 48539/99, 5 November 2002, [52].

[193](ECtHR) *Allan v. The United Kingdom*, Application No. 48539/99, 5 November 2002, [52].

[194](ECtHR) *Allan v. The United Kingdom*, Application No. 48539/99, 5 November 2002, [52]. In its judgment the Court referred to a number of relevant judgments from Australia and Canada.

[195]See further Jackson and Summers (2012), p. 246, where it is argued that the ECtHR's case-law restricts the protection of a suspect or accused from the inherently coercive atmosphere of police questioning.

[196]Jackson and Summers (2012), pp. 177–179, referring to the decision in *Portmann v Switzerland*. See further Sect. 2.3.2.5.3, about the use of listening devices in police or prison cells after an accused has asserted the right to silence.

of police custody with little or no access to legal assistance, and being repeatedly questioned by the police.

The ECtHR's three pronged test for assessing whether the essence of right to silence and the right against self-incrimination has been diminished, formulated with reference to the Canadian and Australian cases, does not allow for the gravity of the offending to be taken into consideration. Therefore, in relation to confessional evidence, art. 6 primarily promotes fairness to the accused rather than pursuing a community interest in bringing those who commit crime to justice, which is greater in serious cases.

2.3.2.3 Drawing Adverse Inferences of Guilt from the Silence of the Suspect of Accused

Is guaranteeing the right to silence only effective when silence may not be taken into account in the determination of guilt or innocence?[197] There are many reasons why a suspect or accused person may chose to remain silent: intoxication; memory problems; protecting another person or a secret relationship; fear of reprisal; fear of sounding guilty; desire to avoid pre-trial custody; or fear of revealing guilt itself.

In the ECtHR case of *John Murray v. the United Kingdom* the significant question to be decided was whether permitting adverse inferences to be drawn from the silence of the accused, when asked to answer questions by the police and again at his trial, amounted to an indirect form of compulsion that shifted the burden of proof and effectively removed the right to silence and the right against self-incrimination.[198] A man had been allegedly held captive by the IRA and forced to make a taped confession. When the police arrived, his blind-fold was removed and he saw the applicant at the top of some stairs pulling tape out of a cassette. The police entered the house where they found the applicant coming down the stairs. The Court found that the accused person's right to silence was not absolute, in that it did not prevent the accused's silence being taken into account on the question of guilt, in circumstances that clearly called for an explanation of innocence from the accused.[199] The Court stated that the question of whether adverse inferences from silence violate art. 6 should be determined in light of all the circumstances of the case, "...having particular regard to the situations where inferences may be drawn, the weight attached to them by the national courts in their assessment of the evidence and the degree of compulsion inherent in the situation."[200]

---

[197]This view is reflected in art. 67(1)(g) of the Rome Statute of the International Criminal Court.

[198](ECtHR) *John Murray v. The United Kingdom*, Application no. 18731/91, 8 February 1996, [41–44, 46].

[199](ECtHR) *John Murray v. The United Kingdom*, Application no. 18731/91, 8 February 1996, [47].

[200](ECtHR) *John Murray v. The United Kingdom*, Application no. 18731/91, 8 February 1996, [47].

The Court found that a system which warns the accused (who is possibly without legal assistance) that adverse inferences may be drawn in support of other evidence of guilt involves "...a certain level of indirect compulsion."[201] However, in deciding that there had been no breach of art. 6 on the basis of the adverse inferences, the Court emphasised the following factors: the applicant was permitted to and did in fact remain silent during interrogation and at trial; the adverse inference was a matter of common-sense in light of the evidence against him; and, as required by law, the evidence of his silence was persuasive only and there was other evidence which strongly supported a finding of guilt.[202] The Court referred to the existence of other systems, where evidence is freely assessed, that may have regard to all the circumstances of the case and all the conduct of the accused. The Court commented, as described by the Commission, that this was a formalised system which allowed for common-sense inferences to play an open role in the assessment of evidence.[203]

Nevertheless, the Court found that the applicant was faced with a fundamental dilemma about whether or not to remain silent and the impact that this could have upon his trial. In such circumstances, by remaining silent during police questioning, an accused sets himself on a particular course. Whether he chooses to give evidence and call witnesses at a trial or if he maintains his silence, either way, strong adverse inferences may be drawn in support of other evidence against the accused.[204] Therefore, the Court found that a restriction on the applicant's right of access to a lawyer during the first 48 hours of his police detention, during which time he was interrogated, deprived him of a fair hearing.[205] Thus, the Court has clearly connected the consequences of what happens in the pre-trial phase to the fairness of the proceedings as a whole, especially to what happens at the trial.

In *Condron v The United Kingdom* the ECtHR found that in systems where the decision about whether to draw an adverse inference of guilt from the silence of the accused is ultimately left to the discretion of a lay jury (who has no obligation to provide reasons for their verdict), the trial judge has an obligation to direct the jury of the proper manner in which to draw such inferences.

The applicants' solicitor had advised them to answer "no comment" when they were being interrogated by the police. This advice had been given on the basis that they were having withdrawal symptoms from their heroin use which, in the solicitor's opinion, rendered them unfit to be interviewed. The trial judge directed

---

[201](ECtHR) *John Murray v. The United Kingdom*, Application no. 18731/91, 8 February 1996, [50].

[202](ECtHR) *John Murray v. The United Kingdom*, Application no. 18731/91, 8 February 1996, [48, 51, 52].

[203](ECtHR) *John Murray v. The United Kingdom*, Application no. 18731/91, 8 February 1996, [54].

[204](ECtHR) *John Murray v. The United Kingdom*, Application no. 18731/91, 8 February 1996, [66].

[205](ECtHR) *John Murray v. The United Kingdom*, Application no. 18731/91, 8 February 1996, [63, 66, 70]. See further in relation to the right to access a lawyer at Sect. 2.3.2.2.2 above.

the jury that they were at liberty to draw an adverse inference of guilt, even if they were satisfied of the plausibility of the applicants' explanation for remaining silent that they acted on the strength of the solicitor's advice. Finding that this jury direction was deficient, the ECtHR concluded:

> ...as a matter of fairness, the jury should have been directed that it could only draw an adverse inference if satisfied that the applicants' silence at the police interview could only sensibly be attributed to their having no answer or none that would stand up to cross-examination.[206]

The Court in Condron discussed the use of safeguards in criminal proceedings that allow adverse inferences to be safely drawn from the silence of the accused.[207] In the circumstances of the case the Court referred to the requirement for a judge to give proper directions, which appropriately limit the reliance on adverse inferences of guilt being drawn from silence, and that such directions are subject to review on appeal.

2.3.2.4 Questioning Under Compulsory Powers

2.3.2.4.1 Obtaining Statements Under Compulsory Questioning Powers

In 1996, in the case of *Saunders v. the United Kingdom,*[208] the Grand Chamber ECtHR examined the validity of using administrative compulsory questioning powers to obtain documents and statements from officers or agents of a company and using this evidence to support a conviction for fraud in relation to share dealing. Department of Trade and Industry (DTI) inspectors relied on section 434(3) of the Companies Act 1985, to examine the applicant on oath in an information-gathering investigation and did not have the power to investigate or deliberate on criminal liability. According to s. 436(3), it was contempt of court, punishable by a fine or imprisonment of up to 2 years, to refuse to produce documents or to answer the DTI inspectors' questions. Pursuant to s. 434(5) of the legislation the applicant's answers were used against him by the prosecution authorities at a subsequent criminal trial. The applicant had been faced with the dilemma of either potentially incriminating himself or facing charges and punishment for contempt of court, and had ultimately answered the inspectors' questions. On the face of it, the statements in question appeared to be non-incriminating, such as exculpatory remarks or non-controversial information about questions of fact. However, at the trial the statements were used by the prosecution to cast doubt upon other statements made by him and evidence given at his trial.

The Grand Chamber did not object to the use of compulsory questioning powers *per se*. The Court observed that an administrative investigation is capable of

---

[206](ECtHR) *Condron v. The United Kingdom*, Application no. 35718/97, 2 May 2000, [61–62].

[207](ECtHR) *Condron v. The United Kingdom*, Application no. 35718/97, 2 May 2000, [61–62, 66].

[208](ECtHR) *Saunders v The United Kingdom*, Application no. 19187/91, 17 December 1996.

involving the determination of a criminal charge, in which case the protections under art. 6 arguably should apply. However, the investigation under consideration in the particular case did not and, therefore, fell outside the protection of art. 6.[209] The Court classified the DTI investigators as being involved in a 'preparatory investigation', which was conducted in the legitimate public interest of the effective regulation of financial and commercial activities. The question posed by the Court was whether, in the light of the circumstances of the case, the use at trial of the pre-trial statements made by the applicant under threat of contempt of court was incompatible with art. 6.[210] The Court found that the various procedural safeguards referred to by the Government, including judicial discretion to exclude evidence on the ground of unfairness, could not be relied on "...in the present case since they did not operate to prevent the use of the statements in the subsequent criminal proceedings."[211] In coming to this conclusion, the Court found that the protection of the right against self-incrimination was not limited to admissions of guilt or statements that are directly incriminating.[212] Furthermore, in the Court's view, the 'fairness requirement' of art. 6 meant that the gravity of the offending was not to be taken into account.[213]

In the case of *Heaney and McGuiness v. Ireland*[214] the ECtHR found that anti-terrorist legislation compelling suspects of terrorism offences to account for their movements was in violation of the right to silence and the right against self-incrimination. Prior to being interrogated, the applicants were cautioned about their right to remain silent. They were further advised that if they did not account for their movements when questioned, they risked facing up to 6 months imprisonment. At the relevant time, the impugned measure was silent as to the use that could be made of any statements compulsorily obtained. The applicants asserted their right to remain silent. They were acquitted of the substantive offence but were imprisoned for failing to account for their movements. On the basis of the information that was provided to the applicants in the course of the interrogation, the Court held that the degree of compulsion imposed on them by the special measure destroyed the very essence of their right against self-incrimination and the right to

---

209 (ECtHR) *Saunders v The United Kingdom*, Application no. 19187/91, 17 December 1996, [67]; see further (Commission—admissibility) *Abas v The Netherlands*, Application no. 27943/95, 26 February 1997; (ECtHR) *IJL, GMR and AKP v The United Kingdom*, Application no. 29522/95, 30056/96 and 30574/96, 19 September 2000; Emmerson et al. (2012), p. 617.

210 (ECtHR) *Saunders v The United Kingdom*, Application no. 19187/91, 17 December 1996, [69].

211 (ECtHR) *Saunders v The United Kingdom*, Application no. 19187/91, 17 December 1996, [75].

212 (ECtHR) *Saunders v The United Kingdom*, Application no. 19187/91, 17 December 1996, [71–72].

213 (ECtHR) *Saunders v The United Kingdom*, Application no. 19187/91, 17 December 1996, [74]. This is consistent with the decision of the Court in (ECtHR GC) *Gäfgen v Germany*, Application no. 22978/05, 1 June 2010, where the Court held that the gravity of offending was not relevant to an assessment of fairness under art. 6 in relation to coercion by threats to torture a suspect.

214 (ECtHR) *Heaney and McGuinness v Ireland* (Application no. 34720/97), 21 March 2001.

remain silent.[215] Further, in response to the United Kingdom's claim, that the terrorist nature and complexity of the alleged offences justified the provision, the Court stated that the public interest could not be relied on to this extent.[216]

In *O'Halloran and Francis v. the United Kingdom*[217] the central issue was whether the right to silence and the right against self-incrimination were violated when the registered keeper of a car was required under the relevant legislation to provide the identity of the driver of the car when caught speeding on a camera. The intention was that this information could then be used in a prosecution. Failure to give the information was an offence which was punishable by a fine and by penalty points against the owner's driving license. The Court asserted that direct compulsion requiring a suspect or potential suspect to provide information which contributes or might contribute to his conviction will *not* automatically result in a violation.[218] The Court stated:

> While the right to a fair trial under Art. 6 is an unqualified right what constitutes a fair trial cannot be the subject of a single unvarying rule but must depend on the circumstances of the particular case.[219]

Based on previous case-law, the Court confirmed that the factors that would be taken into account, in determining whether the essence of the right against self-incrimination and the right to silence had been violated, included: the nature and degree of the compulsion used; the existence of relevant safeguards in the procedure; and the use to which the material obtained was put.[220]

In applying this approach, the Grand Chamber emphasised two aspects of the case: firstly, by owning and driving a car, a person subjects themselves to the regulatory regime in issue, which has been imposed in recognition of the fact the use of cars have the potential to cause grave injury; and, secondly, the information required by the scheme was limited to the identity of the driver, which was not incriminating of itself and could not on its own found a conviction for the offence of speeding.[221] Accordingly, the Grand Chamber held that there was no violation of

---

[215](ECtHR) *Heaney and McGuinness v Ireland* (Application no. 34720/97), 21 March 2001, [55]; but see (ECtHR) *King v The United Kingdom*, Admissibility decision, 8 April 2003.

[216](ECtHR) *Heaney and McGuinness v Ireland* (Application no 34720/97), 21 March 2001, [57–58].

[217](ECtHR) *O'Halloran and Francis v The United Kingdom*, Application nos. 15809/02 and 25624/02, 29 June 2007; see also, (ECtHR) *Weh v. Austria*, Application no. 38544/97, 8 April 2004.

[218](ECtHR) *O'Halloran and Francis v The United Kingdom*, Application nos. 15809/02 and 25624/02, 29 June 2007, [53].

[219](ECtHR) *O'Halloran and Francis v The United Kingdom*, Application nos. 15809/02 and 25624/02, 29 June 2007, [53].

[220](ECtHR) *O'Halloran and Francis v The United Kingdom*, Application nos. 15809/02 and 25624/02, 29 June 2007, [55].

[221](ECtHR) *O'Halloran and Francis v The United Kingdom*, Application nos. 15809/02 and 25624/02, 29 June 2007, [57–60]; (ECtHR) *Weh v. Austria*, Application no. 38544/97, 8 April 2004, [54].

art. 6.[222] The context in which the road traffic scheme exists, coupled with the limited nature of the information sought, which cannot be said to have fully shifted the burden of proof, distinguishes this case, on the facts, from the other examples of compulsory questioning powers. This decision may be a discrete exception to the right against self-incrimination that relies on the European consensus that the immunities do not apply in this situation.[223] This type of case may otherwise be difficult to investigate.[224]

As referred to in Chap. 1 (Sect. 1.2.2), where the person being questioned under compulsory questioning powers is not a suspect, then these administrative powers may be used; so long as the answers provided are not used in future substantive criminal proceedings against him. However, where criminal proceedings are pending or anticipated then compulsory powers should not be used to obtain any form of self-incriminating information. This was the conclusion of the ECtHR in the case of *Weh v Austria*. Here the majority of the Court found that a registered car owner who was asked the identity of the driver of his car when it was caught speeding, was not the subject of criminal charges, within the autonomous meaning under art. 6.[225]

It is possible that the key to resolving cases that deal with the use of compulsory powers to obtain incriminating statements from suspects or accused persons, is to identify on a case-by-case basis whether the essence of the right to silence and the right against self-incrimination has been destroyed or extinguished to such an extent that a state measure is in violation of art. 6.[226] However, the contracting states have to be able to draw some general principles about minimum standards of investigative fairness. At the same time, each case should be reviewed for exceptions to those general principles.

Compulsory questioning powers can be extremely powerful, particularly if they lead to individuals supplying incriminating statements that may ultimately be used for the purposes of a criminal investigation. Even if a suspect is granted a direct use immunity, which prohibits the use of incriminating testimony obtained under compulsory powers from being used in subsequent criminal proceedings, new lines of enquiry may be opened up, leading the way to real derivative evidence that can spur on an investigation. However, this type of measure needs to be viewed in the context of the balance between effectiveness in seeking the truth and fairness to the suspect or accused in each legal system.

---

[222](ECtHR) *O'Halloran and Francis v The United Kingdom*, Application nos. 15809/02 and 25624/02, 29 June 2007, [61–63].

[223]Emmerson et al. (2012), p. 622; see also *Antonio Sánchez López v. Spain*, Communication No. 777/1997, U.N. Doc. HR COMMITTEE/C/67/D/777/1997 (25 November 1999), where the Human Rights Commission considered that a similar complaint by an owner of a vehicle, who was caught speeding, was inadmissible because he was punished for non-cooperation with the authorities and not for traffic offences.

[224]Jackson and Summers (2012), p. 274.

[225](ECtHR) *Weh v Austria*, Application no. 38544/97, 8 April 2004 [40–54]; see further Jackson and Summers (2012), pp. 251–252.

[226]Emmerson et al. (2012), p. 622.

2.3.2.4.2 Obtaining Derivative Real Evidence Using Compulsory Questioning Powers

In the case of *Gäfgen v. Germany*, the ECtHR found that the use at trial of statements obtained directly through torture or other forms of ill-treatment contrary to art. 3 would be contrary to the principles of a fair trial in art. 6, irrespective of their probative value.[227] The use of real evidence that was located as a result of self-incriminating statements (derivative real evidence) made by the use of torture may need to be excluded as a rule, if the evidence is decisive to the outcome of the case.[228] Although the Court acknowledged that there were competing interests on either side in the question of admissibility of evidence obtained under *compulsion*, due to the absolute nature of art. 3, balancing these interests will be irrelevant to the use of derivative real evidence obtained as a result of statements made under other forms of prohibited ill-treatment.[229] However, the Court did not rule out the use at trial of derivative real evidence obtained as a result of statements made under other forms of compulsion not covered by art. 3.

2.3.2.5 Covert Surveillance to Obtain Self-Incriminating Statements

2.3.2.5.1 Unregulated Covert Recordings Containing Self-Incriminating Statements

The use of covertly recorded conversations in a criminal trial has attracted complaints about the fairness of the proceedings, where there has been no regulation of the particular surveillance measures used. Covert recordings can be a method by which the police can obtain evidence of self-incriminating statements without engaging the right to silence possessed by a suspect or accused who is to be questioned under caution.

In the case of *Khan v. The United Kingdom*[230] the police had placed a listening device on the private premises. By chance, the applicant visited his friend and a conversation was recorded during which the applicant admitted his involvement in the importation of drugs. The applicant had already been interviewed for this offence but had remained silent and there was no other evidence against him in relation to the offences. Following the trial judge's ruling that the recorded evidence was admissible, the applicant pleaded guilty. His statements were the sole basis for his conviction in relation to drug trafficking. The listening device had been authorised in accordance with investigative guidelines. However, the domestic law

[227](ECtHR GC) *Gäfgen v. Germany*, Application no. 22978/05, 1 June 2010 [166–167].
[228](ECtHR GC) *Gäfgen v. Germany*, Application no. 22978/05, 1 June 2010 [178].
[229](ECtHR GC) *Gäfgen v. Germany*, Application no. 22978/05, 1 June 2010 [175–176].
[230](ECtHR) *Khan v. The United Kingdom*, Application no. 35394/97, 12 May 2000.

at the time did not regulate the use of such devices, and this was the basis of the applicant's complaint.

The central question was whether the criminal proceedings as a whole were fair. The ECtHR found that the relevance of the non-existence of corroborating evidence depends on the circumstances of the case. Where the recorded admissions had been made voluntarily and there was no concern that the statements were unreliable due to coercion or oppression, the need for supporting evidence was correspondingly weaker.[231] The Court found that the applicant had received ample opportunity to challenge the use and the authenticity of the contested recording. Therefore, the Court concluded that although the lack of a scheme regulating the use of listening devices gave rise to a breach of the right to a private life under art. 8, the use at the applicant's trial of the covert recording was not in contravention of the fairness requirements guaranteed by art. 6(1).[232]

In *Bykov v. Russia*,[233] the Grand Chamber decided that where the reliability of evidence is in dispute, the existence of procedures to examine the admissibility of evidence takes on a greater significance.[234] The Federal Security Service of the Russian Federation (FSB) and the police decided to conduct a covert operation to investigate the applicant on suspicion of conspiracy to murder. The police staged the discovery of a dead body, which was reported in the media. Acting on police instruction a third person went to speak to the applicant, with a hidden radio transmitting device. They had a conversation, in which the applicant appeared to participate freely, confirming that the assassination had been carried out. During the conversation, several objects were produced as proof. These objects were later seized by the police from the applicant's estate. The conversation was a small part of a complex body of evidence against the applicant.[235]

The Grand Chamber found that the lack of safeguards against possible abuse in the regulations governing the use of radio-transmitting devices constituted a breach of art. 8. However, in relation to art. 6 and the fairness of the criminal proceedings culminating in the applicant's conviction, the Grand Chamber explained, as follows:

> In the present case, the applicant was able to challenge the covert operation, and every piece of evidence obtained thereby, in the adversarial procedure before the first-instance court and in his grounds of appeal. The grounds for the challenge were the alleged unlawfulness and trickery in obtaining evidence and the alleged misinterpretation of the conversation

[231](ECtHR) *Khan v. United Kingdom*, Application no. 35394/97, 12 May 2000, [37].

[232](ECtHR) *Khan v. United Kingdom*, Application no. 35394/97, 12 May 2000, [38–40], applying the decision in *Schenk v. Switzerland*, Application no. 10862/84, 12 July 1988; see also (ECtHR) *Heglas v. the Czech Republic*, Application no. 5935/02, 1 March 2007, in which the Court similarly found that the lack of a regulatory scheme in relation to dictaphones and lists of telephone numbers gave rise to a breach of art. 8, but their use in criminal proceedings did not violate art. 6, in the circumstances of the case.

[233](ECtHR GC) *Bykov v. Russia*, Application no. 4378/02, 10 March 2009.

[234](ECtHR GC) *Bykov v. Russia*, Application no. 4378/02, 10 March 2009, [95].

[235](ECtHR GC) *Bykov v. Russia*, Application no. 4378/02, 10 March 2009, [103].

recorded on the tape. Each of these points was addressed by the courts and dismissed in reasoned decisions.[236]

In addition, the Court noted that the covert evidence was not the sole basis for the applicant's conviction.[237] The Grand Chamber confirmed that the applicant's defence rights had been respected in the domestic proceedings, and there had been nothing to suggest that the evaluation of the evidence had been arbitrary.[238] Having regard to the safeguards surrounding the admissibility of the evidence, the nature and degree of the alleged compulsion, and the use to which the material obtained was put, the Court concluded that the proceedings as a whole were not contrary to the requirements of a fair trial.[239]

2.3.2.5.2 The Admissibility of Unlawfully Obtained Confessions

Aside from torture, other ill-treatment and entrapment, the definition of what precisely constitutes improper compulsion varies from country to country, with some countries focusing on the need to prevent unreliable confessions and some focusing on preventing improper police practices.[240] In addition, there is a wide variation in the rules of evidence between different legal systems. For example, in a common law trial there are multiple rules governing the admissibility of evidence. However, in civil law systems there are typically few restrictions on the admissibility of evidence.[241] Some systems, such as Denmark, rely more heavily on safeguards in the investigation phase and on disciplinary actions. Nevertheless, the ECtHR has attempted to set some minimum standards of fairness in relation to reliance on unlawfully obtained evidence for conviction that are applicable in all the Convention states.

Article 6 of the ECHR is concerned with the effect of using self-incriminating statements on the fairness of the proceedings as a whole, in the circumstances of each case. Therefore, under the ECHR, the right to a fair trial does not in general regulate whether and how certain types of evidence should be admitted or what weight should be placed on it.[242] Primarily, the ECtHR has examined whether the proceedings as a whole, including the way in which the evidence relied on at trial was obtained, were fair.[243] The decision about whether or not the right to a fair trial

---

[236](ECtHR GC) *Bykov v. Russia*, Application no. 4378/02, 10 March 2009, [95].

[237](ECtHR GC) *Bykov v. Russia*, Application no. 4378/02, 10 March 2009, [98].

[238](ECtHR GC) *Bykov v. Russia*, Application no. 4378/02, 10 March 2009, [98].

[239](ECtHR GC) *Bykov v. Russia*, Application no. 4378/02, 10 March 2009, [104].

[240]Wu (2011), p. 95.

[241]See Chaps. 3 and 4; see also Harris et al. (2014), p. 418.

[242]See further (ECtHR) *Edwards v The United Kingdom*, Application no. 13071/87, 16 December 1992, [34].

[243](ECtHR) *Schenk v. Switzerland*, Application no. 10862/84, 12 July 1988, [46]; see also (ECtHR GC) *Jalloh v Germany*, Application no. 54810/00, 11 July 2006, [94–95].

has been breached may depend on the types of safeguards being used and the existence of other evidence pointing to guilt.[244]

In the case of *Schenk v Switzerland*,[245] a third person made an unlawful tape recording (which was not obtained by consent or with the requisite authority) of a conversation he had with the applicant. This conversation was later admitted in evidence at his trial. The recording had been made after the person had already been in contact with the police and the suggestion was that he had been acting as an agent of the police. No violation of art. 6(1) was found because the applicant had been given the opportunity to challenge the authenticity of the recording, and to examine the maker of the recording and the police officer in charge of the investigation.[246] In addition, the recording was not the only evidence upon which the applicant was convicted.[247]

This rationale was applied more recently in the case of *Chalkley v. the United Kingdom*, where the ECtHR stated, as follows:

> . . .the use at trial of material obtained without a proper legal basis or through unlawful means will not generally offend the standard of fairness imposed by Art. 6 § 1 where proper procedural safeguards are in place and the nature and source of the material is not tainted, for example, by any oppression, coercion or entrapment which would render reliance on it unfair in the determination of a criminal charge.[248]

In order to plant a listening device, the police removed the applicant and his family from his home by arresting him and his girlfriend on old enquiries that had already lapsed without charge. Later, on the basis of recordings of private conversations between the two at their home, they were charged with conspiracy to commit robbery and burglary. The applicant entered a plea of guilty after the recordings were admitted into evidence at the commencement of his trial. There had been no statutory regulation of listening devices at the relevant time and the ECtHR found a violation of the applicant's private life under art. 8.[249] However, when deciding on the admissibility of the case under art. 6, the Court found that the use of the illegally obtained evidence did not constitute a violation of art. 6 because it did not render the trial unfair.[250]

---

[244](ECtHR) *John Murray v. The United Kingdom*, Application no. 18731/91, 8 February 1996, [47].

[245](ECtHR) *Schenk v. Switzerland*, Application no. 10862/84, 12 July 1988.

[246](ECtHR) *Schenk v. Switzerland*, Application no. 10862/84, 12 July 1988, [48].

[247](ECtHR) *Schenk v. Switzerland*, Application no. 10862/84, 12 July 1988, [49].

[248]ECtHR, *Tony Michael Chalkley v. The United Kingdom*, Application no. 63831/00, 12 June 2003.

[249]ECtHR, *Tony Michael Chalkley v. The United Kingdom*, Application no. 63831/00, 12 June 2003.

[250]ECtHR, *Tony Michael Chalkley v. The United Kingdom*, Application no. 63831/00, 12 June 2003; see also ECtHR, *Perry v. The United Kingdom*, Application no. 63737/00, Decision as to Admissibility, 26 September 2002 in which the ECtHR found that the use of unauthorized video surveillance instead of an identification parade, which the applicant had already objected to, did not affect the applicant's defence rights at trial and was therefore not contrary to art. 6.

2.3.2.5.3 Covertly Recording Admissions After a Suspect Has Remained Silent

As referred to at Sect. 2.3.2.2.6 above, in *Allan v. the United Kingdom*,[251] the ECtHR found that covertly conducting the functional equivalent of an interrogation destroyed the essence of the applicant's right to choose whether to speak to the police or to remain silent. An undercover agent had entered the applicant's cell and elicited self-incriminating statements in a series of conversations. The police had previously sought to interview the applicant who had asserted his right to remain silent. The decision of the Court focused on the unfairness of the interrogation contrary to the right to silence. The Court, therefore, did not set any standards in relation to, for example, using a listening device placed in a suspect's cell to obtain spontaneous statements after the suspect has asserted the right to silence.[252]

2.4 Conclusion

This chapter has presented the scope and the reach of the right to silence under the ICCPR and the ECHR, as a significant background to the development of these rights within the national systems that come under the two human rights frameworks. In each framework the minimum standards in relation to the right to silence are primarily concerned with protecting suspects and accused persons against compulsion. In addition, art. 6 has been interpreted to provide protection from the unfair effects of compulsion, when considering the proceedings as a whole. Any evidence obtained in a manner that destroys the right to silence—not only because it has been obtained by torture—should not be relied on for conviction under the ECHR.

The ICCPR creates international legally binding rights for the ratifying states to then implement and apply. In Denmark and England, national law is presumed to comply with international obligations under the ICCPR, although the treaty is not binding in *national* law. Nevertheless, interpretation of national law will usually be consistent with these international minimum standards. The question of audio-visual recording of interviews, however, may be something that is not consistently implemented in the national setting, as will be discussed in the following chapters.

Australia has incorporated the ICCPR rights into national law, through the enactment of legislation that is consistent with the treaty Legislation, such as the uniform evidence law, refers to some of the treaty provisions. However, the ICCPR has not been directly incorporated in its entirety in one act. Nevertheless, all

[251](ECtHR) *Allan v. The United Kingdom*, Application No. 48539/99, 5 November 2002.

[252]See further Jackson and Summers (2012), pp. 182–187, where covert surveillance cases are discussed in light of the decision of *Allan*.

legislation is scrutinised for compliance with Australia's international human rights obligations.

However, by considering the guidelines of the HR Committee, in particular in their general comments, it becomes clear that the minimum standards recommended under art. 14 of the ICCPR apply once a person is 'charged' but are limited in scope in pre-trial proceedings. Uncertainty surrounds the meaning of 'charged', which appears to have a formal meaning. This means that uncertainty also surrounds the application of the right to silence to procedures in the pre-trial phase, such as police questioning. Nevertheless, despite this uncertainty, it appears that the right not to make self-incriminating statements must be seen as applying to pre-trial interrogation to prohibit physical and 'undue' psychological pressure and compelled confessions. This arguably includes the need to caution a suspect and the right to access legal advice. Under the ICCPR and other UN treaties, the right to a lawyer clearly commences immediately after a person is taken into custody. As an additional protection against abusive coercion in the interview room recommendations apply under the ICCPR that audio and preferably video recordings should be made of a police interview.

Denmark and England also come under the framework of the ECHR and have directly incorporated it into national law. Although the ECtHR has not set standards that call for recording of interviews, the right to silence under the ECHR has been defined in more clear terms and with greater scope than under the ICCPR, by having been defined in detail in relation to a larger range of situations. For, example, through its autonomous definition of 'charge', there is a clear right to be informed of an accusation and the right to silence once reasonable grounds of suspicion have been formed.

Similar to the HR Committee, the ECtHR has stated that a suspect or accused should have access to a lawyer from the first interrogation and, clearly, from the time they are taken into custody, in particular before a custodial interview. However, other minimum standards under the ECHR go beyond the protection provided in the ICCPR. The ECtHR has found that the right to access legal advice prior to interrogation is especially important in systems where there are consequences that arise out of the right to silence for the way the defence is conducted at trial. Access to a lawyer may only be waivered by a knowing and intelligent waiver, which is unequivocal and does not run counter to the public interest in having accused persons represented. Therefore, for example, where a suspect or accused risks imprisonment then the need for a valid waiver is all the more important. Access to free legal representation should be available in the interests and the way in which the system for free legal representation works should not adversely affect the fairness of the proceedings. Under the ECHR, a suspect may not be interrogated by an undercover agent—although an agent may covertly engage a suspect in a spontaneous conversation.

If a suspect or accused person does remain silent when asked to give an account of the facts, the ECtHR has found that common sense inferences may be drawn from silence, when the only sensible explanation for remaining silent on the evidence is the guilt of the accused. This should be the subject of a jury direction in systems that rely on lay juries. Such inferences may only be persuasive on the question of guilt and must be accompanied by other evidence in support of the prosecution case. As referred to above, a system that allows such inferences to be drawn should provide a right to access a lawyer prior to interrogation to ensure that the suspect or accused understands the consequences of the right.

The ECtHR has also set minimum standards in relation to the use of compulsory questioning powers. If there are criminal charges pending or anticipated, then the suspect or accused should not be subject to such powers; he or she should receive the benefit of the protection of art. 6. Where the person is not suspected of criminal offences then compulsory questioning powers may be used; however, any statements made should not be used against them in subsequent substantive criminal proceedings. Where the person is not a suspect or accused and other types of compulsory information-gathering powers are used, then any self-incriminating information in the form of real evidence, such as documents or DNA, may be used against them.

Whether certain covert surveillance devices are unregulated or used unlawfully, the ECtHR has stated that, in the absence of evidence of compulsion or oppression, any statements obtained may be used in a criminal trial provided that certain safeguards exist to ensure that the authenticity and reliability of the evidence may be challenged and tested. Further, in the case of unlawfully obtained evidence, the use of the evidence at trial may be a consideration. For example, it may be relevant if the evidence was decisive or supported other evidence in the prosecution case. Similar rules arguably apply to covertly recorded statements that are spontaneous but were made after the accused had asserted the right to remain silent in the pre-trial proceedings.

Under the ECHR, in addition to physical or psychological compulsion, the right to silence means that safeguards must be in place if the functioning of the law creates some type of direct or indirect compulsion to make (self-incriminating) statements. Safeguards, such as the right to access legal advice are therefore of critical importance. Central to the reasoning of the ECtHR about the right to silence is the idea that fairness is something that is evaluated in light of the criminal proceedings as a whole. This implies that in certain situations there may be consequences that result from the right to silence that need to be dealt with or balanced out in the trial phase, for example, if administrative compulsory questioning powers may be used in pre-trial proceedings or if an adverse inference of the guilt of the accused may be drawn from pre-trial silence. The suspect or accused must always be in position to receive the benefit of the protection of art. 6—if not by safeguards that are immediately in place in the pre-trial proceedings, such as access to a legal representative, then by safeguards that are in place in the trial phase.

Despite the fact that the ECtHR's jurisprudence has covered cases coming from different systems with varying legal traditions and legal structures, the Court has nevertheless attempted to fix common minimum standards that are universal. Some would argue that the Court did not go far enough. For example, the EU directive about access to legal representation in pre-trial proceedings has required Member

States to provide a right to access legal representation as soon as a suspicion based on reasonable grounds arises and a person is made aware of the accusations, whether he or she is in custody or not. It is arguable that the ECtHR could have made clearer statements about the right to access a lawyer prior to interrogation, whether taking place voluntarily or while a suspect is in custody. In addition, under the ICCPR, the need to be cautioned about the right to silence and the right to access a lawyer prior to interrogation could well do with being more precise.

This chapter has discussed the international development of the right to silence and the right against self-incrimination as a background to the functioning of those rights in the *national* settings of Denmark, England and Wales and Australia (examined in Part II: Chaps. 3, 4 and 5, respectively) and in *transnational* proceedings (see Part III: Chap. 6). The application of the rights in the national setting are likely to reflect the international development to a significant extent. Where this chapter has identified the international minimum standards that have been set under the human rights frameworks of the ICCPR and the ECHR by the HR Committee and the ECtHR respectively, the following chapters in Part II will consider the value of these sources in the specific national settings.

Further, looking at the operation of the right to silence and the right against self-incrimination from the *international* perspective, by examining the application of the rights in national law, the national chapters will, thereby, disclose whether the international minimum standards are actually being applied in the national normative systems. Arguably, as a result of the different enforcement mechanisms within each framework, it is more likely that the minimum standards set by the ECtHR will be applied uniformly. Political reporting based enforcement of legal obligations under the ICCPR may not be as persuasive as the combined legal-political enforcement under the ECHR, which is backed by the policies and the obligations created under the EU legislative framework. On this basis, it may be understandable why Council of Europe contracting parties, and more particularly those from the EU, have a high level of confidence that other states within the region share the same minimum guarantees for human rights protection. Thus, the uniformity of international and national law is relevant to the discussion about the principle of mutual trust in international cooperation in Part III, Chap. 7.

References

American Civil Liberties Union (2012) Frequently asked questions about the universal declaration of human rights. https://www.aclu.org/files/pdfs/humanrights/udhr60_faq.pdf. Accessed 13 Apr 2012

Barnard C, Peers S (eds) (2014) European Union Law. OUP, Oxford

British and Irish Legal Information Institute (2015) BAILII. http://www.bailii.org/. Accessed 15 Oct 2015

Cape E, Namoradze Z, Smith R, Spronken T (2010) Effective criminal defence in Europe. Intersentia, Antwerpen

Chalmers D, Davies G, Monti G (2014) European Union Law: text and materials, 3rd edn. CUP, Cambridge
Crawford J (2012) Brownlie's principles of public law, 8th edn. OUP, Oxford
D'Amico M (2010) Justice: Article 47 – Right to an effective remedy and to a fair trial. In: Mock W (ed) Human rights in Europe: commentary on the charter of fundamental rights of the European Union. Carolina Academic Press, Durham
Daniel B, Elholm T, Starup P, Steinicke M (2011) Grundlæggende EU-Ret: EU Efter Lisbontraktaten. DJØF Forlag, Copenhagen
De Schutter O (2010) International human rights law. CUP, Cambridge
Elo Rytter J (2013) Individets grundlæggende rettigheder. Karnov Group, Copenhagen
Emmerson B, Ashworth A, Macdonald A (eds) (2012) Human rights and criminal justice, 3rd edn. Sweet & Maxwell, London
European Commission (2011) Green Paper strengthening mutual trust in the European judicial area – a Green Paper on the application of EU criminal justice legislation in the field of detention. Available via EURLEX. http://eur-lex.europa.eu/LexUriServ/LexUriServ.do?uri=COM:2011:0327:FIN:en:PDF. Accessed 15 Jan 2014
European Court of Human Rights, Public Relations Unit (2012) ECHR: Overview 1959–2011. http://www.echr.coe.int/Documents/Overview_2011_ENG.pdf. Accessed 15 Oct 2015
Harris DJ, O'Boyle M, Bates EP, Buckley CM (eds) (2014) Harris, O'Boyle and Warbrick: law of the European convention on human rights, 2nd edn. OUP, Oxford
Jackson JD, Summers SJ (2012) The internationalisation of criminal evidence: beyond the common law and civil law traditions. CUP, Cambridge
Kälin W, Künzli J (2009) The law of international human rights protection. OUP, Oxford
Legal Experts Advisory Panel (2014) Stockholm's sunset: new horizons for justice in Europe. Available via Fair Trials International. https://www.fairtrials.org/wp-content/uploads/Stockholms-Sunset.pdf. Accessed 15 Apr 2014
Moeckli D, Shah S, Sivakumaran S (2010) International human rights law. OUP, Oxford
Office of the High Commissioner for Human Rights (2014) Human Rights Committee – Jurisprudence. http://www.ohchr.org/EN/HRBodies/CCPR/Pages/Jurisprudence.aspx. Accessed 15 Jan 2014
Spronken T (2010) EU-Wide Letter of Rights in Criminal Proceedings: Towards Best Practice. http://digitalarchive.maastrichtuniversity.nl/fedora/get/guid:2ba696f9-993e-42ca-922f-911a9e96e73b/ASSET1. Accessed 15 Jan 2014
Steiner HJ, Alston P, Goodman R (2007) International human rights in context: law, politics and morals, 3rd edn. OUP, Oxford
Trechsel S, Summers S (2006) Human rights in criminal proceedings. OUP, Oxford
United Nations (2015) Treaty collection, databases: status of treaties – chapter IV: human rights. http://treaties.un.org/pages/Treaties.aspx?id=4&subid=A&lang=en. Accessed 15 Oct 2015
United Nations Economic and Social Council (2004) Civil and political rights, including the question of torture and detention – report of the special Rapporteur on the question of torture, Theo Van Boven: Addendum – visit to Spain. UNDoc.E/CN.4/2004/56/Add.2. http://daccess-dds-ny.un.org/doc/UNDOC/GEN/G04/107/16/PDF/G0410716.pdf?OpenElement. Accessed 15 Jun 2012
Van Dijk P, Van Hoof F, Van Rijn A, Zwaak L (eds) (2006) Theory and practice of the European convention on human rights, 4th edn. Intersentia, Antwerpen
Van Hoecke M, Warrington M (1998) Legal cultures, legal paradigms and legal culture: towards a new model for comparative law. Int Comp Law Q 47:495–536
Wu W (2011) European perspective and standards on interrogational fairness. In: Cools M, De Ruyver B, Easton M, Pauwels L, Ponsaers P, Vander Beken T, Vender Lænen F, Vande Valle F, Verhage A, Vermeulen G, Vynckier G (eds) EU criminal justice, financial & economic crime: new perspectives. Governance of Security Research Paper Series, vol 5. Maklu, Antwerp

Part II
National Perspectives on the Right to Silence

Chapter 3
The Right to Silence in Denmark

Abstract The question of whether it is fair to gather confession evidence or evidence of silence in one legal system for use at trial in another commences with an examination of how the balance between effectiveness of law enforcement and fairness to the suspect or accused is maintained from the national perspective. This chapter, therefore, examines the domestic laws of Denmark that limit and protect the right to silence and the right against self-incrimination. This examination demonstrates how there is continuity between the investigative and trial phases in Danish criminal proceedings when viewed as a whole. Thus, this chapter forms part the basis for an overall comparative analysis, which explores the extent to which the national continuity of law becomes disjointed when evidence is transferred across borders. It is also the starting point to a later consideration of the influence of international human rights frameworks, such as the ICCPR and the ECHR, on national law and thereby on the transnational evidence gathering process-due to a lack of or the building of 'trust' between cooperating states.

3.1 Introduction

The Administration of Justice Act (*Retsplejeloven*)[1] is a key legislative instrument in Denmark protecting procedural fairness and balancing this with the need for law enforcement and the traditional focus in Danish law on seeking the truth. Together with certain other enactments, the provisions of the AJA are central in setting out the Danish national law that limits and protects the right to silence and the right against self-incrimination *throughout* the criminal justice process.

An important starting point for considering procedural rights in Danish criminal proceedings is the process of orally communicating to an individual that they are suspected on reasonable grounds of having committed an offence, known as *sigtelse*.[2] As discussed in Chap. 1, Sect. 1.3.5.6, this is equated with the process

[1]Abbreviated to AJA.

[2]For a fuller explanation of the term *sigtelse* see Chap. 1, Sect. 1.3.5.6 and Table 1.1 in Appendix section.

F.M.W. Billing, *The Right to Silence in Transnational Criminal Proceedings*,
DOI 10.1007/978-3-319-42034-9_3

of being 'charged' in other systems; and its meaning lies somewhere between being suspected and being formally accused. Reference here is again made to Table 1.1 in the Appendix section to Chap. 1, which demonstrates the significant conceptual differences in the formation of suspicion and the communication of accusation between the Danish and the English systems. These differences are important in relation to the timing of accruing pre-trial due process rights.[3] Due to the unique meaning of *sigtelse*, the Danish terminology surrounding this process will be maintained throughout.

This chapter will consider a number of provisions of the Administration of Justice Act (AJA) that provide the *sigtet* (charged person) with procedural protection in the form of the right to silence and the right against self-incrimination. The provisions of the AJA are in general to be read on the background of the *sigtet* person's constitutionally protected interests such as dignity, bodily integrity, freedom and privacy. However, the right to silence and the right against self-incrimination do not directly benefit from constitutional protection.

The purpose of a criminal investigation in Denmark is set out in § 743 of the AJA, namely to determine whether the circumstances exist for imposing criminal liability or some other criminal legal consequence. In furtherance of this goal, the AJA regulates the more intrusive investigative methods that may lawfully be used by the police; and the role that may be played by other actors, such as defence lawyers, the prosecution and the court.[4] Given that any police interrogation of a *sigtet* involves seeking statements (*forklaringer*) that are potentially self-incriminating (*belastende*), there are particular rules in the AJA that regulate such a method. A police interview is not considered to have the status of a coercive measure in Denmark. However, the protection of the right to silence and the right against self-incrimination in the AJA is clearly designed to avoid the situation where an interrogation becomes overly coercive.[5] This is especially so where a *sigtet* is at risk of being detained. Administrative authorities may also be involved in questioning a person who is regarded as *sigtet* (charged) in the pre-trial phase and this situation is also relevant to a discussion about the right to silence and the right against self-incrimination. The legislation not only determines the way in which the *sigtet* person's statements can be received in the investigation but also the way they can be used against him at trial (*til skade for tiltalt*).

[3]These conceptual differences about when certain procedural rights accrue can have significant influence in parallel investigations of transnational offending, for example, on the right to access information and the timing of disclosure of co-offender statements.

[4]The AJA does not regulate 'non-intrusive' investigative methods such as surveillance carried out in a public place or making routine or initial enquiries of the public. The work of the police is largely regulated by the Police Act (*Politiloven*).

[5]See, for example, §§ 750 and 752 AJA; a number of provisions in the AJA secure additional rights and safeguards for the benefit of people under 18 and, in particular, under 15 years of age. For example, according to § 750(2), the Ministry of Justice has the responsibility to make rules for the use of independent third persons in police interviews with *sigtet* persons under 18 years. Similarly, § 821c concerns the need for proportionality in relation to the detention of young persons.

An important question is whether these rules that span across the investigative and trial phases build upon and thereby mirror each other. Are they to be considered as interlocking provisions about the right to silence that establish continuity throughout the criminal justice proceedings? For example, the *sigtet* (charged person) or *tiltalt* (accused) may remain silent when asked questions by the police or given the opportunity to speak in court. Does the manner in which the *sigtet* person's silence during an investigation is used at trial on the question of guilt fit into a coherent scheme?

The AJA also covers the use of covert surveillance measures, such as telephone interceptions or listening devices, to obtain confession evidence from a *mistænkt* (suspect) or *sigtet* (charged person). Are such measures considered to be intrusions on the *sigtet* person's effective right to silence? The AJA establishes protective measures that ensure objectivity and proportionality in the way in which these measures are carried out. These rules are to be observed in conjunction with the focus on the court's role in getting to the truth and the manner in which the courts are able to consider all the relevant evidence on the question of guilt, namely through the principal of the *free evaluation of the evidence*. In general, when considering the way in which a *sigtet* person's statements that have been obtained by covert methods are to be used at trial, the emphasis is on 'legality' and proportionality rather than 'admissibility'.

In this chapter, therefore, various provisions in Danish law and the way they are applied in legal practice will be examined to identify how the *sigtet* person's right to silence and right against self-incrimination are simultaneously protected and limited throughout the whole of the criminal proceedings. The focus here is on provisions that may create *regulatory* compulsion, which may have a unique significance in transnational cases. In these cases, evidence not only crosses national borders but also transverses the investigative and trial phases of criminal proceedings. At the 30th Nordic Law Meeting in Oslo in 1984, a Danish Supreme Court judge, Jørgen Gersing, emphasised the importance of ensuring that the foundational idea behind the ECHR, that the *tiltalt* (accused) receives a fair trial, is always in the minds of investigators when they are developing new approaches to the investigation of organised crime.[6] The fact that organised crime, like other serious crimes, frequently takes place across national borders means that the police and prosecutors must find ways to investigate these types of crimes while ensuring that fair trial rights in the domestic law are not lost in the international cooperation process.

[6]Cited by Rørdam (1996), p. 283.

3.2 Human Rights Protection in Denmark

3.2.1 The Danish Constitution (Grundloven)

From 1661, Denmark had a sovereign king with unlimited power. However, following the revolutions in the late 1700s, particularly in the USA and France, liberal-democratic ideas came to Denmark and stirred discontent. In 1848, avoiding a bloody revolution, King Frederik VII created a constitutional law assembly to formulate a democratic constitution for Denmark. The King signed Denmark's constitution on 5 June 1849 incorporating, as a central element, the same broad rights as exist today.[7] When the Constitution was revised for the third time in 1953, the Universal Declaration of Human Rights (UDHR) was taken into consideration. However, constitutional rights were only amended on a few points and today they are considered as somewhat out of date.[8] The constitution refers to the protection of basic freedoms and rights such as personal freedom, private life, religious and political beliefs, association and assembly, property, social welfare and expression. While the right to a fair trial is not protected by the Constitution, the separation of powers, with an independent judiciary to interpret the law, is established in §§ 3 and 64; and the right to financial aid to try an administrative case is covered by § 63.

The Danish Constitution is the most authoritative source of individual rights in Denmark—constitutional rights take precedence over contradictory domestic, EU and international law. Furthermore, the legislature (*Folketinget*) must unconditionally respect constitutional rights.[9] However, in addition to the limited framework provided by the constitution, the Danish positivist tradition of legal interpretation has not encouraged effective protection of human rights by the judiciary. The judiciary is cautious and pragmatic in its interpretation of legislative provisions containing rights.[10] Danish legal tradition has been at odds with the development of human rights in Europe after 1945 and the dynamic and purposive interpretive styles of the ECtHR and the CJEU.[11] The relatively limited scope of human rights protection offered by the Danish Constitution (*Grundloven*) largely explains why the ECHR has a much greater practical significance in Danish law.[12] The ECHR is generally framed more broadly and is developed more deeply by dynamic interpretation in the ECtHR.[13]

[7]Elo Rytter (2013), pp. 29–30.

[8]Elo Rytter (2013), pp. 33 and 65–66.

[9]Elo Rytter (2013), p. 42.

[10]Elo Rytter (2013), p. 33.

[11]Elo Rytter (2013), p. 33.

[12]Elo Rytter (2013), p. 23.

[13]Elo Rytter (2013), p. 23.

3.2.2 The European Convention on Human Rights

Denmark was a part of the Council of Europe's 1950 European Convention for the Protection of Human Rights and Fundamental Freedoms (ECHR) from its beginnings and entry into force following ratification in 1953. The ECHR, therefore, has had (at least) the same binding effect as other international treaties since that time and should be applied in good faith.[14] This is confirmed in art. 1 of the ECHR, which declares that the high contracting parties shall secure convention rights for all people within their jurisdiction.

Denmark incorporated the ECHR into Danish law in 1992 following the first judgment of the ECtHR against Denmark in 1989.[15] Hereafter, the ECHR has been a part of Danish law; however, it has not been incorporated at the level of the Danish Constitution, which will always take precedence in the unlikely event of a conflict.[16]

Although the ECHR is considered to be part of Danish law, the value of the ECHR-block (*EMRK-komplekset*)[17] as a whole, including the ECHR *and* the jurisprudence of the ECtHR, particularly as it relates to Danish legislation and conflict of laws, has been questioned.[18] According to Report 1220 (*betænkning*) of 1991, under the ECHR a state is only obliged to enact or amend legislation in accordance with decisions of the ECtHR (or the Commission as it then was) to which it is a party. However, as the report highlights, a state will be strongly urged to also follow the decisions of the court that relate to third party states to avoid the situation of having a decision made against it in a like situation.[19] Certainly, in conjunction with incorporation, the Danish government has acknowledged the competence of the ECtHR's jurisprudence as binding in Denmark in relation to the 10 years leading up to incorporation.[20]

Nevertheless, the dualistic tradition in relation to interaction between national and international law has led to some controversy about the relevance of the ECtHR's evolving case law. According to a traditional dualistic approach, whereby national law and international law are two separate normative systems, the ECHR-block, as it had been incorporated in 1992 (including the ECtHR's case law up to that time), applies in Denmark with precedence over older Danish law.[21] At the same time, any ECtHR jurisprudence that has come after 1992 has not been

[14]See further arts. 26 and 27 of the 1969 Vienna Convention on the Law of Treaties.

[15]Law no. 285 of 29 April 1992; see further, Toftegaard Nielsen (2004), p. 24.

[16]See UfR 1995.672/2 R and *(ECtHR) Ninn-Hansen v Denmark (1999)* referred to in Elo Rytter (2013), pp. 50–51; but see Elo Rytter (2013), p. 51, where the author also explains the unlikelihood of a conflict between convention and Danish constitutional rights.

[17]See, for example, Vedsted-Hansen (2004), p. 382.

[18]Elo Rytter (2013), pp. 49 ff. Vedsted-Hansen (2004), p. 389.

[19]Inkorporeringsudvalget (1991), p. 196.

[20]Inkorporeringsudvalget 1220 (1991), p. 195.

[21]Elo Rytter (2013), p. 52; see also Inkorporeringsudvalget (1991), pp. 195–196.

included in the incorporation process and therefore has the status of international law, which can only apply in Danish law in accordance with two interpretive rules that apply to international law aimed at harmonisation. Firstly, Danish legislation and other law must be interpreted as consistently as possible with Denmark's international obligations, to avoid a conflict (the interpretation rule). Secondly, Courts and practitioners can presume that the legislature has not intended to violate existing international law obligations, such that a law made after the coming into effect of an international obligation may be overruled to the extent that conflicts with international law obligations (the presumption rule). According to these rules, legislation that comes into effect before the ECtHR establishes a principle of law under the ECHR will not necessarily be bound by it; however, legislation that is passed after the ECtHR establishes a legal principle may be overruled if it cannot be interpreted consistently with the ECtHR judgment. It is arguable that, due the ECtHR's dynamic interpretation, the court's jurisprudence will often be more recent than national legislation and, therefore, not necessarily applicable under the traditional dualistic approach.

As discussed in Chap. 1, Sect. 1.3.3.2.2 above, the reason behind considering the jurisprudence of the ECtHR as being a relevant source of Danish law in connection with the ECHR's incorporation is that the ECtHR uses an 'evolutive' style of interpretation, known as 'dynamic interpretation'.[22] General legal principles are being developed by the court in an ongoing and up-to-date way (akin to the use of common law 'precedent', though not under a formal binding doctrine). One of the main grounds for incorporation of the ECHR into Danish law was to provide a clear foundation for the *judiciary* to ensure the continued relevance of the ECHR as a source of law in Denmark, including where the legislature has not had the opportunity to consider new or altered ECtHR jurisprudence.[23] Thus, the ECHR's incorporation has set in motion a fundamental change in Danish legal and constitutional tradition, including a shift towards the interpretive role of the judiciary in the protection of individual rights.[24]

As a starting point, following incorporation, the ECHR takes precedence over conflicting Danish legislation, although it does not prevent a higher level of

[22]Vedsted-Hansen (2004), pp. 382–383.

[23]Inkorporeringsudvalget (2001), p. 34, referring to the comments to a Bill for an Act (L 230 af 19. february 1992, FT 1991/92, till. A, sp. 5421, afsnit 4.3) which repeated the Committee's findings in Inkorporeringsudvalget (1991), pp. 195–196.

[24]Elo Rytter (2013), p. 34; see further Christensen et al. (2012), p. 42; and see Christoffersen (2015), where the author reflects on the fact that Denmark, along with Norway and the United Kingdom, ratified a treaty that supposedly had already left too much power in the hands of a judiciary (ECtHR): U.2015B.115. But see Inkorporeringsudvalget (2014), p. 37, where the committee said that it was an assumption that incorporation of the ECHR would require ongoing evaluation of Danish law to ensure that the more precise principles developed by the ECtHR were being followed, although this process was not intended to change the balance of power between the legislature and the judiciary.

protection according to national law.[25] According to Elo Rytter and Vedsted-Hansen, the ECHR, taken together with the ECtHR's interpretation of it at any given time, must take precedence over Danish legislation in the event of a conflict.[26] This is regardless of whether the act came into force before or after incorporation of the ECHR; or whether the conflict results from ECtHR's jurisprudence, which is older or newer than the legislation in question.[27] Nevertheless, it should be borne in mind that the Danish judiciary have been conscious not to find a conflict between Danish legislation and the ECHR, unless the ECtHR jurisprudence lays a very clear foundation.[28]

Despite the dualistic approach to international law in general, there are good arguments stated above to say that allowing the ECtHR's case law that has been decided before 1992 to apply in Denmark while more recent decisions may not creates an artificial, impractical and overly formal situation.[29] Such a position does not take into account the unique interpretive style of the ECtHR, which plays an ongoing role in formulating general principles of law consistent with present day conditions under the ECHR. Therefore, the better position under Danish law is that the principles of law developed by the ECtHR may be considered to be applicable law in Denmark. Nevertheless, in connection with the enactment or amendment of legislation, the Danish legislature may indicate clearly that it does not intend to be bound by a certain rule of law under the ECHR-block.[30]

3.2.3 The EU Charter of Fundamental Rights

As discussed in Chap. 2, Sect. 2.3.1.4.4 above, fundamental human rights that are contained in the Charter of Fundamental Rights of the European Union and are considered general principles of law are binding upon EU institutions and the Member States when implementing EU law.[31] Therefore, fundamental rights have precedence over Danish legislation and international law obligations, in relation to the practice of EU law in the Danish national system.[32] However, while the EU Charter has an ever-increasing impact in Danish law in relation to

[25]See art. 53 ECHR; Elo Rytter (2013), p. 51.

[26]Elo Rytter (2013), p. 52; Vedsted-Hansen (2004), p. 384; Inkorporeringsudvalget (1991), pp. 192 and 197; but see Christensen et al. (2012), p. 42; see also Christoffersen (2015), p. 6, where he warns that human rights institutes should strive to ensure that human rights jurisprudence in the future, more so than to date, develops in such a way as to not alter the development of the law according to national democratic processes.

[27]Elo Rytter (2013), p. 51.

[28]Elo Rytter (2013), pp. 55–56.

[29]Elo Rytter (2013), p. 52.

[30]Elo Rytter (2013), pp. 52–53.

[31]See Art. 51, Charter of Fundamental Rights of the European Union OJ 2007 L303/01.

[32]Elo Rytter (2013), pp. 31–32 and 43–44.

EU areas of interest, the content of fundamental rights under the Charter is still essentially the same for equivalent rights as under the ECHR.[33] Indeed, under art. 52(3) of the Charter, though the ECHR is not formally binding on the EU, the scope of Charter rights is at least the same as the equivalent convention rights.[34] This is on the proviso that the Court of Justice of the EU (CJEU) does not amend the content of Convention rights in its decisions regarding equivalent rights. Therefore, it is arguable that the law of the ECHR, as interpreted by the ECtHR, may be regarded as the primary source of fundamental human rights law in Denmark, also in relation to legal practice concerning EU law and national law decisions. The Charter does, however, include a number of additional rights, including economic and social rights.[35]

3.2.4 Denmark's International Human Rights Obligations

In 1972, Denmark ratified the ICCPR and ICESCR.[36] These international human rights instruments, along with the 1948 Universal Declaration on Human Rights (UDHR) and many others, provide the backdrop to human rights protection in Europe and Denmark. However, Denmark has not incorporated the UN human rights treaties into Danish law, and they cannot be directly applied. Nevertheless, Danish law should be interpreted in accordance with the harmonisation interpretation rules stated above (see Sect. 3.2.2). It is through international instruments such as these that Denmark participates in international discourse on human rights obligations.

In 2014, the Committee on Incorporation in the Human Rights Area delivered its report, recommending that Denmark incorporate a number of the UN human rights conventions, including the ICCPR, in order to strengthen citizen's rights.[37]

---

[33]This may be changing, with a preference by the CJEU to consider the EU Charter as the primary source of fundamental human rights in EU legal practice: Craig and De Búrca (2015), pp. 386 (n 30) and 409.

[34]See, for example, Craig and De Búrca (2015), pp. 398–400 and 409.

[35]Elo Rytter (2013), pp. 31 and 34.

[36]Both of these instruments were presented for signature in New York on 19 December 1966. The ICCPR entered into force on 23 March 1976 and the ICESCR entered into force on 3 January 1976.

[37]Inkorporeringsudvalget (2014).

3.3 The Right to Silence and the Right Against Self-Incrimination

Danish law tends to focus on the right against self-incrimination, including where it overlaps with the right to silence (see the second overlapping circle of Fig. 1.1 in Chap 1, Sect. 1.2.2).[38] A number of Danish provisions and legal texts refer only to the right against self-incrimination, in the protection of the individual from abusive coercion. There are a number of theoretical and practical considerations that form the rationale for the right to silence and the right against self-incrimination in Danish law. These are primarily concerned with the principle that the prosecution bears the burden of proof in criminal proceedings, without resort to abusive coercion, and the significance of the court's role in seeking the truth.[39]

The historic development of the rights in Denmark began with the transition from inquisitorial to accusatorial procedure with the entry into force of the AJA on 1 October 1919.[40] Previously, various coercive measures were used to force the *sigtet* (charged person) to participate in the investigation with a legal obligation to tell the truth.[41] The obligation of the *sigtet* (charged person) to answer questions and to tell the truth was abolished by the introduction of reforms to the administration of justice in 1916.[42] Today, § 159(1) of the Criminal Code (CRC) provides that the *sigtet* may not be punished for making a false statement in official questioning.[43]

3.4 The Application of Fair Trial Rights in Pre-trial Proceedings

3.4.1 Sigtelse *(Charging)*

As referred to in Chap. 1, Sect. 1.3.5.6, *sigtelse* is the communication of criminal charges to a *mistænkt* (suspect), after which the person is called the *sigtet*. It is generally a verbal communication and is therefore a more informal process than the

[38] In criminal proceedings, Danish law might be said to describe the right against self-incrimination, including the right to silence, from the point of view of the second overlapping circle illustrated in Chap. 1, Fig. 1.1.

[39] But see Gammeltoft-Hansen (1989), p. 24, in which the author suggests that the *sigtet*'s right not to say anything to the police may be more a matter of principle rather than a practical benefit.

[40] Gammeltoft-Hansen (1989), p. 23; Lindencrone Petersen and Werlauff (2014), 614.

[41] This includes torture, which was abolished in 1837: Lindencrone Petersen and Werlauff (2014), p. 614; Knudsen (1996), p. 1; Frost (1968).

[42] Knudsen (1996), pp. 1–2.

[43] See further §159(2); see also § 158 AJA, which establishes the general rule that a witness cannot lie in court.

charging process known to common law. The point at which a *mistænkt* becomes a *sigtet* is crucial in terms of the rights and obligations possessed.[44] Following from *sigtelse*, fair trial rights include the presumption of innocence, the right to be clearly informed of the accusation against him, the right to silence, the right to legal representation and the right to disclosure. In addition, certain investigative and procedural steps, such as arrest, questioning and some coercive measures require that a person has been given the status of '*sigtet*'.[45] However, the AJA does not explain when a *mistænkt* becomes the *sigtet*, or when a *mistænkt* who is not yet *sigtet*, should be treated as such.

Paragraph 96(2) of the AJA provides that the objective of the prosecution (and the police) is to ensure that the guilty are investigated and prosecuted, but the innocent are not. At the outset, under § 742(2) of the AJA, the state authorities commence an investigation on the basis of a reasonable supposition that a criminal offence has been committed, which includes identifying one or more possible suspects (*mistænkte*). The *mistænkt* should be regarded as *sigtet* at the time when there is enough evidence pointing to a particular *mistænkt* to establish a reasonable suspicion, as the first step in prosecuting and pursuing punishment.[46] It is the first point at which the investigation becomes more focused, on objectively reasonable grounds, on one or more particular persons as the possible guilty party.[47]

For the purposes of art. 6 of the ECHR, in Denmark, the determination of a criminal charge (criminal proceedings) commences when a person is *sigtet* or may be regarded as *sigtet*. The suspect is often *sigtet* early in the criminal proceedings, when evidence gathering in the investigation is still underway. Therefore, the defence lawyer will often be in a position to follow the police investigation and even contribute to the overall fairness of the methods used from an early stage. In some cases a defence lawyer may even become involved before *sigtelse* to represent the individual interests of the *mistænkt* or other persons affected by coercive measures such as search and seizure or telephone interception.[48] *Sigtelse* can also happen independently of arrest, where the conditions of arrest are not present, for example, if the person does not represent a 'flight' risk. Here the person may be notified of the allegations (charges) against him, cautioned about the right to silence, advised of the right to access a lawyer and then questioned or invited to attend the police station for an interview. Some of these steps are described in more detail below.

---

[44]In relation to pre-trial disclosure of the evidence upon which the prosecution intends to rely at trial and pre-trial disclosure by the defence, see §§ 837–839 of the AJA.

[45]See, for example, U2005.2978/2 V where the *mistænkt* should have been regarded as *sigtet* and therefore should have been appointed legal representation, because he had been subject to a property search (*ransagning*), which is a procedural intrusion/coercive measure.

[46]Smith et al. (2008), pp. 184–185; see further Smith (2013), pp. 38–39.

[47]See further Smith et al. (2008), pp. 184–185.

[48]However, this is not necessarily the case; see further UfR 2012.272.H.

3.4.1.1 The Sigtet's Right to Access Legal Representation

As a 'party' in the criminal justice process, the *sigtet* has a number of rights under the AJA that accrue from the moment that charges are initially communicated. This includes the right to choose a defence lawyer under §§ 729 a(1) and 730(1).[49] Alternatively, a public defence lawyer will be appointed where 'necessary' under § 732 of the AJA, particularly if a person is *sigtet* in relation to more serious criminal offences.[50] In either case, the defence lawyer then takes on the rights of the *sigtet*. Amongst other things, this means that the defence lawyer receives a copy of the court book under § 729a(2), and disclosure of the material gathered in the police investigation under § 729a(3).[51] If the *sigtet* is unrepresented, then they receive disclosed documents in their own right.[52] The AJA includes provisions about the types of information or material that may not be disclosed to the *sigtet* person, or to neither the *sigtet* nor his legal representative.[53]

In addition to the right to legal representation (by a lawyer of his or her own choice) and the right to disclosure, the *sigtet* accrues a number of other rights under the AJA throughout the criminal justice process, including the right to silence during a police interrogation and, at trial, the right to be present in court, to be heard and to test the prosecution evidence.[54] Equality of arms (*ligestillingsprincippet*) is established through these fair trial rights.[55]

3.4.2 Arrest

Arrest can technically take place immediately prior to *sigtelse*. Under § 755(1) of the AJA, the conditions that are necessary for arrest are that there are reasonable grounds for suspicion that the person has committed a criminal offence and that the arrest is necessary to prevent further offending, to temporarily secure the person's

[49] Appointment of a public defence lawyer takes place in accordance with §§ 731 and 732 of the AJA. A person under 15 years of age cannot be punished (§ 15 of the Criminal Code) and is therefore not entitled to legal representation. However, a defence lawyer may be appointed pursuant to § 821e. See also § 831 AJA in relation to the appointment of a defence lawyer in cases where the accused pleads guilty.

[50] Smith et al. (2008), p. 275; Smith (2013), pp. 40–41.

[51] On the right to disclosure, see further Smith et al. (2008), pp. 211–224; and Smith (2013), pp. 36–37.

[52] See § 729b, AJA.

[53] Under the AJA, there are some documents that may be disclosed to the defence lawyer, which may not be disclosed or passed on to the *tiltalt* (accused). See § 729c in relation to non-disclosure due to national security, personal security, secret policing methods or other grounds. See further §§ 729a(4) and § 729b, AJA.

[54] See further, Smith et al. (2008), pp. 224–229.

[55] See further Smith (2013), pp. 242 ff.

presence, or to prevent his contact with others.[56] Therefore, in practice, when a person is arrested they will also be *sigtet*, because the level of suspicion that is necessary for both is the same. Under § 758(2), as soon as possible after arrest, the police must make the person aware that they have been *sigtet* in relation to a specified offence or offences. It must be borne in mind that a reasonable suspicion that the arrestee has committed a crime is a safeguard against arbitrary arrest and detention, in accordance with art. 5(1)(c) of the ECHR.[57] Within 24 hours after arrest, the police must take the *sigtet* before a court to determine release, continued arrest or detention on remand in accordance with §§ 760, 762 and 764, AJA.[58]

3.5 Police Questioning and the Right to Silence

3.5.1 Cautioning About the Right to Silence Prior to Questioning

Before a police interview can occur, under § 752(1) AJA, the *sigtet* must be explicitly advised of the allegation against him and told about told that he or she does not have to say anything. If a *sigtet* is not cautioned in accordance with § 752 (1), and is not told that he has the right to remain silent, then any information or statements that are made by him may not be used by the prosecution as evidence against him.[59] On the other hand, in the absence of testimony from the *tiltalt*, if the provisions of § 752(1) are followed prior to interrogation, the prosecution will not generally be prevented at a trial from producing the written record of an interrogation or testimony from the recording officer as evidence of what the *tiltalt* said to the police.[60] However, it appears from U.2003.903 V that the *sigtet*'s account recorded in the police report may not be produced if it is the only or decisive evidence against him at trial.[61] This is an exception to the principle that the evidence relating to the

[56] Arrest must be proportionate and therefore necessary and not just convenient: Feldtmann and Knop Reventlow (2013), p. 67.

[57] See art. 5(1)(c) ECHR and (ECtHR) *Fox, Campbell and Hartley v The United Kingdom* (Application no. 12244/86; 12245/86; 12383/86), 30 August 1990; see further Emmerson et al. (2012), p. 266.

[58] See § 731 of the AJA in relation to appointment of a public defence lawyer prior to a court hearing in relation to arrest.

[59] U.1997.336 V; see also Feldtmann and Knop Reventlow (2013), p. 87.

[60] U.2000.1309H; See further § 871(4) AJA in which production of the *tiltalt*'s statement to the police is exceptional; see also Gomard et al. (2013), p. 436; and see U.2000.1309H; see also § 871 (2)(2), which states that a previous statement of the *tiltalt* which forms part of the court book (being a previous in court statement) can be produced if the *tiltalt* remains silent at trial, or gives an inconsistent account at trial: Jochimsen (2001), p. 3.

[61] See further U.2000.1309H; Gammeltoft-Hansen (1989), p. 355; Jochimsen (2001), pp. 1–3; and Smith (2013), p. 183; see further Smith et al. (2008), p. 667, which refers to the rule in Denmark

question of guilt should usually be directly presented and tested by the defence counsel before the court, so that the court can assess the reliability of the evidence (*bevisumiddelbarhedsprincippet*). The president of the court should comment on the probative value of such a record when he is summing up the evidence prior to deliberations.[62] Unlike other jurisdictions, in Denmark, there is no immediate authority for the prosecution to produce evidence of the police interview. The court must exceptionally grant an order for this to occur. The above is a clear example of the continuity in Danish law between the investigation and the trial phase.

There are standard rules in the AJA that protect the *sigtet* from intrusions on the right to silence when being questioned outside the supervision of the court and ensure that the police and prosecution properly maintain the burden of proof. For example, in addition to the police being prohibited from forcing the *sigtet* to answer police questions, the police may not ask leading questions where something that is denied or not admitted is presumed to be admitted; they may not use threats, promises or false pretences; and they may not prolong an interview with the objective of obtaining a confession.[63] Similar protections apply to questioning in court.[64]

3.5.1.1 Using Undercover Agents to Conduct Police Questioning

The Danish police can use agents (*agenter*) in the course of investigative work. The term 'agent' has a specific meaning under the AJA. The work or activities of an agent (*agentvirksomhed*) is regulated by §§ 754(a) to 754(e) of the AJA and specifically refers to the situation where an undercover police officer encourages a suspect (within regulated limits) to commit or continue to commit a serious crime.

As referred to in Chap. 1, Sect. 1.3.4, the question of fairness in the context of an agent crossing a forbidden line and inciting crime is not directly relevant to the functioning of the right to silence in Danish law. What may be of interest in relation to the right to silence is the use of informers (*meddelere*)[65] and undercover agents to infiltrate the criminal environment (*infiltration*) and obtain information about the

---

that the record of a witness interview contained in the police report may not be used as the decisive evidence in a case, where the defence has not had the opportunity to cross-examine the witness, for example.

[62] U.2000.1309H.

[63] See §§ 750 and 752(3) and (4) AJA.

[64] § 754(1) AJA.

[65] Informers are usually understood to be civilians acting under an agreement with the police to provide information about others' criminality, but may also be police officers: Smith et al. (2008), pp. 410–412, referring to the Ålsgårde case, U.2004.1456H; see further Jochimsen (2003), pp. 155–156; see also Smith et al. (2008), pp. 412–413, about non-disclosure of the identity of an informer.

criminal activities of a particular person or persons by questioning them.[66] The right to silence of a suspect may be breached if a police agent effectively conducts an interrogation outside the protection of the AJA. Rørdam refers to the case of *Allan v The United Kingdom*, to say that the obligations that arise in relation to advising the *sigtet* or arrested persons about the right to silence, and the right to access a defence lawyer, cannot be circumvented by the use of an informant (or other police agent) who is, for example, planted in a cell with the *sigtet* for the purpose of enticing information out of him.[67] However, this issue does not appear to be canvassed in Danish courts.

Paragraph 751(3) of the AJA prohibits secret phonetic recordings to be made when conducting interviews with witnesses (including the *sigtet*). Thus, the provision may be seen as preventing an undercover agent from having a secretly recorded conversation with a suspect, particularly if the conversation takes on the form of an interrogation. However, where there are other covert surveillance tools readily available to gather evidence for trial, such as telephone intercepts, listening devices or covert searches, undercover police agents may not be used as frequently in Denmark as a tool to fight organised crime.[68]

3.5.2 The Sigtet *Person's Access to Legal Advice Prior to a Police Interview*

3.5.2.1 Access to Legal Advice to Ensure a Practical and Effective Defence

As discussed in Chap. 2, in order to comply with art. 6(3)(c) of the Convention, everyone who is charged with a criminal offence has the right to access a lawyer of their own choosing or, if they do not have sufficient means, to be given access to free legal assistance where the interests of justice so require. This right is crucial to ensuring equality of arms and a practical and effective defence under art. 6.[69] Particularly in the custodial setting, the right to consult with a lawyer protects against abusive coercion. In addition, access to a lawyer is fundamental to ensuring that the suspect fully understands the implications of the right to silence and the right against self-incrimination for future criminal proceedings.[70]

---

[66] See Jochimsen (2003), p. 105.

[67] Rørdam (2012), p. 60.

[68] Secret searches under § 799 are most relevant to organised crime investigations.

[69] See (ECtHR) *Artico* (Application no. 6694/74), decision of 13 May 1980, [33]; and *Imbrioscia v Swizerland* (Application no. 13972/88), 24 November 93, [38]. *Salduz v Turkey*, Application no. 36391/02, 27 November 2008 [55] *Pishchalnikov v Russia* (Application no. 7025/04), 24 September 2009 [68, 70, 91].

[70] See (ECtHR) *John Murray v. The United Kingdom*, Application no. 18731/91, 8 February 1996; *Salduz v Turkey*, Application no. 36391/02, 27 November 2008; *Pishchalnikov v Russia* (Application no. 7025/04), 24 September 2009.

In Denmark, the right to access legal assistance in accordance with art. 6(3) (c) and §§ 729 a(1) and 730(1) AJA the will accrue upon *sigtelse*. Therefore, a *sigtet* clearly has the right to access a lawyer of his or her own choice. However, appointment of a public defence lawyer happens in accordance with §§ 731–735 of the AJA. For the *sigtet*, who has not already chosen a defence lawyer or does not have the means to retain the services of a lawyer of choice, the primary situations when a public appointed defence lawyer is considered necessary and obligatory are listed in § 731. Under this paragraph, a public appointed defence lawyer is necessary, for example, when a *sigtet* who is being held in pre-trial detention appears before a court to determine whether he should be released or whether detention on remand should continue; or during the pre-trial examination of a witness or expert, whose evidence will be relied on at trial. Either of these steps will take place prior to the prosecution service filing an indictment against the *sigtet*. Pursuant to § 731, access to a public defence lawyer upon *sigtelse*, arrest, or prior to the first police interrogation, is not a right.[71]

Nevertheless, the provisions of § 732 allow the court to appoint a public defence lawyer in circumstances other than those listed in § 731, if it is considered desirable in the circumstances of the case. In serious cases, therefore, a public defence lawyer is more likely to be appointed early. Paragraph 732(2) obliges the police to advise the *sigtet*, in accordance with Ministry of Justice guidelines, of the possibility to obtain a public defence lawyer and to note this in the police report. In addition, the police should take the necessary steps to bring the question of appointment of a public defence lawyer before the court.

In 1978, under the delegated power in §732(2), the Ministry of Justice issued a binding ministerial order requiring the police to advise the *sigtet* how to request the court to appoint a public defence lawyer, to oversee that the request is placed before the court and to make the requisite notes in the police report under § 732, *in all cases involving offences, other than traffic offences, that carry punishment of more than a fine only*.[72] In addition, the 1978 ministerial order is supplemented by a circular published by the Ministry of Justice on 18 March 2010, in which it is made clear that the authority is bound to ensure that *every person under arrest* should be given the opportunity to contact a legal adviser who can function as a defence lawyer in accordance with the AJA, unless grounds exist to the contrary.[73] The ministerial order states that the police must advise the *sigtet* about accessing a

---

[71] See further Smith et al. (2008), pp. 274–275.

[72] See Ministry of Justice, *Bekendtgørelse om vejledning af sigtede om adgangen til at begære en forsvarer beskikket*, BEK nr. 467 af 26/09/1978, §§ 1 and 4. A ministerial order is a form of delegated power that may create obligations for the individual, as well as placing legal obligations on the authority: Blume (2010), pp. 120–121.

[73] Ministry of Justice, *Cirkulæreskrivelse* nr. 9155, 18 March 2010, § 3.1 and 3.2. A circular is a source of law that creates legal obligations on the part of the authority but is not binding on the citizen: Blume (2010), 121–122. Unlike some jurisdictions, there are no rules in Denmark which allow the police to delay access to a legal adviser in particular types of cases, such as terrorism.

public defence lawyer prior to questioning, in conjunction with issuing a caution about the right to remain silent under § 752(1) of the AJA.[74]

In the event that a *sigtet* wishes to speak to a publicly appointed defence lawyer before being interrogated, the police should delay interrogating. However, under § 4(2) of the ministerial order, if the police consider that the interrogation cannot be delayed, they may call for the presence of a lawyer under § 733(1) who will represent the *sigtet* until the decision about a formal appointment can be determined. In non-trivial cases, a defence lawyer will generally be called early at the request of the *sigtet*, in the expectation that an appointment will be formally made by the court at a later date under § 731, if not § 732.[75] In many cases where access to a lawyer is desired, it is the *sigtet* who initiates contact with a lawyer, who then contacts the court on behalf of the *sigtet* with a view to being appointed as the public defence lawyer in accordance with § 732.[76]

Police interviews with the *sigtet* often occur outside the police station and sometimes take place over the telephone, particularly in relation to less serious offences. In practice, every *sigtet* is offered the right to access a lawyer of their own choosing and the opportunity to access a publicly appointed defence lawyer. The police officer notes the *sigtet*'s response in relation to whether or not he wishes to obtain legal assistance prior to being interviewed in a template document for the police report.

3.5.2.2 Waiver of Access to Legal Assistance Prior to the First Custodial Interview

In accordance with the ECtHR's decision in *Pishchalnikov*, due to the fundamental significance of the right to consult with a lawyer to the effectiveness of other fair trial rights, the right to access a lawyer requires the special protection of the 'knowing and intelligent' waiver standard.[77] Following the opinion of the ECtHR, if a *sigtet* requests legal advice, additional safeguards should be in place.[78] Therefore, he should not be subjected to further interrogation until a defence lawyer arrives, unless he initiates further communication or conversations with the police or prosecution himself.[79] Paragraph 3 of the 1978 ministerial order provides that a request by a *sigtet* for the appointment of a public defence lawyer

---

[74]See Ministry of Justice, *Bekendtgørelse om vejledning af sigtede om adgangen til at begære en forsvarer beskikket*, BEK nr. 467 af 26/09/1978, § 2.

[75]Smith et al. (2008), pp. 275 and 282–283. In relation to the appointment of a public defence lawyer after an indictment has been filed, see Smith et al. (2008), pp. 277–280.

[76]Smith et al. (2008), p. 281; see also Smith (2013), pp. 40–41.

[77][ECtHR] *Pishchalnikov v Russia*, (Application no. 7025/04), 24 September 2009 [77–78]; see also *Trymbach v. Ukraine* (Application no. 44385/01), 12 January 2012, [64].

[78](ECtHR) *Pishchalnikov v Russia* (Application no. 7025/04), 24 September 2009, [78].

[79]*Pishchalnikov v Russia* (Application no. 7025/04), 24 September 2009, [79].

should not necessarily prevent the commencement or continuation of a police interrogation, if the *sigtet* is willing to speak to the police in the meantime. It is not entirely clear whether these two sources of law are consistent. If this question is raised before a Danish court which takes the view, as argued at Sect. 3.2.2 above, that the jurisprudence of the ECtHR is relevant as a source of law to judicial decisions in Denmark, then the court may be required to find that § 3 of the 1978 order must be read consistently with *Pishchalnikov*.

3.5.2.3 The Right to Access Free Legal Assistance

Under § 2(2) of the 1978 ministerial order, in connection with the obligation to advise the *sigtet* about the opportunity to access a public defence lawyer, the police must also make the *sigtet* aware that if found guilty he or she will be obliged to pay for the costs of the public defence lawyer in accordance with § 1008(1) of the AJA.[80] The 2010 circular states that the police may not use the requirement to advise the *sigtet* about the responsibility to pay costs to affect the *sigtet*'s evaluation of the need for the services of a defence lawyer.[81] However, this result may be unavoidable, without any deliberateness or dishonesty on the part the police.

In Denmark, after being advised of the right to silence and of the opportunity to contact a lawyer prior to being questioned, the *sigtet* frequently consents to being interrogated without delay. In less serious cases, where the police do not seek pre-trial detention of the *sigtet*, a defence lawyer will often be contacted late in the proceedings, even after the *sigtet* has been interviewed by the police (and sometimes more than once).[82] Even in cases of serious offending, if the *sigtet* does not choose to assert his right to a defence lawyer sooner, a lawyer may not be contacted until shortly prior to the remand hearing, after the police have already conducted the interrogation.[83]

It is reasonable to assume that the advice about reimbursing costs, in part, accounts for why so many *sigtet* persons are willing to be interrogated by the police without first speaking to a defence lawyer. For re-offenders, the knowledge of the obligation to pay back the state if found guilty in some cases dissuades a *sigtet* to benefit from the protection provided by a defence lawyer throughout the criminal proceedings.[84] Therefore, if a *sigtet* understands the advice about the obligation to reimburse the state for the defence lawyer's fees on conviction, he or she may

[80]If a person is found guilty, they are required to reimburse the state for the necessary costs incurred in the criminal proceedings, including the publicly funded defence lawyer's fees; see further § 1007 of the AJA.

[81]Ministry of Justice, *Cirkulæreskrivelse* nr. 9155, 18 March 2010, § 3.4. The circular also provides guidance in relation to *sigtet* persons who are under the influence of alcohol or who are under 18 years of age.

[82]Rørdam (2012), p. 59.

[83]Rørdam (2012), p. 59.

[84]See further Olesen (2013), pp. 142 and 149–150.

perceive an economic pressure to speak to the police without first seeking proper legal advice. As a money saving strategy, a *sigtet* may confess (even to crimes that he has not committed) in order to avoid the expense of a criminal trial.[85] The requirement to repay the state for defence costs on conviction may create an indirect pressure to participate in questioning without first seeking legal advice. In light of the ECtHR case-law about the fundamental importance of providing access to a lawyer prior to questioning, this is a pressure which arguably adversely affects the fairness of a future trial. This argument could be extended to include a perceived pressure to choose to defend oneself at trial to avoid the economic consequences of being convicted.

The ECtHR examined whether a scheme requiring a convicted person to repay their costs to the state was compliant with art. 6(1) and 6(3)(c) in the case of *Croissant v Germany*.[86] The court considered that such a system would not be compatible with art. 6 if it adversely affected the fairness of the proceedings.[87] Although the court did not find a violation of art. 6 in the circumstances of the case, no consideration was specifically given to the issue of the effect of the existence of the repayment system on the decision about whether or not to seek a defence lawyer prior to interview.[88] Likewise, this question was not raised in the admissibility decision of *Mortensen v. Denmark*, where the Commission found that the fairness of the proceedings had not been adversely affected because the applicant had been provided a proper defence at trial.[89] Neither of these decisions, therefore, has taken into account the significance of pre-trial procedures such as questioning to the overall fairness of the criminal proceedings. The findings in these two cases are arguably inconsistent with the judgments in cases such as *Salduz* and *Pishchalnikov* and lead to a watering down of the fair trial protection provided under § 732(2) of the AJA.[90] Uncertainty surrounds the question of how a Danish court might resolve a case if it was presented with this conflict; although, it is perhaps most likely that they would support the arguments closest to the Danish position in *Mortensen*.

By removing the requirement that a convicted person must repay the state for the costs involved with the attendance of a defence lawyer prior to the first custodial interview, the police would not be required to advise the *sigtet* in accordance with § 2(2) of the 1978 order. This would arguably eliminate the pressure that is felt by some *sigtet* persons not to seek legal advice prior to being interviewed by the police for the first time.

---

[85]Olesen (2013), p. 144; it is an offence for the *sigtet* to admit to crimes that he or she has not committed.

[86](ECtHR) *Croissant v Germany* (Application no. 13611/88) 25 September 1992.

[87](ECtHR) *Croissant v Germany* (Application no. 13611/88) 25 September 1992, [36]. The court emphasised that the German scheme allowed for the opportunity for debt relief based on the inability of the convicted person to pay: [21, 37–38].

[88]See further, (ECtHR) *Croissant v Germany* (Application no. 13611/88) 25 September 1992, [27–38].

[89](ECommHR) *Mortensen v Denmark* (Application no. 24867/94), 15 May 1996.

[90]See the cases about right to access a lawyer in Chap. 2, Sect. 2.3.2.2.2.

3.5.2.4 Interviewing a 'Mistænkt' with the Rights of a 'Sigtet'

When a person is explicitly or implicitly advised that they are *sigtet*, the rights of a *sigtet* accrue. However, although it may not be clear that the strength of suspicion against a person requires the Danish police to proceed with *sigtelse*, in certain circumstances the police will question a *mistænkt* with the rights of a *sigtet*.[91] This is designed to secure the future admissibility of any evidence obtained in the event a witness does end with being *sigtet*. Smith has given a number of examples, as follows: if a 'victim' reports the identity of a guilty person to the police, but it remains unclear whether what has happened is actually a crime, then that person should be questioned with the rights of a *sigtet*; if a crime has been committed but there is doubt as to who, amongst only a few possible suspects, is most likely the guilty person, then all of them should be questioned with the rights of a *sigtet*; if a victim reports a crime and is unsure of who is responsible, but identifies with some justification a particular person as the possible guilty party, then that person should be questioned with the rights of a *sigtet*.[92] Alternatively, in cases where the evidence of guilt is extremely vague, the police can question a *mistænkt* without *sigtelse* but with the right to silence.[93]

The right to silence contained in § 750 is possessed by anyone who is being questioned by the police, other than the requirement to provide a name and address. More specifically, the right of the *sigtet* to information about the accusation and the right to silence when being questioned under § 752, would pass to the *mistænkt* who is being questioned with the rights of a *sigtet*. Beyond this, however, the question of what other rights accrue becomes problematic, particularly in relation to access to a lawyer as a minimum fair trial right.[94] As discussed above, the right to choose a defence lawyer and the appointment of a public defence lawyer is currently dependent on *sigtelse* for offences punishable by imprisonment and arrest. It is arguable, however, that the right of a *sigtet* to access a defence lawyer prior to questioning should also pass to the *mistænkt* who is being questioned under § 752 (1). This would ensure that the use at trial of any incriminating statements inadvertently obtained would not infringe art. 6.

Nevertheless, the rule that requires the *mistænkt* to be *sigtet* where the police form reasonable grounds of suspicion, together with the rules that the *sigtet* should be cautioned and advised of the right to access a lawyer prior to questioning should be borne in mind. These rules point to the requirement that when a *mistænkt* who is being questioned with the rights of a *sigtet* (or any witness) makes self-

[91] See Feldtmann and Knop Reventlow (2013), p. 63.

[92] Smith et al. (2008), p. 186.

[93] Smith et al. (2008), p. 186.

[94] See U.1999.1744Ø; but see Gammeltoft-Hansen (1989), p. 12; and Elo Rytter (2013), pp. 472–473 who state that art. 6 of the ECHR applies to the *mistænkt* who is questioned with the rights of a *sigtet*; see further Smith et al. (2008), pp. 187–188; and Smith (2013), pp. 34 ff.

incriminating statements, the questioning should be discontinued until the individual is *sigtet*, cautioned and offered access to legal representation.

3.5.3 Written Record of Interview

In Denmark, the police are required by § 751(1) to make a written record of the essence of what is said in the police interview. As far as possible, they should record the important statements by using the exact wording. The written record should be completed as soon as possible (§744). Entering the written record into the 24 hour police report may be sufficient under Danish law.[95] In addition, under § 751(2) the *sigtet* must be given the opportunity to examine and comment on the contents of the written report, and the report should be amended accordingly. The *sigtet* is not obliged to sign the report. It has been suggested that often the *sigtet* does not check the report of the interview, either because they choose not to or because they have not been offered the opportunity.[96]

The Berlingske has reported that a majority of criminal lawyers and experts in Denmark would like to see the introduction of audio or video recordings of police interviews as a standard procedure, to avoid having the *sigtet*'s statements recorded incorrectly and to control the use of closed and leading questions, which can affect the reliability of the *sigtet* person's statements.[97] Clearly, an audio recording would increase the accuracy and thereby the reliability of the record of interview. An audio recording could also provide more reliable evidence of the general circumstances of an interview, including whether the police have cautioned the sigtet; advised him or her of his right to see a lawyer prior to questioning; and given appropriate opportunity for a lawyer to arrive or have ensured that the *sigtet* waived the right knowingly and willingly. An audio recording will also provide evidence of the general tone of an interview, which may not be ascertainable only from the words used. For these reasons, an audio recording is a safeguard of the right to silence that is currently missing from the Danish environment. The fact of an audio recording taking place may also in advance strengthen protection for a *sigtet* against the possible use of coercive questioning techniques in the absence of defence lawyers.

Under the present rules in Denmark, if the *tiltalt* chooses to remain silent at trial the prosecution may nevertheless be permitted to produce the written record in the police report as evidence of what the *tiltalt* said in the interview with the police.[98] This rule also applies if the *tiltalt* gives an inconsistent account at trial compared to

---

[95] Strafferetsudvalget (1971).

[96] Dam (2013). This newspaper article is used as a report of a public debate and not as a source of Danish law or evidence of legal culture.

[97] Dam (2013).

[98] See further 871(4) Administration of Justice Act.

what he or she said to the police. Therefore, the record of interview can lay the foundation for the argument that the *tiltalt* has changed his story and is lying—an argument that can have serious consequences for the *tiltalt* at trial. Therefore, the accuracy and reliability of the record of interview is paramount. What was said in the interview is quite often the subject of argumentation in Danish cases.[99] A mistake in the written police report about the interview is a significant matter.

However, representatives of the prosecution and police have suggested that converting to audio or visual recordings of interrogations would require great expenditure compared with the lower evidentiary value of such out-of-court recordings at trial.[100] However, this expenditure may be offset by the cost of the labour involved in recording an interview by hand and disputing its contents at court. Audio recordings also provide the opportunity of making multiple copies simultaneously.

An audio recording could be used in the same way as the written recording is presently used, primarily in the case of disagreement between the parties about what was said or where the *tiltalt* has given an inconsistent account in testimony. Nevertheless, saving an audio-recording of an interview will not entirely rule out the possibility of disputes arising about what was said in the interview—technical faults are possible and a recording may not always be audible, although audibility is likely given advances in technology.

3.6 Adverse Inferences of Guilt and the Right to Silence

3.6.1 Free Evaluation of Evidence on the Question of Guilt

In accordance with Danish legal tradition, deliberations on the question of guilt are based on the free evaluation of the evidence under § 880(2) of the AJA. Questions may be raised before the court on a case by case basis in relation to how evidence is gathered or whether and how particular evidence may be used at trial.[101] However, there are no formal rules that determine admissibility of evidence or the weight that should be attached to certain types of evidence. There is no distinction between direct evidence and circumstantial evidence in Danish law, and no rules that govern how these types of evidence may be used in deliberations. The objective examination of whether a *tiltalt* is guilty or innocent lies within the domain of judicial power, exercised in conjunction with lay judges or jurors. The overall purpose of the judgment of the evidence is to seek the truth.[102]

---

[99]Dam (2013).

[100]Dam (2013); see also Justitsministeriet (2010).

[101]See, for example, §§ 746 and 875 of the AJA.

[102]See further, §§ 867, 874, 884, 925 and 936 of the AJA.

Nevertheless, due to the influence of the jurisprudence of the ECtHR, the evaluation of evidence is no longer completely free. For example, some judges may make guiding comments to lay judges or members of the jury, prior to or during the course of deliberations, about possible issues in relation to unreliability of evidence.[103] For example, as a result of the decision in *Unterpertinger v Austria*,[104] a witness statement has less evidential value if the *tiltalt* or his defence lawyer has not had the opportunity to cross-examine the witness; and the *tiltalt* cannot be found guilty based on this type of evidence alone.[105] Beyond general judicial comment about questions relating to reliability, however, the members of the court may independently determine the weight to be attached to each piece of evidence.

3.6.2 Drawing Adverse Inferences from Silence of the Sigtet *or the* Tiltalt

The *sigtet* and the *tiltalt's* right not to say anything in the AJA is to be found in provisions such as § 752(1), 752(3), 754(1), 861 and 863. Gammeltoft-Hansen states that the centrality of the right to silence to the operation of the AJA involves the general idea that the *sigtet*'s right to silence cannot on its own be used as conclusive evidence of guilt.[106] However, an exception to general principle is found in § 897, which says that in cases that can be dealt with by fine only, if the *tiltalt* remains silent at trial then the court can resolve the case as if the *tiltalt* has pleaded guilty.[107]

There are consequences that follow from remaining silent; and there may be good reasons not to remain silent during an interrogation. For example, if the *sigtet* refuses to participate in an interview with the police, this may influence the court's decision under § 762(1)(3) in relation to remanding the *sigtet* in custody on a presumption that he will attempt to obstruct the prosecution.[108] Even though detention on remand may not be used to encourage a confession, the real possibility exists that by making a confession the *sigtet* may avoid arrest on this basis.[109] In addition, in a case that is found to call for a defence explanation, silence during the interview may also cause a problem in a claim for compensation for unjust

[103] Smith et al. (2008), p. 680; see further Smith (2013), pp. 188–189.

[104] (ECtHR) *Unterpertinger v Austria*, 9120/80, 24 November 1986.

[105] See U.1993.512 V and U.2002.15H.

[106] Gammeltoft-Hansen (1989), p. 23.

[107] See the original exception in § 934 of the AJA, as it applied prior to 1 January 2007, where in cases that were dealt with by fine only, if the *tiltalt* remained silent it was open to the court to make a finding of guilt without hearing further evidence in the case.

[108] Gammeltoft-Hansen (1989), pp. 23–24.

[109] Gammeltoft-Hansen (1989), p. 24.

detention on remand, if the *tiltalt* is later found innocent at trial or a prosecution is discontinued.[110] Furthermore, despite the centrality of the *sigtet* or *tiltalt's* right to silence in Danish law, through the operation of the principle of the court's free evaluation of evidence, it cannot be ruled out that the *tiltalt's* silence in the interrogation or at trial will not be used against him in the evaluation of all the evidence in the case.[111]

Nevertheless, as discussed in Chap. 2, Sect. 2.3.2.3, there may in fact be innocent reasons why a *sigtet* may want to remain silent. Rørdam has described it as inappropriate for the police to inform the *sigtet* that his interests are not served by remaining silent during the interrogation and that the *sigtet* should treat such advice with scepticism.[112] Regardless of the general truth of such a statement, or whether this is true in the circumstances of a particular case, having been told that he does not have to say anything, the *sigtet* should always be in a position to seek independent advice to solve the dilemma about whether he should say something or not.

In principle, the prosecution bears the burden of proof. However, in reality, the evidential circumstances of a case can effectively shift that burden to the *sigtet* or *tiltalt*.[113] For example, if the prosecution have proved a fact which naturally is in the domain of the *sigtet* to address, but he fails to do so, this may lead to an assumption of guilt.[114] The circumstances of the case may, on a practical level, clearly call for the *sigtet* or *tiltalt* to explain his proximity to the evidence, for example, because there is evidence that suggests his presence at the crime scene, his knowledge of the crime, or his possession of items related to the crime.[115] Even if the evidence cannot be said to 'clearly call for an explanation', in a system that is based on the free evaluation of the evidence, it can never be completely ruled out that the *tiltalt*'s silence will not be used against him.[116] In fact, in Denmark whether or not to speak out is considered to be one of the most important tactical issues to be contended with as a defence lawyer and there are various factors to consider.[117]

Although the *sigtet's* silence during the pre-trial phase is likely to weigh against him in the deliberation process, the knowledge of which may encourage some *sigtet*

---

[110]Rørdam (2012), p. 209; see further, § 1018a(3), U.2009.2287Ø and U.1995.426H.

[111]Gammeltoft-Hansen (1989), p. 23; see for example the judgment of the Glostrup City Court dated 4 June 2012 against four co-accused as an example of adverse inferences drawn against one of the accused as a result of remaining silent; see also Toftegaard Nielsen (2011), p. 71.

[112]Rørdam (2012), p. 65.

[113]There are other circumstances in which the burden of proof shifts to the *tiltalt* as a result of legal rules, for example, when proving the use of necessary force in self-defence.

[114]Smith et al. (2008), p. 202; see also U.1990.866Ø, where an adverse inference was drawn in relation to tax fraud, for not declaring the size of funds in Switzerland.

[115]See for example, TfS 1996, 313 where, in the absence of a better explanation, the Eastern High Court rejected the *tiltalt*'s explanation that he had come into possession of a large sum of money that belonged to a union. The *tiltalt* was convicted for not declaring his full income to the tax authority.

[116]Rørdam (2012), p. 208.

[117]Smith et al. (2008), p. 202; Rørdam (2012), p. 62.

persons to speak to the police, the drawing of common sense inferences is not in itself a factor which will affect the overall fairness of a trial.[118] In addition, the court is obliged to provide written reasons for a conviction under § 884(1) and the reasoning of the court is always subject to the safeguard of the appeal process. Nevertheless, the possibility of drawing adverse inferences in the national system does highlight the importance of access to independent legal advice prior to interrogation.[119]

3.6.3 Producing Evidence of the Interview When the Tiltalt *Remains Silent at Trial*

It is most common, in the Danish legal tradition, for the *tiltalt* to give evidence at trial.[120] This applies regardless of whether he or she remained silent during the investigation. The Danish courts want to hear directly from the *tiltalt* that he is innocent.[121] Therefore, where the preference is to hear the evidence first hand in court, evidence of what was said by the *tiltalt* in a police interrogation, such as the record of the interview or evidence from the recording or interviewing officer is not generally used as evidence on the question of guilt. However, if the *tiltalt* does remain silent at trial what is said during the interrogation may be raised as an exception to this rule. Under § 871 of the AJA, even if the *tiltalt* remains silent at trial, the prosecution will usually be able to produce evidence of the statements made by the *tiltalt* to the police. This situation came before the Danish Supreme Court in U.2000.1309H where the court rejected the defence complaint that allowing the prosecution to raise what the *tiltalt* said to the police, after the *tiltalt* had indicated an intention to remain silent at the trial was in violation of the right to silence because it put pressure on him to give evidence at trial.

Where it is only as an exception that evidence of what was said during the interview will be produced and used against the *tiltalt* at trial, it arguably follows that there is no right (or need) to be told that a statement made to investigators may be used in evidence.[122] However, it does highlight the possibility that the way a *sigtet* answers police interrogations, for example, by selective answering, may become part of the evidence against him at a trial. Assistance from legal representation prior to or during an interview may ensure that the *sigtet* participates in his own defence in the most effective way.

---

[118] *John Murray v. The United Kingdom*, Application no. 18731/91, 8 February 1996.

[119] (ECtHR) *Salduz v Turkey*, Application no. 36391/02, 27 November 2008 [52]; *Pishchalnikov v Russia* (Application no. 7025/04), 24 September 2009 [67].

[120] Rørdam (2012), pp. 207–208.

[121] Rørdam (2012), p. 208.

[122] See Feldtmann and Knop Reventlow (2013), p. 91.

3.6.4 *Drawing Adverse Inferences from the Lies of the* Sigtet *or* Tiltalt

There are other circumstances related to suspect evidence, in addition to silence, which may attract an inference of guilt by the court. If the *tiltalt* chooses to give evidence at trial and that evidence varies from an account he or she has given to the police, under § 871(4) the prosecution may exceptionally be given leave of the court to produce the written record of the police interrogation contained in the police report. In determining whether leave should be granted to the prosecution, the court must be satisfied that the *tiltalt* was properly cautioned that he did not have to say anything to the interrogating police in accordance with §752(1).[123] The *tiltalt*'s statement contained in the police report may not be used against him if he had not been cautioned about the right to silence.[124] Further, the statement in the police report cannot be the sole evidence used for conviction, where it has a lower probative value than statements made and tested in court.

There may also be 'innocent' reasons why a *sigtet* lies. Similar to silence, if the tiltalt has lied during the interrogation or at trial, this is a matter that is considered on the question of guilt, together with all the evidence, in accordance with the court's free evaluation of the evidence.

3.7 Administrative Questioning Powers

3.7.1 *The Use of Coercive Measures in Administrative Investigations*

In Denmark, it may be somewhat misleading to suggest that administrative compulsory questioning powers are used in relation to a *sigtet*. In Denmark, such powers should only be used by consent, with the right to silence, in the case of suspicion of criminal activity. This is a different set-up to the use of coercive regulatory powers, for example, in England or Australia, of the kind referred to in the ECtHR's case of *Saunders v The United Kingdom*.[125] However, the manner in which the Danish legislation provides for questioning of criminal suspects under administrative investigative powers is complex and may lead to a watering down of other important pre-trial protections provided for in criminal investigations that are related to the right to silence—most significantly, the right to access a lawyer.

---

[123] Smith et al. (2008), p. 203.

[124] U.1997.336 V.

[125] See the discussion in Chap. 2, Sect. 2.3.2.4.1.

Paragraph 1 of the Danish Coercive Measures Act 2004 (*Tvangsindgrebsloven*; hereinafter CMA)[126] provides guidelines for the use by state administrative authorities (such as tax, company and finance regulation, building industry regulation, workplace health and safety, food safety and product safety authorities) of coercive measures such as house and property searches, documentary examination and seizure and telephone interception.[127] The act provides for the use of coercive measures by the authorities outside the traditional criminal investigative regulation of the AJA. The power of an administrative authority to use these types of methods comes from the regulatory acts that control each individual authority. Among the coercive measures referred to by the CMA are compulsory information-gathering powers, which oblige a person to provide information to an authority (*oplysningspligter*).[128] An authority, such as the tax authority, may oblige persons in a certain category to provide annual information without further request; or compulsory information-gathering powers may include compulsory questioning powers, where the authority has the power to force a person to answer questions in relation to their administrative or business affairs. The obligation to provide information may be enforced by criminal punishment or by coercive fines.[129] Typically, the power to obtain information is used by the authorities in connection with compulsory access to company or business premises to conduct audits, without the authority of a court order such as a warrant. For example, § 6 of the Tax Control Act (*Skattekontrolloven*) allows the Danish tax authority, SKAT, to compulsorily obtain documentary as well as oral information during the course of an audit on company premises. Often the powers will be used to get information from a person in connection with their position held in a company.[130]

3.7.2 Compulsory Powers and the Right Against Self-Incrimination—§ 10

Chapter 4, § 10 CMA, deals with the right not to incriminate oneself in relation to compulsory information-gathering powers (*oplysningspligter)*, such as compulsory

[126]The 2004 *Tvangsindgrebsloven* is commonly known as the *Retssikkerhedsloven* (Procedural Fairness Law) and came into effect on 1 January 2005.

[127]See § 2 CMA and art. 8 ECHR in relation to the right to privacy and the operation of the principle of proportionality to house searches, search and seizure of documentation and other coercive measure, so that they may only be used where less intrusive measures are not sufficient and their use is reasonable in light of the goal of the measure.

[128]See the reference at § 1(3).

[129]But see SKAT (2013), 2.1.1.1, which states that although every citizen is obliged to annually provide information in relation to their personal tax liability, SKAT may not use coercion to obtain this information.

[130]Consideration of the problem of self-incrimination specifically by legal persons is beyond the ambit of the current discussion, which focuses on limitation of a suspect's right to silence and, ultimately, how such limitation affects the balance between effectiveness and fairness in transnational cases. For a discussion of the actual problem of self-incrimination and legal persons in relation to coercive measures in Danish law, see Toftegaard Nielsen (2011), p. 83.

questioning powers. The provisions apply in situations where a breach of the obligation to provide information is normally dealt with by a criminal sanction or where the authority usually has the power to issue administrative fines.[131] In relation to the coercive measures referred to in the act, other than compulsory questioning powers, such as interferences with secret communications, under § 9 (1), (2) and (3), if there are reasonable grounds to suspect that a person (including a legal person) has committed a crime then the use of the coercive measures provided for must generally proceed under the provisions in the AJA, but only insofar as they relate to allegations of criminal liability. However, the *mistænkt* of the criminal allegations can consent for the measures to be conducted under the CMA under § 9(4).

Under § 10(1) if there is a concrete suspicion that a person (including a legal person) has committed a criminal offence, compulsory powers obliging persons to provide information to an administrative authority do not apply to a *mistænkt* (suspect). However, the powers may be applied if it can be ruled out that the information sought does not relate to criminal liability for the suspected offending, or if a *mistænkt* consents to providing the information under § 10(3). Where a *mistænkt* has consented to making a statement, he or she is no longer being compelled to do so.

In accordance with § 10(2) the authorities can compel other persons to provide information, to the extent that the information sought relates to questions other than punishment for a criminal charge. Alternatively, following § 10(4), information against a criminal suspect may be compulsorily sought from another person, with the *mistænkt*'s consent.[132] Paragraph 10 has, therefore, been applied to enable the administrative authorities to maintain control over obtaining information under compulsory powers or during the audit process by consent, even after the authority has formed a concrete suspicion of criminal conduct.[133] Furthermore, an administrative investigation can continue contemporaneously with a parallel criminal investigation.

The consent provisions in § 10(3) essentially allow the *mistænkt* to agree to have criminal charges investigated in the absence of a broad range of defence rights and guarantees under the AJA that prevent an abuse of process. Consent means that any information obtained from the *mistænkt* by consent may be used against him for conviction.

Article 6 of the ECHR will not generally apply to the use of investigative measures under the CMA by administrative authorities who do not have the power to be involved in the determination of a criminal charge.[134] Although, some Danish authorities such as SKAT do have the power to conduct administrative

---

[131]SKAT (2013), 1.6.1; the provisions of § 10 also generally apply to compulsory powers to provide documents.

[132]See further, SKAT (2013), A.C. 1.6.4.

[133]SKAT (2007); TfS 2007, 199.

[134]Some administrative authorities may be capable of being involved in the determination of a criminal charge and will come under the regulation of the ECHR: (ECtHR) *Saunders v United Kingdom* (Application no. 19187/91), 17 December 1996, [67].

prosecutions in minor Significantly, if an interview of a *mistænkt* proceeds by consent under § 10 in relation to a *concrete suspicion* of criminal offences, where the *mistænkt* should be regarded as *sigtet*, then the protections under art. 6 ECHR must apply. Cases and are regulated by art. 6. In any event, if self-incriminating statements are obtained by an administrative authority and used in a criminal trial, then the application of art. 6 will be relevant. This will ensure the future admissibility of any confessional statements obtained under the compulsory powers. In this connection, the provisions of § 10, or the application of them, may be problematic. This may become an issue in a transnational case if the self-incriminating statements are to be used at trial in another system within the ECHR framework and they have been obtained, for example, in the absence of prior legal advice.

3.7.3 Ruling Out That Criminal Law Evidence Will Be Obtained

Where there is a concrete suspicion under § 10, if the authority cannot rule out that the information being sought will not be relevant to a criminal prosecution, then the regulatory investigation can only continue with the *mistænkt's* consent in sub-paragraph 3. Although the legislation does not expressly state what procedure is to be followed in order to rule out that the information sought has the potential to relate to an evaluation of criminal liability, this must be based on an assessment of the anticipated information before it is received, or as it is being received.[135] It may be possible to demand information from a *mistænkt*, which can have no bearing on the question of whether that person has committed a crime.[136] Alternatively, the question of criminal liability may be ruled out if the suspicious conduct is of such low seriousness that there is no question of prosecution of a criminal charge and criminal punishment.[137] However, assessing the potential evidentiary significance of information that will be or is being gathered is not always unproblematic. In a complex case, it may be difficult for the authority to identify allegations of criminal conduct. Sometimes 'innocent' information may be received that becomes more significant from a criminal evidence point of view later. For example, exculpatory comments or statements about non-controversial facts may be used by the prosecution at a later trial to attack a person's credibility as prior inconsistent statements.

[135]Similarly, the principle of proportionality should be applied to the use of a coercive measure to determine whether another, less intrusive, measure should occur either at the point of deciding to use the measure or during the running of the measure.

[136]SKAT (2013), 1.6.3.

[137]See note 52 to the consolidated act (Karnov); see further the ECtHR on the definition of a criminal charge in (ECtHR) *Engel and Others v The Netherlands* (Application no. 5100/71 and others), 8 June 1976.

In practice, the safeguard provisions in § 10 may not be sufficient to overcome the risk that a *mistænkt* will inadvertently slip into making self-incriminating statements, without having been fully advised about the allegations or the right to silence and the right against self-incrimination.[138]

3.7.4 What Is a Concrete Suspicion?

Under § 10 a case must be handed over to the police for investigation and prosecution under the AJA where there is a concrete suspicion ('konkret mistanke') that a person, or a legal person, has committed a criminal offence. However, where the authority is not of the opinion that the level of suspicion in a particular case is 'concrete' and they commence questioning under compulsory powers, if the person being questioned makes admissions in the absence of art. 6 safeguards then these statements may not be admissible in a future criminal trial. Therefore, the ongoing determination of the level of suspicion is important to protecting the future 'admissibility' of self-incriminating statements. This is particularly so if the evidence is to be used in a system with specific rules about access to legal advice prior to interrogation and the admissibility of confession evidence, such as England and Australia.

The distinction, if there is one, between a concrete suspicion in § 10 and reasonable grounds of suspicion ('med rimelig grund') in other paragraphs of the CMA[139] or the AJA is not obvious. Forming a concrete suspicion may mean that there are sufficient traces of evidence which provide objective and reasonable indications of the particular *mistænkt*'s possible guilt. It is the authority with the power to oblige persons to give information that makes the assessment of whether there are grounds for a concrete suspicion. However, as referred to at 3.5.2.4, determining the level of suspicion in a given case is often difficult. Cases have emerged in the ECtHR or national courts, which are centred upon the difficulty for criminal investigators to assess the level of suspicion, for example, when evaluating the need for cautioning about the right to silence or determining whether there is sufficient suspicion for arrest. It is often a decision that must be taken intuitively and on-the-spot.

Note 50 to the revised version of the act in *Karnov* suggests that a determination of whether there is a concrete suspicion will depend on an assessment of whether the authority's suspicion is so strong that either there would be grounds under the AJA for the police to inform the *mistænkt* of a distinct allegation being made against them (*sigtelse*), or, short of *sigtelse*, it would be necessary to question the

[138]Likewise, § 10(2) may ensure that the purpose of questioning third persons is not to obtain information that may relate to punishment, but cannot avoid the situation where a witness coincidentally gives this type of information.

[139]See, for example, § 9(1) CMA; see further § 755(1) AJA.

mistænkt with the rights of a *sigtet*.[140] In connection with questioning under § 10 (3) it is, therefore, clearly arguable that once the level of suspicion can be described as 'concrete', the *mistænkt* should be *sigtet*, regarded as *sigtet* or questioned with the rights of a *sigtet*. This analysis is consistent with the jurisprudence of the ECtHR.[141]

If there is a concrete suspicion of minor offences, the authority may be empowered to carry out *sigtelse* and pursue an administrative solution. For example, under the Tax Control Act and other regulatory legislation about good and services taxes, levies and duties, SKAT is authorised to prosecute minor fraud cases, involving amounts of up to 250,000 Danish kroner. However, in most cases the investigation should perhaps be handed over to the police for *sigtelse* and prosecution under the AJA.

3.7.5 Valid Consent to Provide Self-Incriminating Information—§ 10(3)

Despite the requirement in § 10(1) that an administrative authority must stop using compulsory powers to obtain information from a *mistænkt* as soon as a concrete suspicion arises, § 10(3) may give the administrative authority the opportunity to continue to ask questions and seek information by informed consent, even where it cannot be ruled out that the information sought will be relevant to the question of criminal liability. By the operation of § 10(1) and (3), an interview of the *mistænkt* conducted by consent, which can be withdrawn at any time, does not involve legal compulsion.[142]

Consent to providing information under § 10(3) of the CMA must be in writing, and must be provided on a voluntary, specific and informed basis.[143] In this connection, under the provision the *mistænkt* must be advised that he or she is not obliged to give any information regarding the suspected offences. There are no provisions to prevent any information obtained under § 10(3) from being used in the adjudication of criminal charges at a future trial. Therefore, the consent provisions must be applied consistently with the minimum standards under art. 6 of the ECHR, in relation to questioning suspects in the determination of a criminal charge.

Paragraph 10(3) in consistent with § 752 AJA and art. 6 to the extent that the *mistænkt* is guaranteed and advised of the right to be informed of the nature of the accusations and the right to remain silent and not to incriminate oneself. However,

---

[140]This note may provide evidence of the law, although it is not considered here to be a source of law as such.

[141]See, in particular, the decision of (ECtHR) *Aleksandr Zaichenko v. Russia* (Application no. 39660/02) 18 February 2010; see also *Weh v Austria*, Application no. 38544/97, 8 April 2004.

[142]See §§ 10(3) and 9(4).

[143]See § 9(4) CMA.

under the CMA, the *mistænkt* has no explicit right to access legal advice in connection with being questioned by the authority during an audit under § 10.[144] In practice, the provision of information by an administrative officer about obtaining legal assistance, including the public appointment of a defence lawyer, usually only occurs in connection with administrative prosecutions in less serious cases, after an audit and questioning has already occurred.[145] It would only be in exceptional cases where a defence lawyer may be appointed during an audit in response to a request from a *mistænkt*. However, there is a strong argument that if the *mistænkt* would be regarded as *sigtet* in criminal proceedings but *sigtelse* under the AJA is not the result of the concrete suspicion by the administrative authority, then, at the least, the *mistænkt* should be questioned by consent with the equivalent rights under AJA. This would include, under § 732(2) AJA, a right to legal representation for a person who is at risk of detention due to the seriousness of the offending, for example. In practice, under the AJA, every *sigtet* person is of advised of the right to access a lawyer (as referred to above).[146]

In practice in transnational cases, this deficit may be particularly problematic where self-incriminating statements are to be used as evidence in serious cases (in other legal systems that explicitly provide for legal representation prior to police questioning, even when the suspect is not in custody. Even though the *mistænkt* under § 10(3) is not being questioned in custody, there may nevertheless be a power imbalance between the state and the individual, particularly in investigations of serious offending where the *mistænkt's* liberty may ultimately be at stake. The complexity and implications of substantial and procedural rules may be incomprehensible to an individual being questioned under § 10(3)—particularly a suspect coming from another legal system.[147] In this connection, the ECtHR's judgment in *Aleksandr Zaichenko v. Russia* may arguably be distinguishable on the facts and limited to situations where the suspect the suspect is being questioned in public.[148]

If the consent of the *mistænkt* to being questioned under § 10 was prefaced by a right to seek legal representation, this does not exclude the possibility of an unequivocal, knowing and intelligent relinquishment of the right.[149] Here, the

---

[144]See further, Bjørn (2006). The procedural safeguards under Chapter 2 of the CMA and under the Administration Act (*Forvaltningsloven*) concerning *inter alia* freedom of information, provision of written reasons, notification procedures (including notification of the right to representation), complaints and reporting procedures only apply to house searches and search and seizure of documents: § 1 CMA.

[145]See, for example, § 1 of Bekendtgørelse af lov om adgang til forsvarerbistand under en administrative skatte- eller afgiftsstraffesag (forsvarerbistandsloven), LBK 827 of 25/08/2005.

[146]On the right in Denmark to free legal advice, see Sect. 5.2.3 above.

[147]See *Salduz v Turkey*, Application no. 36391/02, 27 November 2008 and *Weh v Austria*, Application no. 38544/97, 8 April 2004 discussed in Chap. 2, Sects. 2.3.2.2.2 and 2.3.2.4.1.

[148]*Aleksandr Zaichenko v. Russia*, Application no. 39660/02, 18 February 2010, discussed in Chap. 2, Sect. 2.3.2.2.1.

[149](ECtHR) *Pishchalnikov v Russia* (Application no. 7025/04), 24 September 2009 [77–78]; see Chap. 2, Sect. 2.3.2.2.2.

mistænkt would have to be specifically advised of the right to access a lawyer before indicating his or her consent to be questioned by the administrative authority. Consenting to be questioned in accordance with § 10(3), without being specifically made aware of the right to legal representation, may not constitute a valid waiver of the right under art. 6. Furthermore, it would arguably be insufficient to obtain the written acknowledgment that the *mistænkt* agrees in general terms to forego the procedural guarantees provided under the AJA.[150] In the absence of an appropriate waiver, the investigating authority would not be able to commence or continue questioning, unless by the *mistænkt's* own initiative.[151] In the case of compulsory powers to gather information that do not depend on the authority making a specific demand, but instead oblige persons (including legal persons) to send in information on an ongoing basis (for example annually), the authority may arguably be required to write to the *mistænkt* and advise him or her of the concrete suspicion, the right not to say anything in relation to criminal liability for the suspected offences and the right to access (free) legal assistance.[152]

3.7.6 The Obligation of Registered Vehicle Owners to Provide Information

There have been a number of Danish cases, which have examined the obligation of registered vehicle owners to provide information under § 65(1) of the Road Traffic Act (*Færdselsloven*, hereinafter RTA), to identify the driver of the vehicle if it is caught speeding. A registered owner is presumed to be able to inform the police, when requested, the identity of who was driving the vehicle at a given time.[153] It is not assumed that it is purely the work of the police to find this information. However, it appears from the cases that if it cannot be ruled out that the *tiltalt* was the driver of the vehicle at the relevant time, the right against self-incrimination applies to prevent him or her from being punished for refusing to identify who was driving.[154] Similarly, if the *tiltalt* will provide self-incriminating information by identifying the driver of a vehicle, for example, because the information implies that he knowingly lent the vehicle to an unregistered driver, then the right against self-incrimination will apply.[155] However, to avoid punishment under the compulsory information-gathering powers in § 65, it is not enough to argue that it may have

[150]See further (ECtHR) *Pishchalnikov v Russia* (Application no. 7025/04), 24 September 2009 [78–79]; but see note 47 to the CMA (Karnov).

[151](ECtHR) *Pishchalnikov v Russia* (Application no. 7025/04), 24 September 2009, [79].

[152]See note 54 to the CMA (Karnov).

[153]U.2003.1896 V.

[154]U.2002.2729 V.

[155]TfK 2004.760E.

been the *tiltalt*'s loved ones that were driving, and that providing the police with the requested information opens them up to conviction and punishment.[156]

3.7.7 Using Other Coercive Measures in Cases Where There Is a Concrete Suspicion of Criminal Offending—§ 9

If there is a concrete suspicion of criminal activity and an administrative authority does not have the consent of a *mistænkt* to continue an administrative investigation under compulsory information-gathering powers, the question arises whether it can use other coercive measures under the act, such as interception of private communication, to obtain the information being sought.

The general use of coercive measures under § 9(1) in cases where there are reasonable grounds to suspect the *mistænkt* of criminal offending, may only proceed under the AJA, if, under § 9(2), the measures are used with a view to gathering information in relation to questions other than determining criminal punishment. The Danish tax authority, SKAT, have interpreted this to mean that an authority may resort to using other coercive measures that have a 'reasonable administrative law purpose' (*fornuftigt forvaltningsretligt formål*).[157] Therefore, in the absence of consent to be questioned, SKAT will apply coercive measures, other than compulsory questioning powers, whenever this is reasonable and proportionate in relation to the purpose of obtaining information in order to make a correct assessment of tax liability. This will apply even in cases where there are reasonable grounds of suspicion (or a concrete suspicion) that a person has engaged in fraud, for example. In the opinion of the SKAT, the fact that it may come into possession of information that could also be used as evidence of a criminal offence is not relevant to the *implementation* of a coercive measure for the purpose of tax assessment.[158]

Other commentators have suggested that § 9(2) should be read to mean that § 9 (1) should always be followed, and coercive measures to gather evidence should always be implemented under the AJA, in cases where there are reasonable grounds of suspicion.[159] In accordance with the latter interpretation, any information received in a parallel administrative investigation can only ever amount to 'information' in the criminal law investigation. The question of whether to proceed under 9(1), in cases of suspected criminal offending, is then a tactical one between the

[156] U.2003.1029H; the Danish courts have not upheld the exception to the right against self-incrimination found by the ECtHR in relation to registered car owners and compulsory questioning powers.

[157] SKAT (2007); TfS 2007, 199; SKAT (2013), 1.6.1.

[158] SKAT (2007); TfS 2007, 199; SKAT (2013), 1.6.1. The same argument applies to the application of § 10(2) and obtaining information under compulsory powers from third persons.

[159] Bjørn (2006); enacting an explicit 'use immunity' for the benefit of the *mistænkt* under § 9 of the CMA may resolve the debate about the application of § 9(2).

police and the authority.[160] If information is gathered under an administrative investigation that is relevant to criminal proceedings, it cannot be used as evidence for conviction. Alternatively, the administrative authority can wait several years until after the completion of criminal proceedings and use the results as the basis for any administrative decisions—an option which may not necessarily be in the community's best interests.[161]

Under art. 6, the same arguments apply to § 9 as the ones that apply to § 10 of the CMA. Following *Funke v. France* and *Weh v. Austria*, the compulsory powers of an administrative authority should not be used to circumvent fair trial rights under art. 6 of someone who in reality is the subject of a criminal investigation. Therefore, in cases where there is a concrete suspicion, other compulsory measures, such as telephone interception, should not be used to obtain information which may incriminate the *mistænkt*. Alternatively, compulsory powers under the administrative investigation should allow the *mistænkt* the same procedural guarantees as provided to the sigtet under the AJA.

3.8 Covert Surveillance and Confessions

3.8.1 Invasions of Secret Communications

Privacy of communication is protected by § 72 of the Danish constitution (*Grundloven*) as well as art. 8 of the ECHR. However, under §§ 780–789 of the AJA, the police can lawfully interfere with a person's secret communications in the investigation of criminal offences, where use of the measure is proportionate.[162] Under the supervision of the court provided for in § 783 of the AJA, § 780 allows the police to invade the secrecy of communications of a *mistænkt* by a number of methods, such as telephone and postal interception (including e-mail), information about connections between phones and listening devices.[163]

[160]This type of tactical decision, about the extent to which parallel investigations could run in order to benefit from information gathered outside the AJA, might be occurring under the current law. However, there are cases that emphasise the view that it would be unlawful for the administrative authority to continue obtaining information under compulsory powers after a case has been transferred to the police and informal charges had been communicated to the relevant person: Toftegaard Nielsen (2004), p. 78.

[161]See further Smith et al. (2008), p. 210.

[162]See, in particular, § 782; these measures may be used in relation to offences carrying a penalty of at least 6 years imprisonment; on necessity and proportionality of police activities generally see Henricson (2012), pp. 206 ff.

[163]See further § 781(1) for the conditions required to implement a measure under § 780; see also § 791 b in relation to data copying.

3.8.2 Appointment of Intrusion Lawyers—§ 784 AJA

In the process of an investigation, to ensure that the measures encompassed by § 780 are used appropriately, § 784 of the AJA contains a safeguard by requiring the public appointment of a defence lawyer to represent the interests of the persons affected by the proposed measure, including the *mistænkt*. This lawyer is commonly known as an intrusion lawyer (*indgrebsadvokat*). The measure remains covert and the *mistænkt* is not made aware of an application before the court or the use of an approved measure. Even if the person who is the subject of the measure has already been appointed a public defence lawyer, an additional lawyer is appointed. Although, a defence lawyer, if there is one, should be made aware of the court hearing.[164] The intrusion lawyer may be barred in the future from representing the *sigtet* in the case.[165] For security reasons, intrusion lawyers are not used with all types of secret measures, for example, covert agents.

Many organised crime investigations will start with the use of measures that allow the police to hear the private conversations of the persons falling under suspicion. Therefore, the appointment of intrusion lawyers means that some form of defence representation, and thereby supervision of the fairness of the police investigation, is often in place from very early on.[166]

3.8.3 Urgent or Out of Time Measures—§§ 783(4) and 746(3) AJA

Paragraph 783(1) stipulates that using a coercive measure to invade private communication must occur under the supervision of the court, by way of a court order (similar to a warrant). If the police need to use a secret communication measure urgently they can implement the measure and thereafter place the matter before the court within 24 hours (§ 783(4)).[167] The court must then decide whether or not to approve the past and ongoing use of the measure.[168]

---

[164] See § 784(2).

[165] § 785(2).

[166] See also the safeguard provision in § 788 that requires notice of the measure to be given to the person concerned at the completion of the measure.

[167] § 806(3) AJA where the decision to conduct an urgent search (ransacking) must be examined by the court within 24 hours. If the court decides not to approve the search, the seized items, and any copies, shall be returned immediately: see the fourth note to § 806(3) of the act (Karnov); and see further U.2006.2895H; U.2005.3287V.

[168] See also § 783(2) which provides the court must send a letter to the Ministry of Justice if it finds that an intercept measure should not have been use in broad connection with a person in addition to specific telephone numbers.

If a judge finds that the police did not seek the court's approval of a measure before the expiration of 24 hours, disputes about the lawfulness of the measure can be dealt with under the general provisions of § 746(3). The same applies if the police did not return an intercept, listening device or similar warrant before the expiration of a time limit previously set by the court.[169] Consistent with § 783(4), the court must decide whether to uphold or to nullify the measure.

Beyond the requirement in §746(3) that the court should seek a statement from the police, neither § 783(4) nor § 746(3) provides the judge with guidelines as to how to determine whether or not to approve the measure, or what the relevant criteria are. Both provisions may require the court to make a retrospective examination of the merits of execution of a secret communication measure by an assessment of the conditions contained in §§ 781 and 782(2) and of proportionality in § 782(1).[170] When upholding a limitation on an individual's right to privacy, for example, by a technically unlawful telephone intercept, the court must consider that the measure is necessary in a democratic society to achieve the legitimate aim of law enforcement in the circumstances of the particular case.[171] Evaluating proportionality will involve a weighing and balancing of the various competing interests, including the seriousness of the breach of individual rights by the police, the seriousness of the offending and the importance of the evidence to the prosecution of the case.[172] It is very rare that a court will override the investigative decision use a measure against secret communication.[173]

In accordance with § 783(4), if the court finds that there were insufficient grounds for using a measure against secret communication, then a letter is sent to the Ministry of Justice.[174] Significantly, however, if the court decides not to approve or to overturn a measure, neither § 783(4) nor § 746(3) sets out what will happen to any evidence that has been 'unlawfully' obtained.

---

[169] See further U.1972.164V.

[170] In relation to proportionality, see for example Tfk 2008 209 V relating to an unlawful but proportionate seizure under the similar provisions of § 806(3) AJA; see also Gomard et al. (2013), pp. 314–315; see further the following provisions of AJA which directly engage the principle of proportionality: §§ 755(4), 762(3), 770b, 782, 791a(5), 791b(2), 791c(2), 792e(1), 797, 805(1), 816, 817(4), 818(3), 819(3), 821c; and see also the following provisions which indirectly observe proportionality: §§ 765, 805(2) and 821a(1).

[171] See further, Gomard et al. (2013), pp. 165 and 252.

[172] See Smith et al. (2008), p. 31.

[173] Smith (2013), p. 93.

[174] Similarly, a letter should be sent by the court to the Ministry of Justice in relation to the unlawful use of an *agent provocateur* or an emergency controlled delivery under § 754c(3) or secret searches under § 799(2).

3.8.4 Evidence Obtained Coincidentally—§789 AJA

Paragraph 789 is of narrow scope and concerns the random discovery of evidence during the use of a secret communication measure to gather evidence about an unrelated offence. This paragraph is most relevant if the police locate evidence that is not related to the implementation of the measure, and which could not have been the basis of implementation due to the fact that the offence is punishable by a term of imprisonment of less than 6 years.[175] Under § 789(1), during the use of a coercive measure against secret communication, if the police obtain information about an offence that was not the basis and could not have been the basis for (lawfully) implementing the measure, that information can nevertheless be used in the course of an investigation. In accordance with ØLK-1993-06-01,[176] the police can use the information obtained as a result of an unlawful intrusion into secret communication to guide the investigation and indirectly obtain other (derivative) evidence, which can be used as evidence in criminal proceedings.[177] However, in accordance with § 789(2), such information cannot be used directly as evidence in criminal proceedings—such evidence (of an offence which is unrelated to the measure and is not punishable by at least 6 years) is *prima facie* excluded.[178] The admissibility of evidence under § 789(3) is discussed in detail at Sect. 3.9.3 below.

3.9 The Principle of Material Truth and Admissibility of Confession Evidence

3.9.1 The Principle of Material Truth and the Administration of Justice Act

In Danish criminal proceedings, like in most democratic systems, the community's moral right to sanction wrongdoers is pursued by the prosecution.[179] In order to ensure that this process does not arrive at delivering criminal sanctions against the innocent, at the same time as ensuring that the guilty are made to account, the Danish courts are bestowed with the primary function of getting to the truth of a

[175]It also covers the coincidental gathering of evidence for an offence punishable by less than 6 years that involves endangerment of human welfare or life (in relation to communication by methods other than telephone or other telecommunications).

[176]See FM 1993 154.

[177]§ 789(1) therefore is an acceptance of the legitimacy of using what is known as *fruits of the poisonous tree*.

[178]See also § 800(1) and (2) in which evidence randomly obtained in a property search for an unrelated offence may only be used as information in an investigation, or may be used as criminal law evidence if certain conditions apply (similar to the conditions in § 789(3)).

[179]Smith et al. (2008), p. 26.

matter, and this is described as the principle of material truth (*den matrielle sandheds princip*).[180]

Paragraphs 837 and 839 AJA, requiring pre-trial disclosure by the prosecution and the defence (to the court and to each other), support the principle of material truth. Of significance in the running of the trial, certain provisions of the AJA carve out a function that obliges the court to ensure that all the relevant evidence is presented. Under § 867, the court's president is empowered and obliged to ask questions of witnesses when the interests of truth so requires. Furthermore, § 874 (4) provides that the court can conduct examinations of people, objects or locations where necessary and desirable. Paragraph 874(3) provides the court with the power to order that particular evidence must be presented in court, where necessary to be fully informed. This paragraph holds even if neither the prosecution nor the defence wish for the evidence to be admitted. It is a power that is rarely used and is generally executed in the *tiltalt*'s favour—due to the general understanding that the court is not required to relieve the prosecution of the burden of proof.[181] Other examples of rules that support the principle of material truth are § 873(1) of the AJA, which provides for a prior inconsistent statement to be read from the court's book if a *tiltalt*, witness or expert changes his or her testimony; and § 936, which allows for the rehearing of witnesses in the Danish High Court when necessary for the court's information.

This strong focus on finding the truth is considered to be one of the most important fair trial guarantees.[182] The principle of material truth is usually understood as a defence right, where the court has an obligation to ensure that all the evidence in a case is brought forward, including evidence of innocence.[183] For example, paragraph 746(1) may be used by the defence during the investigation to formally request the police to make further enquiries or take additional investigative steps. This is a clear example of the principle of material truth working to promote the right to equality of arms. However, it is obvious that a rule which calls for as much relevant information about the charges as possible, will work in two directions—both in favour of the prosecution and in favour of the defence. In principle, seeking the material truth only serves to protect the innocent *tiltalt*, and provides for the conviction of the guilty.[184]

In a system which places strong weight on the principle of material truth, less importance attaches to formality and formal rules.[185] For example, under § 925, if a particular rule of court procedure is not followed or is used incorrectly, a subsequent judgment will not be overturned and returned to a lower court unless,

[180]See further Smith et al. (2008), pp. 26 ff. Smith (2013), pp. 16 ff.

[181]Smith et al. (2008), p. 28; Smith (2013), p. 18; see further § 96, U.1968.68H and U.1976.236.Ø.

[182]Smith et al. (2008), p. 26.

[183]Smith et al. (2008), p. 27; Smith (2013), p. 17.

[184]Smith et al. (2008), p. 27, referring to U.1979.912 V; Smith (2013), p. 17.

[185]Smith et al. (2008), p. 29; Smith (2013), pp. 18–20.

exceptionally, a different result would have followed had the rule been complied with.[186]

The focus on the truth rather than formality is apparent in cases that involve unlawfully obtained evidence. Often the courts will hear evidence, despite a lack of compliance with the evidence-gathering rules under the AJA. On the other hand, cases that have typically resulted in a judgment being overturned include cases where judicial bias is objectively proved, where a *tiltalt* has lacked legal representation (and advice about the right to legal representation) or lacked adequate services of an interpreter.[187] Therefore, it appears that formal rules which are central to fairness and equality of arms under art. 6 of the ECHR are generally being followed.

The principle of material truth in Danish criminal procedure, in so far as it is permissive of the use of unlawfully obtained evidence or evidence that involves inherent dangers, is supported and counter-balanced by another extremely important fair trial principle stated in § 96(2) of the AJA—the principle of objectivity. The prosecution must remain objective in their assessment of the evidence and in taking the decision that a criminal prosecution should be pursued.[188] In the preparation of the indictment, the prosecution is required to consider all the alleged facts in the case, including those that point to innocence. The principle of objectivity means that criminal proceedings should not be pursued where there is no prospect of a conviction.[189]

In addition, through the close working relationship between the prosecutors and the police in Denmark, § 96(2) implicitly requires the police to be objective in the discharge of their investigative duties.[190] Although, it is commonly acknowledged that it is not possible for an investigator to be completely objective in their approach to solving a case.[191] Smith has suggested that § 92(2) places the prosecutor and the police investigator in a similar role as the inquisitorial judge who must investigate a case as well as come with suggestions as to its resolution.[192] With a binding focus on the objectivity principle in the investigation of a case and in the preparation of an indictment, the problem of the admissibility of unlawfully or unfairly obtained evidence may not arise so frequently.

Once a case reaches the trial stage, however, the objectivity of the prosecution cannot be equated with that of the court or the trial judge.[193] The accusatorial nature

---

[186] See, for example, U.1991.255H and U.2005.753H.

[187] See further, FM 1989 17; U.1992.300H; U.1996.6 V; U.1996.699 V; U.1996.1384 V; U.1996.1434/2Ø; U.1996.1488 V; U.1997.1422/1 V; TfK 2003.435/1Ø; TfK 2008.710Ø.

[188] Smith et al. (2008), p. 27 and see 132–135; Smith (2013), p. 17.

[189] § 721(1)(2) AJA. In very limited situations, the prosecution authority may find a case so uncertain that it is necessary to indict a person and pursue a court resolution on the question of guilt: Smith et al. (2008), p. 135; this situation is similar in common law systems, where such a case may be pursued as a "jury question."

[190] Smith et al. (2008), p. 27; Smith (2013), p. 17.

[191] Smith et al. (2008), p. 27; Smith (2013), p. 17.

[192] Smith et al. (2008), p. 27; Smith (2013), p. 17.

[193] Smith et al. (2008), pp. 27–28.

of the Danish criminal justice system is clearly felt in the bipartisan trial which is heard by an objective court. Once the decision has been made to proceed on indictment, and provided that the probative value of the evidence does not change in the course of the trial, the prosecution's role is to present the evidence that supports a finding of guilt. At trial, the *tiltalt* is armed with certain rights, including the right to legal representation. A defence lawyer presents the case from the *tiltalt*'s point of view and has the right to challenge the prosecution evidence, including cross-examination of prosecution witnesses. Not only is the pursuit of material truth the primary procedural focus, but the accusatorial trial also gives rise to an expectation that the truth will emerge through the clash of opposing parties.

In the Danish system, the interests of the *mistænkt* or *sigtet* may be represented by a defence lawyer from very early on in the process. For example, a large fraud investigation may begin with a search of premises for documentary evidence and a defence lawyer may be involved in evaluating the police investigation from that early stage. Alternatively, individual interests, including the interests of the *mistænkt* or *sigtet*, may also be protected from an early stage in an investigation of a large narcotics case by the use of court appointed 'intrusion' lawyers (*indgrebsadvokater*) to assess the proportionality of an application to the court for the use of secret coercive measures. Therefore, the use of defence lawyers or intrusion lawyers in the investigative phase provide another layer of protection of the fairness of criminal proceedings that are in addition to the requirement that coercive measures must have the court's approval and supervision. These types of safeguards may limit the scope of the problem of admissibility of unlawfully obtained evidence, even though the principle of material truth is inclusive by nature.

3.9.2 Judicial Decisions About the Legality of Investigative Measures—§ 746(1) AJA

If a *sigtet* or *tiltalt* wishes to raise an issue about the legality of an investigative step, the matter may be brought before the court under § 746(1). Disputes may arise, for example, in relation to the covert recording or interception of private communications or pre-trial questioning of a *mistænkt* or *sigtet*. There is a tendency to narrowly construe the judge's power to settle a dispute about the legality of a measure. The primary consideration is whether the rules in the AJA were followed. Generally, the court does not settle the dispute by a wider examination of criteria such as the appropriateness of a measure.[194] However, if the dispute concerns the lawfulness of

[194]See, for example, U.2001.661 about the legality of putting together a photo folder for photo identification, where the court did not accept the defence lawyer's protest about the manner in which it was put together.

an investigative measure that normally requires court supervision under provisions that include discretionary powers, then the court may also consider the appropriateness of the measure in the circumstances of the case.[195]

Under § 746(1) a dispute about the legality of investigative measures may be raised during the course of the investigation, or after the investigation is completed[196]; and the issue may be raised by the defence or brought before the court by the police or the prosecution (for guidance). For example, in U.2005.3007 V the court examined the question of whether photographs captured during an unlawful search of an apartment should be excluded from an ongoing investigation. The court found that § 746(1) could not be used to require the police to destroy them. It was further decided that it would be up to the trial court to determine the extent to which the photographs could be relied on as part of the prosecution case. This would encompass both the questions of whether the photographs could be relied on at all, and what weight should be attached to them.

3.9.3 Admissibility of Accidentally Discovered Evidence Under § 789(3) AJA

Under § 789 of the AJA, evidence of an offence which is punishable by less than 6 years that is randomly discovered during the implementation of an invasion of secret communications measure in relation to an unrelated offence may be used as information in an investigation but is *prima facie* inadmissible as evidence in the court proceedings. However, under § 789(3) the court has a discretion to expressly override the prohibition in sub-paragraph (2). The court may allow the prosecution to rely on unlawfully obtained secret communication evidence where: other investigative steps will not be suitable to secure evidence in the case; the offence is punishable by imprisonment of 18 months or more; and the court otherwise finds it appropriate to admit the evidence. In this way the provisions in § 789 ensure that the prosecution may only benefit from the unlawfully obtained evidence in sanctioned circumstances in more serious cases.

Under the first ground for allowing randomly but 'unlawfully' located evidence to be used by the prosecution, it must be shown that the police took the necessary steps to try and obtain evidence by other means.[197] The second ground requires the court to ensure that the offending is sufficiently serious so that it is fair ('*rimeligt*') to use the evidence. The third ground for allowing the unlawful evidence requires the court to consider the fairness ('*rimeligheden*') of using the evidence, with reference to the circumstances in which the warrant was originally obtained. All these conditions were met in U.2001.945Ø, in which the court found that it was

[195] U.1965.410H.

[196] U.1986.383.

[197] U.2001.945Ø; U.2010.1250Ø.

appropriate to use listening device evidence located in a narcotics case against a person accused of fraud in a position of trust, attempted theft and breaching weapon laws.

Under § 789(3)(2), in addition to the evidence that was located by chance, the court will *inter alia* examine whether evidence was found in relation to the offending which was the basis for the warrant. In such cases, it is often considered appropriate to admit the evidence that was discovered by chance.[198]

3.9.4 Judicial Discretion to Exclude Evidence

Paragraph 875 of the AJA, which provides that a disagreement between the parties about the use of evidence is to be settled by the trial court with a formal decision (*kendelse*), implies that the court has discretion to determine the admissibility of evidence. Principally, however, § 875 indicates that a decision under this paragraph (by a court other than the trial court) does not constitute a final resolution in the proceedings.

As discussed above, 789(3) explicitly provides a judicial power to override mandatory exclusion in sub-paragraph (2) and admit evidence that has been randomly uncovered by the police while executing a warrant permitting an invasion of secret communications in relation to unrelated matters. This is the only provision in the AJA that explicitly provides for a judicial discretion to admit or exclude unlawfully obtained evidence from a criminal trial.

Paragraph 746(1) requires a court to determine whether an investigative measure is legal or illegal. Similarly, §§ 746(3) and § 783(4) call for a determination by the court whether to approve (or uphold) or overturn a particular investigative measure, which may not have been 'lawful' because it did not have the requisite supervisory order from the court. However, if a measure is overturned, these provisions are silent as to what should happen to any unlawfully obtained evidence. Therefore, acceptance of unlawfully obtained evidence remains open, which clearly illustrates a legislative preference for truth over formality, even in the more controversial area of coercive measures. This silence may presuppose that any evidence gathered, even under a technically unlawful measure, is generally admissible.

In 2001, Peter Rørdam suggested that from the mid-1990s judicial practice in relation to unlawfully obtained evidence developed to deal with the dilemma facing Danish courts about how to balance the competing interests between prosecuting crime (and making sure that wrongdoers are not acquitted for purely technical reasons) and securing convictions in accordance with the rule of law.[199] Rørdam suggested that there are four guiding principles (*sætninger*) that have surfaced[200]:

[198]U.2001.945Ø.

[199]Rørdam (2001), *Mangler ved bevisets tilvejebringelse*, U.2001B.337, 338.

[200]Rørdam (2001), p. 338; see also Brøbech (2003), pp. 401–402.

I. Evidence must be excluded if there is a significant possibility that using the evidence will result in a miscarriage of justice[201];
II. On the other hand, if the methods used to gather the evidence were unlawful for purely technical reasons and the reliability of the evidence remains intact, then the evidence will, as a rule, be admitted;
III. Evidence may be excluded, if the unlawful methods used were deliberate or planned, even if the evidence is reliable[202]; and
IV. If the evidence has been obtained contrary to the prohibition on self-incrimination then the evidence should, as a rule, be excluded.

In a PhD thesis from 2003, with some uncertainty, Brøbech has broadened the scope of these guiding principles referred to above. In relation to the last criterion, Brøbech reported a practice of exclusion of evidence that has been obtained in violation of any fundamental procedural protection, referring, for example, to cases of entrapment and non-compellability of family members (not without exception).[203] Brøbech has added a fifth criterion, that if the unlawfully obtained evidence is favourable to the accused, as a rule, the evidence should be admitted—an example of the rule *favor defensionis*.[204]

In addition, Brøbech highlighted that in cases where the error complained of is purely technical, practice shows that the significance of the evidence to the case,[205] whether it could have been obtained lawfully[206] and the gravity of the crime are

[201]Rørdam (2001), pp. 338–339; Garde (2003), p. 1; Brøbech (2003), p. 401; Smith et al. (2008), pp. 641–642.

[202]See U.1995.752H where the Danish Supreme Court allowed the use of documentary evidence and U.2009.2640 V where the Western High Court allowed the stolen items to be confiscated, despite the use of technically unlawful measures, on the basis *inter alia* that there was no reason to assume that the police deliberately set out to avoid the legal rules in relation to ransacking; but see the Copenhagen City Court decision reported by Rørdam (2001), p. 341 where the prosecution was refused permission to rely on the *tiltalt*'s statement to the police made after he had been told by the police incorrect information about the strength of the fingerprint evidence against him.

[203]Brøbech (2003), p. 402; at 197–205, Brøbech highlights a number of Danish decisions that support the exclusion of evidence obtained by the police during interviews conducted with arrested persons who were being held in isolation without the permission of a defence lawyer; Brøbech (2003) also relies on the jurisprudence of the ECtHR to conclude that entrapment is another example of an unlawful method that will lead to exclusion of evidence in Danish courts: 402.

[204]Smith et al. (2008), p. 643 referring to Brøbech (2003), pp. 401–402. For example, in the trial following the case of U.2005.3108H, the witness evidence was admitted with the consent of the *tiltalt*, even though it had been obtained in breach of § 750 of the AJA. The principle *favour defensionis* is also known as *in dubio pro reo*; see Lindencrone Petersen and Werlauff (2014), pp. 649–651; see further § 216 of the AJA in relation to finding a verdict that is favourable to the *tiltalt* if there are equal votes for and against.

[205]U.1993.512 V; see also an unreported Danish Supreme Court decision of 25 February 1994 about a misunderstanding by the police of the need for a search warrant to take photographs of livestock, referred to in Rørdam (2001), p. 340.

[206]U.1993.512 V; see also an unreported Danish Supreme Court decision referred to in Rørdam (2001), p. 340.

also relevant factors.[207] Therefore, within the framework of art. 6, proportionality is not only relevant criterion in relation to the initial decision about the lawfulness of a coercive measure, but is also relevant to the decision whether to admit or exclude the evidence that was obtained.

However, exclusion of evidence is not a common outcome in Denmark. Due to the elevated status of the principle of material truth, an assessment of proportionality has generally still resulted in a conclusion that the unlawfully obtained evidence is admissible.[208] Despite the argument that using unlawfully obtained evidence may encourage the police to use unlawful means, the courts' traditional tendency to make use of unlawfully obtained evidence is apparent in serious cases that involve clandestine offending which is otherwise difficult to investigate.[209] For example, there are cases in the context of organised crime where the investigative methods are found to be in violation of the rules in the AJA but the courts proceed to admit the evidence.[210] Instead, there is a clear preference for admitting unlawfully obtained evidence and attaching it appropriate weight in the context of the other evidence in the case.[211] The court may ascribe lower weight to evidence on account of the unlawful methods used to obtain it. The weight to be attached to unlawfully obtained evidence, like any type of evidence, is a matter for the discretion of the court, in accordance with the principle of the free-evaluation of evidence enshrined in § 880. The ECtHR has frequently asserted that decisions about the probative value of certain types of evidence and laying down rules of admissibility are beyond the scope of supervision under art. 6.[212]

Recently, however, there appears to be an increased readiness to attach significance to compliance with formal rules, as a result of the influence of the ECtHR.[213] For example, Danish courts have followed the jurisprudence of the ECtHR and held that the defence should have the opportunity to cross-examine a witness on their

[207]U.1993.512 V; U.2005.3108H; see further Brøbech (2003), pp. 401–402; Smith et al. (2008), pp. 30–31 and 642; Smith (2013), pp. 20–21; and Rørdam (2001), p. 340 referring to the Danish Supreme Court decision 25 February 1994.

[208]Smith et al. (2008), pp. 29–30 and 641; Smith (2013), pp. 20 ff.

[209]Smith et al. (2008), p. 32; the tendency of the court to make pragmatic decisions by allowing unlawfully obtained evidence to be considered in the deliberations on guilt is similarly apparent in cases relating to entrapment and whether the accused was incited by the police to engage in illicit conduct. For example, in U.1997.82H, the jury was left to decide on the legal question of whether entrapment had occurred in conjunction with deliberating on the question of guilt: Smith (2013), p. 164.

[210]Smith et al. (2008), p. 31; Smith (2013), p. 20; and see, for example, U.1991.881Ø in relation to a controlled delivery.

[211]In cases of entrapment, evidence obtained by the actions of an *agent provocateur* that do not have the court's approval under the AJA may lead to automatic acquittal, discontinuance of the prosecution or, upon conviction, a reduced sentence depending on the extent to which the agent's conduct had expanded on the *tiltalte*'s criminality: U.1997.82H.

[212]See (ECtHR) *Edwards v The United Kingdom*, Application no. 13071/87, 16 December 1992; *Miailhe v. France* (No. 2), Application no. 18978/91, 26 September 1996; *Stewart v The United Kingdom*, decision of the Commission, 1984; see also Emmerson et al. (2012), pp. 129–130 and 597–599.

[213]Smith et al. (2008), p. 30; Smith (2013), p. 20.

statement made to the police, in preference to the statement being introduced directly into the proceedings, particularly if that evidence is the only or the decisive evidence in the case.[214] It has also been suggested that in recent years there is an increasing tendency in the Danish courts to exclude unlawfully obtained evidence.[215] However, beyond the criteria of reliability and *favor defensionis*, it may be difficult to draw conclusions about legal practice in Denmark in connection with unlawfully obtained evidence.[216] According to Smith, it is unfortunate that the Danish courts will set aside formal rules, which often have an important protective function, in cases where unlawfully obtained evidence is highly significant or where the *tiltalt* is accused of serious crime more readily than in cases that are 'trivial'.[217] However, this is the nature of decisions about the admissibility of evidence that are based on proportionality and a weighing and balancing of competing interests. Smith concludes that this pragmatism builds upon an ingrained unwillingness to let apparently guilty persons go free.[218]

3.9.5 Exclusion of Evidence Obtained in Breach of the Right to Silence and the Right Against Self-Incrimination

Rørdam and Brøbech have referred to a developing practice in Danish courts to exclude evidence that has been obtained contrary to the right to silence and the right against self-incrimination, and the fundamental principle that a person should not be compelled to provide evidence against himself.[219] To protect these rights, § 752 (1) of the AJA requires that the *sigtet* is made aware that he is not obliged to say anything when questioned by the police. In Denmark, evidence that is obtained in absence of this caution should not be used against a *tiltalt* on conviction. This is the case, even if the reliability of the evidence is intact.[220]

Rørdam and Brøbech have highlighted a number of cases which support the view that evidence obtained from a *sigtet* person in breach of § 752(1) of the AJA, where the *sigtet* was not advised that he does not have to say anything, will be excluded as a matter

[214] See, for example, the decision of the Western High Court in U.1996.6. Under § 745d, if the police are going to conduct an interview of a particular witness who may not be available to give evidence at trial, or conduct an identity parade, a photo-identification procedure or some other similar investigative step then the defence lawyer should be advised and given the opportunity to be present.

[215] Rørdam (2001); Smith et al. (2008), pp. 30 and 641; Greve and Cornils (2011), p. 51; Brøbech (2003), pp. 401–402.

[216] See Garde (2003).

[217] Smith et al. (2008), pp. 643–644.

[218] Smith et al. (2008), p. 644.

[219] Rørdam (2001), p. 340; Brøbech (2003), pp. 163 ff..

[220] Brøbech (2003), p. 402.

of general principle.[221] This result was arrived at in the 1970 decision of U.1970.901 V. The *tiltalt* was indicted for punching another person in the face and the Western High Court found that an interrogation, which was conducted by the police without cautioning the *tiltalt* that he did not have to say anything, could not result in a discontinuation of a prosecution but could not be used as evidence against the *tiltalt* at a future trial.

In U.1997.336 V, a tractor driver was observed driving a tractor in Horsens in such a way that indicated that the tractor's brakes were faulty. At trial, on the basis of this observation, the Western High Court found that the driver should have been regarded as *sigtet*. Therefore, he should have been cautioned in accordance with § 752(1) of the AJA that he did not have to say anything prior to being interrogated by the police. It did not appear from the police report, or from the evidence at the hearing, that the driver had been cautioned either prior to the immediate questioning by the police at the abattoir or over the phone on a later date. The court held, as a general rule, that the consequence of not following the provisions of § 752(1), which were to be regarded as providing significant fair trial protection, was that the information given by the *sigtet* could not be used by the prosecution against the *tiltalt*.[222] In other words, the evidence was to be excluded entirely from the trial.

In both cases, the *tiltalt* was indicted for less serious offences and both trials were able to proceed on the basis of other evidence. These are factors that in Denmark would usually speak for exclusion of evidence. It is, therefore, uncertain whether the courts would have arrived at a different result had the cases involved decisive (or only) confessional evidence in relation to very grave crimes.[223] Nevertheless, the written judgments emphasise the significance of the *sigtet*'s rights under § 752(1). It appears from the formulation of the decisions that fairness to the *tiltalt* was the primary consideration on the question of the admissibility of confessional evidence obtained in a police interrogation in the absence of a proper warning about the *sigtet*'s right to remain silent.[224] No reference was made, in

[221]See Rørdam (2001), pp. 340–342; Brøbech (2003), p. 181 and following, in which the author presents two reported decisions and a number of unreported decisions that support the proposition; this position, as a general rule, was accepted by Garde (2003).

[222]This conclusion is also supported by four city court unreported decisions handed down between 1990 and 2001, in each of which evidence of police interviews conducted contrary to 752(1) is excluded, regardless of the significance of the evidence or the seriousness of the offending. In three of the cases, the court also explicitly referred to the fact that the *tiltalt* had not been advised in relation to accessing a legal adviser: see Brøbech (2003), pp. 185–187.

[223]See the decision of the Western High Court on 9 October 1998, reported in Brøbech (2003), pp. 183–185, in which the *tiltalt* was questioned by the regulatory fishing authority without being advised about the right against self-incrimination, where the pre-trial statement was admitted in part due to the significance of the evidence and the organised nature of the offending.

[224]See also U.1995.856H where evidence of an interview conducted by the police of a detainee in isolation was excluded because the *sigtet*'s defence lawyer did not give permission as required under the AJA as a fair trial protection; but see the unreported decision from 1998 of the Western High Court, in which the Court admitted the evidence of an administrative interview conducted by the fisheries control authority without the rights of a *sigtet* despite the court's view that the police should have been in control of the investigation, described by Brøbech (2003), pp. 183–185.

either case, to whether either *tiltalt* had been advised about the possibility of accessing a defence lawyer or of being appointed a public defence lawyer prior to questioning; or whether this issue was even raised as a factor before the court.

The above findings by the Western High Court are also supported by four city court unreported decisions handed down between 1990 and 2001, in each of which evidence of police interviews conducted contrary to 752(1) was excluded, without reference to the significance of the evidence or the seriousness of the offending.[225] In three of the cases, explicit reference was additionally made to the fact that the *tiltalt* had not been advised in relation to accessing a legal adviser in accordance with the provisions of § 732(2). However, it is not clear from the description of the cases whether the confessional statements were made during the course of custodial interviews (and therefore whether they had a stronger right of access).

The importance of protecting the rights of a *sigtet* person under interrogation may arguably be reflected by a practice in the Danish courts to exclude evidence following a breach of § 752(1)—although this result is not yet conclusive. A breach of § 732(2) in relation to the need to advise a *sigtet* person about the opportunity to have a public defence lawyer appointed, may also be a factor which leads to exclusion. Therefore, it appears that courts may exercise judicial power to exclude evidence that has been obtained in violation of the right to silence and the right against self-incrimination and possibly other fundamental procedural guarantees under the AJA. Beyond this, however, it is unclear to what extent there is a general trend against the traditionally accepted position in Denmark that relevant and reliable unlawfully obtained evidence should be admitted.[226]

Smith highlights three clear examples of recent decisions where unlawfully obtained evidence has been ruled admissible. Firstly, in U.2005.1967Ø a person who claimed to be the victim of prior instances of car arson set up a camera on a public road, contrary to the prohibition on CCTV recording. The Eastern High Court allowed the prosecution to use footage from the camera against an alleged arsonist, referring to the seriousness of the offending, the opportunity for the defence lawyer to challenge and test the evidence and the fact that the footage showed only a small section of the road and was, therefore, not a serious breach of public integrity. Secondly, in U.2005.3007 the police were not required to destroy photographs that were taken without consent during an unlawful search (ransacking) of a private apartment. Thirdly, in the case of U.2005.3108H, the Supreme Court found that witness evidence which was compulsorily obtained under the threat of punishment under English rules was strictly in violation of the fundamental principles in relation to the right to remain silent of any person under § 750 of the AJA. Nevertheless, in coming to that conclusion, the court emphasised that it had not formed a view about how the evidence should be handled at trial. Had this case concerned self-incriminating evidence the result may have been different. This decision, when compared to the decisions outlined above in relation to

[225]Reported in Brøbech (2003), pp. 185–187.
[226]Garde (2003).

evidence obtained in violation of 752(1) and 732(2), emphasises the particular importance of the fair trial guarantees protected by these provisions in Danish law.

3.9.6 *The Relevance of Disciplinary and Criminal Actions to Exclusion of Evidence*

Aside from the issue of fundamental procedural guarantees under the AJA, the reliance on unlawfully obtained evidence in the search for the truth is to be understood in light of the importance in Denmark of using disciplinary and criminal actions against the police. It is anticipated that these types of actions will force compliance with formal rules[227] and overcome the concern that misuse of coercive measures may be easy to conceal[228]. A deliberate unlawful intrusion into secret communications in order to obtain evidence may give rise to conviction and punishment, under §§ 147 and 148 of the Criminal Code (*Straffeloven*), by a fine or a term of imprisonment of up to 3 years. Disciplinary proceedings against an officer are also possible, as well as other disciplinary actions such as notifications from the court to the Ministry of Justice. To what extent the possibility of disciplinary or even criminal action may protect investigative fairness will depend on the policing environment, in particular the attitude of the leadership and whether officers are prepared to cover for each other.[229] An examination of the actual use of disciplinary and criminal actions against the police in Denmark is beyond the scope of the present discussion, other than to identify the fact that this aspect of criminal justice may impact on the perceived necessity for the Danish judiciary to react to unlawfully obtained evidence in the forum of the criminal proceedings. However, reliance on disciplinary or criminal proceedings have not factored as criteria in judicial decisions about exclusion of evidence. In the ECtHR, the use of disciplinary or criminal proceedings against investigative misconduct may be relevant to whether an applicant (for compensation) can still claim victim status at the time an application is first heard.

3.10 Conclusion

This chapter has presented the Danish leg of the journey, mapping out the national law that limits and protects the right to silence and the right against self-incrimination throughout the criminal justice process. Beside the incorporation of the ECHR into Danish national law, in discovering what the Danish applicable law

[227]Greve and Cornils (2011), pp. 51–52; Smith (2013), p. 21.

[228]Smith et al. (2008), p. 32.

[229]See further Smith et al. (2008), pp. 644–645; Brøbech (2003), pp. 359 and 404–405.

is in relation to the rights, the provisions of the Administration of Justice Act[230] (AJA) have been an important focus. In addition, the questioning of individuals suspected of wrongdoing by administrative authorities has also been considered under the provisions of the Coercive Measures Act (CMA). Further, ministerial guidelines (based on delegated power) in relation to questioning of *sigtet* persons have also been taken into consideration. The rules in relation to the right to silence and the right against self-incrimination have been examined out from three particular situations in which self-incriminating statements or evidence of silence may be obtained from a *mistænkt* or *sigtet* in the pre-trial phase: police questioning; questioning by an administrative authority; and using covert surveillance measures that invade secret communications, such as interception and listening devices. In addition, reference has been made to the evidentiary value of the *tiltalt* remaining silent at trial.

Paragraph 752(1) AJA is central to the protection of the *sigtet* person's right to silence and the protection of the *mistænkt* who is being questioned with this right. In addition to advising on the accusations in the case, as the investigative authority under the AJA, the police are required to tell to *sigtet* that he does not have to say anything prior to asking him questions. This step is important for continuity of the proceedings, in determining the manner in which any self-incriminating statements made may be used at a future trial. What the *sigtet* says to the police is not *prima facie* to be used as evidence at trial where it cannot be tested in the same way as in-court testimony. However, exceptionally, the prosecution may be given permission to lead evidence of the written record in the police report or from the recording officer of what was said during a police interview, where the *tiltalt* does not give evidence or has made an inconsistent statement in evidence. Where the *sigtet* has not been cautioned about the right to silence then the Danish courts are likely to exclude the confession evidence from the trial. This is an interesting and important exception to the otherwise strong Danish legal tradition of the free evaluation of evidence.

If the *sigtet* refuses to cooperate when asked questions and decides to remain silent, it cannot be ruled out that this factor will be taken into account (by the court and possibly the jury) on the question of guilt. However, the fact of remaining silent cannot be the only or decisive evidence of guilt. Similarly, if the *tiltalt* remains silent at trial this may be relevant as supporting evidence of guilt in some cases.

The right to access a lawyer prior to police questioning, including a publicly appointed defence lawyer where necessary, is also granted under the AJA in § 732 (2), which is to be read in conjunction with a ministerial order and a circular on the matter. However, the operation of the rules about access to a lawyer makes the effective availability of this right uncertain. In addition, judicial decisions that deal with exclusion of confessions obtained in absence of the caution do not consistently refer to significance of whether the *sigtet* was or was not advised of the right to legal

[230] *Retsplejeloven*, hereinafter AJA.

representation. This further brings the effective availability of the right to legal representation into question.

In cases where the *sigtet* is at risk of detention the police are obliged to inform him or her of the right to access a lawyer. In practice, every *sigtet* is offered the right. However, the pathway to the right is not obstacle free. Along with being advised of the right to obtain legal representation, including for free, the *sigtet* is also warned that upon conviction he or she will be required to repay the defence lawyer's fees. This warning arguably discourages the *sigtet* to obtain the benefit of legal advice that enables him or her to fully understand to consequences of the right to silence and to participate effectively in the ongoing criminal proceedings. Although the likelihood of a trial is less, the *mistænkt* who is being questioned with the right to silence does not have the benefit of a full right to access legal advice, which includes a publicly defended lawyer where necessary. Nevertheless, the police will usually warn a *mistænkt* in this situation that he may wish to have his own lawyer present. The significance of the fact of remaining silence to the possibility of remand, future claims for unjust detention and to the possibility of adverse inferences being drawn at trial cannot be underestimated. It appears that in Danish law a connection between the right to access a lawyer and the right to silence is not yet fully developed. However, taking into account the fact that what happens during the interview is not *prima facie* evidence at trial, it may be concluded that this is the way in which the Danish legislature has determined the balance between effectiveness and fairness in relation to the right to silence.

In transnational cases, whether a suspect of foreign offences will gain or has gained access to legal representation prior to questioning may be relevant to issues related to international cooperation. Not only may this right be decisive in the admissibility of confession evidence in certain systems and, therefore, significant to maintain continuity in transnational cases, it is also relevant to the question of whether applicable common minimum standards are actually being applied from an international perspective.

Administrative authorities may also conduct pre-trial questioning of a person about whom there is a concrete suspicion of criminal wrongdoing and who may be, therefore, regarded as *sigtet* (charged). Although such questioning takes place under the Coercive Measures Act, it occurs with the consent of the individual involved. The right to be cautioned that the suspected person does not have to say anything is clearly set out in § 10(3) of the Act. However, again no clear pathway to a right to access a lawyer prior to questioning has been established, despite the fact that any statements obtained (or evidence of silence) can be used against the person in a future criminal trial—even in relation to serious charges. In fact, under the legislation it may be permissible for an authority to request the suspected person to sign away the rights under the AJA that he would normally benefit from during pre-trial questioning by the investigating police. Nevertheless, although by consent, this practice (and the use under the CMA of other coercive measures, such as interception) may be contrary to the jurisprudence of the ECtHR where it establishes the principle that administrative powers should not be used to circumvent fair trial guarantees where criminal charges are pending or are anticipated. This is the type of consideration that may be relevant when another state is considering whether certain minimum standards of fairness are being applied in Danish investigations. This may be relevant, for example, because it may affect a decision about whether to have a suspect or even a witness (where there is a concern that the witness may be involved in criminality) questioned in Denmark.

In Danish law, while there is a direct connection between cautioning and the right to silence and an unclear connection between access to a lawyer and the right to silence, there appears to be no direct connection between using covert surveillance measures to obtain confessional statements and the right to silence. Self-incriminating statements obtained in breach of the right to be cautioned under § 752 (1) will lead to exclusion of the evidence. This rule may even apply regardless of the seriousness of the offending. This reflects the fundamental significance of the right to silence and the right against self-incrimination as a principle of Danish law. However, cases dealing with evidence of self-incriminating statements obtained through the use of unlawful covert surveillance measures are not dealt with in the same way. In the limited situation of evidence gathered by chance by the use of a measure in unrelated matters under § 789 AJA the court will consider the admissibility of evidence. However, in general the provisions of the AJA focus on determining the proportionality (and thereby the legality) of a measure used without the authority of the court by balancing competing interests such as the seriousness of the unlawfulness (for example, the breach of privacy), the seriousness of the offending and the importance of the evidence to the case. In general, Danish courts do not consider the admissibility of this type of technically unlawful evidence and instead place emphasis on the principle of seeking the material truth.

In the event that there is no direct connection between the right to silence and covert surveillance in Danish law, the treatment of these types of covert measures must be viewed in light of other important protective safeguards that are built into the investigative phase. Prosecution lawyers work closely with the police in the investigative phase of Danish prosecutions. In addition, 'defence' lawyers become involved in cases early, by virtue of *sigtelse* (particularly in serious cases) and by the use intrusion lawyers in applications by the prosecution in relation to the use of covert measures—applications which involve the court's authority and ongoing supervision. These are all pre-trial factors that are relevant to the way in which the Danish criminal system allows trial court to exercise its freedom in evaluating all the relevant evidence and attaching weight as it sees fit—with minimal general consideration of the 'admissibility' of evidence.

From the above it can be observed that there are important connections between the different ways in which self-incriminating statements are obtained from a *mistænkt* or *sigtet* in the investigative phase of Danish prosecutions and the ways in which these statements may then be relied on at trial. These connections establish continuity in the ongoing process of criminal proceedings—and fairness in the whole of the proceedings. For example, where the right to be cautioned is breached in the pre-trial proceedings, the evidence may not be used at trial. The lack of a clear right to access a lawyer and the absence of a requirement that police interviews should be audio or video-recorded must be viewed in light of the fact that the police interview is not prima facie evidence in a criminal trial. Familiar to many civil law systems, the tendency to seek the truth by relying on unlawfully obtained evidence without first examining its admissibility must be viewed in light of the additional legal safeguards and oversight of investigative procedures in Denmark. Therefore, obtaining a confession in a police interview in Denmark and then directly admitting the written record of the interview in another system may create unfairness. Likewise, gathering self-incriminating statements by use of a covert surveillance measures in another system may not be accompanied by the same built-in investigative safeguards. Confessions obtained by administrative authorities in another system may have been gathered by compulsion rather than by consent. In transnational cases, these types of disruptions to the continuity in national criminal proceedings may create unfairness.

In the next chapter, the exploration of the rules that limit and protect the right to silence continues across the water and into the national system of England and Wales. In that system, on the background of the common law tradition, the way the balance between effectiveness and fairness is maintained in criminal proceedings may be structured very differently—even though this jurisdiction has also incorporated the minimum standards of fairness that have been established under the ECHR.

References

Bjørn O (2006) Retssikkerhedsloven og skatteforvaltningsretten. Karnov Group Online. SR.2006.130

Blume P (2010) Juridisk Metodelære, 5th edn. DJØF Forlag, Copenhagen

Brøbech B (2003) Ulovligt tilvejebragte beviser i straffeprocessen. DJØF Forlag, Copenhagen

Christensen JP, Albæk Jensen J, Hansen Jensen M (2012) Dansk Statsret. DJØF Forlag, Copenhagen

Christoffersen J (2015) Menneskerettens proportionalitetsprincip. Karnov Group Online. U.2015B.115

Craig P, De Búrca G (2015) EU law: text, cases and materials, 6th edn. OUP, Oxford

Dam P (2013) Advokater oplever fejl i politiets afhøringer. Available via Berlingske. http://www.b.dk/nationalt/advokater-oplever-fejl-i-politiets-afhoeringer. Accessed 15 Jan 2014

Elo Rytter J (2013) Individets grundlæggende rettigheder. Karnov Group, Copenhagen

Emmerson B, Ashworth A, Macdonald A (eds) (2012) Human rights and criminal justice, 3rd edn. Sweet & Maxwell, London

Feldtmann B, Knop Reventlow S (2013) Denmark. In: Ligeti K (ed) Towards a prosecutor for the European Union, vol 1, A comparative analysis. Hart, Oregon

Frost JL (1968) Anmedelse om bogen: Tysk Strafferets Indtrængen Karnov Group Online. U.1968B.130

Gammeltoft-Hansen H (1989) Strafferetspleje II. DJØF Forlag, Copenhagen

Garde P (2003) Birgitte Brøbech: ulovligt tilvejebragte beviser i straffeprocessen, Jurist og Økonomforbundets Forlag. Karnov Group Online. FM 2003.166

Gomard B, Møller J, Talevski O, Thønnings P (2013) Kommenteret Retsplejelov Bruxelles I-forordning: Bind III, 9th edn. DJØF Forlag, Copenhagen

Greve V, Cornils K (2011) Studies in Danish criminal law. DJØF Publishing, Copenhagen

Henricson I (2012) Politiret, 5th edn. DJØF Forlag, Copenhagen

Inkorporeringsudvalget (The incorporation committee) (1991) Den europæiske menneskerettighedskonventionen og dansk ret. Betænkning 1220-1991 (The European Convention on Human Rights and Danish Law Report 1220). Ministry of Justice, Copenhagen

Inkorporeringsudvalget (The incorporation committee) (2001) Inkorporering af menneskerettighedskonventioner i dansk lovgivning, Betænkning 1407-2001 (Incorporation of human rights conventions in Danish Law, Report 1407-2001), Ministry of Justice, Copenhagen

Inkorporeringsudvalget (The incorporation committee) (2014) Betænkning om inkorporering mv. inden for menneskeretsområdet. Betænkning 1546-2014 (Report on incorporation and other matters in the area of human rights 1546). Ministry of Justice, Copenhagen

Jochimsen J (2001) Dokumentation af forklaringer i straffesager. Karnov Group Online. TfK2001.197

Jochimsen J (2003) Anonyme Vidner og Hemmelige Agenter. Forlaget Thomson, Copenhagen

Justitsministeriet (Denmark) (2010) Spørgsmål og Svar nr 1562 fra Folketingets Retsudvalg (Alm. Del). Available via Folketinget. http://www.ft.dk/samling/20091/almdel/reu/spm/1562/svar/748715/906995/index.htm. Accessed 15 Jan 2014

Knudsen ML (1996) Passivitet som bevis i straffesager. Justitia (1)

Lindencrone Petersen L, Werlauff E (2014) Dansk Retspleje, 6th edn. Karnov Group, Copenhagen

Olesen A (2013) Løsladt og gældsat. DJØF Forlag, Copenhagen

Rørdam T (1996) Bekæmpelse af den organiserede kriminalitet. Paper presented with Bay Larsen L, Thiesen H at Gruppdiskussion Bekämpandet av den organiserade brottsligheten – 34th Nordiske Juristmøde, Stockholm, 21–23 August 1996

Rørdam P (2001) Mangler ved bevisets tilvejebringelse. Available via Karnov. U.2001B.337

Rørdam T (2012) Forsvareren. Karnov Group, Copenhagen

SKAT (2007) Tvangsindgreb og retssikkerhedsloven, SKM2007.91.SKAT. http://skat.dk/SKAT.aspx?oID = 1613857. Accessed 15 Jan 2014

SKAT (2013) Den juridiske vejledning 2013-1 af 2013-01-24, A.C. Kontrol og straf. Available via Karnov Group online. Accessed 15 Jun 2013. Now available in version: 2.4 – 31.07.15. http://skat.dk/SKAT.aspx?oId = 1919749&chk = 211047. Accessed 15 Oct 2015

Smith E (2013) Straffeprocess, 7th edn. Forlaget Thomson, Copenhagen

Smith E, Jochimsen J, Kistrup M, Poulsen JL (2008) Straffeprocessen, 2nd edn. Forlaget Thomsen, Copenhagen

Strafferetsudvalget (1971) Betænkning om Efterforskning i Straffesager m.v. Betænkning nr. 622 af 1971 (Report no. 622 of 1971 on Investigation in Criminal Cases etc.). http://www.statensnet.dk/betaenkninger/0601-0800/0622-1971/0622-1971_pdf/searchable_622-1971.pdf. Accessed 5 Jan 2016

Toftegaard Nielsen G (2004) Centrale straffeprocessuelle principper. Karnov Group Online U.2004B.77

Toftegaard Nielsen G (2011) Straffesagens Gang. DJØF Forlag, Copenhagen

Vedsted-Hansen (2004) Inkorporering af menneskerettighedskonventioner – realitet og symbol. In: Toftegaard Nielsen G, Vedsted-Hansen J, Haagen Jensen C (eds) Festskrift om Menneskerettigheder til Carl Aage Nørgaard. DJØF Forlag, Copenhagen

Chapter 4
The Right to Silence in England and Wales

Abstract This chapter commences with a description of the human rights framework in England and Wales, which visibly comes under the framework of the ECHR with the enactment of the Human Rights Act 1998 (UK). In examining how the balance between effectiveness of law enforcement and fairness to the suspect or accused is maintained in the domestic law of England and Wales, attention is focused again in this chapter on the limitation and protection the right to silence and the right against self-incrimination. Similarly to the national law in Denmark, there is continuity between the investigative and trial phases in national criminal proceedings in England and Wales. This is demonstrated by looking into the connections in the law between, for example, the manner of advising a suspect of the right to silence, the availability of pre-trial legal advice and the drawing of inferences of guilt from the accused's silence at trial. The law relating to confessions and administrative questioning powers as well as covert surveillance are also examined. Together with Chaps. 3 and 5, this chapter lays the foundation for a comparison with the national laws of Denmark and Australia in Chap. 6 showing the how the differences in procedural rules about the right to silence may cause admissibility problems in transnational cases. It also becomes part of the vertical perspective in Chap. 7, looking at the question of uniformity of trust-building minimum standards.

4.1 Introduction

The utility of international cooperation in the prosecution of transnational criminal cases requires that evidence may be gathered in one national legal system in a manner and form that renders it functional in another national jurisdiction. Admissible evidence must not only be relevant[1] and reliable, but its use at the trial on the

[1]The primary test of admissibility in a common law jurisdiction is relevance to a material fact—relevant evidence proves or supports a fact in issue between the prosecution and the defence. To be admissible relevant evidence must also pass a number of more specific tests, such as reliability, fairness etc.

F.M.W. Billing, *The Right to Silence in Transnational Criminal Proceedings*,
DOI 10.1007/978-3-319-42034-9_4

question of guilt by the national trial court must be fair, taking all the circumstances of the case into account. The previous chapter looked at the national law of Denmark to identify the ways in which the right to silence and the right against self-incrimination are limited and the safeguards that are in place to ensure overall fairness in Danish criminal proceedings. In this chapter, a similar path will be followed through the law in England and Wales. In relation to fairness in English criminal trials, the House of Lords in *R v Johnstone*[2] held that the state has a responsibility to ensure that legislation maintains a reasonable balance between the general community interest in obtaining a conviction and the individual interest in a fair trial. This balancing of interests is equally relevant to limiting and protecting the right to silence in national and transnational criminal proceedings.

In England, the purpose of a criminal investigation is to ascertain whether a person should be charged with an offence and whether he is guilty of it.[3] Procedural safeguards are built into English criminal proceedings that protect the position of the suspect or accused—this includes the right to silence and the right against self-incrimination, the latter often referred to in English law as the 'privilege against self-incrimination'. The operation of the right to silence and the 'privilege', to protect suspects and accused persons throughout the course of criminal proceedings, has been influenced by the enactment of legislation such as the Police and Criminal Evidence Act 1984 (PACE) and the Human Rights Act 1998.[4] The Human Rights Act is an incorporation of the ECHR and the minimum standards set by the ECtHR.

This chapter examines the ways in which confession evidence (admitted to committing a crime or it can be inferred) and evidence of silence may be gathered in the investigative phase and used in the trial phase.[5] In particular, the law that limits and protect the right to silence will be considered relation to the following situations: police questioning, including covert questioning of a suspect by an undercover police agent; the drawing of adverse inferences by the tribunal of fact from

[2]*R v Johnstone* [2003] UKHL 28; [2003] 1 W.L.R. 1736, [48], which deals with presumptions of fact against innocence in a compact disc bootlegging case.

[3]Sections 1(4) and 22, Criminal Procedure and Investigations Act 1996.

[4]At an international level, the United Kingdom Parliament is the legislature that represents England and Wales, as well as Northern Ireland and Scotland. However, by the powers of devolution, England and Wales, Scotland and Northern Ireland form three separate legal jurisdictions. The parliaments of Northern Ireland and Scotland each make the laws in relation to national criminal justice and law enforcement that apply within those jurisdictions. On a national level, the UK Parliament makes laws in relation to the legal jurisdiction of England and Wales. The Supreme Court of the United Kingdom (which is at the same hierarchical level as the former House of Lords) is the highest appellate court for all three jurisdictions.

[5]Section 82(1) of the Police and Criminal Evidence Act 1984 (UK) defines a confession as including any statement, which is wholly or partly adverse to the person who made it. The section adds that such a statement is a confession, "...whether made to a person in authority or not and whether made in words or otherwise." English and Australian law draws a distinction between a confession and a formal admission, made on behalf of the defence at the beginning of a trial to narrow the issues to be tried.

pre-trial silence of the accused or the accused's failure to testify; the drawing of adverse inferences as a result of statements made by the accused to the police in the pre-trial phase; questioning of suspects by the use of regulatory compulsory questioning powers; and obtaining confession evidence by the use of covert surveillance.

In addition, judicial discretion to exclude evidence on the ground of unfairness in the proceedings as a whole will be considered. This is an important procedural safeguard in English criminal proceedings, which allows the court to assess the effect of the way in which evidence is gathered on the fairness of using the evidence at trial. The functionality of other provisions in English law that aim at establishing a balance between effectiveness and fairness, must be examined in the light judicial discretion to exclude evidence. The aims of the criminal trial and accompanying rules of evidence in England and Wales are to maintain fairness and establish the truth through accurate fact finding in order to convict the guilty and acquit the innocent.[6] Investigation and prosecution is a continuum; and it is impossible to divorce, completely, questions about the fairness of the manner in which evidence is gathered from the question of admissibility of evidence at trial.[7]

The examination of the national balance between effectiveness and fairness in England concentrates on serious cases, which are the cases that are most likely to have transnational aspects to them.[8] Although today approximately 95 % of criminal cases are dealt with in the Magistrates' Court by a judge sitting alone, or by a panel of three lay magistrates, serious crimes are still tried in the Crown Court by a judge and jury of 12 citizens. All cases that are punishable by a minimum level of imprisonment of 12 months will be heard in the Crown Court.[9]

Much of the law of evidence in England and other common law jurisdictions is based on the assumption that the normal method of criminal trial in relation to indictable offences[10] is by judge and jury. The rules of evidence are constructed around this traditional distinction between law and fact and their operation ensures that admissibility and the merits of relying on the evidence (questions of law) are determined prior to assessing credibility or the weight to be attached to it (questions

---

[6]Malek (2010), pp. 3–4.

[7]Malek (2010), pp. 1–2.

[8]Proportionality and the reality of budgetary considerations encourage states not to use the mechanics of international cooperation to enforce the law in relation to minor offences. For example, in accordance with art. 2(1) of the European Arrest Warrant Framework Decision, surrender under an EAW in the EU may only take place in respect of offences that carry a minimum level of imprisonment of 12 months.

[9]The Magistrates' Court has jurisdiction to hear criminal cases where the maximum level of imprisonment is 6 months, and no more than a total of 12 months for multiple offences.

[10]In England and Wales, an indictable offence is usually a serious offence, which, if committed by an adult, will be tried on indictment by a judge and jury in the Crown Court. A summary offence is a less serious offence, which is triable in the Magistrates' Court before a single judge or a panel of three lay magistrates. Some indictable crimes can also be tried summarily and are called 'either way' offences. See further the Interpretation Act 1978, Criminal Law Act 1977 and the Magistrates' Courts Act 1980.

of fact). Even in the case of a trial which is conducted before a judge sitting alone or three lay magistrates, the distinction between the functions of the tribunal of law (the judge) and the tribunal of fact (the jury) assists in delineating the different considerations involved in each function.[11] In jury trials, the legally trained and experienced judge can filter out evidence that is considered improper and that may distract the lay jury from the material issues, when deliberating on the question of guilt.[12] Further, the judge can use judicial comment to advise the jury on the different ways that evidence can be used properly. These reflections on the different functions of the tribunal of law and the tribunal of fact are important to understanding the role of judicial discretion to exclude unlawfully and unfairly obtained confession evidence and directions to the jury about how to handle certain types of evidence, such as evidence of an accused's silence. The role of judge and jury, as well as the rules of evidence, are among the distinctive features of English law that may impact on the transfer of evidence in transnational cases involving England and Wales.

4.2 The Human Rights Framework in England and Wales

4.2.1 The Human Rights Act 1998 and Incorporation of the ECHR

The Human Rights Act 1998 (HRA)[13] was enacted by the United Kingdom Parliament to incorporate the European Convention on Human Rights (ECHR)[14] into the domestic law of England, Wales, Scotland and Northern Ireland, albeit indirectly. The primary provisions of the HRA came into force on 2 October 2000. The HRA gave "further effect" in domestic law to the rights and freedoms guaranteed under the ECHR, which already had an influence on the decisions of domestic courts.[15]

Under s. 3 of the HRA,[16] legislation made within the UK must be read and given the effect of being compatible with Convention rights so far as possible.[17] All

[11]Malek (2010), p. 2.

[12]See further, Malek (2010), p. 2.

[13]The Human Rights Act 1998 (UK) applies in all countries of the UK.

[14]Convention for the Protection of Human Rights and Fundamental Freedoms, opened for signature 4 November 1950, entered into force 3 September 1953, CETS 005 (as amended by Protocols 11 and 14).

[15]See the statement of purpose in the Human Rights Act 1998 (UK) and see further Emmerson et al. (2012), p. 151.

[16]In England and Australia a legislative provision is a 'section', the abbreviation for which is 's.' The plural abbreviation is 'ss.'

[17]This includes past and future (primary and subordinate) legislation. See further s. 3(2)(a) of the Human Rights Act 1998 (UK).

legislation (primary and subordinate) must be interpreted, if possible, in a way that is compatible with the Convention.[18] If a court finds that domestic legislation is not compatible with the ECHR, it may issue a 'declaration of incompatibility' under s. 4 of the HRA.[19] However, pursuant to s. 4(6), this does not mean that the act is annulled or that the provision overturned.[20] The incompatibility remains (and the provision is enforced with the incompatibility) until the legislature rectifies the problem.[21]

In addition, s. 6(1) of the HRA prohibits any public authority from acting in a way that is incompatible with Convention rights.[22] Thus, the police, the prosecution and judicial officers must not unjustifiably infringe a suspect or accused person's human rights under the ECHR. Section 8(1) of the HRA requires courts to "grant such relief or remedy, or make such order, within its powers as it considers just and appropriate." In addition, the courts are empowered to exclude otherwise admissible evidence where its admission may infringe a Convention right without justification.

Pursuant to s. 2(1) of the HRA, when interpreting the Convention rights, a domestic court or tribunal is bound to take into account the judgments and opinions of the European Court of Human Rights (ECtHR) or other bodies under the Convention (also in cases involving third parties). However, the domestic courts are not bound to follow the precedence of the ECtHR.

This situation of being under an obligation to follow the ECtHR and yet not being strictly bound by the court's decisions has been contentious.[23] A suggestion was made during the political debates that it is possible that the domestic courts may even successfully 'lead' the ECtHR on an issue; and that certain ECtHR cases may not be of assistance to the domestic courts, for example, where the case originated in another system or relied on a 'margin of appreciation', or due to the passage of time.[24] On the other hand, in *AF v Secretary of State for the Home Department* the House of Lords found that "... the authority of a considered statement of the Grand Chamber is such that our courts have no choice but to accept it and apply it."[25]

[18]Sections 3(1) and (2)(a) of the Human Rights Act 1998 (UK). If primary legislation is found not to be incompatible then the courts have the power to make a 'declaration of incompatibility': ss. 53 (2)(b) and 54(2). Incompatible subordinate legislation is generally not applied: s. 4(4)(b).

[19]See sections 4(1)–4(4). In the case of secondary legislation, a declaration will be made if the primary act does not remove the incompatibility.

[20]English courts cannot annul legislation.

[21]See section 4(6). There is a fast track system for returning the incompatible provision to Parliament under s. 10 of the Human Rights Act.

[22]See further ss. 7, 8 and 10 of the HRA, about the rights for victims of violations to bring proceedings against the state.

[23]See the blog by Harris et al. (2014).

[24]Emmerson et al. (2012), p. 173.

[25][UKHL] *AF v Secretary of State for the Home Department* [2009] UKHL 29 at [108] (per Lord Carswell); see further, for example, *R v Horncastle* [2009] UKSC 14 at [120].

Emmerson et al have described the statement of Lord Bingham in *R. (Ullah) v Special Adjudicator* as the "... decisive explanation" of the relevance of the ECtHR in domestic law, where his Lordship said:

> [a court] is required by section 2(1) of the Human Rights Act 1998 to take into account any relevant Strasbourg case law. While such case law is not strictly binding, it has been held that courts should, in the absence of some special circumstances, follow any clear and constant jurisprudence of the Strasbourg court ... This reflects the fact that the Convention is an international instrument, the correct interpretation of which can be authoritatively expounded only by the Strasbourg court. From this it follows that a national court subject to a duty such as that imposed by section 2 should not without strong reason dilute ... the effect of the Strasbourg case law. It is indeed unlawful under section 6 of the 1998 Act for a public authority, including a court, to act in a way which is incompatible with a Convention right. It is of course open to member states to provide for rights more generous than those guaranteed by the Convention, but such provision should not be the product of interpretation of the Convention by national courts, since the meaning of the Convention should be uniform throughout the states party to it. The duty of national courts is to keep pace with the Strasbourg jurisprudence as it evolves over time: no more, but certainly no less.[26]

In addition to this statement, the argument also applies in the UK that not following the judgments of the ECtHR may lead to a success by an individual in that Court.[27] Therefore, for the above reasons, following a successful application to the ECtHR, the UK courts are under a general duty to apply a clear decision of at least the Grand Chamber.

Of importance, the HRA has, for the first time, enabled citizens of the UK to directly challenge the decisions and actions of the state that are in violation of convention rights in the national courts. Thus, the application of the Convention in the domestic courts captures the essence of the notion of subsidiarity.[28] An additional factor in the UK undertaking to be bound by the supervisory jurisdiction of the ECtHR means that unsuccessful litigants have a right of individual petition to that Court (as discussed in Chap. 2, Sect. 2.3.1.3.2).[29]

4.2.2 The Interrelationship Between Convention Rights, Statute, Codes of Practice and the Common Law

Prior to, and in addition to, the implementation of convention rights with the enactment of the HRA, the Police and Criminal Evidence Act 1984 (PACE) codified many of the procedural rights encompassed by articles 5 and 6 of the

---

[26] *R. (Ullah) v Special Adjudicator* [2004] 2 A.C. 323 at [20]; see further Emmerson et al. (2012), pp. 175–177.

[27] Emmerson et al. (2012), p. 173.

[28] See the blog by Harris et al. (2014).

[29] Article 34, Convention for the Protection of Human Rights and Fundamental Freedoms, opened for signature on 4 November 1950, entered into force 3 September 1953, CETS 005.

ECHR, such as the right to be given reasons for arrest (s. 28); time limitations that apply in relation to pre-trial detention (ss. 41–46); and the right to access legal advice when in police custody (s. 58). There are also provisions that regulate the admissibility of evidence, which are based on fair trial principles and are derived from common law. For example, s. 78 provides for a judicial discretion to exclude evidence based on unfairness. Today, statutory provisions such as those contained in PACE are the primary source of law in relation to these rights in England and Wales. The language of these provisions is to be interpreted in light of the jurisprudence of the ECtHR and the common law (as far as it is compliant with convention rights).

Section 66 of PACE provides for a delegated power allowing the Secretary of State to issue binding codes of practice in relation to the exercise of police powers in areas such as the arrest, detention and questioning of criminal suspects and accused persons.[30] These codes are important in determining the valid laws. For example, the relevant code of practice in relation to police interviews is the Code of Practice for the Detention, Treatment and Questioning of Persons by Police Officers (Code C). Under s. 67(11) of PACE, if any provision of the Code is relevant to a question arising in criminal proceedings, the court must consider that provision in resolving the question concerned. However, a breach of the provisions does not necessarily amount to oppression or mean that a confession is unreliable. Section 67 (9) applies the codes of practice to police as well as to other investigative agents, such as customs officers, Companies Act inspectors, and military police and Inland Revenue investigators. Most policing activities are covered by codes of conduct.

In addition, there are specific practice codes, which regulate the way in which covert techniques are used and covert operations are conducted, such as the Covert Human Intelligence Sources Code of Practice[31] made under s. 71 of the Regulation of Investigatory Powers Act 2000 (UK). One of the purposes of this code is to emphasise proportionality in relation to using this form of covert technique, as well as to encourage compliance with convention rights, in particular articles 6 and 8. It applies to operations conducted by the police, national crime squads, the national intelligence services, Revenue and Customs and other government agencies.

A number of the most fundamental fair trial rights such as the presumption of innocence and the duty of the prosecution to prove the guilt of the accused are deeply embedded in the English common law. Therefore, although the HRA directly incorporates art. 6 of the ECHR into domestic law,[32] even today discussions in English law about issues that relate to these essential principles often start with the common law.[33]

[30]The Code of Practice for the Detention, Treatment and Questioning of Persons by Police Officers (Code C) does not apply to persons being questioned under the Terrorism Act 2000, which applies its own requirements: Code C [1.11].

[31]Covert Human Intelligence Sources Code of Practice.

[32]Article 6 of the ECHR is reproduced in Article 6 of Schedule 1, Human Rights Act 1998 (UK).

[33]See, for example, on reverse burden of proof, Emmerson et al. (2012), p. 667.

Under s. 11(1) of the HRA (and art. 53 of the ECHR), an individual's reliance on a convention right does not restrict the application of any other right or freedom conferred by domestic legislation or common law. Therefore, the ECHR provides "a floor of rights" and does not detract from any different, more developed or superior rights within the law of the UK or other contracting states.[34] However, as stated at Sect. 4.2.1, past and present legislation must be compatible with convention rights; and pre-existing common law rules that infringe convention rights must give way to the ECHR by virtue of the judiciary's duty of compliance under s. 6 (1) HRA (referred to at Sect. 4.2.1 above).[35]

4.2.3 The EU Charter of Fundamental Rights and Domestic Law in the UK

In accordance with art. 51(1) of the Charter of Fundamental Rights of the European Union,[36] the provisions of the EU Charter apply to EU institutions and other bodies and to the Member States, including the United Kingdom, when they are implementing Union law. Article 1(2) of Protocol (No. 30) on the application of the Charter of Fundamental Rights of the European Union to Poland and to the United Kingdom[37] does not constitute a general opt-out of the Charter by these two countries (later joined by the Czech Republic).[38] Article 1(2) reaffirms the principles contained in art. 51(1) of the EU Charter to the effect that the Charter does not create justiciable rights as between private individuals.

The content of the EU Charter may, in general, be considered at least the same as the content of the equivalent convention rights (as discussed in Chap. 2, Sect. 2.3.1.4 above). In addition, the ECHR was a significant source of inspiration in the development of fundamental rights as general principles of EU law.[39] Furthermore, the EU Charter applies in the UK only in relation to implementation of EU law. Therefore, the ECHR as provided for by the enactment of the HCA 1998 may be considered the most significant source of international human rights law in the UK.

---

[34]See the reference to the Lord Chancellor's speech during parliamentary debates referred to in Emmerson et al. (2012), p. 178.

[35]Specific provisions of the HRA "insulate" incompatible primary legislation from judicial override, so that legislative incompatibility must be dealt with on a political level: Emmerson et al. (2012), p. 183.

[36]Charter of Fundamental Rights of the European Union, OJ 2007 L303/01.

[37]Article 1(2) of Protocol (No. 30) on the application of the Charter of Fundamental Rights of the European Union to Poland and to the United Kingdom, OJ C 306/156.

[38]However, Art 1(2) of Protocol (No. 30) appears to rule out new EU rights being derived from Articles 27 to 38 (social rights) of the Charter: (CJEU) *N.S. v Secretary of State for the Home Department*, case C-411/10, 12 December 2011, [116–122].

[39]See, for example, Craig and De Búrca (2015), pp. 385 ff.

4.2.4 England and Wales' International Obligations Under the ICCPR

The United Kingdom ratified the ICCPR on 20 May 1976, and has made no relevant reservations in relation to art. 14 ICCPR.[40] Therefore, as discussed in Chap. 2, Sect. 2.2.1.2 above, art. 14 of the International Covenant on Civil and Political Rights creates international obligations in relation to the right to a fair trial. However, the ICCPR has not been incorporated into the domestic law of England and Wales and, thus, the international obligations are not legally binding in domestic law (though they arguably have political and moral effect).

4.3 The Right to Silence in England and Wales

In Chap. 3 above (Sect. 3.3) reference was made to the fact that Danish law views the right to silence from the perspective of the right against self-incrimination. However, in English common law, the tradition approach to the accused's right to silence and right against self-incrimination is that both these rights fall under a broad notion of the *right to silence* (see the first overlapping circle of Fig. 1.1 in Chap. 1, Sect. 1.2.2).

In 1993, the right to silence was examined by Lord Mustill in *R v Director of the Serious Fraud Office, Ex p. Smith*.[41] Lord Mustill observed that there is no general single right to silence at common law. The term, "the right of silence," was described as compendious but inaccurate. It was found that the term comprises a disparate group of 'immunities', which are all concerned with the protection of citizens from abuse of investigative powers, but which differ in nature, origin and incidence. Rather than being a general right that can always be asserted positively, Lord Mustill described the immunities as functioning to protect a person from the generally applied process. This analysis of the composite right to silence, which is made up of various immunities, has been referred to by the ECtHR in the 1996 case of *Murray v. The United Kingdom* (referred to in Chap. 2, Sects. 2.3.2.3 and 2.3.2.2.2).[42]

Lord Mustill identified at least six varieties of the immunity:

"(1) A general immunity, possessed by all persons and bodies, from being compelled on pain of punishment to answer questions posed by other persons or bodies.

[40] International Covenant on Civil and Political Rights (ICCPR), opened for signature in New York on 19 December 1966, entered into force 23 March 1976, UNTS 999/171

[41] *R v Director of the Serious Fraud Office, Ex p. Smith* [1993] A.C. 1, 30–31.

[42] (ECtHR) *John Murray v. The United Kingdom*, Application no. 18731/91, 8 February 1996, 32.

(2) A general immunity, possessed by all persons and bodies, from being compelled on pain of punishment to answer questions the answers to which may incriminate them.
(3) A specific immunity, possessed by all persons under suspicion of criminal responsibility whilst being interviewed by police officers or others in similar positions of authority, from being compelled on pain of punishment to answer questions of any kind.
(4) A specific immunity, possessed by accused persons undergoing trial, from being compelled to give evidence, and from being compelled to answer questions put to them in the dock.
(5) A specific immunity, possessed by persons who have been charged with a criminal offence, from having questions material to the offence addressed to them by police officers or persons in a similar position of authority.
(6) A specific immunity (at least in certain circumstances . . .), possessed by accused persons undergoing trial, from having adverse comment made on any failure (a) to answer questions before the trial, or (b) to give evidence at the trial."[43]

The rationale for the different immunities varies, as well as the level of public concern at statutory or judicial encroachment on them.[44] Lord Mustill identified four different motives for the above immunities. First, the court referred to the common view in civilized states that everyone is entitled to some level of personal liberty and privacy. This rationale probably lies behind the common law position from the late middle ages that all testimony was voluntary; and may also in part lie behind the non-compellability rule, which still applies today in relation to accused persons.[45] Although the extent to which the entitlement to liberty and privacy is limited may vary between states, individuals or generations, ". . .few would dispute that some curtailment of the liberty is indispensable to the stability of society."[46] In England, the modern privilege against self-incrimination in the context of witness privilege, insofar as it applies to the accused who chooses to give evidence, only applies in relation to incriminating matters that go beyond the facts in issue in the trial.[47]

[43] *R v Director of the Serious Fraud Office, Ex p. Smith* [1993] A.C. 1, 30–31.

[44] *R v Director of the Serious Fraud Office, Ex p. Smith* [1993] A.C. 1, 31–32.

[45] Helmholtz et al. (1997), p. 47; See further Criminal Evidence Act 1898 (UK), which made the accused a competent but non-compellable witness. It was found that by implication the 1898 act also removed the right of the accused (who had previously been disqualified as a party to the case from giving sworn evidence) to make an unsworn statement: Bentley (1998), p. 147; Brawn (2006), p. 190.

[46] *R v Director of the Serious Fraud Office, Ex p. Smith* [1993] A.C. 1, 32.

[47] Richardson (2011), [12.4]; the civil witness privilege rule was extended to non-party criminal witnesses in the late 1700s and was given exclusionary effect in the 1847 case of Garbett: Helmholtz et al. (1997), p. 89 referring to the Treason Act 1696, 7 Will. 3 c. 3, §§1 and 7 in relation to cases of treason and 1 Anne st. 2, c. 9, §3 (1702) in relation to all felonies, see also 107, 171 and 180; Langbein (2003), pp. 254, 279 and 281.

The second motive to which Lord Mustill referred is the long history of reaction against abuse of judicial power, specifically in relation to judicial interrogation. From the 1300s the ecclesiastical courts, and the prerogative courts of Star Chamber and High Commission, borrowed the *ex-officio oath* from Continental inquisitorial procedure, by which a person swore by god to tell the truth. Thus, the fear of god was used to conduct fishing expeditions and press confessions out of individuals under interrogation.[48] As referred to in Chap. 1, Sect. 1.2.2, the Latin maxim, *nemo tenetur prodere seipsum* (that an accused person could not be obliged to bear witness against himself by providing self-incriminating evidence) developed in Continental and English ecclesiastical courts in response to the oath. This was an attempt to curb intrusive inquisitorial powers.[49] Even after the abolition of the prerogative courts and the prohibition of interrogation under oath by the English Parliament in 1640, magistrates and justices of the peace (in the absence of a police force) had power to compulsorily interrogate an accused prior to trial in preparation of the prosecution case.[50] It was not until 1848 that the St John Jervis's Act formally required the committing magistrate or justice to make the accused aware that he might decline to answer questions in the pre-trial examination and warn him that anything he or she said could be used in evidence at the trial.[51] Even today, although abuse of pre-trial judicial interrogation is outdated, practices such as the use of quasi-judicial compulsory investigative questioning powers serve as reminders. However, modern criminal procedure has regulated pre-trial interrogation by the police with proper procedures for cautioning suspects and persons who have been charged.[52]

Lord Mustill similarly raised the third motive for the immunities as "...the instinct that it is contrary to fair play to put the accused in a position where he is exposed to punishment whatever he does."[53] In the distant past, compulsory judicial interrogation on oath was said to create a 'trilemma'—a person was forced to either not answer questions, and face harsh penalties for contempt of court; confess against the natural duty to the self; or give false testimony under oath and face serious and final judgment before god.[54] In 1721, an English treatise was written about the 'confession rule' dealing with the 'involuntariness' of confessions made on oath and in fear of eternal damnation.[55]

---

[48]Helmholtz et al. (1997), 14–15, 18, 21 and 101; see further Helmholtz (2004), pp. 607–608.

[49]Helmholtz et al. (1997), pp. 6–7, 17, 23, 29 and 44; the witness privilege was loosely based on the *nemo tenetur* maxim: Helmholtz et al. (1997), p. 157.

[50]Langbein (2003), pp. 40–43 and 278; Helmholtz et al. (1997), pp. 8, 32, 92 and 101.

[51]Helmholtz et al. (1997), p. 92.

[52]The Judges' Rules of 1912 were the forerunners of the Police and Criminal Evidence Act 1984 and the practice codes.

[53]*R v Director of the Serious Fraud Office, Ex p. Smith* [1993] A.C. 1, 32–33 (Lord Mustill).

[54]Helmholtz et al. (1997), pp. 27–28 and 31; see further *Murphy v Waterfront Commission* 378 US 52 (1964); and *EPA v Caltex* [1993] HCA 74, [33].

[55]Hawkins, *A Treatise of the Pleas of the Crown*, (1721) bk 2 ch 31, [2] referred to by McHugh J in *Azzopardi v The Queen* [2001] HCA 25.

In the absence of immunity against prosecution, for example, modern compulsory questioning powers could expose an accused to punishment regardless of whether he remains silent or not. In addition, current practices in English trials such as the drawing of adverse inferences from the accused's pre-trial silence or failure to testify at trial can place a suspect or accused in the position where he may not have a real choice—either he confesses and faces conviction; remains silent and risks adverse inferences being drawn; or he lies and risks a conviction for perjury. Recent critics of the 'trilemma' problem argue that it only applies in relation to guilty accused and therefore cannot justify additional safeguards in the criminal justice process.[56]

Finally, the fourth rationale for the immunities that make up the right to silence in English law is the desire to minimise the risk that an accused person may be convicted on the basis of an untrue out of court confession. Following the early beginnings of adversary procedure in the 1730s, when judges began to allow the (albeit limited) use of defence counsel, and when extra-judicial confessions to prosecutors was prominent, the confession rule was extended in 1783 by the judgment in the case of *Warickshall*. In the words of Lord Summer, the court held:

> A confession forced from the mind by the flattery of hope or by the torture of fear comes in so questionable a shape, when it is to be considered as evidence of guilt, that no credit ought to be given it.[57]

Accordingly, a court could not be satisfied that an extra-judicial confession was voluntary if the confession had been obtained by the use of force or inducement. However, the 'fruits' of the condemned confession could be admitted.[58] The English common law rules that preconditioned admissibility on proof that a confession was made voluntarily were the precursor for s. 76 of PACE, which provides for exclusion of confessions obtained through oppression or other circumstances affecting reliability (discussed at Sect. 4.8.1 below).

In Lord Mustill's opinion, the six immunities that fall under the umbrella of the right to silence, share the general purpose of protecting an accused person from an abuse of power in relation to their will to speak, but offer protection, which is circumstantially diverse and has been influenced by a variety of factors over centuries of English law.[59] Helmholtz shares this view.[60] Based on these opinions, it is inaccurate to speak of a general right to silence in English law or under art. 6 of the ECHR (which is the floor of protection). Further, with the exception of the right of a suspect to refuse to answer questions under caution or the non-compellability of the accused, Lord Mustill and the ECtHR have held that the composite right to

---

[56]See further Jackson and Summers, 268, *inter alia* referring to the dissenting judgment of Martens J in (ECtHR) *Saunders v. The United Kingdom*, Application no. 19187/91, 17 December 1996.

[57]*R v Warickshall* (1783) 1 Leach 263 [168 ER 234].

[58]See Helmholtz et al. (1997), p. 154.

[59]*R v Director of the Serious Fraud Office, Ex p. Smith* [1993] A.C. 1, 33 (Lord Mustill).

[60]Helmholtz et al. (1997), p. 12; see also *Azzopardi v The Queen* [2001] HCA 25, [168] (McHugh J).

silence is otherwise not absolute. Therefore, it is not possible for a legislature to generally abolish or amend the right to silence by any one measure or provision. Likewise, it would not be possible to assess the necessity of a limiting measure, without first identifying the variety of the right in question and the underlying rationales.

4.4 Police Questioning and the Right to Silence

The Police and Criminal Evidence Act 1984 (PACE)[61] and the PACE codes of conduct contain a number of provisions which regulate the treatment of suspects by the police, including the applicable arrest, detention and questioning procedures (referred to at Sect. 4.4 above).[62] By the process of regulating investigative actions in obtaining evidence for trial, these provisions define the ways in which fair trial rights impact on the pre-trial phase in England and determine the sequence of events that lead to a lawful and fair police interrogation. Because a suspect, even unwittingly, can provide confession evidence at any time of the investigation, this framework is important to protect the integrity of confession evidence by guarding against what might be an abuse of investigative powers.

4.4.1 Initial Questioning Without Arrest

The police have general powers which entitle them to question any person whom they think may have information about the commission of a crime (PACE Code C, Note 1 K). Although all citizens have a civic duty to assist the police to prevent crime and identify offenders, there is no obligation to answer police questions (beyond giving a name and address).[63] With consent, questioning of witnesses may occur in a public place, a private location or at a police station. A witness may be asked to attend a police station voluntarily to make a witness statement in relation to an investigation. Voluntary attendance means that the individual may leave at will (Code C, 3.21 and Note 1 K). Anyone assisting the police voluntarily in their own home may withdraw their consent and require the police to leave at any time (Code C, 3.22).[64]

[61]Police and Criminal Evidence Act 1984 (UK).

[62]Code of Practice for the Detention, Treatment and Questioning of Persons by Police Officers (Code C); and Code of Practice for the Statutory Power of Arrest by Police Officers (Code G).

[63]For stop and search powers in England and Wales, in order to allay or confirm suspicions, see PACE Code A.

[64]If a signed witness statement turns out to be false, this may lead to criminal charges (making a false statement).

A suspect may initially be questioned as a witness. However, in order to obtain usable evidence from a suspect, as soon as a constable of police forms the opinion that there are grounds to suspect that an individual has committed an offence the officer must 'caution' the person before any questions, or further questions, are asked. The suspect must be cautioned in the following terms:

> You do not have to say anything. But it may harm your defence if you do not mention when questioned something which you later rely on in Court. Anything you do say may be given in evidence.[65]

This caution must be used before a suspect is arrested, advised of a potential prosecution being brought against him or charged (Code C, 10.1 and 10.5).[66]

If a suspect has been cautioned but is not under arrest, they must be informed of that fact and that they are not obliged to remain and answer questions (or may request the police to leave). In accordance with the PACE codes of practice, they should also be informed that they may receive free, independent legal advice and how to obtain it (Code C, 10.2 and 3.21–3.22). However, this is not a 'right' and lack of access to a lawyer while not in custody and when being questioned by the police does not necessarily render any confession evidence that follows inadmissible—the fairness of non-custodial questioning is an issue that will be decided taking all the circumstances of the case into account.[67] The admissibility of confession evidence may be considered under ss. 76 and 78 PACE (as discussed at Sect. 4.8.1 and 4.8.2 below).

The fairness of a suspect unwittingly being questioned by an undercover agent (and whether this was the functional equivalent of an interrogation in derogation of the right to silence and the right to access legal representation) is a question that will be dealt with by the courts under s. 78 PACE (referred to at Sect. 4.8.2 below).[68]

4.4.2 *Arrest and Questioning*

A suspect who voluntarily participates in police questioning may only be prevented from leaving the police station, or requesting the police to leave private premises, if their arrest as a suspect becomes necessary (s. 29 PACE, Code C, 3.21 and Code G, Note 2G).

---

[65]Code C, [10.5]. This caution must be used before a suspect is arrested, advised of a potential prosecution being brought against him or charged. The terms of the caution change if an accused is exceptionally questioned after being charged—in this situation no adverse inferences can be drawn (Code C, 16.4).

[66]For a more detailed discussion about cautioning and drawing adverse inferences at trial, see 4.5.1.1 below. A juvenile, mentally disordered or vulnerable person must be cautioned in the presence of an independent person: Code C, 10.12.

[67]*Ambrose v Harris, Procurator Fiscal, Oban (Scotland)* [2011] UKSC 43.

[68]About the use of covert human intelligence sources (CHIS), see 4.7 below.

A constable of police may execute an arrest warrant in connection with the commission of an offence.[69] Alternatively, section 24 of PACE, in addition to PACE Code G, 2.3, provides that a constable of police may arrest a person without a warrant if, *inter alia*, he or she is about to commit an offence, is in the act of committing an offence, or there are reasonable grounds for suspecting the person is guilty of an offence.[70] In addition, the power of arrest is conditional upon necessity (Code G, 2.4). The constable must have reasonable grounds for believing that arresting the person is necessary, for example, to prevent physical injury or damage to property, to allow the prompt and effective investigation of the offence, or to prevent a prosecution from being hindered by the disappearance of the person (s.24 (5) PACE and Code G, 2.9).[71] Arrest to allow prompt and effective investigation is a broad ground, which may include the right of the police to arrest a suspect for the purpose of interrogation.[72] Most common law jurisdictions do not allow the police the power of arrest to interview and in England and Wales the power under s. 24 PACE is the result of a recent enactment.

In accordance with s. 28 PACE and Code C, 3.21, if it becomes necessary to arrest a suspect, for example during an interview, he or she must be informed immediately (or as soon as practicable) of the fact of arrest; and the grounds and reasons for the arrest, provided for in Code G.[73] Upon arrest, a person must also be cautioned, unless it is impracticable or they have already been cautioned immediately prior (Code C, 10.4).[74] A person who is arrested must also be taken immediately, or as soon as practicable, before the appropriate custody officer at a police station (ss. 30, 35 and 36 PACE, Code C, 3.21 and Code G, 2.1A), unless they are released on bail (s. 30A and 47).

In addition to being told the grounds and reasons for arrest and being cautioned, as above, an arrested person must be told of their rights in relation to their detention (Code C, 3.1). Importantly, s. 58 PACE (and Code C, 3.1 and 6.1) ensures that a person arrested and held in custody has the continuing right to communicate with a lawyer in private. According to the provisions of Code C, this right includes the right to free, independent legal advice (Code C, 3.1, 6.1 and Note 6B).[75] At the investigative stage (prior to formal charges being laid) legal advice is not means

[69]See, for example, s. 125 of the Magistrates' Court Act 1980 (UK); see also s. 33 PACE, which allows a constable to execute an arrest warrant, even if he or she is not in possession of it.

[70]See PACE Code C, [10A] about how reasonable grounds for suspicion may be formed; Section 24A of PACE deals with arrest without warrant by persons other than a constable of police.

[71]See also PACE Codes C, [10.3] and G [2.2 and 4.3]. See further s. 32(5) and (6) of PACE in relation to the power to search the person or location for evidence; and see s. 32(8) for the power to seize and retain any items located.

[72]See Howse (2013), p. 133.

[73]See also PACE Code C, [10.2, 10.3 and 10B].

[74]For a discussion of the terms of the caution, see 4.5.1.1.1 below. For search on arrest, refer to s. 32 PACE.

[75]This rule is subject to certain restrictions in relation to terrorism offences (see Code C, Annex A and C). The same rules apply for vulnerable persons, juveniles and children.

tested and is provided free to all detainees.[76] If a detainee does not wish to consult with a legal representative there must be a clear waiver of the right recorded in the custody record (Code C, 6.5). Amongst other rights, there is a right not to be held incommunicado (Code C, 5). In addition, an arrested person who comes from another state has the right to contact and consult with their consular offence, embassy or High Commission (Code C, 3.3 and 7). There is also a right to an interpreter under Code C. These rights are generally provided in written form to the suspect who is at a police station as part of a Letter of Rights.[77]

There are limitations to the use of police detention and if the grounds for detention no longer apply, an arrested person should be released (s. 34 PACE). Under s. 37 of PACE, where a person is arrested without a warrant and there is insufficient evidence to charge, a custody officer must order the suspect's release, unless continued detention is necessary (based on reasonable grounds) for the purpose of securing or preserving evidence of the offence or obtaining evidence by questioning the suspect.

In general, a person may not be in police detention without charge for more than 24 hours (s. 41(1)). However, pursuant to ss. 42, 43 and 44, if continued detention is required in relation to an indictable offence (for the purpose of securing evidence or questioning) then a person may be detained for up to 96 hours.[78]

4.4.3 Interviewing

Rule 11.1 of the PACE Code of Conduct C[79] defines a 'police interview' as questioning a person under caution. In England, pre-trial custodial interrogation is regarded as coercive and various rules have been devised to regulate it. Following arrest, only exceptionally may a person be interviewed about the relevant offence in a location other than a police station (Code C, 11.1).[80] Exceptions do apply to this rule where the delay of transporting the person to the appropriate station may cause harm to a person or hinder the investigation. When a suspect is interviewed, he or she must be informed about the nature of the offence.[81]

---

[76] Cape et al. (2010), p. 115. The provision of legal advice at the police station is given free in accordance with a duty lawyer scheme.

[77] See Spronken (2010), at Annex 1, Letters of Rights per country, 12 (page 83 of 246).

[78] In most cases, this applies from the time of a person's arrival at the police station or 24 hours after his or her arrest, whichever is the earliest. See also Code C, 15 in relation to reviewing the period of detention. Extensions of detention may be made by a superintendent or higher, or by way of warrant of detention issued by a magistrate.

[79] Other provisions of PACE and Code C deal with other types of measures to establish identity or other evidence, such as fingerprinting (s. 27) and intimate searches (Code C, Annex A).

[80] See Code C, 11.7–11.14 about accurate record keeping.

[81] For the conditions of detention while being interviewed in relation to juveniles or other vulnerable suspects, and the use of independent persons or interpreters at the police station, see Code C, 11.18–11.20, 12 and 13.

Prior to interviewing, or upon recommencement, a suspect should be reminded of the statutory right to access legal advice (PACE, 3.2; Code C, 11.2). In England there is a duty solicitor (lawyer) scheme whereby every person under arrest or in custody may communicate with a defence lawyer over the telephone via a call centre and may speak with a duty solicitor in person. The lawyers who are part of the scheme must undergo specific training and registration. A lawyer may be present in a custodial interview as of right. Detained persons may not be interviewed without a lawyer being present.[82] There is also a right to an interpreter.

For adverse inferences to be drawn in relation to a suspect's failure to answer a question or questions, he should be reminded of the offence being investigated, the fact he is being asked to account for, how this fact may be related to the offending, how the court may draw adverse inferences and that a record is being made of the interview (Code C, 10.11). The suspect should understand the terms of any caution given by the police (Code C, Note 10D).

A suspect who attends the police station voluntarily, and agrees to remain for the purposes of participating in an interview, also has the right to obtain free and independent legal advice, which includes the opportunity to communicate over the telephone with a lawyer from the Defence Solicitor Call Centre (Code C, 3.21). A suspect should be told of the possibility to obtain legal advice prior to being questioned (Code C, 10.2). A suspect attending voluntarily should also be given a written notice about the right to access legal advice and how to obtain it (Code C, 3.2 and Note 6B). A suspect always has the right to contact a lawyer of own choosing if he or she agrees to pay for the costs of such advice. Prior to being charged, a restriction is placed on drawing adverse inferences from the silence of the suspect when being questioned under caution if he has not been offered the opportunity to contact a lawyer (referred to at Sect. 4.5.1.1 below). This restriction does not apply to a person who has attended the police station voluntarily and who cannot be prevented from seeking legal advice at any time (Code C, Note 10C).

4.4.4 *Recording of Interviews*

All police interviews must be adequately and accurately recorded, at least in writing, whether they are conducted at a police station or elsewhere (Code C, 11.7).[83] If practicable, a written record should be made contemporaneously (or as soon as possible) which records an interview verbatim (Code C, 11.8). A written record should be acknowledged in writing by the interviewee (Code C, 11.9, 11.11). Special provisions apply in relation to juveniles, mentally disordered or vulnerable people (11.5–11.20).

[82] Howse (2013), p. 142. The same rules apply for vulnerable persons, juveniles and children.

[83] See also Code C, Note 10E.

An audio-recording shall be made of any interview at a police station in relation to the following: firstly, an interview under caution for any indictable offence (including an indictable offence that is triable on indictment or summarily[84]); secondly, an interview that exceptionally takes place after charge for an indictable offence; or thirdly, to inform an accused person after charge about the contents of a written statement or interview with another person (Code E, 3.1).[85] The tape-recordings shall be made in duplicate or triplicate, the master copy being sealed in front of the suspect or accused. The purpose of tape-recording interviews is to ensure the accuracy and integrity of the interview recording process (Code E, 1.5A and Note 2A).

PACE Code F regulates the use of visual recording of interviews, where an interviewing officer deems it appropriate (Code F, 3.1). In addition to the above three areas, Code F states that visual-recording might be appropriate, for example, in relation to interviewing deaf, blind or speech-impaired persons; in the presence of anyone requiring an appropriate adult; or where the suspect or their representative so requests (3.1).[86]

If a suspect has made a significant statement (or silence) in the hearing of the police in the period prior to the interview commencing, or during a break, this should be put to the suspect at the commencement (or recommencement) of the interview to be confirmed or denied (Code C, 11.4).

An interview must cease after all relevant questions have been asked, allowing the suspect sufficient opportunity to explain any ambiguities or innocence, and when, taking other evidence into account, the officer in charge or the custody officer form the reasonable belief that there is a reasonable prospect of conviction (Code C, 11.6).[87]

4.4.5 Charging and Legal Representation

Generally, once sufficient evidence has been obtained the person must be either charged or released without charge; and if they are not charged within the specified time frames (due to insufficient evidence) they must be released.[88] A person who is detained after charge must be brought before the Magistrates' Court as soon as practicable (s. 46(2) of PACE).

---

[84]For an explanation of the terms 'indictable' and 'summary' offences, see fn 842.

[85]Sections 60 and 60A of PACE obliges the Secretary of State to issue codes of practice in relation to tape-recording and visual-recording of police interviews. See further Code C, 3.2 and Note 6B.

[86]Provisions of the Terrorism Act 2000 independently regulate the practice of visually recording interviews in relation to certain offences under the Act.

[87]See also Code C, 16 about charging.

[88]See ss. 37(8), 41(7), 42(10), 43(15) and 44(7).

An accused person must be cautioned when formal charges are laid and may not be interviewed after this has occurred, unless certain exceptions apply (Code C, 16.2, 16.4 and 16.5). In *Omar v Chief Constable of Bedfordshire*[89] the EWHC considered the formal cautioning process contained in the Home Office Circular 18/1994. The court confirmed that the proper procedures to be followed in the investigation of an offence do not generally allow for the police to seek a confession following formal charges being laid. By this time the accused has had the opportunity to give his account and the police have already formed the view that there is a reasonable prospect of conviction. It is therefore unnecessary and inappropriate after formal charges are laid to seek further incriminating evidence directly from the accused.

A person has a right to legal representation from the time they are arrested and detained; or prior to the first court appearance. Legal aid is still available from the time a person is charged until final verdict. However, after formal charges are laid legal aid becomes means tested and the accused may have to pay for, or make a contribution towards the cost, if he is found guilty.[90]

4.5 Inferences of Guilt and the Right to Silence

4.5.1 Drawing Adverse Inferences from the Silence of the Accused

A suspect has the right to remain silent by refusing to answer the police when they put questions to him about a crime being investigated. Alternatively, a suspect may provide extensive answers to questions, but remain silent in relation to a particular matter that is material to the case. In addition, an accused is a non-compellable witness and may of right choose to remain silent rather than giving evidence at trial. Nevertheless, despite the operation of the right to silent, in certain circumstances English law allows adverse inferences to be drawn when the accused remains silent in the face of accusation.[91]

In 1993 in his judgment in *Ex parte Smith*,[92] Lord Mustill referred to the sixth aspect of the right to silence, which protects against adverse comment being made at trial in relation to the silence of the accused. Lord Mustill referred to this specific immunity as being part of a notion of 'fair play' ensuring that the accused is not

[89] *Omar v Chief Constable of Bedfordshire* [2002] EWHC 3060, [42].

[90] Cape et al. (2010), p. 115; Howse (2013), p. 174. Means testing was abolished in 2001 but was reintroduced in respect of magistrates' court cases in 2006 and is being reintroduced in relation to crown court cases in 2013.

[91] In relation to adverse inferences and the failure to call witnesses or to provide intimate samples, see Starmer et al. (2001), p. 224; and ss. 80(8) and 62(10) PACE respectively.

[92] *R v Director of the Serious Fraud Office, Ex p. Smith* [1993] A.C. 1.

punished for remaining silent. However, it is a well-established idea in the modern history of the common law that "...silence (or even a denial) in the face of accusation is capable, in some situations, of counting as an implied admission..."[93]

This idea received the backing of Parliament by the enactment in 1994 of ss. 34–37 of the Criminal Justice and Public Order Act (CJPOA). In the lead up to the enactment of these provisions, in the late 1980s and early 1990s, there had been a 'sustained attack' on the right to silence by the suggestion that it was misused by a number of hardened criminal groups, such as terrorists. A central theme of the attack was the problem of 'ambush defences', where the suspect remained silent when questioned by the police but later presented a defence at trial which was 'sprung' on the prosecution.[94] A Royal Commission that was set up in the early 1990s to examine the effectiveness of the criminal justice system found that there was no evidence to support the suggestion that ambush defences were resulting in low conviction rates. In addition, the research at the time seemed to suggest that the number of suspects actually exercising the right to silence during pre-trial interrogation was low. The Royal Commission recommended retaining the right silence; though it acknowledged the concern in relation to no-comment interviews and ambush defences and made a number of recommendations aimed at advanced disclosure, such as ss. 5–11 of the Criminal Procedure and Investigations Act (discussed at Sect. 4.5.1.1).[95]

In addition, the UK Parliament went ahead with the limitations on the right to silence that allowed for adverse inferences on the basis of the accused's silence. These provisions effectively removed the immunity referred to by Lord Mustill in a number of cases. Today, in England and Wales, refusal to answer questions or the failure to testify may come at a price.[96] This notion has also received the backing of the ECtHR which has found that provisions which allow the tribunal of fact to draw common sense adverse inferences from the accused's silence do not *per se* infringe art. 6.[97]

4.5.1.1 Adverse Inferences from Pre-charge Silence—s. 34 Criminal Justice and Public Order Act 1994

A suspect may feel compelled to speak to the police by the knowledge that if he remains silent there may be adverse inferences drawn at trial, which can make the prosecution case appear stronger.[98] Section 34 provides that when a suspect is

[93]Mirfield (2003), p. 227; a denial, in certain circumstances, may also be capable of being treated as an admission by the jury: Mirfield (2003), p. 239.

[94]Easton (2014), p. 23.

[95]Easton (2014), pp. 24–26 and 56–59.

[96]Mirfield (2003), p. 227.

[97](ECtHR) *John Murray v. The United Kingdom*, Application no. 18731/91, 8 February 1996, [47].

[98]Malek (2010), p. 1182.

questioned under caution, or informed that formal charges are being brought against him,[99] and fails to mention a material fact, which he could reasonably have been expected to mention *and* which he later relies on in his defence, then the tribunal of fact may draw any inferences of guilt that "appear proper".[100] By putting a suspect on notice of the adverse consequences of remaining silent, the effect of the provision may be to increase the pressure on suspects to state facts to the police that they may not otherwise have disclosed. Because this provision restricts rights which are recognised at common law as appropriate to protect an accused from the risk of injustice, what appears proper in the circumstances of the case it should be narrowly construed.[101]

Section 34 permits adverse inferences to be drawn when a suspect remains silent when questioned or charged by the police, but it also applies if the accused refers in trial testimony to some additional fact that was not mentioned prior to being charged.[102] Further, the section does not exclude the possibility that the trial judge may invite the jury to draw adverse inferences in relation to the accused's silence both at the time of questioning under caution, and when formal charges were being laid.[103] In fact, in some cases the fact that the accused remained silent at both pre-trial stages, may entitle the jury to draw a stronger inference against the accused.[104] However, there is one 'saving' in s. 38(3), which states that a person shall not be convicted solely on an inference drawn from such a failure.[105]

In 1996, the UK Parliament enacted ss. 5–11 of the Criminal Procedure and Investigations Act, which supplements s. 34 by obliging all accused persons awaiting trial on an indictable offence to state the general nature of his defence, including reference to any alibi witnesses the accused intends to call.[106] Failure to

[99]Section 22(9) Counter-Terrorism Act 2008 extends the application of s. 34 to post-charge questioning authorised by a judge under the 2008 Act, which came into force on 16 February 2009. Section 34(2A) disengages the operation of s. 34 in relation to "safety interviews" conducted under Sch. 8(8) of the Terrorism Act 2000, the admissibility of evidence arising from these interviews being dealt with under s. 78 Police and Criminal Evidence Act 1984 (discussed at 4.8.2 below).

[100]For a discussion about the conditions for the operation of s 34 see *R v Argent (Brian)* [1997] 2 Crim.App.R. 27, 32–33, CA (Lord Bingham C.J.).

[101]*R v Bowden (Brian Thomas)* [1999] 2 Crim.App.R. 176, 181, CA (Lord Bingham C.J.); see also (ECtHR) *Averill v The United Kingdom*, Application no. 36408/97, 6 June 2000; see further Richardson (2011), 15-414–15-424.

[102]An accused can rely upon a fact at trial not only by giving evidence, but also through the evidence of a defence witness and by eliciting facts through the cross-examination of a prosecution witness: see Malek (2010), p. 1188; see further, Mirfield (2008), pp. 249–250.

[103]*R v Dervish* [2001] EWCA Crim 2789, [48]; [2002] 2 Cr.App.R. 6.

[104]*R v Dervish* [2001] EWCA Crim 2789, [48]; [2002] 2 Cr.App.R. 6.

[105]If any of the conditions to the application of s. 34 do not apply, then the common law position in relation to the drawing of adverse inferences at trial from the silence of the accused when questioned by the police is preserved by s. 34(5)(b).

[106]Sections 5(6) and 11, respectively, Criminal Procedure and Investigations Act 1996.

comply with these provisions may also attract adverse comment by the trial judge to the jury.[107]

4.5.1.1.1 The Significance of Cautioning and Legal Advice in Relation to s. 34

Pursuant to s. 34(1)(a) the suspect must be cautioned prior to questioning, in order for adverse inferences to be drawn from silence. Following the enactment of s. 34, paragraph 10.4 of PACE Code C provided for a caution to be administered in the following terms:

> You do not have to say anything. *But it may harm your defence if you do not mention when questioned something which you later rely on in court.* Anything you do say may be given in evidence. [emphasis added]

Although minor deviations from this standard caution will not constitute a breach, the sense of the caution must be maintained in order for adverse inferences to be properly drawn.[108] The caution is generally provided in written form at a police station as part of a letter of rights and entitlements provided to suspects arrested by the police or otherwise at the police station.[109]

In addition, under s. 34(2A), in order for s. 34 to apply in relation to a suspect who was questioned at a police station, he or she must have been given the opportunity to consult with a legal representative prior to being questioned or charged. Annex C of the PACE Code C provides that if a person is in custody and asks for but does not receive legal advice then adverse inferences may not be drawn under s. 34. This restriction does not apply if a suspect has attended the station voluntarily for interview.[110]

Section 34 is a truth-seeking tool. It has been designed to promote the public interest in an early disclosure of a suspect's account to the police, by formally allowing common sense adverse inferences to be drawn if the jury is satisfied that tan accused has dishonestly introduced a defence at the trial.[111] As described by the court in *R v Essa*,[112] it is not the silence in the interview that is the focus of s. 34, but what is said at trial. The s. 34 warning puts the suspect on notice that if the nature of the evidence or circumstances in the case clearly calls for an explanation from him, then he may be penalised for not providing it at the earliest opportunity and instead waiting until the trial. Under s. 34, an adverse inference is not drawn purely on the basis that the accused remained silent, but that he remained silent about an

[107]Section 11(5), Criminal Procedure and Investigations Act 1996.
[108]Richardson (2011), [15-416].
[109]Copied in Spronken (2010), at Annex 1, Letters of Rights per country, 12 (page 83 of 246).
[110]See PACE Code C, Note 10 C.
[111]See *R v Webber* [2004] UKHL 1, 33.
[112]*R v Essa* [2009] EWCA Crim 43, [16–17].

explanation that he later gave at trial, which, he could reasonably have been expected to come forward with earlier. However, if the trial defence is simply "I was not there" and there is no evidence to suggest that he was, then the court in *Essa* suggested that a s. 34 direction by a trial judge to the jury may not be of assistance to the prosecution.

There is research that suggests that the current caution may be difficult for suspects to understand; and that eight out of ten people found that the adverse inference element of it was pressuring or threatening.[113] Significantly, other English studies suggest that very few people questioned in police custody actually remain silent, and that the proportion of interviews where a suspect makes a 'no comment' interview may have declined since the enactment of the Criminal Justice and Public Order Act 1994.[114] Some research concluded that the amendments had impacted on the pre-trial and trial defence practices by reducing the reliance on the right to silence, though there was no evidence that this had led to an increase in conviction rates.[115]

4.5.1.1.2 The Right to a Fair Trial and the Drawing of Adverse Inferences

As discussed in Chap. 2, Sect. 2.3.2.3, the drawing of an adverse inference from the accused's silence questioned by the police is of itself compatible with art. 6 of the ECHR.[116] However, in order for adverse inferences to be taken into account without infringing art. 6, it must be fair in all the circumstances of the case.[117] For example, an inference should not be permitted in cases where the safeguard provisions in PACE were not met, such as the requirements of cautioning, the right to access legal advice, or the obligation of the police to refrain from interviewing a suspect and lay charges immediately after forming a view that there is sufficient evidence to establish a reasonable prospect of conviction.[118] Likewise, no adverse inference should be drawn if the circumstances of the case demonstrated that there was bad faith by the police.[119] The jury must be properly directed about the reasons why the explanation put forward by the accused at trial for remaining silent prior to

---

[113]Malek (2010), p. 1182 f. 7 and 1189 f. 67.

[114]Malek (2010), pp. 1182–1183 at fn 11 and 12; see also Easton (2014), pp. 58–59.

[115]Easton (2014), pp. 59–60.

[116](ECtHR) *John Murray v. The United Kingdom*, Application no. 18731/91, 8 February 1996, [47].

[117]*R v Dervish* [2001] EWCA Crim 2789; [2002] 2 Cr.App.R. 6; see also (ECtHR) *Averill v The United Kingdom*, Application no. 36408/97, 6 June 2000 as to the finding of the ECtHR in relation to the conditions that must be met in order to properly draw an adverse inference from silence under s. 34.

[118]*R v Dervish* [2001] EWCA Crim 2789; [2002] 2 Cr.App.R. 6; *R v Pointer* [1997] Crim. L.R. 676. See also (ECtHR) *John Murray v. The United Kingdom*, Application no. 18731/91, 8 February 1996.

[119]*R v Dervish* [2001] EWCA Crim 2789; [2002] 2 Cr.App.R. 6.

trial may be plausible and trustworthy.[120] Recent authority suggests that an adverse inference should only be drawn where the sole sensible explanation for the accused's failure to answer police questions was that the accused had no answer, or no answer that would withstand scrutiny.[121] The jury should be sure that the prosecution's case was so strong at the time of the interview it clearly called for an answer.[122] There may be various reasons why suspects are reluctant to speak to the police, such as fears for personal safety, safety of family members or the risk of being perceived as an informant could account for a reluctance to speak in certain situations.[123]

A controversial issue in relation to s. 34 is whether an adverse inference can be drawn if a suspect has received legal advice to remain silent during questioning.[124] As discussed in Chap. 2, Sect. 2.3.2.3, the ECtHR in *Condron v. The United Kingdom*[125] found that a lawyer's advice to remain silent based on the belief that a suspect was unfit to be interviewed, may be a legitimate reason why a suspect declines to answer questions in the police interview and should be given appropriate weight by the jury.

However, in *R v Howell* the Court of Appeal stated that the fact that a suspect may have genuinely followed legal advice does not necessarily disallow adverse comment:

> [w]hat is reasonable depends on all the circumstances. ...There must always be soundly based objective reasons for silence, sufficiently cogent and telling to weigh in the balance against the clear public interest in an account being given by the suspect to the police. Solicitors bearing the important responsibility of giving advice to suspects at police stations must always have that in mind.[126]

Jury directions need to be carefully formulated to avoid a violation of art. 6.[127] Strengthened by the guidance of the Judicial Studies Board in their specimen directions, s. 34 is said to have "...introduced a degree of consistency and certainty regarding the scope and nature of proper judicial comment on silence, which was absent at common law."[128]

---

[120](ECtHR) *Condron v. The United Kingdom,* Application no. 35718/97, 2 May 2000; see also (ECtHR) Beckles v. The United Kingdom, Application no. 44652/98, 8 October 2002.

[121]*R v Petkar* [2004] 1 Cr. App. R. 270; *R v Benn* [2005] EWCA Crim 2100.

[122]See Malek (2010), p. 1194; see further Richardson (2011), [15-414–15-417].

[123]Easton (2014), p. 173.

[124]See Mirfield (2003), *Silence, Confessions and Improperly Obtained Evidence*, 255–258.

[125](ECtHR) *Condron v. The United Kingdom,* Application no. 35718/97, 2 May 2000; see also (ECtHR) *Averill v The United Kingdom*, Application no. 36408/97, 6 June 2000.

[126]*R v Howell* [2003] EWCA Crim 1, [24]; [2005] 1 Cr.App.R. 1.; [2003] Crim. L. R. 405; but see *R v Betts; R v Hall* [2001] EWCA Crim 224; [2001] 2 Cr.App.R. 16.

[127]*Birchall* [1999] Crim. L. R. 311 CA.

[128]Malek (2010), p. 1185; on new specimen directions see further *R v Armstrong* [2009] EWCA Crim 643.

Section 34(5)(b) preserves the common law in relation to drawing adverse inferences from silence when questioned or charged.[129] Therefore, where the circumstances of the case prevent adverse inferences from being drawn under the section, the trial judge should consider whether adverse inferences from such silence can be drawn at common law. However, the common law position is generally perceived to be more lenient on an accused.[130]

4.5.1.2 Adverse Inferences from the Failure When Arrested to Explain the Presence of Identifying Evidence or Own Presence at the Crime Scene—ss. 36 and 37

The CJPOA also provides for adverse inferences to be drawn against an accused if upon arrest he is questioned by the police and fails to account for his possession of an object, a substance or a mark (s. 36); or for his presence at the crime scene at or around the relevant time (s. 37). These provisions stipulate a 'special warning' must be given by arresting officers in the relevant circumstances to ensure fairness to a suspect that he or she understands the context of the questions being asked and the way in which the replies may be used at trial.[131] Similar to s. 34, it may be necessary for the trial judge to direct the jury as to whether and how to draw adverse inferences that are fair and proper in the circumstances of the case. However, there is no explicit requirement for the jury to consider the circumstances of failing to give an explanation or the reasonableness of the person's failure to do so.[132]

4.5.1.3 Adverse Inferences from the Failure of the Accused to Testify—s. 35

As discussed at part 3 above, the enactment of the Criminal Evidence Act 1898 significantly changed the law in relation to the status of the accused at trial. The accused became a competent but non-compellable witness. In addition, the prosecutor was prohibited from making any comment on the failure of the accused to testify. The provision was silent in relation to comment from other actors in the trial. This was interpreted by the courts to mean that the judge and counsel for a co-accused were not prohibited from commenting.[133] English judges continued to make adverse comments about failure to testify well into the twentieth century.

[129] See also ss. 36(6) and 37(5) that preserve the common law in relation to the other pre-charge failures, which attract adverse comment.

[130] Mirfield (2003), pp. 240–246.

[131] See Malek (2010), p. 1196; see, for example, *R v Compton* [2002] EWCA Crim 2835; (2003) 147 S.J.L.B. 24.

[132] Emmerson et al. (2012), p. 632.

[133] Malek (2010), pp. 1197–1198.

Section 35 of the CJPOA[134] regulates the position in relation to adverse inferences from failure by the accused to testify. At a contested hearing, an accused must be made aware that he or she has the right to give evidence but cannot be compelled to give evidence.[135] However, if an accused chooses to remain silent at trial, the court may draw any adverse inferences that "appear proper".[136] The same applies if an accused is sworn but refuses to answer certain questions without good cause.[137] However s. 35(4) emphasises the non-compellability of an accused and that he will not be in contempt of court as a result of his failure to give evidence.[138]

The Court of Appeal considered the necessary scope of a trial judge's direction under s. 35 in *R v Cowan*.[139] In cases where adverse inferences may be drawn, the trial judge is obliged to direct the jury properly in accordance with the terms of s. 35 and not simply leave it to the jury to decide for themselves what inference can be drawn. The court in *Cowan* accepted the Judicial Studies Board specimen direction, and stated that a jury direction under s. 35 must include reference to the following: first, the prosecution bears the overall burden of proof throughout the proceedings; second, the accused is entitled to remain silent; third, the adverse inference cannot be the sole basis to prove guilt[140]; fourth, the prosecution must have presented a *prima facie* case before any inference can be drawn; and fifth, an adverse inference can only be drawn if silence can be sensibly explained by the accused having no answer to the facts in issue, or none that would stand up to cross-examination. A number of these factors were emphasised by the ECtHR in the case of *Murray v The United Kingdom*.[141]

Where a prosecution case is weak, or contains significant weaknesses, then a proper direction by the trial judge may remind the jury of the inappropriateness of

---

[134]Section 35 was amended by s. 120(2) and Sch. 10 of the Crime and Disorder Act 1998, and by s. 331 and Sch. 36 of the Criminal Justice Act 2003.

[135]Section 35(2) Criminal Justice and Public Disorder Act 1998. The warning must be formulated in accordance with the *Practice Direction (Criminal Proceedings: Consolidation)*, [44]. See also Richardson (2011), [4-306].

[136]Section 35(2), Criminal Justice and Public Disorder Act 1998. The section is premised on the fact that there is no reason due to the physical or mental condition of the accused that makes it undesirable for him or her to give evidence: s. 35(1) and see *R v Cowan* [1996] Q.B. 373 and *R v Friend* [1997] 2 All E.R. 1011; *DPP v Kavanagh* [2005] EWHC 820 QBD (Admin), [16–17] about the wide discretion conferred. See further Richardson (2011), [4-399].

[137]Sections 35(2) and (3) Criminal Justice and Public Disorder Act 1998.

[138]This rule also applies to accused persons under the age of 14 in relation to offences that were committed after September 30, 1998. See further ss. 35 and 120(1) Criminal Justice and Public Disorder Act 1998; and Malek (2010), p. 342.

[139]*R v. Cowan* [1996] 1 Cr. App. R. 1; see also *R v Becouarn* [2003] EWCA Crim 1154.

[140]See s. 38(3) Criminal Justice and Public Disorder Act 1998; and see *R v Birchall* [1999] Crim. L.R. 310.

[141](ECtHR) *John Murray v. The United Kingdom*, Application no. 18731/91, 8 February 1996; see also Mirfield (2003), pp. 247–248.

drawing adverse inferences in certain circumstances.[142] Therefore, the requirement of a direction in the proper form is an important safeguard in the operation of the provision.

4.5.2 Lies by the Accused, Untrue Alibi and Adverse Inferences

Lies told by the accused, even though they may function as a statements made against personal interest, are not treated as confessional statements or direct evidence of guilt.[143] However, they "...may indicate consciousness of guilt and therefore amount to circumstantial evidence of guilt."[144] Evidence of lies can therefore strengthen the case against the accused. If the jury finds beyond reasonable doubt that the accused told a deliberate lie in relation to a material issue; and that the motive to lie was the realisation of guilt and there was no innocent explanation for lying, then this can be used against the accused to corroborate or strengthen the prosecution case.[145] Thus, the jury must bear in mind that some people lie for reasons other than a realisation of guilt, such as to cover up the guilt of another or their own shameful non-criminal behaviour. Like adverse inferences about silence, the prosecution case cannot rely solely on evidence of the accused's lies.

Similarly, pursuant to s. 11 of the Criminal Procedure and Investigations Act 1996 (referred to at Sect. 4.5.1.1 above), if an accused discloses a false alibi the trial judge or other parties may make adverse comments and the jury may draw adverse inferences that appear proper in the circumstances of the case.[146]

Lies by the accused can support evidence of guilt or they may merely reflect on the accused's credibility. If the prosecution can prove that the accused has lied, or made a testimonial statement which is inconsistent with something said prior to trial, then this will reflect poorly on his or her credibility. There is authority to suggest that the credibility of the accused is always an important issue.[147] Therefore, under ss. 101 or 103 of the Criminal Justice Act 2003, an accused lying may even be the gateway for the prosecution to introduce the accused's previous convictions into evidence as evidence of 'bad character' and a propensity to be

[142]See Malek (2010), pp. 1201–1202, referring to the judgment of Lord Slynn in *Murray v DPP* [1994] 1 W.L.R. 1, 10.

[143]*R v Hasan (Aytach)* [2005] 2 A.C. 467. Therefore, evidence of lies can never be excluded under s. 76 PACE, but may be excluded under s. 78, depending on the circumstances of the case: Richardson (2011), 15-354.

[144]Malek (2010), p. 1232; Richardson (2011) [15-414].

[145]*R v Lucas* [1981] QB 720; *R v Goodway* (1994) 98 Cr App R 11.

[146]See further *R v Lesley* [1996] 1 Cr App R 39; *R v Drake* [1996] Crim LR 109.

[147]*R v Baker* [1895] 1 Q.B. 797, 799.

untruthful, particularly if the accused has previous convictions for offences which under English law involve fraud, dishonesty or deception, such as perjury, perverting the course of justice, fraud or deception.[148]

4.6 Administrative Questioning Powers in Serious or Complex Fraud Cases

A number of English statutory provisions create administrative compulsory questioning powers that enable investigators to gather information by questioning individuals about, for example, their business and financial affairs, obliging the individuals to respond. Although they vary in detail, in general, the provisions create offences where the individuals do not respond, or respond dishonestly.

The idea that Parliament can limit the right to silence and the privilege against self-incrimination under compulsory questioning powers is not new.[149] For example, compulsory examination of a bankrupt was allowed under the Bankrupt Law Consolidation Act 1849. Today, the use of compulsory powers in England to question individuals under pain of punishment is often seen in the commercial and corporate sectors.[150] Recent illustrations of such powers include provisions under the Insolvency Act 1986, which allow for the examination of the bankrupt, the bankrupt's spouse, other persons with knowledge of the bankrupt's affairs, as well as individuals involved in the formation of or employed by insolvent companies.[151] Department of Trade and Industry (DTI) inspectors have compulsory questioning powers under ss. 431(5), 432(2) and 436(3) of the Companies Act 1985.

In *Brown v Stott (Procurator Fiscal, Dunfermline)*,[152] Lord Bingham considered the enactment of compulsory questioning powers in relation to road safety offences. It was highlighted that, while the overall fairness of criminal proceedings must be maintained, specific fair trial rights such as the right to silence and the privilege against self-incrimination are not absolute. The decision also referred to the need to establish a fair balance between the general interest of the community and the rights of the individual when considering whether Parliament's enactment of compulsory questioning powers in relation to serious social problems was compatible with art. 6.

Part 2, Chapter 1 of the Serious Organised Crime and Police Act 2005 (SOCPA) creates compulsory questioning powers to be exercised by the Director of Public

[148]See further Emmerson et al. (2012), pp. 609–610; and see, for example, the Theft Act 1968 (UK).

[149]Mirfield (2003), p. 228.

[150]See further Richardson (2011), [16-69, 30-10].

[151]See further ss. 366(1) and 235(3) of the Insolvency Act 1986.

[152]*Brown v Stott (Procurator Fiscal, Dunfermline)* [2001] 2 All ER 97; [2003] 1 A.C. 681.

Prosecutions, the Director of Revenue and Customs Prosecutions and the Lord Advocate under ss. 60–70 of the Act in relation to 'lifestyle' offences (such as drug trafficking, money laundering and prostitution),[153] terrorism offences,[154] customs, tax and theft offences.[155] The Criminal Justice Act 1987 (CJA) also contains similar powers in relation to serious and complex fraud investigations, and provides a useful statutory example of the ways in which these powers can operate in England and Wales.[156]

Section 1 of the Criminal Justice Act 1987 (CJA) allows for the appointment of the Director of the Serious Fraud Office (for England, Wales and Northern Ireland) to investigate and prosecute "any suspected offence" which appears to involve serious or complex fraud (subsection 3).[157] This takes place under the superintendence of the Attorney-General. The investigation may be conducted independently or in conjunction with the police or some other appropriate person. Under s. 2, by written notice, the Director of the SFO can order a person to supply information by answering questions or producing documents. Section 2, sub-sections (2) and (13) of the CJA together make it a criminal offence (punishable by a fine or by imprisonment of up to 6 months) for a suspect or other relevant person to fail to answer questions or disclose information that may be relevant to an investigation under the Act.[158]

This provision was the subject of the case before Lord Mustill in *R v Director of the Serious Fraud Office, Ex p. Smith*.[159] Lord Mustill referred to the decision of the Divisional Court in the case to illustrate the rationale behind such provisions:

> Faced with the myriad opportunities for the concealment of fraudulent activities which companies and trusts provide, Parliament has given the Serious Fraud Office the power to call upon a suspected person to come into the open, and to disclose information which may incriminate him.[160]

The question that has been raised in cases such as this is how far down the road had Parliament actually intended to go? To what extent did the CJA remove the protection afforded by the Police and Criminal Evidence Act 1984 (PACE) and the PACE Code C, in particular the right to refuse to answer police questions? Criminalising the non-provision of self-incriminating information in serious fraud

---

[153] For a full description of lifestyle offences, see Schedule 2 of the Proceeds of Crime Act 2002.

[154] Sections 15–18, Terrorism Act 2000.

[155] For a full description of lifestyle offences, see Schedule 2 of the Proceeds of Crime Act 2002; see also ss. 15–18 Terrorism Act 2000, s. 170 Customs and Excise Management Act 1979, s. 72 Value Added Tax Act 1994 and s. 1 Criminal Attempts Act 1981.

[156] See also Financial Services Act 1986, the Banking Act 1987 and the Youth Justice and Criminal Evidence Act 1999.

[157] The Director may also institute criminal proceedings (subsection 4).

[158] See also s. 2(1), CJA.

[159] *R v Director of the Serious Fraud Office, Ex p. Smith* [1992] 1 All ER 730.

[160] *R v Director of the Serious Fraud Office, Ex p. Smith* [1992] 1 All ER 730, 736–737; see further *The Fraud Trials Committee Report* (London, 1986).

cases, involves three major issues: firstly, do the relevant authorities have the right to demand answers to questions on pain of punishment in the pre-trial phase; secondly, if incriminating answers are obtained, are they admissible in a criminal trial; and thirdly, is any real evidence that has been obtained by the investigating authority either directly under compulsory production powers or indirectly, on the basis of the answers given under the compulsory pre-trial examination (derivative real evidence),[161] admissible at trial? Each of these questions will be examined in detail to the extent they relate to the non-provision of statements and the right to silence.

4.6.1 The Legitimacy of Applying Administrative Questioning Powers in the Pre-trial Phase

The respondent in *R v Director of the Serious Fraud Office, Ex p. Smith*, a company chairman and managing director, was charged, cautioned and admitted to bail in relation to being a party to a company that carried on business with intent to defraud its creditors, contrary to s. 458 of the Companies Act 1985. Subsequently, he was served a notice by the Director of the Serious Fraud Office under s. 2 of the CJA obliging him to attend and answer questions or furnish information to the investigating officers. Since the respondent had been compelled (prior to trial) to provide answers that were potentially self-incriminating, the decision particularly raised the immunity described by Lord Mustill at 3 above.

In addition to compelling the accused to answer incriminating questions, it was found that the penalty provided for in s. 2(13) of the CJA, over-rode the general position in Code C, paragraph 16.5, that the police have a duty to refrain from questioning after forming the opinion that there is sufficient evidence against the suspect for there to be a reasonable prospect of conviction. In coming to the conclusion that it was irrelevant that the respondent had been compulsorily examined after charge, Lord Mustill referred to the judgment of Windeyer J in *Rees v Kratzmann*:

> If the legislature thinks that in this field the public interest overcomes some of the common law's traditional consideration for the individual, then effect must be given to the statute which embodies this policy.[162]

Lord Mustill concluded that Parliament had clearly intended to establish an inquisitorial regime in relation to serious or complex fraud. Although the aim of one aspect of the right to silence is to prevent self-incriminating material from becoming available in the first place,[163] it was found that the legislature had chosen

---

[161] Starmer et al. (2001), pp. 209 ff. refer to all real evidence obtained under compulsory powers as 'derivative' evidence. However, the author defines derivative evidence as only referring to real evidence that has been obtained as a result of the information gathered from the answers given under compulsory questioning powers.

[162] *Rees v. Kratzmann* (1965) 114 CLR 63, 80.

[163] Mirfield (2003), p. 226.

to limit the protection of the right to silence to this extent, in the context of investigating serious or complex fraud.[164]

4.6.2 *The Admissibility of Incriminating Answers Obtained Under Compulsion*

As discussed in Chap. 2, Sect. 2.3.2.4, in the case of *Saunders v. The United Kingdom*, the ECtHR considered the legitimacy of the English compulsory questioning powers under the Companies Act 1985. The court gave its *imprimatur* on the use of compulsory questioning powers, provided that any statements made by the suspect or accused could not be used against him in subsequent criminal proceedings. The ECtHR indicated their approval of the law creating the compulsory questioning powers used by the Serious Fraud Office, which also provided the suspect with 'use immunity'. Such immunity prohibits the state from using any testimony gained via the powers as evidence against the suspect in subsequent criminal proceedings.

As an English statutory example, s. 2(8) of the CJA ensures that any incriminating answers that are obtained under compulsory questioning procedures are not admissible in evidence against a person in subsequent proceedings. This is with two exceptions. Firstly, any incriminating statements obtained under compulsory examination may be used against an accused in a subsequent prosecution for an offence under s. 2(14) of knowingly or recklessly making a false or misleading testimonial statement, which is punishable under subsection (15) by up to 2 years imprisonment or a fine or both. Secondly, ss. 2(8) and (8AA) together provide that a statement obtained by virtue of s. 2 may be used against a person in subsequent proceedings for some other offence as a prior inconsistent statement, if contradictory evidence has been adduced by or on behalf of the accused. Therefore, the prosecution may not use the statements as evidence of a prior inconsistent statement of an accused.

An investigator or inspector under compulsory questioning powers is often charged with the duty to investigate and report the business affairs of an individual or a company. Therefore, the duty is to ascertain facts, rather than to investigate and prosecute criminal offences. However, under s. 1 of the CJA the Director of the Serious Fraud Office has powers to investigate suspected offences and to conduct criminal proceedings. Thus, while it is of concern that the protective provisions of PACE and the PACE codes of conduct do not generally apply to compulsory

[164]Legal professional privilege may never be overridden under s. 2(9) and confidentiality agreements may not be broken without the consent of the other party under s. 2(10). See further *Roche Registration Limited, R (on the Application of) v Secretary of State for Health (Rev 2)* [2014] EWHC 2256 (Admin) (9 July 2014) in which a claim of unfairness under common law about lack of provision of information about the relevance of a Medicines and Health Care Regulatory Agency re-inspection to separate proceedings being brought by the European Medicines Agency was dismissed.

examinations,[165] the 'use immunity' is necessary to ensure that these powers may be viewed as proportionate. Likewise, under s. 65 of the SOCPA a statement made by a person under compulsory questioning powers in response to a 'disclosure notice' may not be used against him in criminal proceedings.

Section 2 of the CJA defines the extent to which Parliament says the right to silence can be reasonably limited, in view of proportionality and the aims of the power in question. While Parliament has allowed this aspect of the right to silence to be limited, by compelling suspects to give sworn testimony which may be self-incriminating, it has restricted the subsequent use that can be made of that testimony.[166] The compulsory questioning procedures may be used to generate information which uncovers the existence of company fraud and 'informs' an investigation, but the state has been prevented from relying on such (self-incriminating) testimony as the basis for a fraud conviction.

If a person is questioned by an officer of the Serious Fraud Office not using the compulsory powers, then s. 67(9) of PACE obliges persons other than police officers to also have regard to the codes of conduct in the investigation of offences or charging of offenders. Under PACE Code C a suspect being questioned by an authority should be offered access to legal advice prior to being questioned (as referred to at 4.4.5 and 4.5.1.1.1 above). However, if the person is being questioned outside of the custodial setting then a lack of access to a lawyer may not necessarily render the questioning unfair or inadmissible. Nevertheless, despite the fact that Code C protections may not apply to self-incrimination under compulsory questioning powers, a trial judge always has discretion to exclude any statements under s. 78(1) or 82(3) of PACE (discussed in detail at Sects. 4.8.2 and 4.8.3 below).

One anomaly has arisen due to the interaction of the provisions under the Insolvency Act 1986 and those under the CJA in relation to serious fraud investigations. In 1994, the House of Lords in *Hamilton and Others v Naviede* found that nothing expressly or impliedly prevents the Director of Serious Fraud from obtaining incriminating statements made in an examination under the Insolvency Act for use in a fraud prosecution.[167] Mirfield suggests:

> In principle, subject to the precise terms of the statutory language used, the same would seem to apply to statements indicating guilt of serious fraud which have been collected under any other statutory power precluding the privilege against self-incrimination, at least where the statute in question grants full admissibility at a criminal trial.[168]

However, in light of the duty under the Human Rights Act 1998 of public authorities to act in a way that is compatible with human rights and the jurisprudence of the ECtHR, it is unlikely that the Director of Serious Fraud would use evidence obtained under compulsory powers in this way.

---

[165] See *R v Seelig* [1992] 1 WLR 148, 157–158.

[166] See further Hansard, HC vol. 305, cols 639–640 cited in Starmer et al. (2001), p. 215.

[167] *Hamilton and Others v Naviede* [1995] 2 AC 75.

[168] Mirfield (2003), p. 231.

Following the ECtHR's decision in *Saunders v. The United Kingdom*,[169] the English courts have traditionally confined the application of the right to silence and the privilege against self-incrimination to evidence of statements being a product of the accused's will. There are a number of statutory examples of regulatory compulsory powers in England and Wales where suspects (or accused persons) are required to produce real evidence that is incriminating. Frequently, such evidence may also be acquired under search warrants, or even by force in some circumstances. By contrast to statements, in line with the decision in *Saunders*, when real evidence is obtained directly from a suspect or accused by compulsory powers in England and Wales there is usually no 'use immunity' to prevent its admissibility against the suspect or accused in future criminal proceedings.[170]

4.6.3 The Admissibility of Real Derivative Evidence

Compulsory questioning powers are used to directly obtain information (usually oral statements) about suspected criminal offences. From that information criminal investigators may be able to locate real evidence, for example documentary evidence or evidence of a clandestine laboratory, which is also relevant to a prosecution. This real evidence may be called derivative evidence because it has been obtained *indirectly* as a result of the information obtained directly under the compulsory questioning powers.

As an example, the CJA is silent as to the admissibility of derivative real evidence, which is located as a result of the information that is directly obtained through the use of compulsory questioning powers. Although the answers given under compulsory examination may not generally be used by the prosecution in future criminal proceedings (other than in the limited ways referred to above and specified in s. 2(8) and (8AA)), the CJA does not restrict the use of the *information* contained in that testimony. This approach is consistent with the doctrine of the 'fruits of the poisonous tree', which was first asserted at common law in relation to the confession rule in the 1783 case of *Warickshall*.[171] In this case the evidence of a

[169](ECtHR) *Saunders v. The United Kingdom*, Application no. 19187/91, 17 December 1996; and see the discussion about compulsory powers in Chap. 2, Sect. 2.3.2.4.

[170]See further ss. 2(3) and 2(4) of the Criminal Justice Act 1987 in relation to compulsory powers and warrants to obtain documents or knowledge of the whereabouts; s. 6(4) of the Road Traffic Act 1988, making it an offence to refuse to provide breath specimen without reasonable excuse; s. 78 Police and Criminal Evidence Act 1987; and *R v S(F)* [2008] EWCA Crim 2177 where the Court of Appeal found that while the 'privilege against self-incrimination' was engaged to a limited extent, compulsion to obtain memorized keys to unlock encrypted data was proportionate; but see the decisions in (ECtHR) *Funke v. France*, Application no. 10828/84, 25 February 1993; (ECtHR) *Weh v. Austria*, Application no. 38544/97, 8 April 2004, [42–43].

[171]*R v Warickshall* (1783) 1 Leach 263 [168 ER 234]; in relation to the 'fruits of the poisonous tree' doctrine, see further the judgment of the US Supreme Court no. 82-1651, *Nix v. Williams*,

confession that was induced by 'promises of favour' was inadmissible. However, the stolen property that had been found as a result of the confession was deemed admissible. Questions surrounding the admissibility of derivative real evidence obtained as a result of an unlawful interrogation may be dealt with by an application for discretionary exclusion of evidence on the ground of unfairness to an accused under s. 78 PACE (see Sect. 4.8.2 below).

As discussed in Chap. 2, the ECtHR has not directly commented on the fairness of using derivative real evidence obtained as a result of using compulsory questioning powers.

4.7 Obtaining Confession Evidence by Covert Surveillance Measures

English legislation establishes a range of covert surveillance measures such as interceptions, listening devices and covert human intelligence sources (CHIS), also known as undercover operatives, which are available to the police and other investigative agencies for secretly recording conversations and obtaining self-incriminating intelligence and evidence. It is well established that powers enabling private transactions to be recorded and observed constitute an interference with the right to a private life in art. 8 of the ECHR. However, these interferences are frequently accepted as being necessary in a democratic society in the prevention or detection of crime, provided they are accompanied by safeguards against abuse.[172] Neither suspicion nor offence type is a criterion for covert surveillance. However, covert surveillance must be necessary and proportionate in relation to permitted purposes, such as national security and the prevention and detention of crime.[173]

In England and Wales the Regulation of Investigatory Powers Act 2000 (RIPA) governs the use of covert surveillance, together with Part III of the Police Act 1997 (which regulates surveillance powers that involve interference with property) and the Intelligence Services Act 1994 (which regulates surveillance warrants for use by MI6 and GCHQ).[174] Part I of RIPA provided for the issuing of warrants by the Secretary of State in relation to most interceptions. However, s. 37 of the Protection of Freedoms Act 2012 (UK) now requires judicial approval for obtaining or disclosing communications data. The remaining provisions of RIPA that involve

11 June 1984, 467 U.S. 431 (1984), 441 ff. and the discussion of the doctrine in (ECtHR) *Gäfgen v. Germany*, Application no. 22978/05, 1 June 2010, [73].

[172](ECtHR) *Klass and Others v. Germany*, Application no. 5029/71, 6 September 1978. See further Emmerson et al. (2012), pp. 315–319; Starmer et al. (2001), pp. 34–35 and 37–39.

[173]See further Howse (2013), pp. 148–152.

[174]In relation to offences carrying a term of imprisonment of at least 3 years imprisonment; see further Emmerson et al. (2012), pp. 319 and 325–329. GCHQ stands for Government Communications Headquarters.

other types of secret surveillance and the relevant Police Act provisions rely on a system of authorisations by an appropriately high ranking officer under the scrutiny and supervision of Commissioners, who hold or have held high judicial office. However, s. 38 of the Protection of Freedoms Act requires judicial approval for directed surveillance and CHIS that is considered intrusive surveillance.[175] In addition, the regulatory scheme is supported by a number of practice codes, being the Interception Code of Practice, the Surveillance Code of Practice and the Covert Human Intelligence Sources Code of Practice (CHIS Code).

RIPA and the Covert Human Intelligence Sources Code of Practice (CHIS Code) regulate the use of both informants and undercover officers. There is no special authorisation procedure for the use of infiltration. However, where a breach of art. 8 is at risk then the procedures laid down should be followed.[176] Due to the expense and difficulty in carrying out infiltration, the measure is reserved for cases involving serious crimes.[177]

Section 17 of RIPA prohibits the disclosure or adducing of any interception evidence in legal proceedings, which discloses the use of an interception of communications. Therefore, interception evidence generated in the UK can only be use as information in a criminal investigation and not as evidence in a prosecution.[178] Under s. 18, this prohibition also applies to interception data being used outside the UK. Section 15(6) provides that the Secretary of State will determine the conditions under which interception data may be surrendered for use in investigations outside the UK.

An accused may attempt to have confession evidence excluded where it was obtained unlawfully or unfairly through the use of covert surveillance. However, evidence obtained in breach of art. 8 is not automatically inadmissible and its admission will not necessarily render a trial unfair. The judicial discretion to exclude evidence is an important safeguard in the use of covert surveillance evidence and is discussed at 4.8 below.

---

[175]Directed surveillance is covert surveillance to obtain personal information that is not intrusive, in that it does not involve an entry onto private property: Home Office (UK), Covert Surveillance and Property Interference: Revised Code of Practice (Home Office, London, 2010).

[176]See Howse (2013), pp. 164–166. See also s. 80, RIPA 2000.

[177]Howse (2013), p. 165.

[178]Section 15 of RIPA provides for safeguard arrangements to ensure the security of interception data by limiting or prohibiting its disclosure. Under s. 15(6) these safeguard arrangements do not apply to data, which is surrendered to a country outside the UK. In this event, pursuant to s. 15(7), the Secretary of State determines the appropriate safeguard arrangements. However, there is no exception to s. 17 in relation to the use of UK interception evidence outside the UK.

4.8 Judicial Discretion to Exclude Unlawfully or Unfairly Obtained Confession Evidence

The issue of 'admissibility' of evidence and the use of judicial discretion to exclude improperly, unlawfully or unfairly obtained evidence is an extremely important fair trial safeguard in English criminal law. Frequently, where limitations have been placed on procedural rights during gathering evidence in the pre-trial phase, this problem is addressed, at least in part, by the question of admissibility. PACE and the PACE codes of practice contain protective provisions in relation to the treatment of a suspect by the police, for example during arrest, detention or questioning. A breach of the act or a code may be relevant to a determination under s. 76(2) or s. 78 but will not necessarily result in rejection of a confession.[179] Thus, judicial discretion to exclude evidence is an important element in the overall legal structure of English criminal proceedings, balancing the effectiveness of law enforcement with fairness to the accused. Judicial warnings may limit the use that can be made of evidence as an alternative to outright exclusion in the case of evidence that has multiple uses, some of which infringe art. 6.

There is a distinction in English law between confessions that are inadmissible because they are likely to be unreliable[180] and confessions that, although otherwise admissible, may be excluded from a trial in the exercise of judicial discretion because they were obtained in a manner that was unfair or their use causes unfairness.[181] Today, ss. 76 and 78 of the Police and Criminal Evidence Act 1984 (PACE) encapsulate this distinction. The former type of confession will be dealt with briefly, as will the common law discretion to exclude evidence, where the probative value of the evidence is outweighed by prejudicial effect of admitting it.

The focus of the following discussion is on the judicial discretion provided by s. 78 to exclude (wholly or partly) confession evidence that has been obtained unlawfully or unfairly. This discretion can be applied to restore balance when some aspect of the right to silence has been unfairly restricted by the operation of the law. As illustrated by the case of *R v Konscol* in the opening to chapter 1, a transnational case that turns on the fairness of relying on a confession that has been obtained according to the law of another country may be dealt with in an English trial under s. 78.

---

[179] *R v Samuel* [1988] QB 615 CA; in relation to decisions under ss. 76(5), 76(6) and 76A see *Lam Chi-Ming v The Queen* [1991] 2 A.C. 212, PC; *A v Secretary of State* [2005] 1 W.L.R. 414 at [241]; but see the judgment of the ECtHR in (ECtHR) *Gäfgen v. Germany*, Application no. 22978/05, 1 June 2010, [175–176] in relation to the potential inadmissibility of derivative evidence obtained in breach of Art. 3; s. 77 deals with confessions by mentally handicapped persons made in the absence of an independent person.

[180] A confession may be considered unreliable, for example, because it was obtained in oppressive circumstances or because the police offered some kind of inducement.

[181] Malek (2010), p. 1205.

The admissibility of evidence is a question of law to be decided by the trial judge. If the evidence is deemed to be admissible, the questions of the weight, credibility and sufficiency of evidence, although sometimes guided by legal principles supplied by the trial judge, are questions of fact and in the domain of the jury (or a judge sitting alone or as a panel of three magistrates).[182] Therefore, when the admissibility of evidence is challenged, the normal course in English jury trials is that the evidence which is sought to be excluded is heard and the question of admissibility (such as unfairness) determined on a *voir dire*[183] in the absence of the jury.[184] The jury is not required to hear the evidence and the legal arguments raised, and then ignore that in their deliberations on the question of guilt. Thus, the merits of the evidence are determined prior to the facts in issue—and not at the same time. This distinction is important in English criminal law. The right of an accused person to give evidence on a *voir dire*, without affecting his right to silence in the main trial is absolute.[185] After an admissibility ruling and the trial has recommenced before the jury, the prosecution may not lead evidence of what the accused said on a *voir dire*.[186] This is so even if, in the course of cross-examination about the admissibility of a confession, the accused confesses to a material fact in the main trial—a rule which has led to criticism of the English rules of evidence as not being suited to an enquiry which should arrive at the truth.[187] Where a single judge in summary proceedings decides to exclude unfairly obtained evidence under s. 78, it is thought that the judge should seek the views of the parties and consider whether the substantive hearing should be heard by a differently constituted bench (judge).[188]

---

[182]On questions of law and questions of fact, and the roles of judge and jury, see Malek (2010), pp. 19 and 196 ff.

[183]A *voir dire* is a 'trial within a trial' about a question of law such as the admissibility of evidence, where the judge hears the evidence in question and makes his or her legal ruling in the absence of the jury.

[184]There are a few exceptions to this rule, such as an examination of a witness to determine whether he or she understands the nature of an oath or the duty to tell the truth: Malek (2010), p. 14. The standard of proof on a *voir dire* is usually the criminal standard of proof: Malek (2010), p. 150.

[185]*R v Brophy* [1982] A.C. 476.

[186]See further Malek (2010), pp. 1223–1225.

[187]See for example, Malek (2010), p. 1225.

[188]*DPP v Lawrence* [2007] EWHC 2154 Admin; [2008] 1 Cr.App.R. 10. See further Malek (2010), pp. 14–17; but see Malek (2010), pp. 1217–1218, where it is argued that the line between law and fact has become blurred in relation to mandatory exclusion of confession evidence obtained by oppression under s. 76(2).

4.8.1 Admissibility of Confessions—Mandatory Exclusion Under s. 76 Police and Criminal Evidence Act 1984

Though the question of obtaining evidence by oppression, inducement or other circumstance affecting reliability is a national problem, and is not relevant to the transnational perspective of irregularly obtained evidence, the general framework for considering the admissibility of confessions in England and Wales begins with s. 76 Police and Criminal Evidence Act 1984.

Prior to the enactment of PACE in 1984, the admissibility of confessions was determined by the question of 'voluntariness'. The court in *Ibrahim v R* stated the common law test, as follows:

> It has long been established as a positive rule of English criminal law that no statement by an accused is admissible in evidence against him unless it is shown by the prosecution to have been a voluntary statement in the sense that it has not been obtained from him either by fear of prejudice or hope of advantage exercised or held out by a person in authority.[189]

Likewise, it became well established in English law that a confession was involuntary and therefore inadmissible if it had been obtained by oppression.[190]

Now the admissibility of confession evidence in England and Wales is regulated in the first instance by s. 76 of PACE. Section s. 76(2) provides that a confession is inadmissible against the accused, "...notwithstanding that it may be true ...", if it was or may have been obtained (a) by oppression; or (b) as a result of any statement or act which was likely, in the circumstances, to render unreliable any confession that might have been made. Section 76(8) defines oppression as including torture, inhuman or degrading treatment, and the use of or threat of violence.[191]

4.8.2 Judicial Discretion to Exclude Evidence of a Confession on the Ground of Unfairness—s. 78 Police and Criminal Evidence Act 1984

Section 78 of PACE introduces a wide exclusionary discretion to be exercised if it appears to the court that the exclusion of prosecution evidence is required in order to safeguard the fairness of the proceedings.[192] The court must consider all the

---

[189] *Ibrahim v R* [1914] A.C. 599 at 609.

[190] Malek (2010), pp. 1205–1206, referring to the obiter dictum of Lord Parker CJ in *Callis v Gunn* [1964] 1 Q.B. 495, 501 and paragraph (e) of the Judges' Rules (1965).

[191] These acts are prohibited by Art. 3 of the ECHR, and some such acts are prohibited by the criminal law; for a more detailed analysis of the word "oppression", see Malek (2010), pp. 1207–1208.

[192] See *Loosely v R* [2001] UKHL 53, [42] (Lord Hoffman); in relation to criminal proceedings, the breadth of the discretion is discussed by the Court of Appeal in *R v Smurthwaite* [1994] 1 All

circumstances surrounding the admissibility of the evidence, including the circumstances in which the evidence was obtained.

There are a number of ways in which the circumstances surrounding the making of a confession may adversely affect fairness at trial. If a confession has been obtained unlawfully, this may affect the fairness of proceedings. Statute often sets out important safeguards, such as the requirement for supervisory or judicial oversight in the use of coercive measures. Similarly, codes of practice set important limits on the use of investigative power. Impropriety that involves a breach of a practice code may lead to the exclusion of evidence on the ground of unfairness. In addition, if the circumstances in which a confession was obtained make it unreliable, for example, where the prejudicial effect of the evidence outweighs its probative value, this may also lead to exclusion under s. 78.

Looking beyond unlawfulness, there is no exhaustive definition of what s. 78 'fairness' entails. It is clear from the wording of s. 78 that guaranteeing the accused's right to a fair trial via art. 6 ECHR is central to the provision. In addition, the exclusion of evidence is justified where the admission of evidence will bring the administration of justice into disrepute. As Choo puts it, the judiciary will become an accomplice to improper action.[193]

The decisions of the English courts also reveal that s. 78 functions as a safeguard ensuring balance in criminal proceedings, in the interaction between investigative and prosecutorial powers, and defence rights. The court in *R v Smurthwaite* explained that 'fairness of the proceedings' is a policy which operates in two directions, involving *fairness to the public* and *fairness to the accused*.[194] On the one hand, the court must *inter alia* ensure that an accused is not acquitted on a mere technicality, which will be contrary to the good reputation of the court. 'Fairness to the public' in this sense could be interpreted to mean 'effectiveness' in convicting the guilty. On the other hand, in the application of the discretion, the English courts have frequently applied s. 78 so that investigative agents are not able to obtain a forensic advantage by breaking the rules. Therefore, using the maintenance of balance and the reputation of the court as rationales to explain s. 78 leads to a broad discretion to exclude evidence, based on weighing and balancing of the competing interests of the public and the accused.

In the event that s. 78 is concerned with balancing competing interests, factors such as the gravity of the crime, the seriousness of the impropriety by the police and the probative value of the evidence are all relevant to the exercise of discretion.[195] The weight to be attached to one circumstance may vary depending on the existence and importance of other relevant circumstances. This requires the trial judge to look at the preponderance of evidence to establish whether admitting the confession

E.R. 898; and by the House of Lords in *R v Khan* [1997] A.C. 558. It also applies to non-confession evidence: *R v Mason* (1988) 86 Cr. App. R. 349.

[193] Choo (1989), p. 276. On exclusion of evidence under s. 78 see also Stone (1995), pp. 10–17.

[194] *R v Smurthwaite* [1994] 1 All ER 898, 902–903.

[195] See further *R v Bailey* (1993) 897 Cr.App.R. 365, 375; *R v Walsh* (1990) 91 Cr.App.R. 161.

would have such an adverse effect on the fairness of the proceedings as to require its exclusion.

Some unreliable statements may not be caught by s. 76(2)(b) but will fall to be considered under s. 78. For example, if a mentally disordered person waives the right to legal advice, in the absence of assistance from an independent adult, this may raise questions about the fairness of using the confession at trial.[196] Failure to observe the codes of practice in relation to tape-recording of interviews (Code E), where significant voluntary statements made outside of the interview room are not subsequently repeated on tape, may raise an issue under s. 78.[197] A failure of this type may give rise to allegations or findings that the police have fabricated confessional statements. This is especially true if the suspect was not allowed access to a solicitor (in accordance with s. 58 of PACE and Code C, 6), who could have acted as an independent witness.[198] Although, there is authority to suggest that the issue of whether or not a suspect made an unrecorded confession is probably a question to be determined by the jury,[199] the Court of Appeal has found that the impropriety of not keeping a proper record of an interview with a suspect, or statement made by him, is 'significant and substantial' and may on its own justify the exclusion of the confession under s. 78. This is because of the potential for disadvantage to the accused in challenging the making of the statement at the trial.[200]

As discussed at 4.2.2 above, access to confidential, independent legal advice at the police station is a statutorily conferred right under s. 58 of PACE. Not only does the presence of a legal representative protect against fabrication of confessional statements, but it is central to safeguarding the suspect's right to silence in the face of accusation by the police, and to protecting against oppression during the interrogation.[201] Section 78 has been relied on for the exclusion of confessions, for example, where a denial of access to legal advice has led to a suspect making a confession he or she may not otherwise have made.[202] In *R v Kirk*, the Court of Appeal held that where the accused had been arrested on one offence, about which

[196]See *R v Aspinall* [1999] 2 Cr.App.R. 115; and *R v J* [2003] EWCA Crim 3309.

[197]See Code E, [4.6] and Code C, [11.4] in relation to the obligation to put any significant statement made by the suspect in the hearing of the police, which has not previously been the subject of an interview, subsequently to the suspect during an audio recorded interview. See also *R v Scott* [1991] Crim. L. R. 56; and *R v Dunn* (1990) 91 Cr.App.R. 237; but see *R v Van Gelderen* [2008] EWCA Crim 422.

[198]*R v Hoyte* [1994] Crim. L. R. 215.

[199]Malek (2010), p. 1213.

[200]*R v Keenan* [1989] 3 All E.R. 599; [1990] 2 Q.B. 54, 66 and 70.

[201]See (ECtHR) *Magee v. The United Kingdom*, Application no. 28135/95, 6 June 2000; see also (ECtHR GC) *Salduz v. Turkey*, Application no. 36391/02, 27 November 2008 discussed in Chap. 2, Sect. 2.3.2.2.2 above, where the ECtHR has clearly stated that a suspect has an indispensable right to consult with a lawyer prior to the first interrogation, particularly in the custodial setting.

[202]*R v Samuel* [1988] Q.B. 15; see further *R v Walsh* (1989) 91 Cr.App.R. 161, 163.

he did not seek legal advice, and was then questioned about a more serious offence, the critical admissions made by him should have been excluded in the exercise of judicial discretion under s. 78.[203] Section 78 has also been applied where a confession has been made on the basis of a lie told by the police to the suspect or the suspect's legal advisor about the prosecution evidence available.[204]

4.8.2.1 Section 78 and Covert Surveillance

Preventing an abuse of process and maintaining the rule of law is an inherent power and duty of every English court.[205] Therefore, s. 78 has often been used in connection with cases of entrapment by undercover agents, where in English law entrapment is not a substantive defence. However, in such cases the application to exclude evidence is based on the premise that the accused should not be tried at all.[206] This is to be distinguished from a case requiring "...a determination about the forensic fairness of admitting evidence."[207]

In the case of *R v Smurthwaite*, in considering the operation of s. 78 to cases of entrapment, the Court of Appeal also indicated the following factors may be generally relevant to the admissibility of covert conversations:

> How active or passive was the officer's role in obtaining the evidence? Is there an unassailable record of what occurred, or is it strongly corroborated? ...[Has the officer] abused his role to ask questions which ought properly to have been asked as a police officer and in accordance with the codes?[208]

In the case of *R v Christou and Wright*,[209] the Court of Appeal considered the fairness of admitting evidence gathered in an undercover sting, in which a London jewellery shop was set up and staffed by the police. The aim of the operation was to recover stolen goods and obtain evidence against suspects of stealing and dishonestly handling stolen items. The appellants were convicted of handling stolen goods. They argued on appeal that the evidence ought to have been excluded under s. 78 on the basis that they had not been cautioned according to Code C or allowed access to legal advice before the recorded conversations took place.[210] The Court of Appeal found that the undercover police at the shop spoke to the appellants

---

[203] *R v Kirk* [2000] 1 W.L.R. 567, 572.

[204] *R v Mason* [1987] 3 All E.R. 481; (1988) 86 Cr. App. R. 349.

[205] *Loosely v R* [2001] UKHL 53, [1] (Lord Nicholls).

[206] The appropriate course in a case of entrapment would be to apply for a stay of proceedings before the commencement of the trial. If the stay is refused it would still be open to apply for exclusion of evidence under s. 78, but the applicable principles are those in relation to a stay: *Loosely v R* [2001] UKHL 53, [104].

[207] *Loosely v R* [2001] UKHL 53, [18] (Lord Nicholls).

[208] *R v Smurthwaite* [1994] 1 All E.R. 898, 903; (1994) 98 Cr.App.R. 437.

[209] *R v Christou and Wright* [1992] 1 Q.B. 79.

[210] *R v Christou and Wright* [1992] 1 Q.B. 79, 99.

on equal terms and not as persons who the appellants believed were interrogating officers.[211] There was no intimidation or pressure and the appellants voluntarily applied themselves to the trick.[212] Therefore, the Code had no application in such a context. However, the court left it open that there might be certain undercover situations aimed at circumventing the Code that would fall foul of s. 78.[213]

Indeed, in *R v Bryce*, the Court of Appeal held that evidence of conversations between the appellant and an undercover officer should have been excluded because the conversations went directly to the central issue of guilty knowledge and there was no contemporaneous recording to eliminate the possibility of fabrication.[214] The existence of a record of what occurred is clearly a factor to be taken into account by the court in assessing the admissibility of covertly obtained confessions.

A number of other judgments indicate that the seriousness of the offence charged may also be relevant to the exercise of discretion under s. 78.[215] In *R v Latif and Shahzad*, concerning the use of an undercover agent to investigate the appellants' involvement in heroin importation worth £3.2 million, the Court of Appeal referred to the Parliament's intention in enacting s. 78, which was to protect the innocent and not to allow for the exclusion of cogent evidence in relation to serious crimes.[216] Similarly, in the case of *R v Khan*, the appellant had been released without charge when questioned by the police, having declined to answer most of the questions put to him. Six months later, he was recorded at a friend's house by a covertly planted recording device admitting his involvement in heroin importation. The recording was the sole evidence against the accused. Nevertheless, the Court of Appeal dismissed the appellant's appeal against conviction. The court found that although the planting of the device may have involved a breach of privacy and a civil trespass, this was outweighed by the gravity of the offending. In addition the Court pointed to the fact that the surveillance operation was conducted in accordance with Home Office guidelines and was not specifically designed to obtain evidence against the appellant.[217] The House of Lords highlighted the opportunity that had been available to the appellant to contest the authenticity of the tapes and the fairness of using them.[218] The court also reiterated that a breach of the right to privacy in art. 8(1) of the ECHR may be a relevant factor in the exercise of the

[211]*R v Christou and Wright* [1992] 1 Q.B. 79, 99.

[212]*R v Christou and Wright* [1992] 1 Q.B. 79, 84.

[213]*R v Christou and Wright* [1992] 1 Q.B. 79, 91.

[214]*R v Bryce* [1992] 4 All E.R. 567, 572; (1992) 95 Cr.App.R. 320.

[215]See for example *R v Latif and Shahzad* [1994] Crim. L.R. 750; *R v Khan* [1994] 4 All E.R. 426; *R v Cadette* [1995] Crim. L.R. 229.

[216]*R v Latif and Shahzad* [1994] Crim. L.R. 750; see also *R v. Latif and Shahzad* [1996] 1 All ER 353, 360 (Lord Steyn).

[217]*R v Khan* [1994] 4 All E.R. 426, 429 and 437–438. For a discussion of the ECtHR's judgment in the same case see Chap. 2, Sect. 2.3.2.5.1 above.

[218]*R v Khan* [1997] A.C. 558 (House of Lords).

judge's discretion to exclude evidence. However, the admission of the evidence obtained did not automatically render the trial unfair under art. 6.[219]

The cases referred to above are examples of covertly recorded conversations that do not involve interceptions. Section 9 of the defunct Interception of Communications Act 1985 was replaced by s. 17 of the Regulation of Investigatory Powers Act 2000 (RIPA), discussed at 4.4.7 above. In accordance with these provisions any evidence which discloses that a lawful interception warrant has been used, or that a lawful interception with the consent of one party has been conducted or that an unlawful interception has been carried out is explicitly rendered inadmissible.[220] This rule is aimed at protecting the use of interceptions as a covert method of surveillance from public knowledge and scrutiny. However, in *R v Sargent*[221] the House of Lords held that s. 9 in itself did not prohibit the police from using the transcript of an inadmissible intercept (whether lawful or not) during an interview with a suspect. Any confession obtained in the interview, as a result of raising the intercept transcript with the suspect, was not automatically rendered inadmissible, provided that a jury could properly convict on the confession evidence obtained in the interview after all references to the actual intercept were excluded from the interview transcript.[222] The question of whether the properly edited transcript had been admissible, and whether it had been open to the jury to convict the accused solely on the basis of his confession in the police interview, was dealt with under s. 78.

4.8.3 Common Law Discretion to Exclude Evidence—Probative Value Versus Prejudicial Effect

Discretionary exclusion of relevant evidence at common law has been expressly preserved by the operation of s. 82(3) of PACE. This includes exclusion of confession evidence. In English common law, the exercise of a trial judge's discretion to exclude evidence is based on an old principle that evidence is intrinsically unreliable where its probative value[223] is outweighed by the prejudicial effect to the accused of allowing the evidence to be adduced. Hence, at the outset, the assessment is weighted in favour of admission.

---

[219]See also *R v P (Telephone Intercepts: Admissibility of Evidence)* [2002] 1 AC 146; [2001] W.L.R. 463 (House of Lords); but see *PG and JA v United Kingdom* [2002] Crim. L.R. 308 and (ECtHR) *Condron v. The United Kingdom,* Application no. 35718/97, 2 May 2000.

[220]See further *Morgans v DPP* [2001] 1 A.C. 315.

[221]*R v Sargent* [2001] UKHL 54; [2003] 1 A.C. 347.

[222]*R v Sargent* [2001] UKHL 54; [2003] 1 A.C. 347.

[223]The probative value of evidence is defined as the extent to which the evidence can rationally affect the assessment of the probability of the existence of a fact in issue (Uniform evidence law, Australia). Therefore, probative value determines weight.

The common law discretion is aimed at ensuring fact-finding by the jury that is *rational*. Judges are unlikely to concede a real risk of being misled by prejudicial evidence.[224] The prejudicial effect of evidence is determined by the way in which the tribunal of fact may attach to it more weight than it rationally warrants.[225] The fact-finder may use prejudicial evidence on a basis that is logically unconnected with the issues in the case.[226] In other words, prejudicial evidence may not prove any fact in issue. The 1980 House of Lords decision of *Sang*[227] is frequently cited in England as the authority for this principle. However, the rule is rarely relied upon because its scope is imprecise and more limited than s. 78.[228] Further, the principle is not easily applicable to cases involving the admissibility of confessions, which are inherently highly probative.

4.9 Conclusion

The purpose of international cooperation in gathering evidence in criminal matters is that any evidence gathered is admissible in the national system where the trial is taking place. Therefore, the question of admissibility involves ensuring that it is *fair* to take that evidence into account in deliberations about guilt. This chapter has examined the law in England and Wales that limits and protects the right to silence in order to determine how the balance between effectiveness and fairness is maintained in the national system. The chapter looked to whether there are rules that limit the right to silence in the investigative phase, which are counteracted in the trial phase and vice versa. This helps in confirming whether there is continuity in the way the national law allows confession evidence and evidence of silence to be gathered and used. The chapter will assist in determining whether there are rules in English law that might impinge on fairness in transnational cases. This may be the case where another state is prosecuting an English accused in an indirectly related matter and wishes to make use of sworn testimony of the accused in an unrelated prosecution. For example, some states may find that using previous testimony of the accused that was obtained under compulsion to tell the truth to be contrary to the right to silence. Alternatively, some states consider that it is contrary to the right to silence to draw adverse inferences from pre-trial silence in any circumstances. Other systems within the EU need to be confident that England is applying the minimum standards set by the ECtHR in the international cooperation process; and, more specifically, what minimum standards are being applied in relation to the right to silence.

[224] Gans and Palmer (2014), p. 327.

[225] Gans and Palmer (2014), p. 327.

[226] Gans and Palmer (2014), p. 327 citing the Australian Law Reform Commission (1985), [644].

[227] *R v Sang* [1980] AC 402.

[228] Malek (2010), p. 1212.

The right to silence was first considered in the context of police questioning. The procedural rules that apply to police questioning can be divided into three general categories: initial questioning, when suspects are questioning voluntarily outside the police station; voluntary attendance at a police station for the purposes of questioning; and custodial interrogation, when the suspect is questioned while under arrest or in police custody. As soon as the police form a suspicion of criminal offending on reasonable grounds then the suspect must be cautioned about the right to silence prior to being questioned (whether voluntarily or as a detainee). This applies regardless of whether the police question a suspect as a witness first, he is questioned in public, he is attending the police station voluntarily or whether he is under arrest or in custody.

In order to allow for adverse inferences to be drawn at trial about the accused's silence prior to being formally charged, where the circumstances call for an explanation and accused then raises a defence first time in trial testimony, the accused must have been warned by the police of this possibility. The standard caution used in England may have the effect of taking away with one hand what was given with the other: having been told that he does not have to say anything, the suspect is then told that his defence at trial may suffer if he does remain silent. It is accepted that this type of warning indirectly compels a suspect to speak and the English law has therefore placed emphasis on the need for legal representation for a suspect being questioned under caution. A suspect who is willing to be questioned by the police should therefore be advised of the opportunity to receive legal advice prior to questioning and has the right to refuse to answer questions until he does. However, it is not a right as such to have access to legal representation at this initial stage and evidence will not necessarily be excluded because the suspect did not receive legal advice prior to questioning in this situation. The fairness of questioning which takes place on the spot, outside the atmosphere of the police station is matter that is determined by judicial discretion under s. 78 PACE. For example, in addition to cautioning and the opportunity to receive prior legal advice, the cases that relate to covert police questioning indicate the factors that may be relevant to the exercise of the trial judge's discretion are: the active or passive nature of the officer's role; whether the provisions of PACE and the PACE codes of conduct should have been followed; whether there was an unassailable record of the questioning; and the strength of the corroborating evidence. These types of factors indicate that the reliability of the confession evidence received and the possibility for the accused to test the evidence in court are important considerations. The rules in relation to initial police questioning clearly demonstrate the relationship between pre-trial and trial proceedings.

Most often police questioning takes place at a police station; and the desire for tape-recording of interviews in relation to indictable offences (to protect the integrity of the record of interview as an item of evidence in ongoing proceedings) demands this. A suspect always has the right to engage a lawyer of their own choosing, including having a lawyer present during a police interview. In addition, a suspect who is voluntarily being questioned at a police station has the opportunity to obtain free legal advice that includes access to a free duty solicitor (publicly appointed defence lawyer) at a call centre. The suspect should be told about the right to access legal advice and how to obtain it, including receiving a written notice of the right. However, again, the suspect who attends for questioning voluntarily and who may stop the interview at any time of their own initiative and even leave the premises does not have a 'right' to receive legal advice *immediately* prior to questioning. In England, in the situation of a voluntary suspect interview the police may continue questioning, even if the suspect has indicated that he wishes to remain silent and wait until he has received legal advice, so long as he willingly continues answering the questions. However, many suspects may lack the capacity to 'put their foot down' and resist. In light of the protective features of the trial judge's discretion to exclude unfairly obtained evidence under s. 78 PACE, the English codes of practice in relation to police interviews have allowed for an element of 'fairness to the public', by permitting the police to encourage spontaneous confessions in non-custodial situations.

On the other hand, in England custodial interviews are recognised as being coercive. Therefore, the English law includes a number of rules related to the right to silence that enable a suspect or accused to participate knowingly in the proceedings and that ensure the right has ongoing effect. Every detainee has a statutory right to access free legal representation prior to charge, after which legal aid becomes means tested. This includes the right to receive advice over the telephone and in person when necessary.[229] There is also a right to have a publicly appointed legal representative present during the interview. Due to the coercive atmosphere of a custody interview together with the element of indirect compulsion involved in the terms of the caution, there is a restriction on the drawing of adverse inferences of silence at trial if the detainee has not been given the opportunity to speak with a lawyer prior to be asked questions. Again the connection between the law in relation the right to silence in pre-trial proceedings and the law limiting and protecting the right in the trial phase is apparent.

Although the provisions directly dealing with the custodial questioning of suspects are stringent, the question of what might happen to real evidence that is located as a result of unlawfully obtained confession evidence is a matter that has not been sufficiently dealt with. What will become of the forensic evidence, for example, that is obtained from a crime scene, the location of which only became known to the police as a result of a custodial interview where the suspect had been unlawfully denied access to a legal representative in advance? This problem also arises in the case of questioning of suspects under compulsory questioning powers. Although the use in England of these types of compulsory information-gathering powers is structurally very different to administrative questioning powers in Denmark, they do not raise a problem in relation to unfairness and the admissibility of evidence. Although the right to silence is abrogated by Parliament in authorising the use of compulsion to question suspects, for example, in relation to serious fraud cases, the suspects are provided with immunity from having the answers that they

[229]Detainees also have the right to the service of an interpreter as necessary.

give being used against them in future criminal proceedings. This may raise questions about the appropriateness of using regulatory powers to question criminal suspects; however, the immunity removes concerns about the unfairness of relying on self-incriminating statements obtained under compulsion as part of the prosecution case. This is particularly so given the protective nature of provisions that operate at trial such as s. 78 PACE. Nevertheless, what is left behind is the fact that the statements obtained under compulsion may contain significant information that can lead the police to further incriminating material. Section 78 could provide some protection of an accused at trial. However, this may be particularly problematic (and may result in inadmissibility of evidence) in transnational cases where the English compulsory questioning powers were used to obtain suspect statements in an investigation in a system that has more strict rules for dealing with the 'fruits of the poisonous tree'.

This chapter looked finally at the use of covert surveillance to obtain confession evidence, particularly in the context of the judicial discretion to exclude prosecution evidence on the ground of unfairness. Self-incriminating statements may be obtained by interception of communication; however, such evidence may not be directly presented in a criminal trial. This type of surveillance is carried out under the control of the warrants issued by the Secretary of State. Other types of secret surveillance, such as listening devices, require warrants issued by high ranking officers under the supervision of Commissioners who hold or have held judicial appointments. There are also practice codes that relate to secret surveillance and may become relevant in relation to the fairness of relying on the evidence. Consistent with the ECtHR jurisprudence and the law and practice in Denmark, these types of measures may be used after a suspect has asserted their right to silence.

A consideration of the case-law about covert surveillance highlights the importance in the English jury system of the s. 78 fairness safeguard, as a final gate through which confession evidence must pass before it is allowed to be considered on the question of guilt. However, the English courts have clearly approached the fairness discretion as operating in two directions and requiring a weighing and balancing of all the competing interests involved in effectiveness in law enforcement and fairness to an accused—including the seriousness of the offending in question. This connection between the pre-trial and trial phases strengthens the argument that criminal proceedings may be seen as a continuum. All other English rules that regulate investigations function with the final check and balance of judicial discretion in sight. The significance of s. 78 PACE and similar provisions may be lost in transnational cases, where evidence is gathered in England for use in a foreign trial.

In the following chapter the journey into the right to silence takes us to the criminal law in Australia, where the minimum standards under the ICCPR are applicable. This is with a view to ultimately discovering whether there are fundamental differences in the way the national balances between effectiveness and fairness are maintained in Denmark, England and Australia and whether any differences will hinder the fairness of using confession evidence in transnational criminal cases.

References

Bentley D (1998) English criminal justice in the nineteenth century. The Hambledon Press, London

Brawn D (2006) The Court of Queen's Bench of Manitoba 1870–1950: a biographical history. University of Toronto Press, Toronto

Cape E, Namoradze Z, Smith R, Spronken T (2010) Effective criminal defence in Europe. Intersentia, Antwerpen

Choo A (1989) Improperly obtained evidence: a reconsideration. Legal Stud 9:261–283

Craig P, De Búrca G (2015) EU law: text, cases and materials, 6th edn. OUP, Oxford

Easton S (2014) Silence and confessions: the suspect as a source of evidence. Palgrave Macmillan, Hampshire

Emmerson B, Ashworth A, Macdonald A (eds) (2012) Human rights and criminal justice, 3rd edn. Sweet & Maxwell, London

Gans J, Palmer A (2014) Uniform evidence. OUP, Sydney

Harris DJ, O'Boyle M, Bates EP, Buckley CM (2014) UK withdrawal from the convention? A broader view. Available via the UK Constitutional Law Association. http://ukconstitutionallaw.org/2014/07/24/david-harris-michael-oboyle-ed-bates-and-carla-buckley-uk-withdrawal-from-the-convention-a-broader-view/. Accessed 4 Sept 2014

Helmholtz RH (2004) The Oxford history of the laws of England, vol 1, The canon law and ecclesiastical jurisdiction from 597 to the 1640s. OUP, Oxford

Helmholtz RH, Gray CM, Langbein JH, Moglen E, Smith HE, Alschuler AW (1997) The privilege against self-incrimination: its origins and developments. University of Chicago Press, Chicago

Howse T (2013) England. In: Ligeti K (ed) Toward a Prosecutor for the European Union, vol 1, A comparative analysis. Hart, Oxford

Langbein JH (2003) The origins of adversary criminal trial. OUP, Oxford

Malek HM (gen. ed) (2010) Phipson on evidence, 17th edn. Thomson Reuters (Legal) Ltd, London

Mirfield P (2003) Silence, confessions and improperly obtained evidence. Clarendon, Oxford

Richardson PJ (ed) (2011) Archbold 2011: criminal pleading, evidence and practice. Sweet and Maxwell, London

Spronken T (2010) EU-wide letter of rights in criminal proceedings: towards best practice. http://digitalarchive.maastrichtuniversity.nl/fedora/get/guid:2ba696f9-993e-42ca-922f-911a9e96e73b/ASSET1. Accessed 15 Jan 2014

Starmer K, Strange M, Whitaker Q (2001) Criminal justice, police powers and human rights. Blackstone Press Limited, London

Stone R (1995) Exclusion of evidence under Section 78 of the Police and Criminal Evidence Act: practice and principles. Web JCLI 3:10–17

Chapter 5
The Right to Silence in Australia

Abstract The Australian legal system has its roots in the English common law legal tradition. Nevertheless, the majority of the states of Australia have uniformly codified the law of evidence so that this legislation covers many of the issues relating to procedural rights in Australian criminal proceedings. Following an outline of the human rights framework in Australia, which largely draws on the incorporation of provisions of the ICCPR into various Australian enactments, this chapter considers how the Australian national law balances the limitation and protection of the right to silence and the right against self-incrimination. Again, it can be seen that the investigative and trial phases of the Australian criminal justice system (at federal and state levels) form a coherent whole. The different settings of the right to silence, such as the police interrogation, the approach to adverse inferences being drawn at trial from pre-trial silence, administrative compulsory questioning powers and the use of covert surveillance illustrate the continuity that exists in the national criminal proceedings. Therefore, this chapter is also referred to in Chap. 6 to examine what happens to this continuity when confession evidence or evidence of silence is obtained and used in transnational cases. Falling outside the framework of the ECHR, the chapter serves as a point of reference in Chap. 7 when assessing the force of influence of the ICCPR and the ECHR.

5.1 Introduction

This chapter moves beyond the immediate reach of the ECHR to the Australian criminal justice system. Although Australia does not belong to a regional human rights framework, the laws of Australia must comply with Australia's international obligations under seven core UN human rights treaties. One of these is the ICCPR, which has been incorporated into the provisions of national legislation and may be regarded as the relevant international framework (and a legal source) for the right to a fair trial in Australian criminal proceedings.

There is no direct constitutional protection of fair trial rights in Australia, other than the right to trial by jury, or a single, compendious human rights enactment. Nevertheless, constitutional separation of powers has promoted the independence

F.M.W. Billing, *The Right to Silence in Transnational Criminal Proceedings*,
DOI 10.1007/978-3-319-42034-9_5

of the Australian judiciary, which has been a strong force in maintaining the rule of law and protecting rights, such as the right to silence of the suspect or accused. Australia's common law roots are deeply embedded in the English common law tradition and many principles protecting the right to a fair trial have similarly developed through judicial decision-making.

However, today, much of the law that relates to investigating and prosecuting crime is to be found in statutes. Australia is a federation and this chapter will present the applicable law that touches on the right to silence and the fairness of using suspect evidence at federal and state level. Legislation is enacted by the national parliament and by the legislature in each state. Therefore, in addition to commonwealth (federal) law, frequent reference is made to the law of Victoria as an example of state law. Victoria was the first state with a specific human rights charter that was to be applied to the state's enactment of *uniform evidence law* (see Sect. 5.2.4 below)—a significant combination in the context of the legislative protection of procedural rights.[1] The existence of the charter arguably adds an additional dimension to the discussion about fair trial rights at the state level.

As a federation, Australia has independent law enforcement agencies at the federal level (the Australian Federal Police) and in each state and territory. Commonwealth legislation provides for rules about investigations and prosecutions of federal offences. State legislation regulates the investigation and prosecution of state offences. There is a Commonwealth Director of Public Prosecutions and also each state has its own prosecutorial service. Federal criminal law encompasses offences that relate to national borders, national systems and national security, such as customs, tax, welfare or security. The states prosecute other serious (and minor) offences that generally can take place within state borders and do not involve national systems. After criminal charges have been laid, a brief of evidence will be sent from the federal or state police to the corresponding prosecution office.

Both federal and state offences are tried within the same overall judicial court systems. Each state has a magistrates' court as well as one or two higher level criminal courts, such as county and supreme courts. State supreme courts will have jurisdiction to try the most serious offences such as murder and, for example, serious and complex fraud. State supreme courts will also have an appellate jurisdiction to hear appeals against conviction and sentencing. These appeals are not retrials of the facts. If an appellate court finds that a conviction is unsatisfactory, then the case may be remitted back to the trial court for retrial. The High Court of Australia is the highest appellate court in the land and hears appeals on questions of law of general relevance from all federal and state courts.

Similar to English criminal law, Australian law draws a distinction between *indictable* offences, which are serious offences tried by jury (if guilt is contested) in the superior courts and *summary* offences that are tried by a single judge in the

[1] In other words, Victoria was the first Australian state or territory to have both a human rights act and a uniform evidence law act. The Australian Capital Territory's uniform evidence law came after the Victorian enactment (as discussed below).

Magistrates' Court.[2] This chapter considers the law in relation to the right to silence in investigations and prosecutions of indictable offences. The role of the judge as the tribunal of law and the jury as the tribunal of fact has a resonating significance in the way criminal proceedings are conducted, a reflection of the English common law ancestry. This is apparent in the vital role of judicial discretion to determine the issue of balance between effectiveness and fairness and the admissibility of evidence.

This chapter begins by outlining the way in which federal and state statute law in general provides for the protection of human rights such as the right to a fair trial. More specifically, an explanation will be given of federal and state uniform evidence law legislation, which plays a central role in the protection of procedural rights. Section 5.3 of this chapter, sets out the specific human rights framework operating in Australia; and Sect. 5.4 explains how the right to silence is conceived in Australia. In Sect. 5.4–5.7, similar to the previous chapters dealing with Danish and English law, this chapter considers the way in which the law in Australia limits and protects the right to silence in relation to the following situations: police questioning; drawing adverse inferences from the silence or statements of the suspect or accused; using compulsory questioning powers to question the suspect or accused; and using covert surveillance to obtain self-incriminating statements. In addition, judicial discretion to exclude confession evidence will be examined in Sect. 5.8, as a significant safeguard of the fairness of criminal proceedings.

5.2 The Human Rights Framework in Australia

The Australian human rights framework relies principally on political enforcement of human rights, with an emphasis on political commitment to seven core UN human rights treaties,[3] community education and consultation, legislative compliance and administrative review. The Australian Human Rights Commission promotes political enforcement of human rights along these lines. In addition, procedural rights are legally enforced through the law of evidence found in legislation and at common law. The Australian state of Victoria and the Australian

---

[2]An indictable offence is punishable by a term of imprisonment of at least 12 months and guarantees the accused a trial by judge and jury in a higher court, unless it is a less serious indictable offence that may be heard summarily; and a summary offence is punishable by up to 12 months imprisonment must be tried in the Magistrates' Court by a judge sitting alone: see ss. 4G and 4H, Crimes Act 1914 (Cth).

[3]These are the International Covenant on Civil and Political Rights; the International Covenant on Economic, Social and Cultural Rights; the Convention on the Elimination of All Forms of Racial Discrimination; the Convention on the Elimination of All Forms of Discrimination Against Women; the Convention Against Torture and other Cruel Inhuman and Degrading Treatment and Punishment; the Convention on the Rights of the Child; and the Convention on the Rights of Persons with Disabilities.

Capital Territory (ACT) enforce human rights through human rights statutes.[4] Specific human rights, such as procedural rights in criminal law, have traditionally been protected in the common law. The Australian courts have frequently been the champions of human rights in Australia pulling these various threads together. Many of the common law (judge-made) principles have now become the subject matter of legislative enactment and amendment.

5.2.1 National Obligations Under the ICCPR

As discussed in Chap. 2, Sect. 2.2.1 above, Article 14(3) of the International Covenant on Civil and Political Rights (ICCPR) guarantees the right to a fair trial, including a right not to be compelled to testify against oneself or to confess guilt (after charges have formally been laid). In addition, the Human Rights Committee (HR Committee) has found that this right includes protection from pressure applied by investigative authorities with a view to obtaining a confession, possibly from the first interrogation. Australia ratified the ICCPR on 13 August 1980, and has made no relevant reservations in relation to Article 14 ICCPR.[5]

5.2.2 Federal Institutions and Legislation Establishing the Rule of Law

In Australia, in the absence of constitutional rights or any specific federal human rights charter, the rule of law is supported by strong democratic institutions and legislation dealing with specific issues such as non-discrimination and privacy. The central features of Australia's constitutional system are the doctrines of responsible government and the separation of the executive and legislative branches of government from the judiciary. The Australian Parliamentary Joint Commission on Human Rights has the function of ensuring that all Australian federal legislation and proposed legislation complies with Australia's human rights obligations and answers any inquiries referred to it by the Australian Attorney-General's Department.

---

[4]See Sect. 5.2.3 below.

[5]International Covenant on Civil and Political Rights (ICCPR), opened for signature in New York on 19 December 1966, entered into force 23 March 1976, UNTS 999/171.

The Australian Constitution does not guarantee a right to a fair trial, other than trial by jury.[6] Instead, at the federal level, the right to a fair trial is safeguarded through federal legislation and by the common law.[7]

5.2.3 State Protection of Human Rights in Specialised Human Rights Legislation

Two states have enacted statutes[8] that set out the human rights that are applicable in that jurisdiction, and require that all statutory provisions are compliant with human rights and interpreted in accordance with those rights. They are the ACT's Human Rights Act 2004 and Victoria's Charter of Human Rights and Responsibilities Act 2006. Hence, the scope of protection of the right to a fair trial, generally, as well as the right to silence and the right against self-incrimination, may vary to some extent within Australia.[9] Under both acts, the primary source of rights is the ICCPR. However, in relation to interpretation of the rights contained in each, explicit reference is made to the potential relevance of international law and the judgments of international courts and tribunals. In the Explanatory Memorandum of the ACT act refers directly to the importance of the ICCPR and the ECHR, including the decisions of the HR Committee and the ECtHR. Consistent with international human rights law, both the Australian Capital Territory (ACT) and Victorian laws provide that it is unlawful for public authorities (such as the police, prosecutors and judges) to act in a way that is incompatible with human rights.[10]

Both the ACT act and the Victorian charter enact the right to a fair trial and provide for a number of minimum guarantees that may flow from a criminal charge, including the right not to be compelled to testify against oneself or to confess guilt.[11]

---

[6]Trial by jury in specified circumstances (Section 80) is the only procedural guarantee provided for in the Commonwealth of Australia Constitution Act (The Constitution). Sections 71 and 73, respectively, fully invest judicial power in the courts, independent of government, and establish the appellate jurisdiction of the High Court (allowing the process of judicial review).

[7]See *Dietrich v R* [1992] HCA 57; (1992) 177 CLR 292 in which the majority of the High Court held that although there is no right at common law to have publicly provided legal representation, in some cases representation is appropriate to ensure a fair trial. The majority found that trial judges should use their power to adjourn a case in the interests of fairness to encourage legal aid agencies to provide counsel. Deane J and Gaudron J went further to imply a right to a fair trial in chapter three of the Constitution, as part of the requirement of judicial process and fairness that may be implied from the doctrine of the separation of powers.

[8]Human Rights Act 2004 (ACT); Charter of Human Rights and Responsibilities Act 2006 (Vic).

[9]Gans and Palmer (2014), p. 20.

[10]Section 40B(1) Human Rights Act 2004 (ACT); s. 38(1) Charter of Human Rights and Responsibilities Act 2006 (Vic).

[11]Sections 24 and 25(2)(k), Victorian Charter on Human Rights and Responsibilities Act 2004 (Vic); Sections 21 and 22(2)(i), Human Rights Act 2004 (ACT). These acts also include provisions that deal with arrest and lawful detention, such as the right to be informed promptly of the reasons

5.2.4 *Uniform Evidence Law Harmonising Procedural Rights at Federal and State Levels*

The operation of the right to silence in Australia presented here is based in part on a consideration of procedural rights as they function in the Australian *uniform evidence law* (UEA). The uniform evidence law is the result of efforts to harmonise the rules of evidence and procedure in Australia, involving *federal* and *state* enactments. In the mid-eighties the Australian Law Reform Commission (ALRC) foreshadowed the need for harmonisation, and it was the ALRC's recommendations in 1987 that led to the enactment of the uniform evidence law.[12] The uniform evidence law currently comprises of seven Australian statutes that resemble each other closely[13]: Evidence Act 1995 (Commonwealth); Evidence Act 1995 (New South Wales); Evidence Act 2001 (Tasmania); Evidence Act 2004 (Norfolk Island)[14]; Evidence Act 2008 (Victoria); Evidence Act 2011 (Australian Capital Territory); and Evidence (National Uniform Legislation) Act 2011 (Northern Territory).[15] In the first instance, it is generally the numbering system from the Commonwealth Act that will be referred in the course of the present discussion.

The uniform evidence law has attained 'majority status' in Australia based on the number of jurisdictions that have enacted it, the number of court systems (including the Federal Court system) affected by it and the percentage of population reached by it.[16] Therefore, the principles relating to confessions and the immunities that make up the right to silence will not be discussed at common law, except to the extent that the common law is relevant to interpreting and supplementing the uniform evidence law.

The Australian definition of what constitutes a confession is similar to the English and Danish definitions. However, Australian uniform evidence law talks about 'admissions' rather than confessions. Thus, any previous representation (statement) will be an admission, if it supports some inference of fact that is adverse to the party who made it.[17] Therefore, the words 'admission' and 'confession' are

---

for arrest and of any charges to be laid or proceedings to be commenced. See for example Sections 21–23 of the Victorian Charter and Sections 18–20 of the ACT Act.

[12]See further Australian Law Reform Commission, *Evidence*, Report No. 26 (Interim), 1985; Australian Law Reform Commission, *Evidence*, Report No. 38, 1987.

[13]The following states are the three Australian jurisdictions that have not enacted uniform evidence law: Queensland, Western Australia and South Australia.

[14]Norfolk Island is an external territory of Australia.

[15]There are differences in the initial enactments and amending acts to be found in some of these statutes. However, the provisions that are the subject of this research may be considered 'uniform'.

[16]Gans and Palmer (2014), p. 1.

[17]See the uniform evidence law dictionary; see further Gans and Palmer (2014), pp. 156–159. Admissions may be made before or after criminal prosecutions are commenced in conversations with others, including interactions or interviews with criminal investigators. Admissions may also be in the form of testimony at pre-trial hearings or a guilty plea that is later withdrawn, but does not include oral testimony at a trial. In accordance with s. 81(1) of the uniform evidence law, the hearsay rule and the opinion rule do not apply to admissions on the grounds that they are reliable—because people are unlikely to lie against their own future legal interests—and necessary, to

largely interchangeable and this work will generally refer to confessions rather than admissions.

The uniform evidence law is not a 'code' in the sense that it does not purport to contain all the law on the various issues to the absolute exclusion of others, such as the common law. A number of the state uniform evidence acts include a provision to the effect that the uniform evidence law is not exhaustive and does not affect other principles or rules of evidence, unless the uniform evidence law explicitly or by necessary implication provides otherwise.[18] The uniform evidence law must, therefore, be understood and interpreted in the context of the laws that may influence it. Local statutes may supplement it regulating, for example, police powers. It may be interpreted by reference to common law, and supplemented by the common law of evidence that fills in any gaps. Judges considering the law may be influenced by similar overseas law; and the uniform evidence law may be modified by human rights law. Nevertheless, in certain areas, such as the admissibility of evidence, there is some debate as to whether the uniform evidence law is a code.

The Human Rights Act 2004 (ACT) and the Charter of Human Rights and Responsibilities Act 2006 (Vic) applies to the Evidence Act 2011 (ACT) and the Evidence Act 2008 (Vic) respectively, which may also have an impact on the scope of protection in these states—even compared to other uniform evidence law states.

5.2.5 *The Common Law as a Source of Domestic Law About the Right to a Fair Trial*

At common law the main function of the court in the administration of justice is to ensure that a fair trial according to law is secured.[19] The right against self-

directly contradict things a person previously said to protect their legal interests; and because people are likely to be familiar with factors that are adverse them. However, the hearsay and opinion rules are not prevented from applying to the admission in relation to proceedings against a third party, such as a co-accused, unless by consent (ss 81 and 83 uniform evidence law). Certain statements made by a third party under the authority of another may be permitted to be used as admissions under s. 87.

[18]See, for example, ss. 8 and 9 of the uniform evidence law. Hearsay evidence is an example of the abolition of a common law rule. Hearsay evidence is inadmissible unless an exception is available. Evidence of a previous representation of another person is hearsay when it is used to prove the existence of a fact that it can reasonably be supposed that the person intended to assert by the representation (s. 59(1) uniform evidence law). However, the uniform evidence law has removed the application of the hearsay rule to confessions, which effectively means that a suspect who confesses to a crime is speaking directly with the court, without the benefit of the court-room protections ensuring fairness of proceedings. See further Gans and Palmer (2014), p. 161.

[19]Halsbury's Laws of Australia (2012), [80-1630].

incrimination, also known by the old common law term as the 'privilege against self-incrimination', is a rule of common law and a procedural 'right'.[20] The protection offered may be claimed against any demand for compulsory disclosure, unless a statute conferring extra-curial powers of compulsion formally abrogates the 'privilege', expressly or by necessary implication.[21] However, the right against self-incrimination is not available to an accused person (who chooses to give evidence) in criminal proceedings in relation to the facts in issue.[22]

5.2.6 The Interrelationship Between Federal Law, State Law and the Common Law

Both the Australian Federal Police (AFP) and the police forces of a particular state could have an interest in obtaining evidence from overseas. However, it is primarily the AFP that is concerned with foreign evidence, in investigating crime that traverses Australia's national maritime borders. Therefore, in relation to transnational crime, the Commonwealth (federal) investigative rules relating to the gathering of evidence are highly relevant. Nevertheless, even federal investigations will be pursued at trial in the state court, under state evidence and procedural laws. The Commonwealth uniform evidence law will apply to federal offences being tried in uniform evidence law states and to cases being held in Federal Courts. However, uniform evidence law will not apply in cases dealing with commonwealth crimes that are being tried in non-uniform evidence law states such as Western Australia.

The common law is always in the background—it is a source of authority in relation to issues that have not been included in the uniform evidence law and in relation to interpreting legal provisions at both the federal and state levels, unless its application has been explicitly abolished.

5.3 The Right to Silence in Australian Law

5.3.1 The Composite Right to Silence

The evolution of the right to silence and the right against self-incrimination in Australian criminal law shares its path with the meandering development of the principles in English common law, influenced by changing procedures and statutory

[20]In this context and under s. 187 uniform evidence law, because the privilege is treated as a human right, it is not available to employees of a corporation if it is claimed on the basis that their evidence may incriminate the *corporation*: Gans and Palmer (2014), p. 293.

[21]See, for example, *Controlled Consultants Pty Ltd v Commissioner for Corporate Affairs* [1985] HCA 6; (1985) 156 CLR 385; and *Hamilton v Oades* [1989] HCA 21; (1989) 166 CLR 486.

[22]See generally, Gans and Palmer (2014), pp. 291–293.

provisions, and the establishment of the adversarial system. Historically, the right to silence may have been incidental to the development of the privilege against self-incrimination and may have come into being to ensure effective protection of the privilege. The fullest expression of the right to silence in criminal proceedings is tied to the non-compellability of an accused at trial and the right to silence of a suspect under caution. There has been some confusion in Australian case law as to whether a general right to silence existed beyond these two situations. Nevertheless, the Australian common law today may be influenced by constitutional norms and the principles reflected in international and regional human rights treaties, so that the emerging law is not only dictated by English and local legal history.[23] Today Lord Mustill's analysis of the composite right to silence, of which the right against self-incrimination in various situations is a (major) component, has been accepted by the Australian High Court.[24]

In England in 1912, the Judges' Rules were drawn up to regulate the questioning of suspects in relation to obtaining evidence to be used at trial. They were revised for the last time in 1964 and were re-issued in 1978 by way of a Home Office circular.[25] In England, they were ultimately replaced by the provisions of the Police and Criminal Evidence Act 1984 (PACE) and by the PACE Codes of Conduct (see Chap. 4, Sect. 4.4). The Judges' Rules significantly affected the way in which certain procedural rights unfolded in Australian common law and legislation. They continue to influence criminal law in non-uniform evidence law states.

5.3.2 *The Right Against Self-Incrimination in the Uniform Evidence Law*

Section 128 of the uniform evidence legislation operates as a rule of evidence law that is connected to the trial process.[26] A witness's 'testimonial evidence', or sworn answers to questions, is covered by the 'privilege against self-incrimination'[27]; and it can be claimed both during the trial and during the pre-trial stage, in relation to a court order (summons) for the production of documents as part of disclosure.[28] The

[23] *Azzopardi v The Queen; Davis v The Queen* [2001] HCA 25, [36–36].

[24] *RPS v The Queen* (2000) 199 CLR 620, 630; see also *Azzopardi v The Queen; Davis v The Queen* [2001] HCA 25, [7] and [160–161].

[25] Judges' Rules and Administrative Directions: Home Office Circular 89/1978 (UK).

[26] Gans and Palmer (2014), pp. 283–284. Section 128A creates a mechanism for determining claims of privilege against self-incrimination with respect to information a court has ordered to be disclosed in civil proceedings for a freezing (Mareva) or search (Anton Pillar) order: Gans and Palmer (2014), p. 284.

[27] Gans and Palmer (2014), p. 289.

[28] Gans and Palmer (2014), pp. 282–284 and 289–290.

right applies where the evidence sought to be adduced, either directly or circumstantially, implicates a witness or an accused in a criminal offence.[29] Where the right under the uniform evidence law does not apply, the common law 'privilege' may be claimed and this is not restricted to judicial proceedings.[30]

Section 128(3) of the uniform evidence law allows a witness to voluntarily exchange the right against self-incrimination for a use immunity certificate; and s. 128(4) allows the court to order the witness to provide incriminating evidence in the interests of justice, by substituting the right for a use immunity certificate.[31] The certificate may grant both 'direct-use' and 'derivative-use' immunity, so that neither the answers nor any information or evidence derived from those answers may be used in future Australian proceedings.[32] However, this procedure is not available if the offence arises under the law of a foreign country.[33] If a witness or an accused are being examined (by consent) in court in relation to foreign offences the right against self-incrimination will under s. 128 will apply.

Similar to the position at common law, in accordance with s. 128(10) of the uniform evidence law, the right against self-incrimination does not protect an accused in criminal proceedings in relation to the facts in issue. If an accused chooses to give evidence he or she cannot only be required to answer questions that relate directly to the events that are the subject of the offence or offences charged, but also other matters being relied upon by the prosecution to prove the commission of the offence. For example, in *Cornwall v The Queen*[34] the accused was charged with conspiring to import a large quantity of cocaine. The prosecution also relied on evidence that the accused was in the business of supplying drugs. The trial judge found that the right against self-incrimination did not apply to the accused person's testimony about supplying drugs and it was open to the jury to find that the accused participated in the conspiracy, in order to carry out the illicit trade.[35] The High Court held that the words of s. 128(10) were not limited to direct evidence of the mental or physical elements of the crime charged, but extended to indirect or circumstantial evidence that establishes the facts in issue.[36]

---

[29]See *Application Concerning Section 80 of the Supreme Court Act and Sections 119 and 128 of the Evidence Act* [2004], NSWSC 614, [35–36] (Brownie JA). The privilege against self-incrimination also applies to civil court proceedings.

[30]See *Mann v Carnell* [1999] HCA 66; (1999) 201 CLR 1; and *Esso Australia Resources Ltd v Federal Commissioner of Taxation* [1999] HCA 67; (1999) 201 CLR 49.

[31]See further Gans and Palmer (2014), p. 291.

[32]Gans and Palmer (2014), p. 292.

[33]Section 128(4)(a).

[34]*Cornwall v The Queen* [2007] HCA 12; (2007) 231 CLR 260.

[35]*Cornwall v The Queen* [2007] HCA 12; (2007) 231 CLR 260, [7].

[36]*Cornwall v The Queen* [2007] HCA 12; (2007) 231 CLR 260, [84] (Gleeson CJ, Gummow, Heydon and Crennan JJ).

5.4 Police Questioning and the Right to Silence

5.4.1 Investigative Authority to Make Initial Enquiries

The majority of investigative acts neither have nor require a specific legislative basis, being acts that can equally be performed by ordinary people.[37] There are virtually no legal impediments to passive police investigation or conversing with people, and the gathering and processing of information is almost entirely unregulated.[38] While there is a civil duty to assist a police investigation, the police have no power to force a person to make a witness statement, other than to give their name and address.

5.4.2 Cautioning and Persistent Questioning

Section 23F of the Crimes Act 1914 (Cth) states a general rule that a person who is under arrest or a person who is not under arrest but is accompanied by an official who believes there to be sufficient evidence to establish that the person has committed an offence should be cautioned that they do not have to say or do anything, but anything they say or do may be used in evidence. Similar provisions apply under state legislation, such as s. 464A of the Crimes Act 1958 (Vic) in relation to cautioning a person who is suspected of committing a criminal offence, but only if he or she is under arrest.

Section 139 of the uniform evidence law provides that, in certain circumstances, if the investigating officers do not caution a suspect or accused (whether under arrest or not), that a person does not have to say or do anything but if they do say or do anything then that may be used in evidence, prior to questioning then evidence of anything that is said or done by the suspect or accused during the course of questioning may be taken to have been obtained improperly for the purposes of the public policy discretion to exclude evidence under s. 138.[39]

However, even if the suspect has been cautioned, there is no rule of law that an interview conducted despite objection by the suspect, or despite an indication that a suspect does not wish or intend to participate, is not admissible.[40] It is a question of degree on the facts of each case as to whether the investigator's persistence has crossed the line so as to render using any evidence of the answers against the

---

[37] Gans and Palmer (2014), p. 339.

[38] Gans and Palmer (2014), p. 339.

[39] *R v Sharp* (2003) 143 A Crim R 344; [2003] NSWSC 117, [12] (Howie J). See also Gans and Palmer (2014), p. 344.

[40] Odgers (2014), [1.3.5800]; *R v Clarke* (1997) 97 A Crim R 414, [21] (Hunt CJ); *R v Phan* (2001) 123 A Crim R 30, [54]–[56] (Wood CJ).

defendant unfair.[41] Here, the issue will be whether the confession evidence should be excluded in the exercise of the trial judge's discretion.[42] The defendant may have a particular vulnerability, such as age, which will be relevant to the exercise of discretion.[43] In Victoria and the ACT, the exercise of the discretion to exclude confession evidence obtained in the absence of cautioning may be influenced by the human rights statutes that apply. At common law, in *R v Swaffield; Pavic v R*, following Australian common law precedent, the High Court concluded that the absence of a caution triggers the trial judge's discretion to exclude an admission, but does not require it.[44]

5.4.3 Arrest

Police powers allow the police to undertake measures that the ordinary person is unable to do, such as detaining people for investigative purposes. Although police powers were initially based in the common law, they are now primarily regulated by numerous statutes.[45] Investigative or coercive powers typically involve an intrusion into individual rights. Therefore, the statutes that create such powers are interpreted narrowly.[46]

Police powers are also concerned with the administration of justice, aimed at ensuring compliance with the law and the effectiveness of the criminal justice system. Powers of arrest and detention include powers to: prevent ongoing or future unlawful behaviour, such as preventative arrest; ensure the court attendance of those charged with a criminal offence; ensure the preservation of evidence for prosecution and, where appropriate, lawful disposal. There may be collateral uses of the power of arrest, so long as the primary purpose is still being served. For example, detention of a suspect while preparing a charge and waiting for the court to open may provide the opportunity for questioning. However, in the absence of specific statutory authorisation, detaining a person for questioning beyond the period of time required to bring that person before court is unlawful.[47] South

[41]Odgers (2014), [1.3.5800]; *R v Clarke* (1997) 97 A Crim R 414, [21] (Hunt CJ); *R v Phan* (2001) 123 A Crim R 30, [54]–[56] (Wood CJ); *R v Quach* (2002) 137 A Crim R 345; [2002] NSWCCA 519, [100]–[104]; see also *R v Swaffield; Pavic v R* [1998] HCA 1; (1998) 192 CLR 15, [93–95] (Toohey, Gaudron and Gummow JJ); and *Em v The Queen* [2007] HCA 46; (2007) 232 CLR 67.

[42]See Sect. 5.8.2 below.

[43]*R v Phan* (2001) 123 A Crim R 30, [54]–[56] (Wood CJ). See also *R v Medcalfe* [2002] ACTSC 83, [24] in relation to the questioning of suspects suffering from a mental disability or mental disorder.

[44]*R v Swaffield; Pavic v R* [1998] HCA 1; (1998) 192 CLR 15, [94–95] (Toohey, Gaudron and Gummow JJ).

[45]Gans and Palmer (2014), p. 341.

[46]See, for example, *Coco v R* [1994] HCA 15, (1994) 179 CLR 427 in relation to planting listening devices.

[47]Gans and Palmer (2014), p. 341, citing *Williams v R* [1986] HCA 88; (1986) 161 CLR 278.

Australia is the only Australian jurisdiction that has amended the common law rule that a person may not be detained for questioning. In SA, a person may be detained for questioning in relation to serious offences.[48]

The most important pre-condition to the exercise of nearly all police powers is a suspicion[49] or belief[50] based on 'reasonable grounds' that the suspect has committed a crime and that the exercise of the power will produce a particular benefit.[51] Therefore, police cannot use coercive or intrusive powers unless there has been an initial investigation.[52] A suspicion or belief does not have to be based on admissible evidence, but may be based on hearsay, opinion or tendency evidence.[53] Thus, police powers are often used when an investigation moves beyond initial inquiries to an attempt to gain more concrete evidence of a crime.[54] However, this does not preclude relying on evidence of an unrelated offence, accidentally found while lawfully excising a power in relation to a different investigation.[55] Often the pre-conditions for the exercise of coercive or intrusive police powers, including powers of arrest, are supplemented by a warrant procedure. Thus, an independent person examines the existence and legitimacy of the statutory pre-conditions, before the power is exercised.[56]

In general, Australian federal and state legislation allows a person to be arrested and detained for a limited period of time to enable investigation of an offence, for example between 4 and 12 hours.[57] The person should usually be released unconditionally, released on bail or brought before a judge to determine bail in that time.

5.4.4 Access to Legal Advice

Similar to the rules about cautioning, under s. 23G of the Crimes Act 1914, prior to questioning of a suspect who is under arrest or believed to have committed an offence, the suspect must be offered the opportunity to access a legal practitioner of his own choosing and be given assistance and facilities to arrange receiving legal

[48]See s. 78, Summary Offences Act 1953 (SA). The Commonwealth and the Victorian Crimes Acts give no power to arrest a suspect for the purpose of questioning: see further s. 3 W Crimes Act 1914 (Cth); s. 458 Crimes Act 1958 (Vic).

[49]A 'suspicion' is 'more than a mere idle wondering': *Queensland Bacon Pty Ltd v Rees* [1996] HCA 21; (1996) 115 CLR 266, 303 (Kitto J).

[50]A belief is an 'inclination' assenting to a proposition: *George v Rockett* [1990] HCA 26; (1990) 170 CLR 104, 116.

[51]Gans and Palmer (2014), p. 342.

[52]Gans and Palmer (2014), p. 342.

[53]Gans and Palmer (2014), p. 342.

[54]Gans and Palmer (2014), p. 342.

[55]Gans and Palmer (2014), p. 342.

[56]Gans and Palmer (2014), p. 342.

[57]See for example s. 23C of the Crimes Act 1914 (Cth); see also s. 464A Crimes Act 1958 (Vic).

advice. An interview of a suspect must be deferred for a reasonable length of time in order to allow a suspect the opportunity to contact a legal practitioner. The legal representative is entitled to be present during the interview and to give advice during the interview, so long as this does not unreasonably interfere with questioning. Exceptions to this rule apply in accordance with s. 23L. These exceptions include: where it is believed on reasonable grounds that the access to legal advice may lead to an accomplice avoiding apprehension; the destruction of evidence; and compromising the safety of others by deferral of questioning. However, in these situations access to another legal adviser must be offered.

Some state legislation provides similar entitlements to prisoners and individuals who are in custody awaiting trial.[58] For example, s. 464C of the Crimes Act 1958 (Vic) provides a right to communicate with a legal practitioner prior to custodial questioning of a suspect and questioning must be deferred to enable the suspect to receive legal advice, unless it is believed on reasonable grounds that this may lead to an accomplice avoiding apprehension, the destruction of evidence, or compromising the safety of others. This entitlement includes the provision of facilities to enable private communication to take place. At common law there is no right to legal representation. However, there is a right not to be tried unfairly and an accused person cannot be forced to go to trial unrepresented.[59]

Under s. 25(2)(b) of the Charter of Human Rights and Responsibilities Act 2006 (Vic) a person who has been *charged* with a criminal offence in Victoria has the right to communicate with a lawyer of their own choosing in the preparation of their defence; and under ss. 25(2)(d), (e) and (f) a person who has been charged has the right to defend themselves through a legal representative of his own choosing or, if eligible, through a legal representative in accordance with the Legal Aid scheme, to be told about the right, if eligible, to legal representation and to have Legal Aid representation if the interests of justice so require. Legal Aid is the scheme by which suspects and accused persons throughout Australia have access to free publicly appointed legal representation. Legal Aid is means tested and becomes available after formal charges have been processed, marking the beginning of the trial phase. This is in stark contrast to the availability of free legal representation to those suspected of criminal conduct in England and Wales and in Denmark, both of which apply the minimum standards set by the ECtHR within the ECHR framework.

Provisions which lay down rules relating to access to a legal practitioner are rules that the legislature has determined must be followed in order for a confession or admission to be regarded as fairly obtained. A breach of the rules will not automatically render a confessional statement inadmissible. However, a breach of

---

[58]Halsbury's Laws of Australia (2012), [80-1630]. However, this may be limited in some circumstances by anti-terrorism legislation: Halsbury's Laws of Australia (2012), [80-1635].

[59]*Dietrich v R* (1992) 177 CLR 292, 299–304; 109 ALR 385 (Mason CJ and McHugh J), where a statutory right was interpreted negatively, so that an accused could not be forced to go to trial unrepresented by the trial judge exercising a discretion.

these types of procedural protections makes it prima facie unfair to the accused to admit the evidence. The evidence of the circumstances in which the confession was obtained may displace a prima facie case of unfairness, where the breach was insignificant or irrelevant to the making of the confession.[60]

5.4.5 *Interviewing*

As stated above, the statutes authorising police powers, such as the power to interview, are not detailed. The law of evidence therefore deals with issues of impropriety. In particular, the law of evidence may determine when coercion or deception renders an otherwise lawful interview inadmissible.

5.4.5.1 Unfair Interrogation

According to Australian law, it would not necessarily be unfair for the police to continue questioning a suspect, with a view to obtaining relevant evidence, even after the suspect has indicated that he or she wishes to remain silent.[61] For example, an officer is not bound to terminate an interview where the first answer given indicates a desire not to comment.[62] It is a question of degree and each case will turn on its own facts in relation to what extent unfair pressure was placed on a suspect; or whether an unfair forensic advantage has been gained by the prosecution from using the answers against an accused at trial.[63]

Personal characteristics that make a suspect vulnerable, such as their age or lack of familiarity with the English language, have been found to give rise to an unfair disadvantage when faced with persistent interrogation by the police.[64] Other examples where the courts have found unfairness due to persistent questioning are where the police put allegations of other admissions to a suspect during an interview, so as to allow those admissions to be introduced into evidence without the need to call the people who witnessed them[65]; where multiple interviews caused the defendant to make inconsistent admissions[66]; and where the suspect reacted to new allegations in a way that was damaging to his credibility.[67]

[60] *R v Thomas* [2006] VSCA 165, [113].

[61] Odgers (2014), [1.3.5800].

[62] *R v Clarke* (1997) 97 A Crim R 414, 419 [Hunt CJ); *R v Phan* (2001) 53 NSWLR 480; 123 A Crim R 30, [54–55] (Wood CJ).

[63] See further *R v Clarke* (1997) 97 A Crim R 414, [21] (Hunt CJ); *R v Quach* (2002) 137 A Crim R 345; [2002] NSWCCA 519, [100–104].

[64] *R v Phan* (2001) 53 NSWLR 480; 123 A Crim R 30, [56] (Wood CJ).

[65] *McDermott v R* (1948) 76 CLR 501.

[66] *R v Amad* [1962] VR 545.

[67] *Van Der Meer* (1988) 35 A Crim R 232, 258.

5.4.6 *Recording Police Interviews Between the Suspect and the Investigating Authorities*

Under s. 23V of the Crimes Act 1914 (Cth), where it is reasonably practical to tape-record a confession, this must be done or any confession evidence that results will be inadmissible, unless admission of the confession evidence would not be contrary to the interests of justice. Where not practical, then a written record should be made as soon as possible which is read and corrected by the suspect.

State law usually requires audio or video-recording. For example, under ss. 464F and 464H of the Crimes Act 1958 (Vic) an audio or video-recording must generally be made of a confession by a suspect (or someone who should be regarded as a suspect) to an investigating official. If the confession was made before the commencement of questioning, or during a break or at a location without recording facilities, the confession must be put to the suspect and confirmed on tape. The court may admit confession evidence that has not been taped if satisfied on the balance of probabilities that the circumstances are exceptional and justify the reception of the evidence.

In common law jurisdictions, statements allegedly made by the defendant to the investigating authority may be of great significance at his or her trial. In Australia, an audio or audio-visual recording of the interview is typically tendered in evidence via the testimony of the investigating officer, as evidence of what was said. Having been cautioned about the right to remain silent, evidence of what was said by the defendant to the investigating officers is generally considered relevant and admissible at trial, unless there are grounds to show otherwise. The common law prosecutor may seek to adduce evidence of what was said by the accused to the investigators, for a number of purposes: as confession evidence or evidence of admissions that are relevant to the facts in issue; in very limited circumstances, as evidence of remaining silent or failing to answer to certain questions;[68] as evidence of lies, showing a consciousness of guilt; or as a prior inconsistent statement.

In 1991, in the case of *McKinney v R*,[69] in order to counter the occurrence of 'verballing' (where police witnesses falsely claim that the defendant has freely confessed to a crime), the High Court made a requirement that, where the prosecution relies on police testimony of an admission by a suspect, the jury should be warned of the dangers of convicting the defendant based on police testimony that is uncorroborated by a contemporaneous audio or video recording. Given the importance of admissions and the words chosen by the defendant in the pre-trial phase, all Australian jurisdictions, including uniform evidence law jurisdictions, have enacted exclusionary rules for certain admissions, unless an electronic recording is

[68]But see the uniform evidence law provisions described at Sect. 5.5.1 below.

[69]*McKinney v R* [1991] HCA 6; (1991) 171 CLR 468, 476.

available at the trial.[70] These rules, and the wide-spread practice that has followed, generally require as a minimum that any formal questioning of a person suspected of a serious crime by an investigating officer must be corroborated by an electronic (or digital) recording. Though not completely uniform with each other, the rules preserve the accuracy and integrity of the evidence that flows from an interview. Where the correct procedure has not been followed the defendant may also object to the admissibility of evidence of admissions, on the grounds of involuntariness, unreliability, unfairness or public policy.

Despite the recognised advantages of electronically recording admissions,[71] there remain a number of difficult issues. Some of the statutes do not cover the situation where a suspect makes a spontaneous admission outside the formal interview setting. In a significant number of cases, police officers have claimed that suspects have made off-the-record admissions before or after formal questioning, or during 'tea breaks'.[72] In *Kelly v R*,[73] the police testified that about 30 min after a formal interview had ceased, in which the defendant had retracted an earlier confession, he said, 'Sorry about the interview—no hard feelings, I was just playing the game'. The police neither asked the defendant what he meant, nor requested that he repeat what he said on tape. Nevertheless, the majority of the High Court held that the police evidence was admissible, because Tasmania's statute is limited to formal interviews.

In all Australian jurisdictions the legislation allows an admission to be adduced despite the absence of an electronic recording where there is reasonable excuse,[74] or 'special'[75] or 'exceptional'[76] circumstances. The situations that may be covered by the exception may include mechanical failure of the equipment; the refusal of the suspect to allow the questioning to be recorded; or the reasonable absence of the recording equipment. In *Nicholls v R; Coates v R*,[77] the police testified that the defendant, with the encouragement of the police, made admissions to an attempted murder while trying to strike a deal for a reduced sentence. The defendant denied

[70]See for example *Crimes Act 1914*, s. 23 V (Cth), applicable in the ACT (s23A(6)); *Criminal Procedure Act 1986* (NSW), s. 281; *Crimes Act 1958*, s. 464H (Vic). In Victoria, for example, the practice is that three simultaneous recordings are made: one for the prosecution's brief of evidence, one for the defendant's copy of the brief of evidence; and one to be used as the trial exhibit. See also s. 86 of the uniform evidence law, which provides that a document prepared by an investigating official as a record of an oral admission made by a suspect will not be admissible unless it has been signed or otherwise marked by the defendant, acknowledging its truth. In Victoria, by virtue of the operation of s. 464H(1) of the Crimes Act 1958 requiring any evidence of an admission to be accompanied by an electronic recording, s. 86 will only be relevant if there are exceptional reasons to allow the adducing of an admission that has not been recorded.

[71]*Kelly v R* [2004] HCA 12; (2004) 218 CLR 216, [37]. See also McHugh J at [96]–[108].

[72]Gans and Palmer (2014), p. 174.

[73][2004] HCA 12; (2004) 218 CLR 216.

[74]*Criminal Procedure Act 1986* (NSW), s. 281(2)(a)(ii).

[75]*Crimes Act 1914*, s. 23 V(5) (Cth).

[76]*Crimes Act 1958*, s. 464H (Vic).

[77][2005] HCA 1; (2005) 219 CLR 196.

the conversation, which allegedly came during a 45 min toilet break during a formal police interview, which was recorded. The police failed to mention the 'admissions' when recording resumed, did not take contemporaneous notes and claimed to have lost the written notes that were made subsequently. The majority of the High Court held that there was no reasonable excuse for the absence of an electronic recording in the circumstances.[78] The majority were concerned about opening up a clear risk of 'verballing' which would defeat the purpose of the recording requirements. Most Australian jurisdictions require the police to make a later recording of an off-the-record admission in which the admission is read out to the suspect who is invited to comment on it.[79]

Another issue that has been raised by the relevant case law is the ineffectiveness of the recording legislation to address the issue of unreliability. In the case of *Mallard v R*, Mallard made a bizarre recorded confession recounting details of a murder that only the true murderer could have known. However, it was shown that these details may have been fed to Mallard in a series of earlier interviews and the High Court subsequently quashed Mallard's conviction. Fresh evidence in the form of forensic tests in fact implicated another man in the murder. To avoid such improprieties, most Australian statutes require not only the admission to be recorded, but also all the surrounding questions and answers. The High Court has interpreted these provisions narrowly, finding that the recording requirement does not extend to informal 'pre-interviews', before the tape is turned on, or to separate interviews that are distinct in time and place.[80] One approach, mooted by the Court of Appeal of Western Australia in one of Mallard's earlier unsuccessful appeals,[81] is the exclusion of all unrecorded admissions to the police due to unfairness on the basis that their use in a trial will put the defendant at an unfair forensic advantage.[82]

[78]*Nicholas v R; Coates v R* [2005] HCA 1; (2005) 219 CLR 196, [106] (McHugh J, [152]–[157] (Gummow and Callinan JJ), [218]–[219] (Kirby J), see further in dissent [11] (Gleeson CJ) and [341]–[344] (Hayne and Heydon JJ).

[79]*Crimes Act 1914*, s. 23 V(1)(b) (Cth); *Criminal Procedure Act 1986* (NSW), s. 281(2)(a)(ii); *Crimes Act 1958*, s. 464H(1)(c), (e) (Vic). See further the report of the Coldrey Committee 1986, which recommended changes to Crimes Act 1958 (Vic) to introduce tape-recording of interviews and confirmation of confessions on tape as part of a package of changes aimed at striking a balance between effective law enforcement and protecting individuals from unlawful and unfair treatment. In *R v Simmons; R v Moore (No. 3) [2015] NSWSC 189* the Supreme Court of NSW excluded evidence of unrecorded statements by an accused under ss. 85(2) and 90 of the uniform evidence law that were made in circumstances where the accused had been under some pressure, also from family members, to say something and the truth of the statements may have been affected.

[80]*Pollard v R* [1992] HCA 69; (1992) 176 CLR 177; *Heatherington v R* [1994] HCA 19; (1994) 179 CLR 370.

[81]Cited in Gans and Palmer (2014), p. 176.

[82]In uniform evidence law jurisdictions, this objection would be raised under s. 90. In *The Queen v Willis (Ruling No. 1)* [2015] VSC 261, an unrecorded caution did not result in exclusion of the accused's statements on public policy grounds under s. 138 uniform evidence law.

5.5 Adverse Inferences of Guilt and the Right to Silence

5.5.1 Drawing Adverse Inferences When the Accused Remains Silent

5.5.1.1 Adverse Inferences from the Accused's Silence Prior to Trial

Section 89(1) of the uniform evidence law prohibits an inference that is adverse to the accused from being drawn at trial in relation to evidence of the silence of the accused, or another person, when being questioned by an investigating official about the commission of a crime (unless this is a fact in issue in relation to the specific charge).[83] The section prohibits adverse inferences from being drawn if the accused, or another witness, remained silent throughout an entire interview, or if they remained silent in response to one or some of the questions.[84] According to s. 89(4), the prosecution may not use such evidence to prove that the accused was conscious of his or her guilt; or to discredit the accused, for example, by suggesting that an account of the facts given at trial is a 'recent invention'.[85] These provisions acknowledge that there may be reasons other than guilt that lead a suspect to remain silent or answer police questions selectively.

Section 89 is consistent with the information that a suspect receives when he or she is cautioned by the police, namely that he is *entitled* to remain silent and that it is only the statements that he makes that may be introduced as part of the case against him.[86] In addition, as it relates to criminal defendants, s. 89 essentially corresponds to the rule as it was expressed by the majority of the High Court in *Petty and Maiden v The Queen*:

> A person who believes on reasonable grounds that he or she is suspected of having been a party to an offence is entitled to remain silent when questioned or asked to supply information by any person in authority about the occurrence of an offence, the identity of the participants, and the roles which they played... An incident of that right to silence is that no adverse inference can be drawn against an accused person by reason of his or her failure to answer such questions or to provide such information. To draw such an adverse inference would be to erode the right of silence or to render it valueless.[87]

---

[83]If the silence is a fact in issue, for example, for not answering a summons or contempt of court, evidence may be led: s. 89(2) uniform evidence law.

[84]*King v The Queen* (1986) 15 FCR 427, 436; *R v Barrett* [2007] VSCA 95; (2007) 16 VR 240. The explanatory memorandum to Victoria's s. 89 specifically refers to the maintenance of this position.

[85]See further the common law case of *Petty and Maiden v The Queen* [1991] HCA 34; (1991) 173 CLR 95, [3] (Mason CJ, Deane, Toohey and McHugh JJ).

[86]The effect of s. 89 in combination with the 'caution' means that it is unlikely that the prosecution could rely on silence in response to police questioning as evidence of post-offence conduct for the purposes of ss. 18–24 of the Jury Directions Act 2013 (Vic); see further Sect. 5.5.2.1 below.

[87]*Petty and Maiden v The Queen* [1991] HCA 34; (1991) 173 CLR 95, [2] (Mason CJ, Deane, Toohey and McHugh JJ).

Thus, at common law, adverse inferences may not be drawn in relation to the defendant's silence in earlier court proceedings, such as a committal hearing.[88] Section 89(1) is arguably narrower than the common law, because it only relates to questioning by an investigator rather than any person in authority.[89] Nevertheless, decisions that have interpreted the section have held that the common law rule in *Petty and Maiden* still applies to prohibit adverse inferences being drawn in relation to silence at pre-trial hearings as well.[90]

In Victoria, a number of provisions of the Criminal Procedure Act 2009 (Vic) require pre-trial disclosure by the prosecution and the defence, such as: response to the prosecution summary of the opening statement and pre-trial admissions; notice about calling expert witnesses; and notification of alibi witnesses.[91] These types of the disclosure requirements may impinge on the accused's right to silence prior to trial. Under s. 237 of Act, with leave of the trial judge in limited circumstances a party may comment, for example, on a failure to comply with or an evidentiary departure from pre-trial requirements. However, in light of the right to silence and due to the wording of the provision, it is unlikely that a comment could be made in relation to such a failure or departure by the accused in a criminal case.

A significant protection of the right to silence is to be found in s. 41 of the Jury Directions Act 2013 which prohibits any judicial comment allowing the jury to draw an inference that the accused is guilty if he or she remained silent when questioned by the police.

5.5.1.2 Adverse Inferences from Pre-Trial Silence About Serious Indictable Offences in NSW

In 2013, s. 89A was inserted into the New South Wales (NSW) uniform evidence law allowing inferences of guilt, as appear proper, to be drawn from evidence of a suspect's silence during official questioning about a serious indictable offence, if he then raises a defence for the first time at trial.[92] This type of inference suggests that the trial defence is a recent invention. The adverse inference involved may only be drawn if the suspect was given prior access to an Australian lawyer and the police, in the presence of an Australian lawyer and in addition to the usual caution, give the suspect a *special warning* that it may harm the suspect's defence if he does not

---

[88]Where the uniform evidence law is narrower than the common law, there may be an argument for the preservation of the common law rule under s. 9 uniform evidence law.

[89]Section 89(1) is also significantly broader in application than the common law in other respects. Most notably, the section applies in relation to the silence of a party to the proceedings or 'another person', including, for example, a defence witness: *Jones v R* [2005] NSWCCA 443; see further Gans and Palmer (2014), p. 160.

[90]*R v Anderson* [2002] NSWCCA 141; *R v Stavrinos* (2003) 140 A Crim R 594.

[91]See ss. 186, 189 and 190, respectively, of the Criminal Procedure Act 2009 (Vic).

[92]A serious indictable offence is an indictable offence, which is punishable by life or a term of imprisonment of 5 years or more: Crimes Act 1900 (NSW).

mention anything which he later relies on in court. This provision functions similarly to the drawing of inferences from pre-trial silence (and the suggestion of recent invention) in England and Wales as referred to in Chap. 4.[93]

5.5.1.3 Adverse Inferences from the Failure of the Accused to Testify

The decision whether to testify must be one of the most important tactical decisions to make in the course of a criminal defence. Often there are a number of reasons why an accused does not testify. Equally, there will be different implications leading from this decision. Above all, an accused is entitled not to testify—and this is not an admission of guilt or innocence.[94]

In a division of the uniform evidence law dealing with the 'competence and compellability of witnesses', s. 20(2) provides that in a criminal proceeding for an indictable offence,[95] a prosecutor is prohibited from making any comment on the accused person's failure to testify.[96] This includes an indirect comment suggesting that the accused could have given evidence and did not do so, for example, by emphasising the fact that a co-accused testified.[97] However, the prohibition does not preclude the prosecutor from commenting that the questions put to prosecution witnesses in cross-examination are not evidence and suggesting that the prosecution case has not been contradicted.[98]

A trial judge may comment upon the failure of the accused to give evidence, so long as the comment does not suggest that this failure was due to the accused person's guilt, or the accused person's belief in his or her own guilt.[99] Being a 'comment' denotes that a jury is entitled to ignore what the trial judge says about the potential significance of the accused person's failure to testify.[100] However, pursuant to s. 20(2) a co-accused may make an adverse comment about the failure

[93]See Dixon and Cowdery (2013), p. 9 in which they are highly critical of the complexity and low value of the new caution, based on empirical evidence from Northern Ireland and England and Wales.

[94]*Weissensteiner v The Queen* [1993] HCA 65; (1993) 178 CLR 217, [33].

[95]See s. 20(1) uniform evidence law. An indictable offence is a more serious offence, which guarantees the accused a trial by judge and jury in a higher court. A summary offence carries a lower penalty range and must be tried in the Magistrates' Court by a judge sitting alone.

[96]Likewise, the prosecutor is not allowed to comment on the failure of the accused's spouse or de facto partner, parent or child not to give evidence: 20(3).

[97]*R v Villar; R v Zugecic* [2004] NSWCCA 302, [119].

[98]*R v Yammine and Chami* [2002] NSWCCA 289, [56–58].

[99]The same rule applies to the failure to give evidence of the accused's spouse or de facto partner, parent or child not to give evidence: 20(3) and (4).

[100]This is in contrast to a judicial *direction* about an impermissible path of reasoning, which the jury may not ignore: *Azzopardi v The Queen* [2001] HCA 25, [49–50].

of another accused to testify, as an indication of the guilt of that accused.[101] In the event, s. 20(5) permits the judge to comment on the co-accused person's comment. It is not clear from the section whether a trial judge's comment under s. 20(5) is limited by the rule in s. 20(2) as it applies to the judge.

The scope of the rule in s. 20(2) is a matter of some debate. Does the restriction in s. 20(2), on judges' comments about the failure to testify, preclude the trial judge from inviting the jury to use the accused person's failure to testify in evaluating the probative value of the prosecution case? This is particularly relevant in cases which clearly call for an explanation from the accused. For example, the accused may have been found at the crime scene.

The 1993 common law decision of the High Court in *Weissensteiner v R*[102] has been reconsidered by Australian courts on numerous occasions in light of the enactment of s. 20 of the uniform evidence law and other similar provisions. In *Weissensteiner*, the majority focused on the failure of an accused to *explain or contradict* evidence.[103] The accused had been found in sole possession of a boat belonging to two others. The boat still contained numerous personal items belonging to the two missing people, including certain possessions which they always took with them. The accused told others numerous inconsistent and unsubstantiated stories about the whereabouts of the two. The accused never explained to the police or at the trial his knowledge of the circumstances of the case.

In *Weissensteiner*, the High Court confirmed that the accused was entitled to put the prosecution to its proof by remaining silent. His failure to testify could not be an implied admission of guilt.[104] That is why, as the High Court reasoned, the accused person's failure to testify was not evidence of itself and could not be used to add to the prosecution case.[105] However, the High Court unanimously found that the jury was entitled to use the accused person's failure to explain or contradict the prosecution's case when the failure to testify "clearly assisted" in its evaluation of the probative value of the existing circumstantial evidence.[106] Therefore, doubts about the reliability of circumstantial evidence, or of the inferences that need to be drawn, can more readily be discounted where the accused fails to testify.[107] The majority observed that:

---

[101]The same rule applies to the failure to give evidence of the accused's spouse or de facto partner, parent or child not to give evidence: 20(3) and (4).

[102]*Weissensteiner v The Queen* [1993] HCA 65; (1993) 178 CLR 217.

[103]*Azzopardi v The Queen* [2001] HCA 25, [13].

[104]*Weissensteiner v The Queen* [1993] HCA 65, [33]; (1993) 178 CLR 217 (Mason CJ, Deane and Dawson JJ).

[105]*Weissensteiner v The Queen* [1993] HCA 65, [33]; (1993) 178 CLR 217 (Mason CJ, Deane and Dawson JJ).

[106]*Weissensteiner v The Queen* [1993] HCA 65, [28, 33]; (1993) 178 CLR 217 (Mason CJ, Deane and Dawson JJ).

[107]*Weissensteiner v The Queen* [1993] HCA 65, [28, 30, 33]; (1993) 178 CLR 217 (Mason CJ, Deane and Dawson JJ).

> [I]n a criminal trial, hypotheses consistent with innocence may cease to be rational or reasonable in the absence of evidence to support them when that evidence, if it exists at all, must be within the knowledge of the accused.[108]

Therefore, following *Weissensteiner*, there may be cases where the only explanation of the accused person's innocence would have to lie peculiarly within his or her knowledge, so that the accused person's failure to provide the details of this explanation may provide 'especially strong support' for formulating the inference, based on the prosecution's circumstantial case, that the defendant is guilty.[109] As the majority in *Weissensteiner* stated, in relation to the right to silence:

> ...it is not to deny the right; it is merely to recognize that the jury cannot, and cannot be required to, shut their eyes to the consequences of exercising the right.[110]

In *RPS v The Queen*[111] and *Azzopardi v The Queen; Davis v The Queen*,[112] the majority of the High Court pointed to the substantial limitations on interpreting provisions such as s. 20 imposed by fundamental accusatorial principles, for example, that the prosecution bears the onus of proving the accused person's guilt beyond reasonable doubt.[113] Furthermore, the question of the effect in a criminal trial of the failure of an accused to testify must be "...answered against the background of the common law principle that an accused person should not be compelled to incriminate himself or herself."[114] Such principles, the majority said, had become "deeply embedded in the common law of Australia" and had become some of the most important features of the contemporary criminal trial in this country.[115] Therefore, the majority in *RPS* observed that the prohibition in s. 20 (2), on suggesting guilt as the reason for refusing to testify, should not be narrowly constructed.[116] Otherwise, the opportunity to give exculpatory evidence becomes an obligation to self-incriminate.[117]

On the background of these important accusatorial principles, in *Azzopardi* the majority of the High Court said:

> [I]f an accused does not give evidence at trial it will almost always be desirable for the judge to warn the jury that the accused person's silence in court is not evidence against the

---

[108] *Weissensteiner v R* [1993] HCA 65, [28]; (1993) 178 CLR 217 (Mason CJ, Deane and Dawson JJ). See also [32].

[109] Gans and Palmer (2014), p. 34; see also *Weissensteiner v The Queen* [1993] HCA 65, [30]; (1993) 178 CLR 217 (Mason CJ, Deane and Dawson JJ).

[110] *Weissensteiner v R* [1993] HCA 65, [33]; (1993) 178 CLR 217 (Mason CJ, Deane and Dawson JJ).

[111] *RPS v The Queen* (2000) 199 CLR 620.

[112] *Azzopardi v The Queen; Davis v The Queen* [2001] HCA 25; (2001) 205 CLR 50.

[113] *RPS v The Queen* (2000) 199 CLR 620, [22]–[27] (Gaudron ACJ, Gummow, Kirby and Hayne JJ); *Azzopardi v The Queen; Davis v The Queen* [2001] HCA 25, [34]; (2001) 205 CLR 50.

[114] *Azzopardi v The Queen; Davis v The Queen* [2001] HCA 25, [35]; (2001) 205 CLR 50.

[115] *Azzopardi v The Queen; Davis v The Queen* [2001] HCA 25, [38]; (2001) 205 CLR 50.

[116] *RPS v The Queen* (2000) 199 CLR 620, [20] (Gaudron ACJ, Gummow, Kirby and Hayne JJ).

[117] *Azzopardi v The Queen; Davis v The Queen* [2001] HCA 25, [56]; (2001) 205 CLR 50.

> accused, does not constitute an admission by the accused, may not be used to fill gaps in the evidence tendered by the prosecution, and may not be used as a make-weight in assessing whether the prosecution has proved its case beyond reasonable doubt.[118]

In adapting the decision in *Weissensteiner* to determine when the failure to offer an innocent explanation may be the subject of comment by the trial judge, the majority held:

> It is to be emphasised that cases in which a judge may comment on the failure of an accused to offer an explanation will be both rare and exceptional. They will occur only if the evidence is capable of explanation by disclosure of additional facts known only to the accused.[119]

Therefore, in assessing the probative value of the prosecution evidence, the jury may take into account the failure by the accused to offer an innocent explanation only in cases where that explanation would have been given by way of additional facts "…to those already given in evidence by the witnesses who were called."[120]

While acknowledging that there may be cases where a trial judge is permitted under s. 20(2) to comment on the failure of an accused to offer an innocent explanation based on matters peculiarly within his or her knowledge, the majority remarked:

> However, as with all judicial comments on the facts in a jury trial, it will often be better (and safer) for the judge to leave the assessment of the facts to the determination of the jury in the light of the submissions of the parties. Unnecessary or extensive comments on the facts carry well-recognised risks of misstatements or other errors and of blurring the respective functions of the judge and the jury.[121]

Reading these paragraphs together, although a *Weissensteiner* type comment has not been completely ruled out, there is little scope for its application in future Australian trials—under both common law and uniform evidence law. However, the cases dealing with this section have been controversial. In the opinion of Callinan J of the High Court, s. 20(2) was enacted for the protection of an accused who chooses not to give evidence, and now makes a *Weissensteiner* type comment wholly impermissible.[122] Arguably, following the reasoning in *Azzopardi*, even on a broad reading of the prohibition in s. 20(2), a *Weissensteiner* comment by a trial judge may be given in cases where it would completely defy common sense, in the circumstances, to tell the tribunal of fact that the failure of the accused to give evidence may not be taken into account in any way in the evaluation of the evidence. Likewise, it may be nonsensical for a co-accused who testifies, and

[118] *Azzopardi v The Queen*; *Davis v R* [2001] HCA 25, [51]; (2001) 205 CLR 50, 70; 119 A Crim R 8, [51].

[119] *Azzopardi v The Queen*; *Davis v R* [2001] HCA 25, [68] (Gaudron, Gummow, Kirby and Hayne JJ); (2001) 205 CLR 50.

[120] *Azzopardi v The Queen*; *Davis v R* [2001] HCA 25, [64] (Gaudron, Gummow, Kirby and Hayne JJ); (2001) 205 CLR 50.

[121] *Azzopardi v The Queen*; *Davis v R* [2001] HCA 25, [52]; (2001) 205 CLR 50, 70; confirming the position of Gaudron ACJ, Gummow, Kirby and Hayne JJ in *RPS v The Queen* (2000) 199 CLR 620, [42].

[122] *Azzopardi v The Queen*; *Davis v R* [2001] HCA 25, [189]; (2001) 205 CLR 50.

subjects himself to cross-examination by the prosecution and the accused, to not be allowed to make the suggestion that the accused did not give evidence because he or she is guilty.

In Victoria, s. 42 of the Jury Directions Amendment Act 2013[123] will ensure that if the accused fails to testify the trial judge, the prosecution and the defence counsel will be prohibited from saying to the jury that they may draw an inference that is adverse to the accused from his failure to testify.[124] The common law rules in *Weissensteiner* and *Azzopardi* have been abolished.[125] Further, if defence counsel requests the trial judge to warn the jury that adverse inferences from failure to testify are not permitted, this should include when necessary a direction that the evidence of failure to testify cannot be used to strengthen the prosecution case.[126]

5.5.2 *Drawing Adverse Inferences from Lies Told by the Accused*

5.5.2.1 Lies as Admissions Showing a Consciousness of Guilt

Although evidence of silence is not readily admissible as part of a prosecution case, what a suspect positively says to the police is directly admissible through the evidence of the interviewing officer and the presentation of the 'record of interview', usually by playing an audio or video-recording to the jury. Therefore, a suspect is cautioned about the right to remain silent and the fact that what he does say may be used against him. In general, this may apply whether the suspect tells the truth or lies to the police.

A suspect may engage in a number of different types of post offence conduct from which an inference of guilt may be drawn. For example, a suspect may be seen fleeing from the scene of a crime, may conceal a weapon or other object associated with a crime, or may tell lies in relation to their involvement with the crime. The difficulty with post offence conduct as a form of circumstantial evidence is that inferences about what motivates certain conduct are very uncertain.[127]

Under the uniform evidence law, any previous statement by an accused will be a relevant and admissible *admission* (not subject to the hearsay or opinion evidence rules) if it supports an adverse inference of guilt.[128] In addition to confessions and admissions about facts, the truth of which can be proved by the prosecution, the

[123]The current version includes amendments up to 1 July 2015.

[124]Section 42.

[125]Section 44.

[126]Section 41(2)(d)(iv).

[127]Victorian Law Reform Commission (2009), p. 48.

[128]See the Dictionary of the Evidence Act 2008 (Vic), for example, which refers to the Commonwealth Act.

prosecutor may attempt to prove a lie. Thus, controversy has arisen as to whether Part 3.4 of the uniform evidence law, about admissibility of confession evidence and judicial discretion to exclude evidence (described below), applies to evidence of the commission of a crime that is based on a lie told by the accused to the police.[129] Legal authority suggests that the definition of 'admission' extends to a lie, relied upon by the prosecution as an implied admission.[130] Therefore, evidence of a lie will only be admissible as an admission if, for example, the circumstances in which it was made do not affect its reliability and its admission would not be unfair.[131]

Although defining 'admission' broadly characterises a lie as an 'admission' of guilt, a narrow definition would have benefited the prosecution, because the evidence would not have to comply with the proof requirements[132] or the exclusionary rules of evidence in relation to confessions.[133] At common law, if the defendant tells a lie about a fact in issue it will not be classified as an 'admission'.

There are a number of reasons why a suspect might lie to the police about their criminal involvement. A suspect may 'boast' about their criminal endeavours in order to improve their status within criminal circles.[134] Alternatively, a suspect may panic, lie in order to protect another person, or to avoid an unjust accusation or certain consequences that are unrelated to the crime.[135] However, a suspect or accused may lie to cover-up their crimes—the spin-off being that the prosecution may prove the lies and present the cover-up of lies as part of their case. In addition, the jury may be directed that they can use evidence of the accused telling lies as evidence that he was aware of his own guilt. Although, following the High Court decision in *Edwards v R*,[136] a judge's warning about drawing inferences of guilt from post offence conduct should also point out the dangers involved in jumping to conclusions about what has motivated the accused.[137]

As referred to at Sect. 5.5.1.1 above, it is unlikely that the silence of a suspect can be used as evidence post-offence conduct under ss. 18–24 of the Jury Directions Act 2013 (Vic)[138] by virtue of the combined effect of s. 89 of the uniform evidence

[129]See *Lee v R* [1998] HCA 60; (1998) 195 CLR 594, [21]; *R v Horton* (1998) 45 NSWLR 426; *R v GH* [2000] FCA 1618; (2000) 105 FCR 419, [79] and [82]; *DPP v Cook* [2006] TASSC 75 [31] and [120].

[130]Gans and Palmer (2014), pp. 162–164.

[131]*R v Esposito* (1998) 105 A Crim R 27, 43–44 (Wood CJ). See for example s. 84, dealing with oppressive conduct, s. 85 regulating potential unreliability of admissions, or s. 90 which considers the admissibility of admissions on the basis of unfairness to the defendant.

[132]For example, pursuant to s. 82 of the uniform evidence law, an admission must be proved first-hand.

[133]See further, Gans and Palmer (2014), p. 163.

[134]*Weininger v R* [2003] HCA 14; (2003) 212 CLR 629.

[135]*Edwards v R* (1993) 178 CLR 193, 210–211.

[136]*Edwards v R* (1993) 178 CLR 193.

[137]Victorian Law Reform Commission (2009), pp. 47–50. These types of warnings have attracted much criticism from law reformers such as VLRC.

[138]As amended from 1 July 2015.

law and the standard caution. Nevertheless, in certain circumstances evidence of lies told to the police may be used as incriminating post-offence conduct, requiring a trial judge to direct the jury that they must be satisfied that the only reasonable explanation for the conduct was the accused's own belief that he or she was guilty.

5.5.2.2 Lies, Prior Inconsistent Statements and the Character of Criminal Defendants

In common law legal systems, a coercive atmosphere created, for example, by persistent questioning may cause a suspect or accused to give inconsistent accounts of the defence position, or to make a statement to the police that is later contradicted by a testimonial statement at trial. In certain circumstances, the accused person's credibility can then be damaged at the trial stage, when the credibility of the accused is always a key issue.

In *Dawson v R*, Dixon J referred to the "thesis in English law" that the elements of a crime are to be established by direct and circumstantial evidence of the events comprising the crime, and are not to be inferred by evidence that relates to the character or tendencies of the accused person.[139] This distinction is based upon the notion of fairness to the accused and upon the need to confine the trial process.[140] The free admission of evidence which is only relevant to the credibility of the accused (credibility evidence),[141] may be time-consuming and may lead to many issues being tried with no direct bearing on the facts in issue.[142] Therefore, as a general rule stated in ss. 102 and 108A of the uniform evidence law, credibility evidence is prima facie inadmissible. However, there a number of exceptions in the uniform evidence law, which allow the prosecution to attack the accused person's credibility.

In criminal proceedings, s. 103 provides that the credibility rule does not apply to evidence adduced during cross-examination, which the court agrees could substantially affect the assessment of the accused person's credibility.[143] In deciding whether to allow the exception in s. 103 to apply, pursuant to s. 103(1)(a) the court may consider *inter alia* whether the evidence shows that the accused has knowingly or recklessly made a false statement while under an obligation to tell the

[139] *Dawson v R* [1961] HCA 74; (1961) 106 CLR 1, 16.

[140] *Palmer v R* [1998] HCA 2; (1998) 193 CLR 1, [52] (McHugh J).

[141] See further s. 101A uniform evidence law.

[142] Gans and Palmer (2014), p. 209.

[143] See *Dupas v The Queen* [2012] VSCA 328 [21 December 2012), [151–152], for a discussion of the terms 'credibility' and 'reliability'. Unreliable evidence may be false, untrustworthy and not credible; but the term unreliability may also mean that there is a risk that the evidence will be misleading or false or inaccurate, due to the nature of the evidence or the position of the witness. Section 108A applies this exception in the case of an accused who is not testifying.

truth.[144] Thus, if the court allows, the prosecution could cross-examine an accused about telling lies to the police, about having a history of lying, or even about having prior convictions for offences such as fraud or perjury. Section 103 could be a 'gateway' to introducing evidence of the accused person's 'bad character' into a trial—this may be particularly damaging if the accused has a previous criminal history for dishonesty type offences.[145] The prejudicial effect of such evidence would have to be assessed and potentially minimised under ss. 136 or 137 uniform evidence law discussed below at Sect. 5.8.1.

In addition, under s. 104(3)(c), one of the few ways that the prosecution may freely impugn an accused person's credibility is by cross-examining him or her about a prior inconsistent statement.[146] The prosecution is free to cross-examine about any prior inconsistent statement, including a statement that goes to credibility only. The prosecution may cross-examine an accused about a testimonial statement which contradicts a previous sworn statement or even evidence of a prior inconsistent statement made to the police or another witness.

Under s. 106, if an accused denies a credibility matter which has been raised in cross-examination under ss. 103 or 104, the prosecution may prove the lie, prior inconsistent statement, prior conviction or a previous false sworn statement by leading evidence from another source.

5.6 Compulsory Questioning Powers in Australia

5.6.1 Compulsory Questioning Powers and the Right Against Self-Incrimination

As a rule, a witness (including a suspect or accused person) must not be compelled to make self-incriminating statements in judicial or quasi-judicial settings.[147] However, the right against self-incrimination may be abrogated by an act of Parliament so that a person may be compelled to make *testimonial* self-incriminating statements or produce documents that tend to be incriminating. In 1944 in *R v Kempley*, the Court of Criminal Appeal of NSW concluded that the common law privilege against self-incrimination can be excluded by the clear

[144]Leave (permission) of the court may be granted to a prosecutor under s. 104(2) and (4) to cross-examine about a matter that goes only to the accused's credibility, if evidence was raised on behalf of the accused attacking the credibility of a prosecution witness.

[145]See further Gans and Palmer (2014), p. 250.

[146]The rules that apply under s. 104(3)(c) to the prosecution cross-examining an accused about a prior inconsistent statement are the same as the rules about cross-examination about motive to lie (s. 104(3)(a)) and the accused's inability to recall matters to which his or her evidence relates (s. 104(3)(b)); see also s. 108B.

[147]*Michael George Valentine v Technical and Further Education Commission & Anor* [2007] NSWCA 208.

intention of an act of Parliament. Furthermore, the court found that admissions made under compulsory interrogation pursuant to certain regulations could properly be received in evidence in a later prosecution.[148]

Witnesses summonsed to give evidence in an independent commission, royal commission or a similar regulatory inquiry may be compelled to give truthful sworn testimony or be liable for punishment for refusing to answer questions or giving false sworn testimony.[149] The various provisions creating compulsory questioning powers vary from authority to authority. However, in Australia today, like in England, such powers are usually accompanied by use immunity, such that the witness's answers may not be used in future criminal proceedings against him, and an entitlement to legal representation of the witness's own choosing. In general, however, the answers may be used in proceedings for offences against the statutory instrument establishing the commission, such as giving false testimony or refusing to answer questions.[150] In such proceedings it will not necessarily be unfair to an accused to use a statement which has been lawfully compelled against him or her.[151]

More recently, following a majority decision of the High Court of Australia in *X7 v Australian Crime Commission*,[152] it was found that the general provisions of the Australian Crime Commission Act 2002 could not be interpreted to allow an examiner to summons and question a person who has already been charged but not yet tried for offences which are the subject of an examination under the Act. The accused had been charged though not yet tried for conspiracy to traffic drugs and to deal with money being proceeds of crime, about which the accused was to be examined under compulsion in forfeiture proceedings. The relevant questions were: firstly, whether the provisions of the act permitted the accused to be compelled to answer questions; and, secondly, in the circumstances of the case, whether there was a real risk of interference with the administration of justice. In the leading judgment it was said:

> Permitting the Executive to ask, and requiring an accused person to answer, questions about the subject matter of a pending charge would alter the process of criminal justice to a marked degree, whether or not the answers given by the accused are admissible at trial or kept secret from those investigating or prosecuting the pending charge.
>
> Requiring the accused to answer questions about the subject matter of a pending charge prejudices the accused in his or her defence of the pending charge (whatever answer is given). Even if the answer cannot be used in *any* way at the trial, any admission made in the

[148]*R v Kempley* (1944) 18 ALJR 118, 122; (1944) 44 SR (NSW) 416; see further Gans and Palmer (2014), p. 289, where the authors state that statutes concerned with extra-curial powers of compulsion fall outside the scope of the law of evidence and the privilege against self-incrimination under the uniform evidence law and the common law.

[149]See, for example, *Pyneboard Pty Ltd v Trade Practices Commission* [1983] HCA 9.

[150]See, for example, the case of *Director of Public Prosecutions (NSW) v Attallah* [2001] NSWCA 171 in which the accused was prosecuted under the Independent Commission Against Corruption Act 1988 for making a false statement.

[151]*Director of Public Prosecutions (NSW) v Attallah* [2001] NSWCA 171, [27]; see also *Director of Public Prosecutions (NSW) v Alderman* (1998) 45 NSWLR 526; 104 A Crim R 115.

[152]*X7 v Australian Crime Commission* [2013] HCA 29 (delivered on 26 June 2013).

> examination will hinder, even prevent, the accused from challenging at trial that aspect of the prosecution case. And what would otherwise be a wholly accusatorial process, in which the accused may choose to offer no account of events, but simply test the sufficiency of the prosecution evidence, is radically altered. An alteration of that kind is not made by a statute cast in general terms. If an alteration of that kind is to be made, it must be made by express words or necessary intendment.[153]

In obiter, of persuasive effect, the Court suggested that had the Act expressly allowed a person subject to pending criminal charges to be compulsorily examined then this would have been contrary to s. 80 of the Australian Constitution, which guarantees a right to trial by jury and rules out inquisitorial hearings to determine criminal responsibility for indictable offences.[154] Under s. 28(1A) of the 2002 Act, before issuing a witness summons an examiner must now be satisfied that "...it is reasonable in all the circumstances to do so."

Following *X7*, the High Court of Australia revisited these questions in *Lee v NSW Crime Commission* (*Lee 1*),[155] in which the accused was compelled to answer questions by a court in confiscation proceedings related to offences of money laundering and trafficking in drugs and firearms about which the applicant had been charged. Here the majority took a more pragmatic approach to find that the act in question clearly authorised the use of compulsory questioning powers by a court, even when the accused had already been charged with related offences. In assessing the risk of interference with the administration of justice, the High Court emphasised that the court had made a non-publication order in relation to the testimony to protect the accused in future proceedings. In *Lee v The Queen* (*Lee 2*), the prosecution had extraordinarily come into possession of the transcripts from the compulsory examination of the accused in their criminal trials and, therefore, the HCA found unanimously that the trial had been unlawful.[156] In *Zhao*, the Victorian Court of Appeal placed some emphasis on practical considerations in terms of prejudice and unfairness, in an attempt to reconcile *X7*, *Lee 1* and *Lee 2* in assessing the risk of interference with the administration of justice, in a case where the accused was to be questioned in relation to common material facts.[157]

Like in England and Denmark, a number of administrative and regulatory authorities in Australia use compulsory information-gathering powers, such as the Australian Securities and Investments Commission (ASIC), the Australian

---

[153] *X7 v Australian Crime Commission* [2013] HCA 29, [70–71] per Hayne and Bell; see also [124–125 and 158].

[154] *X7 v Australian Crime Commission* [2013] HCA 29, [160].

[155] *Lee v NSW Crime Commission* [2013] HCA 39.

[156] *Lee v The Queen* [2014] HCA 20.

[157] *Zhao v Commissioner of the Australian Federal Police* [2014] VSCA 137. In a case against the CFMEU (a legal person), a single judge of the Federal Court of Australia found that the accepted principles to determine on balance whether to stay civil proceedings in the event of concurrent criminal charges had not been sufficiently amended by the decisions in *X7*, *Lee 1*, *Lee 2* or *Zhao*: *Director of the Fair Work Building Industry Inspectorate v Construction, Forestry, Mining and Energy Union* [2015] FCA 47.

Competition and Consumer Commission (ACCC), the Australian Prudential Regulation Authority, the Australian Taxation Office, Centrelink and Medicare Australia. For example, s. 20 of the Australian Securities and Investments Commission Act 2001 provides ASIC with the power to use compulsory questioning powers and s. 68 provides direct use immunity.

Compulsory questioning powers may also be used in relation to intelligence-gathering investigations about criminal offences. For example, the Australian Crime Commission Act 2002 (Cth) allows such powers to be used in connection with the investigation of federal offences other than by questioning a person who has already been charged with a criminal offence. Under the 2002 Act, an examiner may require a person to attend an examination to answer questions and produce documents. Section 30 of the 2002 Act provides that the punishment for failing to attend or failing to answer questions is up to 5 years imprisonment. In addition, a person may be liable for punishment for making a false statement. This is a high level of punishment by imprisonment by comparison with, for example, 6 months of imprisonment in relation to offences under the English compulsory questioning powers of the Serious Fraud Office (SFO). If the right against self-incrimination is claimed, use immunity is available so that the answer or the document may not be used in criminal or pecuniary penalty proceedings, other than confiscation proceedings or proceedings in relation to the falsity of the evidence.

In Victoria, the Victoria Police may apply under s. 5(1) of the Major Crime (Investigative Powers) Act 2004 to the Victorian Supreme Court to use compulsory powers if it is suspected on reasonable grounds that an organised crime offence has been, is being or is likely to be committed.[158] This is so even if there are proceedings on foot in another court or tribunal in relation to the subject matter.[159] The person being examined has the right to legal representation (of own choosing) and the presence of an interpreter, independent person, parent or guardian, where appropriate. Section 36 provides that an examiner (who has the power to administer the oath) questions a person on oath in private proceedings. In accordance with s. 3D, a Public Interest Monitor may be involved in the proceedings. The operation of the Act falls under the supervision of the Victorian Inspectorate. In addition, people who have been summonsed or who have an interest in examination proceedings may apply through 'special counsel' to have the order for the use of the powers revoked.[160] Supreme Court supervision, with the involvement of a Public Interest Monitor, oversight by the Victorian Inspectorate and the possibility of revocation of the orders on application by special counsel are all safeguards that counteract the coercive nature of the proceedings.

[158]See further s. 4 Major Crime (Investigative Powers) Act 2004 (Vic); the order is called a coercive powers order.

[159]S. 29(2) Major Crime (Investigative Powers) Act 2004 (Vic). Under sub-section (3) the Examiner must take reasonable steps to ensure that the examination does not prejudice any other proceedings in relation to the offences.

[160]See ss. 12 and 12D.

Under ss. 36, 37 and 38, if the witness fails to attend, refuses to give sworn evidence, fails to answer questions, fails to give evidence and produce documents or other real evidence or makes a false statement he or she may be liable to imprisonment of up to 5 years. The right against self-incrimination is generally abrogated by s. 39 and a use immunity is available so that any answer given, or document or thing produced during the examination may not be used in criminal or pecuniary penalty proceedings, other than in proceedings for breaches of the Act itself, confiscation proceedings or proceedings in relation to the falsity of the evidence.[161]

In circumstances where compulsory information-gathering powers are not accompanied by a use immunity, the fairness of using statements made under compulsion may be determined by the provisions of the uniform evidence law dealing with exclusion of evidence, in particular s. 90 (as discussed at Sect. 5.8.2 below).[162] Where criminal proceedings are pending, it has been held at common law that it is contrary to the due administration of justice to then initiate a parallel administrative investigation into the criminal liability of a person in relation to the same subject matter.[163]

The use of compulsory questioning powers in Australia has been controversial in some areas, such as industry—particularly the use of such powers in cases where there is no suspicion of any wrong doing.[164] Administrative inquiries in the absence of suspicion have been regarded as undemocratic fishing expeditions. It may be preferable to reserve the use of compulsory information-gathering powers to investigations involving serious suspected criminal conduct such as fraud, organised crime and terrorism.

> A fundamental aspect of the rule of law is that legislation conferring a power on the executive or one of its agencies which may be used to breach individual rights and liberties should not be granted 'at large'. The power must be carefully constrained and its exercise subjected to safeguards that ensure the accountable and appropriate use of the power.[165]

This reservation in relation to the use of compulsory questioning powers in Australia must be seen in light of the high level of punishment that attaches—particularly, as stated above, compared to compulsory questioning powers in England. In Denmark, the wider use of regulatory questioning powers is on the basis of the consent of the suspect, rather than the use of compulsion. These differences may be significant in a transnational context.

---

[161]It appear that the statements may not even be used to prove a prior inconsistent statement in criminal proceedings.

[162]See *R v Frangulis* [2006] NSWCCA 363 in relation to an obligation to answer questions of a private detective in order to pursue an insurance claim.

[163]*Hammond v The Commonwealth of Australia* (1982) 152 CLR 188, 206 (Deane J, referring to the unlawfulness of an 'extra-curial inquisitorial investigation').

[164]*Thorson v Pine* (2004) 139 FCR 527 (Marshall J).

[165]Williams and McGarrity (2008), p. 247.

The use of compulsory powers by the Australian Building and Construction Commission (ABCC) in the building and construction industry, under s. 52(1) of the Building and Construction Industry Improvement Act 2005 (Cth) was questioned for fairness.[166] Under the Act a person being examined was entitled to legal representation of their own choosing, even though not everyone would avail themselves of that right. In addition, s. 53(2) provided a direct and a derivative use immunity so that neither the statements nor the information gathered from the statements could be used against a person in future criminal (or civil) proceeding.[167] However, the powers did not recognise confidentiality agreements and were not accompanied by provisions for judicial review. This regime was replaced in 2009 with the enactment of the Fair Work Act 2009 (Cth), which created the Fair Work Commissioner, a workplace relations tribunal, and the Fair Work Ombudsman. The Fair Work Ombudsman may employ Fair Work Inspectors who are able to respond to complaints by conducting audits during which they may interview persons present at a workplace and require the production of documents. The right against self-incrimination is not abrogated by the 2009 Act in relation to interviews so that such persons may not be compelled to make self-incriminating statements. The right is abrogated in relation to the powers to obtain documents. However, in this respect a use immunity has been granted under s. 713 of the 2009 Act.

At common law and under the uniform evidence law, the 'privilege' against self-incrimination only protects against compulsory *testimonial* disclosure, which would include the production under summons of self-incriminating documents or things.[168] Such a summons is usually a summons to give evidence and produce, for example, documents. However, in *Sorby v Commonwealth*,[169] Gibbs CJ commented that the common law privilege does not prohibit a witness from being compelled to provide a fingerprint,[170] to show his or her face, speak or provide a sample of writing so that the jury or another witness may identify him or her.[171] Similarly, the protection afforded by the privilege against self-incrimination does not extend to providing a sample of breath for analysis under motor vehicle legislation.[172]

---

[166]See Williams and McGarrity (2008), pp. 244–279.

[167]A person could not be liable to prosecution for contravening any other law under s. 54 of the BCIIA 2005.

[168]In this situation, a summons to produce documents is usually a summons to give evidence and produce documents.

[169][1983] HCA 10; (1983) 152 CLR 281.

[170]See also Halsbury's Laws of Australia (2012), which states that there is no rule of law that a person cannot be made to incriminate him or herself by the giving of fingerprints: [195-770]. See also *R v Carr* [1972] 1 NSWLR 608 at 611 from the NSW Court of Criminal Appeal.

[171]*Sorby v Commenwealth* [1983] HCA 10, [8]; see also Halsbury's Laws of Australia (2012), [325-8245] and *R v Smith* [1984] 1 NSWLR 462, 482 in relation to the relevance of the privilege against self-incrimination to the right to refuse to participate in an identification parade.

[172]*King v McLellan* (1974) VR 773.

5.6.2 *Compulsory Questioning Powers and Derivative Real Evidence*

As discussed above, the Major Crime (Investigative Powers) Act 2004 applies a (direct) use immunity in respect of statements, documents and things made or produced under summons during an examination under the Act. However, the Act was silent about the use that could be made of any real evidence that was retrieved as a result of testimony during an examination (derivative real evidence). In *Das v Victorian Human Rights § Equal Opportunity Commission*[173] the Chief Justice of the Supreme Court of Victoria examined whether a derivative use immunity should be implied under the Act as part of the right against self-incrimination, including the right not to be compelled to testify against oneself or to confess guilt and the right to a fair hearing under the Charter of Human Rights and Responsibilities Act 2006 (Vic).[174] Warren CJ found that once the common law privilege against self-incrimination is abrogated, an individual's right to claim the privilege falls away unless immunities are provided by the legislature.[175] The privilege against self-incrimination in English and Australian common law extends to protection of direct and indirect use of statements.[176] This would also be true of the right under s. 25(2)(k) of the Charter.[177] Warren CJ found it is unlikely that the ordinary rules about exclusion of evidence would work to exclude derivative evidence, even though it can be just as damaging as the direct incriminating statement. Therefore, it was held that the way in which the 2004 Act is currently applied is in breach of the right under the Charter.[178]

The judge considered the scope of the right against self-incrimination in relation to derivative evidence in international human rights law, where there is no clear position, and other national jurisdictions.[179] Warren CJ found assistance in Canadian law, which, like the common law in Australia, recognises derivative use immunity and applies a purposive approach to interpreting the right.[180] Warren

[173]*Das v Victorian Human Rights and Equal Opportunity Commission* [2009] VSC 381. This case provides a detailed examination of the right to a fair trial and the right against self-incrimination under the Charter of Human Rights and Responsibilities Act 2006 (Vic) and at common law in Australia: [37–49].

[174]See ss. 7(2), 24(1), 25(2)(k) and 32.

[175]*Das v Victorian Human Rights and Equal Opportunity Commission* [2009] VSC 381, [44].

[176]*Das v Victorian Human Rights and Equal Opportunity Commission* [2009] VSC 381, [46–48, citing Australian and English authority].

[177]*Das v Victorian Human Rights and Equal Opportunity Commission* [2009] VSC 381, [38, 40].

[178]*Das v Victorian Human Rights and Equal Opportunity Commission* [2009] VSC 381, [84].

[179]*Das v Victorian Human Rights and Equal Opportunity Commission* [2009] VSC 381, [97].

[180]*Das v Victorian Human Rights and Equal Opportunity Commission* [2009] VSC 381, [98].

CJ referred to the majority opinion of the Canadian Supreme Court, which classified derivative evidence into four categories as follows[181]:

(1) Evidence that could not have been discovered *but for* the testimony;
(2) Evidence that was discovered as a result of the testimony, but that could have been discovered without such testimony;
(3) Evidence that would, or would probably, have been discovered even without the testimony; and
(4) Evidence that was discovered after the testimony was given, but independently of the testimony.

As a safeguard, this system of classification is able to inform a court on the question of admissibility of evidence in connection with maintaining a fair trial. The nature of the connection between the statement and the evidence will determine the extent to which the right against self-incrimination is being limited and the scope of protection that will be necessary in a given case. Evidence that would not have been obtained *but for* the compelled testimony of the accused should generally be excluded.[182] A decision about admissibility of derivative evidence contrary to the right against self-incrimination would need to be based on an evaluation of what is proportionate in the circumstances of the case with respect to the nature of the evidence or the offending.[183] However, as an alternative to the judiciary, it is arguable that the legislature could justify subjecting this general rule to special considerations in certain types of cases.

5.7 Covert Surveillance to Obtain Confession Evidence

5.7.1 Covert Surveillance and Deception

In the case of *Tofilau v The Queen; Marks v The Queen; Hill v The Queen; Clarke v The Queen* the High Court stated:

> All forms of covert surveillance, many of them authorised (subject to safeguards, such as a requirement for judicial approval) by statute, involve a kind of deception. Interception and recording of telephone conversations often produces evidence of admissions tendered at a criminal trial, as well as circumstantial or direct evidence of criminal activity. The parties to

[181] *Das v Victorian Human Rights and Equal Opportunity Commission* [2009] VSC 381, [133], referring to *S(RJ)* [1995] 1 SCR 451; *Branch* [1995] 2 SCR 3; *Phillips* [1995] 2 SCR 97; and *Re Application* 2004 SCC 42.

[182] *Das v Victorian Human Rights and Equal Opportunity Commission* [2009] VSC 381, [135].

[183] *Das v Victorian Human Rights and Equal Opportunity Commission* [2009] VSC 381, [145–148]. About the principle of proportionality in Australian law, for example in relation to sentencing law, anti-terrorism legislation or the exclusion of evidence for public policy reasons, see the Encyclopaedic Australian Legal Dictionary and Halsbury's Laws of Australia (2012), [80-2325].

> those conversations speak in the erroneous belief that they are not being overheard. They have no opportunity to consult a lawyer, or to take advice on what they should or should not say. They are not given any warning that what they say may be used against them. They do not waive any right to silence. Yet, if a suspect, in an intercepted and secretly recorded conversation, makes an admission, that admission is ordinarily and rightly regarded as voluntary. At least, it is not regarded as involuntary simply because the person making the admission is the victim of a form of deception.[184]

The impropriety of using covert surveillance to obtain confession evidence will depend on the circumstances of each case and whether there was a causal connection between the deception and the admission.[185] However, like 'coercion', what precisely constitutes unacceptable forms of deception is not clear under the uniform evidence law.[186] In general, the question of whether confession evidence has been obtained unlawfully, improperly or unfairly will be governed by the provisions of the uniform evidence law dealing with exclusion of evidence and exclusion of confessions. Part 1AC commencing at s. 15K of the Crimes Act 1914 (Cth) regulates the use of covert human intelligence sources, termed 'assumed identities' at the federal level.

5.7.2 Warrants

Today, most police powers are to be found in Australian statutes, including powers for the use of listening devices and telephone intercepts. The powers generally involve a breach of individual rights, namely privacy, and the statutory provisions creating the powers will usually be interpreted narrowly. The power to use a covert surveillance measure involving private communications, such as a telephone intercept or a listening (and recording) device is the requirement that the investigators believe on reasonable grounds that the use of the measure will provide the investigation with additional relevant information and evidence. This question must be determined by judicial authority in the form of a warrant.[187] For example, under s. 46 of the Telecommunications (Interception and Access) Act 1979, a judge or a member of the Administrative Appeals Tribunal may issue a warrant for a state police agency or the AFP to intercept telecommunications where there are reasonable grounds for suspecting that a particular person is likely to use a service and the anticipated information would be likely to assist with the investigation of a serious offence as defined in the Act, including an offence that is punishable by at least 7 years imprisonment. Section 46(2) sets out the matters which the issuing judge must take into account, including the extent to which the privacy of any persons will be breached,

[184] *Tofilau v The Queen; Marks v The Queen; Hill v The Queen; Clarke v The Queen* [2007] HCA 39, [5].

[185] *Hamberger (Employment Advocate) v Williamson and CFMEU* [2000] FCA 1644, [43] (Marshall J). See further Australian Law Reform Commission, *Evidence*, Report 26 Volume 1, [965].

[186] See s. 138(2) uniform evidence law; see further Australian Law Reform Commission, *Evidence*, Report 26 Volume 1, [965].

[187] Gans and Palmer (2014), pp. 342–343.

the gravity of the conduct being investigated, the significance of the anticipated information and the other methods that may be used to obtain the information.

5.8 The Exclusion of Confession Evidence and Fairness

5.8.1 Judicial Exclusion of Confession Evidence

The Australian uniform evidence law follows the common law tradition in terms of its basic goals.[188] Hence, the general rule at trial is that relevant evidence[189] should be admitted; and fact-finding should be a rational process based on all the relevant information being available to the court.[190] Relevant but potentially unreliable or unfairly obtained information should be treated with caution, in one of two ways: admit it, but alert the tribunal of fact to the dangers of relying on it; or exclude it altogether.[191] The parties must feel that they have had a fair hearing, in accordance with predictable and intelligible rules, within the confines of the justice system's resources and the need to protect other public interests such as confidentiality of information and legal professional privilege.[192] Therefore, enhancing the effectiveness of the courts, broad procedural rules about the admissibility of evidence have developed, including exclusionary rules or judicial discretion to exclude evidence on a number of bases.[193]

In relation to confessions, in addition to general rules about the exclusion of evidence, the common law developed a set of rules dealing solely with the exclusion of confessions.[194] Confessions were seen to be worthy of unique rules based on the theory that confession evidence is inherently probative due to the improbability that a criminal suspect or accused will confess to crimes he did not commit.[195] The uniform evidence law has replaced and expanded the common law about the admissibility of confessions by enacting three statutory rules. Two of these rules (found in ss. 84 and 85) have been sourced from the England and Wales *Police and Criminal Evidence Act* 1984 (PACE), reforming the common law in that jurisdiction about confessions that were made 'involuntarily' and were unreliable due to the circumstances in which

---

[188]Gans and Palmer (2014), p. 21.

[189]Section 55(1) of the uniform evidence legislation defines 'relevant' evidence as evidence that, 'if it were accepted, could rationally affect (directly or indirectly) the assessment of the probability of the existence of a fact in issue'; see further *Dupas v The Queen* [2012] VSCA 328 (21 December 2012), [148–149].

[190]Gans and Palmer (2014), pp. 21 and 71.

[191]Gans and Palmer (2014), pp. 6–7.

[192]Gans and Palmer (2014), pp. 9–10.

[193]Gans and Palmer (2014), pp. 9 and 322–323.

[194]The development of the law dealing with confessions has been well documented by the judges of the High Court in the Australian case *R v Swaffield; Pavic v The Queen* [1998] HCA 1.

[195]*Sinclair v King* (1946) 73 CLR 316, 334–335.

they were made.[196] The third rule about exclusion of confessions comes from the common law itself and is known as the unfairness discretion (s. 90).

Section 84(1) of the uniform evidence law provides for the mandatory exclusion of admissions that are made under the influence of oppressive conduct, such as torture, inhuman or degrading treatment, or the threat of oppressive conduct.[197] In addition, under s. 85(2), an admission made to an investigator or as the result of an act of another person who is known or reasonably believed to be capable of influencing a prosecution, must be excluded unless the circumstances in which it was made render it unlikely that the truth of the admission was adversely affected. Mandatory exclusion based on the likelihood of unreliability is required by the provision due to: (1) the nature and manner of the questioning; (2) the nature of any threat, promise or other inducement; or (3) matters internal to the accused, such as age, personality, education or mental, intellectual, or physical disability.[198] Sections 84 and 85 are the gates through which all confession evidence must pass in order to be prima facie admissible. Hereafter, a court will consider whether it would be unfair to the accused to admit the confession (s. 90) or whether the confession should be excluded on public policy grounds (s. 138). Sections 90 and 138 are significant to the discussion of exclusion of confession evidence and the fairness of a criminal trial and are discussed in detail below. The combined effect of these two provisions can be compared to the English broad unfairness discretion in s. 78 PACE.

Finally, there is long established, general common law discretion to exclude probative evidence on the basis that it would be unduly prejudicial to the accused.[199] That discretion is the predecessor to the general discretion contained in s. 135 of the uniform evidence law, whereby evidence may be excluded if the probative value is substantially outweighed by the danger of unfair prejudice to a party.[200] In criminal matters, under s. 137 the court *must* refuse to admit evidence

---

[196]See ss. 76(2)(a) and 76(2)(b) *Police and Criminal Evidence Act* 1984 (England and Wales) discussed briefly in Chap. 4, Sect. 4.8.1, where the act/statement may have happened and the English provisions provide a broader base for exclusion (for example, s. 76(2)(b) is not limited to persons of influence on an investigation); about involuntariness in common law see further the Australian High Court decision in *R v Lee* (1950) 82 CLR 133, 144.

[197]*R v Swaffield*; *Pavic v R* [1998] HCA 1; (1998) 192 CLR 159, [11]; see further *R v L* [1998] NSWSC 225 and *R v Ul-Haque* [2007] NSWSC 1251.

[198]The unlikelihood that the circumstances have affected reliability is to be proven by the prosecution on the balance of probabilities (s. 142): Anderson et al. (2009), p. 320.

[199]See *R v Christie* [1914] AC 545; and *Driscoll v R* [1977] HCA 43.

[200]Australian law is unsettled in relation to whether issues of reliability or credibility are relevant to a determination of the probative value of evidence. The Australian Law Reform Commission in *Uniform Evidence Law* (Report No 102 of 2005) has endorsed the view that questions of credibility and reliability should not be taken into account when determining the probative value of evidence, but should be left to tribunal of fact to determine in the course of a trial, particularly through hearing cross-examination: Gans and Palmer (2014), pp. 324–325. In relation to s. 137 of the uniform evidence law, this view has been established Gaudron J in *Adam v R* [2001] HCA 57, [59–60] and followed in *R v XY* [2013] NSWCCA 121 and *KMJ v Tasmania* [2011] TASCCA 7; but see Papakosmas v R [1999] HCA 37, [86] (McHugh J) followed in *Dupas v The Queen* [2012] VSCA 328, [227–228].

where the probative value is outweighed (not necessarily substantially) by the danger of unfair prejudice *to the accused*. Hence, s. 137 creates mandatory exclusion, though with something akin to 'discretion' in assessing the extent of the probative value of the evidence and the danger of unfair prejudice.[201] In comparing ss. 135 and 137, it is arguable that s. 137 is framed in favour of exclusion of evidence. The types of evidence that may call for the exercise of mandatory exclusion include: highly equivocal evidence, such as where an accused shrugs his shoulders in response to police questioning; accomplice evidence, where the accomplice has received immunity in exchange for a promise to testify against the accused; requiring the accused to dress the same as a robber recorded by a security camera; or evidence suggesting guilt by association.[202] As a general rule, before excluding evidence, the trial judge should always consider the possibilities of admitting the evidence but effectively eliminating the danger of prejudice, for example, under s. 136.[203]

5.8.2 *The Unfairness Discretion to Exclude Confession Evidence*

The third section that deals exclusively with the admissibility of confessions is s. 90, which is based on the common law unfairness discretion. This section provides that evidence of an admission led by the prosecution may be excluded if the trial judge finds that, having regard to the circumstances in which it was made, it would be unfair to an accused to admit it. However, the term 'unfair' is not defined and the scope and purpose of s. 90 is unclear.[204]

Beyond the common law, the origins of s. 90 lie in the 1987 report of the Australian Law Reform Commission (ALRC).[205] Originally, there was no clause corresponding to the common law unfairness discretion, which would allow exclusion of evidence on the ground that it would be unfair to the accused to admit it.[206] Several commentators had suggested the need for an additional discretion to deal with the situation where the accused has chosen to speak with the police but made

[201]Australian Law Reform Commission (2006), [16.37].

[202]See case references in Gans and Palmer (2014), pp. 333–334. Some Australian cases have extended the concept of prejudice to include procedural disadvantage, such as the prejudice that flows from being unable to cross-examine the maker of a statement: Gans and Palmer (2014), p. 328.

[203]See further Gans and Palmer (2014), pp. 334–335, citing case examples such as where an accused was recorded by listening devices and made numerous references to his involvement in unrelated crimes, for which he had not been prosecuted, which showed a system and time-frame behind the crime in issue but also indicated, improperly, possible bad character.

[204]Gans and Palmer (2014), p. 169.

[205]Australian Law Reform Commission (1987).

[206]See further Australian Law Reform Commission (1985).

that choice on the basis of assumptions that were incorrect, due to false representations made by the police or for other reasons.[207] The ALRC conceded that there was a need for an additional discretion to deal more adequately with issues that were not about the 'reliability' of confessions but concerned the choice about whether or not to speak to the investigative authority, and how that affected the fairness of the accused person's trial.[208] The term, 'unfairness', is necessarily vague, allowing the courts to adapt to changing social circumstances.[209] The circumstances of arrest, a compulsive aspect to interrogation, and the absence of electronic recordings may all trigger an examination of the facts under s. 90; and any significant infringement of the accused person's rights must ordinarily result in exclusion.[210] However, the vagueness of s. 90 and its wide scope have created difficulties.

In 1998, two High Court common law cases were held simultaneously in *R v Swaffield; Pavic v R*,[211] dealing with the admissibility of covertly recorded confessions. The majority found that the common law unfairness discretion is concerned with the accused person's right to a fair trial and the unreliability of evidence which has been improperly obtained. In addition, however, relying on the judgment of Mason CJ in the case of *Van der Meer v The Queen*,[212] the majority found that the unfairness discretion has a wider operation and the discretion also encompasses the situation where "...no confession might have been made at all, had the police investigation been properly conducted."[213] Therefore, in addition to unreliability, the wider aspect of the unfairness discretion at common law is to protect against unfair forensic disadvantages.[214] The drafters of s. 90 identified this possibility as the primary reason for enacting the unfairness discretion and ensuring that the uniform evidence law remained as 'protective' of an accused person's right to silence as the common law.[215]

Many admissions are made in response to police questioning in contexts that are insufficiently regulated or completely unregulated, such as questioning outside of police custody. Section 90 may act as a protective barrier in these cases. Australian courts (applying the common law and s. 90) have excluded admissions made to the police in formal interviews, that may not have been made had the relevant laws

207 Australian Law Reform Commission (1987), p. 90.

208 Australian Law Reform Commission (1987), p. 90.

209 R v Swaffield; Pavic v R [1998] HCA 1; (1998) 192 CLR 159, [53, 66 and 68] (Toohey, Gaudron and Gummow JJ).

210 *Foster v The Queen* (1993) 67 ALJR 550, 554–557.

211 *R v Swaffield; Pavic v R* [1998] HCA 1; (1998) 192 CLR 159.

212 *Van der Meer v R* (1988) 62 ALJR 656, 662.

213 *R v Swaffield; Pavic v R* [1998] HCA 1; (1998) 192 CLR 159, [54].

214 *R v Swaffield; Pavic v R* [1998] HCA 1; (1998) 192 CLR 159, [70]; See also s. 38 of the consolidated version of the Jury Directions Act 2013, including amendments up to 1 July 2013.

215 See Australian Law Reform Commission (1987); and see further Gans and Palmer (2014), pp. 169 and 171.

been followed.[216] For example, evidence has been excluded where the police conducted a custodial interview of a detainee in Pakistan, despite being unable to meet his request for legal advice due to Pakistani restrictions[217]; where the police interviewed a statutorily defined 'young person' without the presence of an independent adult[218]; and where the prosecution relied on admissions made at a properly conducted interview, but which had been elicited in an earlier interview that omitted the requisite warnings and did not follow ordinary processes.[219] The protective aspect of s. 90 in one sense aims to protect suspects from themselves; and whether or not a suspect's willingness to go ahead with an interview amounts to a waiver of that protection may be a relevant question.[220]

The High Court considered the effect of s. 90 in the appeal of *Em v The Queen*,[221] which highlights the problem of the provision's wide scope. In this case, the police were investigating a home invasion that had resulted in a murder. The appellant was questioned by police at a police station and was cautioned fully about his right to remain silent. He objected to talking to the police while being recorded and refused to answer any questions. On a later date the two interviewing officers drove the appellant to a park "to have a talk". Each had been fitted with a covert listening and recording device, for which warrants had been issued. They reminded the appellant that he was not under arrest and that he did not have to say anything to the police if he did not want to. But at no stage did the police caution the appellant that if he did say anything it might be recorded and given in evidence. Initially the appellant clearly stated he did not want to talk about the matter. Having denied any knowledge of the crime he then went on to make incriminating and potentially incriminating statements. The critical question in the case was whether the police had been obliged to give a full caution to the appellant, including advice that his answers would be recorded and may be used in evidence, despite the fact that the appellant was not under arrest and that a warrant had been issued for the use of the covert listening device. In other words, did the obligation to give a caution arise; and what was the substance of the caution that was required?

---

[216]Gans and Palmer (2014), p. 171.

[217]*R v Thomas* [2006] VSCA 165, [112–114].

[218]*R v Phung and Huynh* [2001] NSWSC 115; the New South Wales Court of Criminal Appeal refused to extend the wider operation of s. 90 to the situation where a juvenile made an "unguarded incriminating statement" in a private conversation with a support person who had attended to assist the suspect in custody, in *JB v Regina* [2012] NSWCCA 12, [40–41].

[219]*Pollard v R [*1992] HCA 69; (1992) 176 CLR 177. See also the common law cases *Nicholls v R; Coates v R* [2005] HCA 1; (2005) 219 CLR 196, [108] (McHugh J); *R v Schaeffer* [2005] VSCA 306; (2005) 13 VR 337, [6] where police failed to subsequently mention an unrecorded admission in a recorded interview.

[220]Gans and Palmer (2014), p. 171 citing Ashworth (1977); see also *R v Taylor* [1999] ACTSC 47 in relation to the exclusion of a confession by a suspect where there were issues about insobriety, intellectual impairment and cultural issues; see further Anderson et al. (2009), p. 349.

[221]*Em v The Queen* [2007] HCA 46; 232 CLR 67.

Gummow and Hayne JJ held that s. 90 was a final "safety net" provision for cases that could not be dealt with under the other more specific provisions for exclusion of confessions under the uniform evidence law. They found that the reliability of the appellant's statements under s. 85 (or s. 90) was not in question and that the issues in the case were not relevant to the exercise of the discretion under s. 90.[222] Having considered the case under s. 138, in obiter, they concluded that s. 138 was not engaged by the facts and the evidence remained admissible. Exclusion of the confession evidence was neither required through the deeming provisions in ss. 138(2) or 139, which expand on what may be deemed 'improperly' obtained evidence and the need for issuing a caution, nor through the balancing exercise in s. 138(3).[223]

In his dissenting judgment, however, Kirby J considered the application of s. 90 independent of the other uniform evidence law provisions. He concluded that there was clear unfairness and that the admissions should have been excluded.[224] This was due to the lack of a full and proper caution, in violation of accepted police practice and an accused person's right to silence.[225] In his view, in the circumstances of a police interrogation it was necessary not only to warn the appellant about the right to silence but also about the "incidents of that right" (and the consequences of foregoing it).[226] He observed that the police engaged in a deliberate deception, which tricked the appellant out of exercising his right to silence by drawing a distinction between the conversation that took place in the park and an interrogation at a police station, misleading the appellant into believing that the conversation in the park was not being recorded.[227] Kirby J found that a determination pursuant to s. 90 about whether to admit evidence of a confession must be made on the basis that criminal proceedings in Australia normally observe the accusatorial principle by which the burden of proof is on the prosecution and the accused is not usually required to say or prove anything.[228] He also referred to the fundamental obligation of an appellate court to decide whether there has been a miscarriage of justice in a particular case.[229]

The judgment of Gleeson CJ and Heydon J has been described as the 'most authoritative'.[230] Gleeson CJ and Heydon J considered that ss. 84, 85, 138 and 139 were the gates through which the confession evidence in question had already

---

[222] *Em v The Queen* [2007] HCA 46; 232 CLR 67, [111–113, 119–123].

[223] *Em v The Queen* [2007] HCA 46; 232 CLR 67, [118–121]. See the discussion at Sect. 5.8.3 below, in relation to the public policy test under s. 138 of the uniform evidence law.

[224] *Em v The Queen* [2007] HCA 46; (2007) 232 CLR 67, [132, 232 and 238].

[225] *Em v The Queen* [2007] HCA 46; (2007) 232 CLR 67, [212].

[226] *Em v The Queen* [2007] HCA 46; 232 CLR 67, [193–194, 211–214]; referring to *Miranda v Arizona* 384 US 436, 469.

[227] *Em v The Queen* [2007] HCA 46; (2007) 232 CLR 67, [218–224].

[228] *Em v The Queen* [2007] HCA 46; 232 CLR 67, [179].

[229] *Em v The Queen* [2007] HCA 46; 232 CLR 67, [199].

[230] Gans and Palmer (2014), p. 169.

passed.[231] Furthermore, there had been no breach of ss. 135 or 137, or of the New South Wales Police Commissioner's Code of Practice.[232] Therefore, they also applied s. 90 to the facts but, unlike Kirby J, they found in favour of the prosecution. Taking a narrow interpretation of s. 90, they held that the appellant's freedom to speak or not to speak had not been impugned and the admission of the confession at his trial was therefore fair, for the following reasons: first, the appellant was not under arrest and was free to leave at any time; second, while there was no express obligation on the police to caution him, the appellant had been cautioned about his right to silence (in the narrow sense); third, the appellant knew he was speaking to the police and that he was not obliged to do so—he spoke to the officers "knowingly and willingly"; fourth, the appellant possessed no right not to be recorded once the listening device warrants had been obtained; and fifth, the questioning was not overbearing, or unduly persistent, and the police were merely taking advantage of the appellant's ignorance about the fact that the confessions were being recorded and could be used as evidence against him.[233]

Gleeson CJ and Heydon J acknowledged that the appellant's freedom to speak was affected in the sense that he did not know the conversation was being recorded, which was an important factor to him.[234] However, they highlighted that the targeted suspects in virtually all cases of lawfully authorised secret surveillance "...have been deprived of the opportunity to make an informed choice about whether or not to exercise their right of silence."[235] The covert recording of a conversation does not of itself make it unfair (pursuant to s. 90 of the uniform evidence law) to use the evidence of a confession against the defendant at trial.[236]

There are two particular problems in relation to the judgment in the case of *Em*. Firstly, the different approaches taken in the High Court to the relationship between the relevant provisions of the uniform evidence law dealing with exclusion of confession evidence arguably leads to the creation of an unclear precedent for other courts in uniform evidence law jurisdictions. Is s. 90 a safety-net or can it be considered independently of the other provisions? This uncertainty is bolstered by a similar lack of clarity about the relationship between the different judicial discretions in common law; in particular the unfairness discretion and the subsequently formulated public policy discretion.

Secondly, the case of *Em* is factually complicated. However, the differing judicial approaches and opinions highlight the breadth of the discretion in s. 90

---

[231] *Em v The Queen* [2007] HCA 46; (2007) 232 CLR 67, [27–36].

[232] New South Wales Police Commissioner's Code of Practice for Custody, Rights, Investigation, Management and Evidence. See further, *Em v The Queen* [2007] HCA 46; (2007) 232 CLR 67, [37–41].

[233] *Em v The Queen* [2007] HCA 46; 232 CLR 67, [62, 77 and 78].

[234] *Em v The Queen* [2007] HCA 46; 232 CLR 67, [78].

[235] *Em v The Queen* [2007] HCA 46; 232 CLR 67, [78].

[236] *R v Swaffield; Pavic v R* (1998) 192 CLR 159; *Em v The Queen* [2007] HCA 46; (2007) 232 CLR 67, [69].

and the problem in Australian law of a lack of clear common understanding about the minimum standards of conduct that apply to criminal investigators, and the minimum procedural protections to be afforded suspects and accused persons. The question of when the obligation arises to inform a suspect of his or her procedural right to silence, and the extent of the warning that is required, is arguably a rudimentary issue that goes to the heart of a right to a fair trial. Therefore, the decision raises the lack of Australian consensus about minimum standards of investigative conduct as well as the difficulty of achieving such consensus.[237]

5.8.3 Exclusion of Evidence for Public Policy Reasons

In Australian common law, the general public policy discretion to exclude evidence (even where no unfairness to the accused has been found) originated in the judgment from 1970 of Barwick CJ in the case of *R v Ireland*.[238] In that judgment Barwick CJ stated:

> Evidence of relevant facts or things ascertained or procured by means of unlawful or unfair acts is not, for that reason alone, inadmissible. . . .Whenever such unlawfulness or unfairness appears, the judge has a discretion to reject the evidence. . . .In the exercise of it, the competing public requirements must be considered and weighed against each other. On the one hand there is the public need to bring to conviction those who commit criminal offences. On the other hand is the public interest in the protection of the individual from unlawful and unfair treatment. Convictions obtained by unlawful or unfair acts may be obtained at too high a price.[239]

Stephen and Aiken JJ, in *Bunning v Cross*, further explained the principle expressed in *Ireland*, as a discretion that deals with "broader questions of high public policy". They said the principle involves:

> . . .seeking to resolve the apparent conflict between the desirable goal of bringing to conviction the wrongdoer and the undesirable effect of curial approval, or even encouragement, being given to the unlawful conduct of those whose task it is to enforce the law.[240]

Section 138 of the uniform evidence law is largely a codification of the common law public policy discretion, applying generally to all types of unlawfully or improperly obtained evidence. Section 138(1) provides for the discretionary exclusion of evidence that has been obtained improperly, or in contravention of an Australian law, either directly or indirectly (in the case of derivative real evidence), *where the desirability of admitting the evidence is outweighed by the undesirability*.

[237]The decision in Em has been applied in cases such as *R v Frangulis* [2006] NSWCCA 363, [30–31, 33–34, 75–76]; *Sonnet v The Queen* [2010] VSCA 315 and *R v Sean Sonnet (Ruling No. 2)* [2011] VSC 551, [38–41].

[238]*R v Ireland* (1970) 126 CLR 321.

[239]*R v Ireland* (1970) 126 CLR 321, 334–335.

[240]*Bunning v Cross* (1978) 141 CLR 54, 74–75.

Section 138(2) indicates when an admission made in the course of police questioning may be 'improperly' obtained due to coercion or deception. In addition, s. 139 provides that a person who is under arrest or who is suspected of a crime, based on a belief that there is sufficient evidence against them, must be cautioned before being questioned—otherwise any admission made will be considered to have been obtained improperly for the purposes of s. 138(1).

The primary rationales behind the public policy discretion contained in s. 138 are to deter future investigative misconduct; to uphold the legitimacy of the judiciary, by avoiding the taint of overlooking the misconduct in question; and to preserve the rights of those who are the subject of specific investigations.[241] With those aims in mind, investigative conduct that is otherwise lawful can nevertheless be considered improper. A finding of impropriety or that evidence was obtained unlawfully does not automatically lead to exclusion.[242] The decision is based on a balancing of a number of different criteria listed in s. 138(3), in addition to any other relevant factors.

The first three sub-paragraphs feature factors that relate to the cost to society of excluding evidence,[243] namely: (a) the probative value of the evidence; (b) the importance of the evidence in the proceedings; and (c) the nature of the offence, cause of action, defence, or subject matter of the proceedings. It may not always be easy to assess the probative value of the evidence objectively, and even its importance in the proceedings, in isolation from the rest of the evidence in the trial.

Sub-paragraphs (d) to (g) provide the four factors that must be considered by a judge in the exercise of the discretion to exclude evidence, that represent the cost to society of admitting the evidence.[244] These factors focus on the seriousness of the investigative misconduct and aim to ensure the rule of law. They are, as follows: (d) the gravity of the impropriety or contravention; (e) whether the impropriety or contravention was deliberate or reckless; (f) whether the impropriety or contravention was inconsistent with a right recognised by the International Covenant on Civil and Political Rights (ICCPR); and (g) whether there will be any other proceedings taken in relation to the impropriety or contravention. There is general agreement that a deliberate or reckless disregard of the law or a judicially recognised code of conduct—as opposed to an isolated and accidental non-compliance—is a strong factor in favour of exclusion of evidence.[245]

These factors in favour of exclusion have been introduced by the legislature in the interests of the protection of the individual and due administration of criminal justice.[246] In relation to sub-paragraph (f), the more limited or unclear scope of

[241] Gans and Palmer (2014), p. 338.

[242] See *Bunning v Cross* [1978] HCA 22; (1978) 141 CLR 54.

[243] Gans and Palmer (2014), p. 352.

[244] See further Gans and Palmer (2014), pp. 353–354.

[245] *Pollard v R* [1992] HCA 69; (1992) 176 CLR 177, 203, 204 (Deane J); see also Stephen and Aickin JJ (with whom Barwick CJ agreed) in *Bunning v Cross* (1978) 141 CLR 54, 77–79.

[246] *Pollard v R* [1992] HCA 69; (1992) 176 CLR 177, 203, 204 (Deane J).

protection of the right to a fair trial under Article 14 of the ICCPR, in comparison to Article 6 of the ECHR, for example, may undermine the weight to be given to the cost to society of excluding particular evidence.[247] However, when interpreting the provisions of the ICCPR, Australian judgments do occasionally refer to the jurisprudence of the ECtHR, for example, when that court discusses similar or analogous rights.[248] Nevertheless, the question of unfairness to an accused is arguably of peripheral importance in the exercise of the public policy discretion.[249] In addition, the existence of sub-paragraph (g) indicates that if other proceedings, such as disciplinary proceedings, are to be taken in relation to the misconduct, the necessity for the trial court to express its disapproval may not be so great.

The final factor for a judge to examine under s. 138(3) is paragraph (h): the difficulty of obtaining the evidence without impropriety or a contravention of Australian law. If there had been no alternative way in which to secure the evidence, other than by the improper or unlawful methods used, this may provide 'some excuse' for the impropriety or contravention, in favour of admission of the evidence.[250] For example, vital evidence may have been at risk of being destroyed. Therefore, the motivation of investigators who have acted improperly or unlawfully may also be relevant to this factor.[251] Another interpretation of this sub-paragraph, which also supports admission, is that if the evidence could have been obtained anyway, in an alternative and lawful manner, the impact of the improper or unlawful conduct is less. However, admitting evidence on this basis would arguably erroneously condone investigative 'corner-cutting' and a watering down of the rationales of the discretion.[252] In the absence of a clear definition of investigative 'impropriety' in the uniform evidence law (beyond ss. 138(2) and 139), or clearly established guidelines elsewhere about minimum standards of conduct for the gathering of evidence in criminal cases, it is for the courts to set the standards in each case.[253]

---

[247]Compare, for example, Chap. 2, Sects. 2.2.2.2 with 2.3.2.1 and 2.3.2.2.

[248]*Mallard v R* [2005] HCA 68; (2005) 224 CLR 125, [78–80].

[249]*Ridgeway v R* [1995] HCA 66; (1995) 184 CLR 19, [33] (Mason, Deane and Dawson JJ), [14] (Brennan CJ) and [29] (Toohey J).

[250]Australian Law Reform Commission (1985).

[251]See further *R v Pimental* [1995] NSWCCA 401, [24]; (1999) 110 A Crim R 30, 35; Gans and Palmer (2014), p. 355.

[252]Thus, in *Bunning v Cross* [1978] HCA 22; (1978) 141 CLR 54, 80, Stephens and Aiken JJ stated that in cases where a lawful alternative exists, this factor is 'wholly equivocal'.

[253]Gans and Palmer (2014), p. 344.

5.8.4 Overlap Between s. 90 and s. 138

From decided cases, it is apparent that there is a significant amount of overlap between the different grounds for the exercise of judicial discretion to exclude evidence, particularly the unfairness and public policy discretions. This may explain why in England and Wales there is a single, broad unfairness discretion (s. 78 PACE), which involves balancing of competing interests.

In their majority judgment in *R v Swaffield; Pavic v The Queen*, Toohey, Gaudron and Gummow JJ said:

> Unfairness then relates to the right of an accused to a fair trial; in that situation the unfairness discretion overlaps with the power or discretion to reject evidence which is more prejudicial than probative, each looking to the risk that an accused may be improperly convicted. While unreliability may be a touchstone of unfairness, it has been said not to be the sole touchstone. It may be, for instance, that no confession might have been made at all, had the police investigation been properly conducted. And once considerations other than unreliability are introduced, the line between unfairness and policy may become blurred.[254]

The purpose of the unfairness discretion is said to be the protection of "…the rights and privileges of an accused person."[255] However, the case law dealing with investigative misconduct may illustrate the potential for overlap between s. 90, and ss. 138 and 139, where the latter sections also, *inter alia*, aim to protect the individual. For example, impropriety and unfairness may include failure by investigators to inform suspects of their legal rights and the consequences of those rights, or failure to comply with rules of practice or established codes of conduct, such as those dealing with investigative questioning.[256]

In *Foster v The Queen*, the majority of the High Court described the unfairness discretion and the public policy discretion as two independent residual discretions, stating that:

> The first of those discretions exists as part of a cohesive body of principles and rules on the special subject of evidence of confession statements. It is the discretion to exclude evidence on the ground that its reception would be unfair to the accused, a discretion which is not confined to unlawfully obtained evidence… The second of those discretions is a particular instance of a discretion which exists in relation to unlawfully obtained evidence generally, whether confession or 'real'. It is the discretion to exclude evidence of such a confessional statement on public policy grounds… In a case where both discretions are relied upon to support an application for the exclusion of a voluntary incriminating statement obtained by unlawful police conduct, it will commonly be convenient for the court to address *first* the question whether the evidence should be excluded on the ground that its reception and use in evidence would be unfair to the accused. [emphasis added][257]

---

[254] *R v Swaffield; Pavic v The Queen* [1998] HCA 1; (1998) 192 CLR 159, [54].

[255] *R v Swaffield; Pavic v The Queen* [1998] HCA 1; (1998) 192 CLR 159, [52].

[256] Gans and Palmer (2014), p. 345.

[257] *Foster v The Queen* (1993) 67 ALJR 550, 554; 113 ALR 1, 6–7 (Mason CJ, Deane, Dawson, Toohey and Gaudron JJ).

The unfairness discretion deals solely with evidence of confessional statements, and the emphasis is on the effect of investigative impropriety or misconduct on a particular accused at trial.[258] Alternatively, the public policy discretion applies to improperly and unlawfully obtained evidence generally, whether confession or real, and focuses on 'large matters of public policy': whether the trial judge should give more weight to the public requirement of bringing a wrongdoer to conviction, on the one hand; or to the public requirement of upholding the rule of law and protecting individuals (also in future cases), by ensuring that law enforcement officers do not benefit from unlawfully or improperly obtained evidence, on the other.[259] Unfairness to the particular accused is relevant to the public policy discretion, but is not the central point.[260]

In the common law case of *Collins v The Queen*, Brennan J stated:

> When the admission of confession evidence is in question, the material facts are evaluated primarily to determine whether it is unfair to the accused to use his confession against him, and it would be only in a very exceptional case that the residual question would arise as to whether the public interest requires the rejection of the confession.[261]

More recently, in *R v Swaffield; Pavic v R*,[262] the High Court discussed at length the considerable overlapping between the unfairness and public policy discretions at common law. The Chief Justice suggested that the admissibility of confessions should turn first on the question of voluntariness, next on the issue of reliability, and finally on an overall discretion that:

> . . .might take account of all the circumstances of the case to determine whether the admission of the evidence or the obtaining of a conviction on the basis of the evidence is bought at a price which is unacceptable, having regard to contemporary community standards.[263]

The other members of the court agreed with the reformulation of the common law that was suggested by Brennan CJ, with one qualification:

> The qualification is that the decided cases also reveal that one aspect of the unfairness discretion is to protect against forensic disadvantages which might be occasioned by the admission of confessional statements improperly obtained.[264]

---

[258] *R v Swaffield; Pavic v The Queen* [1998] HCA 1; (1998) 192 CLR 159, [18] and [24].

[259] *R v Swaffield; Pavic v The Queen* [1998] HCA 1; (1998) 192 CLR 159, [22] and [24].

[260] *Bunning v Cross* (1978) 141 CLR 54, 74–75.

[261] *Collins v The Queen* (1980) 31 ALR 257, 317.

[262] *R v Swaffield; Pavic v The Queen* [1998] HCA 1; (1998) 192 CLR 159.

[263] *R v Swaffield; Pavic v The Queen* [1998] HCA 1; (1998) 192 CLR 15, [69] (Toohey, Gaudron and Gummow JJ). See also [27] (Brennan CJ). The reference to contemporary community standards has been interpreted by the NSW Court of Criminal Appeal as a reference to "...community standards concerning the maintenance of the rule of law in a liberal democracy, the elements of the proper administration of justice and the due requirements of law enforcement": *R v Suckling* [1999] NSWCCA 36, [40–41].

[264] *R v Swaffield; Pavic v The Queen* [1998] HCA 1; (1998) 192 CLR 15, [70]. See further [74 and 78] (Toohey, Gaudron and Gummow JJ) and [120] (Kirby J). See further Halsbury's Laws of Australia (2012), [195–205].

Such an unfair forensic disadvantage will result in a suspect making an admission that they would not otherwise have made.[265] Together with this aspect of the unfairness discretion, the common law reformulation suggested by the High Court mirrors the provisions of the uniform evidence law, taken in combination.

Therefore, the High Court found that the question of the admissibility of admissions was based on a broad principle about the suspect's freedom to choose whether to speak or to remain silent.[266] The majority Toohey, Gaudron and Gummow JJ, proposed that this principle required the examination of a number of questions: Was the admission involuntary, in the sense that the defendant's will had been overborne? Was the admission unreliable? Should the admission, though voluntarily made, be excluded in the exercise of overall judicial discretion? The High Court envisaged that consideration of this broad judicial discretion would include matters, such as: unfairness to the accused; public policy interests; and whether the prejudicial effect of evidence outweighs its probative value.[267] Thus, rejection of a voluntary admission may occur where the circumstances point to unfairness to the accused if the statement is admitted, or to a situation where there is no such unfairness but the evidence has been obtained at a price that is unacceptable to society.[268]

The *Swaffield* and *Pavic* cases both dealt with the admissibility of voluntary confessions, recorded covertly by an undercover police operative or a police agent *after* the suspects had previously formally declined to answer police questions. In answering the broader question of whether or not the accused person's freedom to choose to speak to the police had been impugned,[269] the majority of the court referred to Canadian authority, which distinguished between undercover agents deceptively *listening to* an admission and deceptively *prompting* an admission[270]:

[265] *R v Swaffield; Pavic v The Queen* [1998] HCA 1; (1998) 192 CLR 15, [155].

[266] *R v Swaffield; Pavic v The Queen* [1998] HCA 1; (1998) 192 CLR 15, [92].

[267] *R v Swaffield; Pavic v The Queen* [1998] HCA 1; (1998) 192 CLR 15, [51–52, 70] (Toohey, Gaudron and Gummow JJ) and [119–120] (Kirby J).

[268] *R v Swaffield; Pavic v R* [1998] HCA 1; (1998) 192 CLR 15, [91] (Toohey, Gaudron and Gummow JJ).

[269] The High Court also considered the 'added question' of whether Rule 2 of the Judges' Rules had been breached and, if so, the consequence for the admissibility of the conversation. The court concluded that the absence of a caution does not require exclusion of an admission at common law: *R v Swaffield; Pavic v R* [1998] HCA 1; (1998) 192 CLR 15, [94–95] (Toohey, Gaudron and Gummow JJ), referred to at Sect. 5.4.2 above.

[270] *R v Swaffield; Pavic v The Queen* [1998] HCA 1; (1998) 192 CLR 15, [91, 97–98, 100–103] (Toohey, Gaudron and Gummow JJ); see also Kirby J, who did not consider that the Canadian approach derived specifically from the Canadian Charter of Rights and Freedoms so as to make it inapplicable in Australian law: [153–155]. The members of High Court referred in particular to the judgments of the Canadian Supreme Court in *R v Herbert* [1990] 2 SCR 151 and *R v Broyles* [1991] 3 SCR 595. The line of reasoning adopted by the Australian High Court and the Canadian Supreme Court in the above cases have been referred to by the European Court of Human Rights, referred to in Chap. 2, Sect. 2.3.2.2.6 above.

> [I]n the absence of eliciting behaviour on the part of the police, there is no violation of the defendant's right to choose whether or not to speak to the police.[271]

Referring to Canadian authorities, the High Court majority in *Swaffield* held that the undercover agent's story that his brother-in-law had started a fire was conduct that elicited the defendant's admissions about his own involvement in arson, in clear breach of Swaffield's right to choose whether to speak.[272] In *Pavic*, although the majority found that his friend was acting as an agent of the State, the evidence of the admissions was properly admitted because there had been no 'interrogation' and the admissions had not been elicited.[273] In the judgment, the majority did not state explicitly whether they were applying the unfairness aspect or the public policy aspect of judicial discretion to the two cases. However, on an examination of the judgment, it appears that they considered the right to silence and unfairness to the accused, without reference to competing public interests, to resolve the issues.[274]

5.8.5 Gravity of Offending and Exclusion of Confession Evidence

In *R v Swaffield; Pavic v The Queen*, in relation to the common law unfairness and the public policy discretions, Brennan CJ commented:

> Of course, the two discretions do overlap and in a sense it is immaterial whether a trial judge considers the facts of a case under one heading rather than another.[275]

However, given that the unfairness discretion does not involve a balancing of the competing public interests in determining what type of investigative misconduct is acceptable in the circumstances of a particular case, the outcome in relation to exclusion of confession evidence in the same case may be different under the two discretions. This is due to relevance or otherwise of the gravity of the crimes being tried.

---

[271] *R v Herbert* [1990] 2 SCR 151, 173.

[272] *R v Swaffield; Pavic v The Queen* [1998] HCA 1; (1998) 192 CLR 15, [98]; see further Gans and Palmer (2014), p. 357, citing *R v Liew* [1999] 3 SCR 227, in relation to the situation where the suspect prompts the conversation leading to a confession.

[273] *R v Swaffield; Pavic v The Queen* [1998] HCA 1; (1998) 192 CLR 15, [100–103]. There are a number of different scenarios that would not lead to exclusion of a confession, such as confessions that have been obtained by prison inmates on their own initiative in order to foster beneficial relations with the authorities or 'real conversations' with persons with whom they have a special relationship, even one with a power imbalance, without police involvement: see (Canadian Supreme Court) *R v Broyles* [1991] 3 SCR 595, 611 (Iacobucci J); *R v Burt* [2000] 1 Qd R 28, 33. See also *R v M* [2002] QCA 486; (2002) A Crim R 324; *R v Lowe* [1997] 2 VR 465 and *R v Walker* [2000] NSWCCA 130; and Gans and Palmer (2014), pp. 358–359.

[274] *R v Swaffield; Pavic v The Queen* [1998] HCA 1; (1998) 192 CLR 15, [98, 100–103].

[275] *R v Swaffield; Pavic v The Queen* [1998] HCA 1; (1998) 192 CLR 159, [28].

If a confession is covertly recorded in a private home by a secretly installed listening device ('bugging' device) in the absence of a warrant, but the confession is apparently reliable and was not elicited or induced by the police in any way, it is likely that the unfairness discretion would not be enlivened.[276] If a court goes beyond the narrower issue of fairness to a particular accused, to consider the public interest in whether highly probative evidence should be rejected due to the improper or unlawful circumstances in which it was obtained, the issue of the gravity of crimes charges becomes highly relevant. In fact, s. 138(3)(c) obliges a court to consider the nature of the offending when exercising the discretion to exclude improperly or illegally obtained evidence. The question of weight to be given to this factor means that the public policy discretion arguably offers less protection to a person accused of a serious crime.

On the other hand, in a case where a confession would not have been made at all but for the improper conduct of the police, the unfairness discretion is likely to offer broader protection to a person accused of a serious crime (such as murder) than the public policy discretion, because the seriousness of the offending will not be a relevant consideration. The case against *Swaffield*,[277] where an undercover officer conducted an (unlawful) interrogation of the accused in an arson case, is on point. Brennan CJ applied the public policy discretion and acknowledged that there had been *strong interests to protect on either side* of the balance. Where the case could legitimately have gone either way, he did not interfere with the decision of the lower court to exclude the evidence. However, the majority of the court applied the unfairness discretion, which in their view clearly called for the exclusion of the confession evidence.[278]

The High Court has also considered the relevance of the seriousness of the offending to the operation of the common law public policy discretion and the exclusion of illegally or improperly obtained evidence, in *Tofilau v The Queen; Marks v The Queen; Hill v The Queen; Clarke v The Queen*.[279] In the case against the accused Clarke, the conduct of undercover operatives was found to be "*not sufficiently unacceptable*" as to cause the evidence in question to be excluded.[280] In these cases, undercover agents created a ruse, which was intended to prompt high level criminals into confessing to a number of unsolved crimes. The undercover officers posed as members of criminal gang and offered membership to the suspects involved on condition that they confessed to their previous criminal activity. The

---

[276]*Ridgeway v R* [1995] HCA 66; (1995) 184 CLR 19, [33] (Mason, Deane and Dawson JJ).

[277]*R v Swaffield; Pavic v The Queen* [1998] HCA 1; (1998) 192 CLR 159.

[278]But see Gans and Palmer (2014), p. 359, in which the authors suggest that the cases of *Swaffield* and *Pavic* must have been decided differently under s. 138 on the basis of the seriousness of the offending in each case.

[279]Cited as *Tofilau v The Queen* [2007] HCA 39, [112]; (2007) 231 CLR 396, 432 (Gummow and Hayne JJ) and [410] and [413], 527-8 (Callinan Heydon and Crennan JJ).

[280]*Clarke v The Queen* [2007] HCA 39, [397].

undercover gang members promised in return that they knew corrupt police officers who would then derail any investigations into those crimes.

The examples of case law discussed above are illustrative of the mechanics of the balancing exercise in s. 138. Investigators of serious crime are not necessarily encouraged to cut corners or take more risks—a deliberate and serious contravention of the law will always be frowned upon, and will carry great weight in the determination of the test under s. 138, and the public interest. However, under s. 138(3)(c), if the facts of the case reveal that the crime is particularly heinous, and great weight is placed on this aspect, then there may be a tendency to ignore the importance of the competing public interest in counteracting serious investigative misconduct. Gans and Palmer argue that investigative misconduct is more harmful when it involves more serious offences, because it ultimately exposes defendants to a higher risk of serious punishment and encourages investigators to be less careful when investigating serious crime.[281] In a dissenting judgment on this point in *R v Dalley*, Simpson J said:

> In my opinion it would be wrong to accept as a general proposition that, because the offence charged is a serious one, breaches of the law will be readily condoned. In my judgment there may be cases in which the fact that the charge is a serious one will result in a more rigorous insistence on compliance with statutory provisions concerning the obtaining of evidence. That a person is under suspicion for a serious offence does not confer a licence to contravene laws designed to ensure fairness.[282]

Above all, on the issue of investigative misconduct, s. 138 must be viewed in the context of the other provisions that deal with the admissibility of confession evidence and judges' exclusionary discretions. For example, torture and other prohibited ill-treatment will never be tolerated under s. 84(1) and clear breaches of a suspect's right to choose whether or not to speak to the police will typically be caught by s. 90, regardless of the crime. The above cases are illustrative of the importance of the other sections of the uniform evidence law. These other provisions draw lines in the sand, which determine the outer-limits of s. 138 in relation to confessions.

From an analysis of the Australian jurisprudence about judicial discretion to exclude evidence, it is arguable that both the common law and the uniform evidence law have treated the question of admissibility of confession evidence differently from other types of evidence. In certain circumstances, the unfairness discretion and s. 90 may quite deliberately remove the balancing exercise, weighing up the costs to society of admitting or excluding evidence, from a decision on the admissibility of confessions. This may be an acknowledgment of the importance of protecting a suspect or accused person's right to silence in the context of police questioning, given the clash between the potential power imbalance between a

[281] Gans and Palmer (2014), p. 353.

[282] *R v Dalley* [2002] NSWCCA 284, [97]; (2002) 132 A Crim R 169, 189.

suspect and the investigating authorities, on the one hand, and the general expectation that confessions are reliable and probative, on the other.[283]

The discretions in ss. 90 and 138 of the uniform evidence law aim to do what the English law does in one broad provision, namely s. 78 PACE. In comparing these three provisions, the significant feature of the Australian law is that the drafters of s. 90 have considered it necessary to remove the gravity of the offending as a factor relevant to the exercise of judicial discretion to exclude confession evidence. It is likely that this is recognition that the right to silence deserves special protection in the overall maintenance of a fair trial. In practice, however, the difference in the level of protection of the right to silence may be more nuanced.

5.9 Conclusion

In this chapter the national law in Australia has been explored to discover the ways in which the balance between law enforcement and the individual's right to silence and right against self-incrimination is established in criminal proceedings. As vast the country, so too is the law. Criminal proceedings in Australia rely on the combination of federal, state and common law under the umbrella of international obligations, which arise under the ICCPR and have been incorporated into national law. Therefore, the majority of Australian states and territories have chosen to join in harmonising the law of evidence in judicial proceedings in Australia with the uniform evidence law. On the whole, the law has enacted the principal features of the common law of evidence, while amending the latter on a number of points. Harmonisation of the rules of evidence used at trial has ensured a more unified approach to the scope of procedural rights in Australian investigations. For example, provisions such as s. 89 (discouraging adverse inferences to be drawn from pre-trial silence), s. 90 (unfairness discretion) and s. 139 (requiring exclusion of evidence on a lack of caution) all encourage the police to respect the right to silence, to caution properly (even a suspect who voluntarily attends for police questioning). These provisions, along with s. 138 (the trial judge's public policy discretion to exclude evidence) otherwise ensure that the suspect is supported by fair procedures, such as access to a lawyer, audio-recording of interviews and overall fairness in the use of covert surveillance measures. These types of enactments enable the accused to participate and engage in the criminal proceedings on a sound and equal footing with the state.

An important distinction between the federal and Victorian state provisions about police questioning is that the Commonwealth rules in relation to cautioning of suspects and access to legal representation (of a suspect's own choosing) apply to suspects who are under arrest or in custody as well as suspects who are voluntarily accompanying a police officer for questioning. The Victorian rules limit these particular protections to custodial situations. This distinction will also be relevant

[283] Gans and Palmer (2014), p. 157.

in transnational cases, where the assumption is that the Australian Federal Police will be applying similar standards as their Danish or English counterparts.

Nevertheless, s. 139 of the uniform evidence law provides that confession evidence will be deemed inadmissible under s. 138 if the police had formed a view that there was sufficient evidence to establish that the suspect had committed an offence and did not caution him. Therefore, this provision ensures that all police in a uniform evidence law jurisdiction are careful in relation to the need to caution suspects about the right to silence after forming reasonable grounds of suspicion. In addition, federal and state provisions emphasise the importance in Australia of audio or audio-visual recording of interviews, as a safeguard against abuse. In Victoria, there is a general rule that confession evidence in relation to indictable offences will be excluded if not audio-recorded or confirmed by audio-recording. These rules also reflect the need to protect the integrity of the record of interview as an 'exhibit' in a trial, which may be directly admitted into a trial, subject to the exclusion of unfair or prejudicial content. Further, s. 90 of the uniform evidence law always serves to provide an extra check and balance of fairness to the accused in relation to the reliance on confession evidence, where all the above safeguards may be relevant to the trial judge's exercise of discretion.

The Victorian rules expressly provide that the questioning of a suspect must be deferred when he or she has indicated a desire to speak with a legal representative until this has been fulfilled. At both federal and state level, the police are obliged to provide a suspect with assistance in facilitating this. However, another potential weakness in the functioning of the right to access a lawyer is that it will not generally be considered 'unfair' for the police to keep questioning a suspect after he has asserted the right to silence but continues to answer questions.

One significant difference in the functioning of pre-charge safeguards in Australia is that free publicly appointed legal advice is in general unavailable. The state-operated means tested Legal Aid schemes are connected to the trial process—remand hearings or preparation for trial may be the first opportunity for an accused to access free legal representation where necessary. This is with the exception in NSW where a suspect of a serious indictable offence who is cautioned that if he remains silent during official questioning about a defence that he raises at trial must be provided with access to an Australian lawyer.

The effect of s. 89 of the uniform evidence law about adverse inferences at trial and the standard caution used in investigations in Australia is that the suspect is entitled to remain silent in response to official police questioning and no adverse inference should be drawn at trial from silence in this situation. This is consistent with the common law on the point and accusatorial principle that it is the prosecution that bear the burden of proof. Under common law no adverse inferences should be drawn against an accused in relation to pre-trial silence in any context. However, this is in contrast to the position in NSW in relation to serious indictable offences, as well as England and even Denmark, where adverse inferences may be drawn in the free evaluation of the evidence.

In addition, as a general rule no inferences of guilt should be drawn from failure to testify, although a co-accused may suggest to the jury to follow this path.

Nevertheless, the trial judge may, with great caution, direct the jury that the accused's failure to testify may be used evidence which supports the prosecution case. But this only applies where the failure to testify may be explained by the accused's consciousness of guilt on account of additional facts known only to the accused and not already introduced into evidence by other witnesses. Therefore, the circumstances in which a trial judge can safely direct the jury along these lines will be extremely rare, if not non-existent.

Throughout this chapter the continuity of Australian law that deals with the right to silence has been illustrated. The way the suspect's confession evidence or evidence of silence is obtained in the investigation phase directly influences the provisions of the uniform evidence law, as well as other statutes and the common law, are enlivened at trial. In particular, this chapter has revealed the interplay between investigative procedures and judicial discretion to exclude evidence under ss. 90 and 138, on unfairness or public policy grounds. Like in England and Wales, Australian criminal proceedings place heavy reliance on the question of admissibility and exclusion of evidence as a final check and balance of the fairness of criminal proceedings and the integrity of the role played by the judiciary in achieving truth and justice.

In conclusion, the provisions of art. 14 ICCPR, which have apparently influenced the drafting of the Victorian human rights charter, have been reflected in the Australian law in relation to police questioning. In particular, for example, the Victorian state provisions about cautioning, access to a lawyer of a suspect's own choosing and audio or video recording of interviews place a similar focus of the protection of suspects in custody from coercive forms of questioning in the criminal investigation. This is reminiscent of the limited scope of protection in pre-trial proceedings identified by the HR Committee in their commentary and case law. Beyond this, however, the Australian law has developed and amended the common law right to a fair trial. This has occurred through the processes of codification and judicial decision-making, frequently seeking inspiration from the judicial authority of other common law countries—with only occasional reference to the guidelines of international human rights decision-making bodies such as the ECtHR. Nevertheless, while some of the detail of the provisions may not apply the same minimum standards being set by the ECtHR, the overall picture is that the Australian law has arrived at a similar place. For example, The Australian High Court's conclusion about the impropriety of using compulsory questioning powers in cases where charges are pending is in accord with the ECtHR's decision in *Weh*. In fact, Australian courts have extended the scope of minimum rights in some areas, such as formulating a test for the use of derivative real evidence that has been located as a result of testimony given under compulsory questioning powers, and have even been a source of inspiration in the ECtHR, for example, in the case of covert police interrogation. In both of these examples, the jurisprudence of the Canadian courts that work out from a combination of common law and constitutional protection of human rights should also be acknowledged.

The right to silence is central to any well-functioning criminal justice system. In the previous chapters, the criminal justice systems in Denmark, England and Wales, and Australia have been examined to determine how the national law limits and protects the right to silence. There is continuity within each of the national systems in the way that the rules in the investigative and trial phases of criminal proceedings relate to each other. Nevertheless, legal rules that are structurally quite different, for example, the form of caution to be used or the way in which administrative compulsory questioning powers operate achieves the balance in each of these systems between effectiveness of law enforcement and fairness in protecting the right to silence. This is so despite the fact that the national law in some systems, for example Denmark and England, should be applying the same common minimum standards that are applicable under common frameworks for human rights protection.

In the next chapter, the divergences in the limitation and protection of the right to silence and the right against self-incrimination between the national systems of Denmark, England and Wales and Australia will be brought into focus. Reflections on the changing scenery in the national systems are necessary to determine whether national balances between effectiveness and fairness are disrupted in a significant way in the international cooperation process.

References

Anderson J, Clegg L, Williams SC (2009) The new law of evidence: annotation and commentary on the uniform evidence acts, 2nd edn. Lexis-Nexis, Sydney

Australian Law Reform Commission (1985) Evidence: Report No. 26 (Interim). http://www.austlii.edu.au/cgi-bin/sinodisp/au/other/lawreform/ALRC/1985/26.html?stem=0&synonyms=0&query=alrc%20evidence. Accessed 15 Jan 2014

Australian Law Reform Commission (1987) Evidence: Report No. 26. http://www.austlii.edu.au/au/other/lawreform/ALRC/1987/38.html Accessed 15 Jan 2014

Australian Law Reform Commission (2006) Uniform Evidence Law: Report No. 102. https://www.alrc.gov.au/publications/report-102 Accessed 15 Jan 2014

Dixon D, Cowdery N (2013) Silence rights. Aust Indigenous Law Rev 17(1):23–37

Gans J, Palmer A (2014) Uniform evidence. OUP, Sydney

Halsbury's Laws of Australia (2012). Lexis-Nexis, Australia

Odgers S (2014) Uniform evidence law, 11th edn. Thomson Reuters, Sydney

Victorian Law Reform Commission (2009) Jury Directions: Final Report. http://www.lawreform.vic.gov.au/sites/default/files/VLRC_JuryDirections_FinalReport.pdf. Accessed 15 Jan 2014

Williams G, McGarrity N (2008) The investigatory powers of the Australian Building and Construction Commission. Aust J Labour Law 21:244–279

Part III
Transnational and International Perspectives on the Right to Silence

Chapter 6
Admissibility of Confession Evidence Across Borders: A Transnational Perspective

Abstract The previous chapters have demonstrated that the question of balancing the effectiveness of law enforcement and the interests of the suspect or accused in relation to the right to silence and the right against self-incrimination relies on in-built continuity that exists between the investigative and the trial phases within the national law. This ensures that any limitations that are placed on the right to silence in the investigative phase are meaningfully counter-balanced at some later point, which may not be until the criminal trial takes place. The opposite is also true, whereby what may be perceived as restrictions on the right to silence in the trial, such as allowing an inference of guilt to be drawn from the accused's pre-trial silence, may have been safeguarded by the circumstances in which the suspect's evidence was obtained, for example, by the protection provided by cautioning about the right to silence, legal representation, a letter of rights, early involvement of the prosecuting authorities etc. Building on this conceptual framework, this chapter examines what happens to the balance at a fundamental level when confession evidence or evidence of silence is transferred across national borders where the equivalent counter-balancing measures are not present.

6.1 Introduction

Chapters 2–5 followed the paths of protection of the right to silence and the right against self-incrimination in the national systems of England and Wales, Denmark and Australia, against the background of the guidance of the Human Rights Committee (HR Committee) and the European Court of Human Rights (ECtHR). Each national system has responded in different ways to the desirability of obtaining an account from a suspect or other self-incriminating evidence, such as silence itself, and the need to ensure that the methods used do not adversely affect the fairness of criminal proceedings when viewed as a whole. Safeguards have been developed within the national systems to balance the legal rules that have the effect of diminishing the rights in the process of obtaining and using confession evidence,

F.M.W. Billing, *The Right to Silence in Transnational Criminal Proceedings*,
DOI 10.1007/978-3-319-42034-9_6

pre-trial and at trial respectively.[1] Safeguards are the legal mechanisms by which the balance between effectiveness of a prosecution and fairness to an accused can be achieved or restored. Sometimes safeguards are in place at the time the evidence is gathered and some safeguards are in place when then evidence is relied on by the prosecution and the fact-finder at trial. Based on the observations and conclusions of the previous chapters, this chapter compares the national responses to protecting and limiting the right to silence in criminal proceedings as a whole and then considers what happens to the '*cross-admissibility*' of confession evidence or evidence of silence, when it is obtained in one of the systems and used at trial in another. Do we disrupt the national continuity of law that exists between the investigative and the trial phases by obtaining the evidence from a suspect abroad? In particular, is the process of gathering confession evidence via international cooperation so disruptive that its admission at trial would result in a denial of fairness, when viewing the proceedings as a whole?

Confession evidence is highly probative and, therefore, it may be desirable for investigative officers to request a foreign agency to seek to obtain an account from a suspect even though he or she may be outside the jurisdiction. It may be preferable that the suspect's account is obtained before extradition or surrender is pursued. In systems where the investigation is carried out by an investigative judge, the authorities may request to have a witness or suspect heard in another national jurisdiction and mutual legal assistance as well as the European Investigation Order (EIO) allow for this possibility.[2] The general observations and conclusions in this chapter will also be applicable to these types of hearings. To avoid the problem of overuse of the European Arrest Warrant (EAW), the EIO also makes it possible for a suspect to be temporarily transferred to the requesting state, though the actual mechanics of such an order are unclear. This means that interrogation of a suspect in the requested state, whether as a result of MLA or an EIO, is still perhaps the better option.

Official police questioning of suspects of foreign offences may assist in transnational cases to eliminate lines of enquiry, for example, where the suspect is already in detention in relation to unrelated matters or does not otherwise present a flight risk. Investigators may wish to have a foreign suspect questioned in order to examine how the person reacts. A suspect of a transnational crime may have already been apprehended and questioned for offences in one national system and the problem may arise concerning the admissibility of any confession evidence already in existence in a parallel prosecution. Alternatively, investigators may seek to have a witness who is outside the jurisdiction questioned and that witness turns out to be a suspect. Frequently in transnational cases, the investigators may wish to use covert surveillance measures to obtain statements by suspects who are outside the

[1]Safeguards such as the right to an interpreter or translator, which are relevant throughout the whole of the criminal justice process, are important rights in relation to enabling the suspect or accused person to participate in the process, though they are not directly relevant to the right to silence *per se*.

[2]The EIO also allows for the witness or suspect to be temporarily transferred to the issuing state for hearing—but this must happen by consent: art. 22 EIO Directive 2014/41/EU, 3 April 2014.

jurisdiction. How do these efforts affect the overall balance between effectiveness and fairness in transnational cases?

Specifically, this chapter will examine the jurisdictional cross-admissibility between the national systems of Denmark, England and Wales and Australia of confession evidence or evidence of silence gathered in the three situations, as follows: statements or evidence of silence obtained through police questioning of suspects or accused persons; statements that are obtained from suspects or accused persons by direct compulsion using compulsory information-gathering powers; and suspect statements obtained by the use of covert surveillance. Obtaining confession evidence in these three situations can all lead to discussions about the fairness of using the evidence and the issue at trial of admissibility of evidence.[3] In particular, the third context leads to a discussion about the exercise of judicial discretion to exclude improperly or unlawfully obtained evidence as a safeguard of fairness in a criminal trial. The comparison of these three evidence gathering methods in the national systems is essential to the analysis of jurisdictional cross-admissibility of suspect evidence and maintenance of fairness in transnational cases. Analysing the differences in the ways in which these investigative methods are used in the national systems lays the foundation for greater mutual understanding between jurisdictions in the international cooperation process.

6.2 Pre-Trial Due Process Requirements in Transnational Proceedings

6.2.1 The Right To Be Cautioned and to Audio-Recording: Denmark, England and Australia

The fundamental significance to fairness of a caution about the right to remain silent prior to questioning is illustrated by the fact that, in principal, Denmark, England and Australia all apply the same minimum standards. Firstly, in all three jurisdictions, a belief that there are reasonable grounds to suspect a person has committed an offence appears to form the basis of the need to caution prior to official questioning.[4] This may be before formal charges have been communicated and is regardless of whether questioning takes place outside the police station, by

[3]This is also true in the Danish setting where a *sigtet* has been questioned without being cautioned about the right to silence, as discussed in Chap. 3.

[4]See further PACE Code C, 10.1 and Note 10A, in England and Wales; § 752 and the reference to *sigtelse* in the Administration of Justice Act in Denmark; for example, s. 23S(a) Crimes Act 1914 (Cth), 464 J(a) Crimes Act 1958 (Vic) and s. 89A uniform evidence law in the Evidence Act 1995 (NSW) in Australia, preserving the common law position, in addition to the jurisprudence of the Australian Courts about s. 90, uniform evidence law, discussed in Chap. 5, Sect. 5.8.2. But see 464A(3,) Crimes Act 1958 (Vic) which provides that there is no statutory right to be cautioned in Victoria unless a suspect is under arrest.

voluntary attendance at a police station or in custody. Second, there appears to be a presumption against the admissibility of confession evidence obtained in the absence of a caution about the right to silence prior to questioning by law enforcement officers.[5] Third, the minimum form of the caution in each jurisdiction is that the suspect should be warned that he does not have to say anything.[6] These three requirements are subject to the following observations about the form of cautioning in each system.

6.2.1.1 Denmark

Pre-trial due process requirements in Denmark should be viewed in light of the fact that, in general, Danish proceedings place great weight on in-court testimony over other forms of evidence. Unlike in England and Australia, there is no procedure that allows evidence of a police interview with the *sigtet* to be directly admitted into evidence at trial. Therefore, the evidence of what a *sigtet* said to the police during official questioning may only be led exceptionally, for example, if the *tiltalt* fails to testify or to prove a prior inconsistent statement. The police are not under any general obligation to warn the *sigtet* that what he says or does during pre-trial proceedings may be used against him.

In addition, in Denmark, there is no express provision permitting or prohibiting the drawing of adverse inferences of guilt from the silence of the accused during police interrogation. Although it cannot be ruled out that the court will place some weight on evidence of silence to support other evidence in the trial, there is no obligation on the police to warn a suspect of the possibility.

Therefore, a fundamental procedural guarantee under § 752 of the Administration of Justice Act is that a *sigtet* must be warned, simply:

- You do not have to say anything.

A practice whereby a prosecutor raises the *tiltalt*'s pre-trial silence with the court when summing up may arguably enliven a requirement that, in particular types of cases, a suspect should be warned about the possibility of adverse inferences being drawn from silence. Although, access to a legal representative who can explain the various consequences that follow from the attitude of the suspect during official questioning could equally place the suspect in an informed position.

The fact that evidence of what a *sigtet* says to the police is not generally admitted at trial may explain why the interview is not audio-recorded as a standard practice.

---

[5]See further PACE Code C, 10.1 in England and Wales; the general practice of the court discussed in Chap. 4, Sect. 4.4.1; and s. 139 of the uniform evidence law in Australia in addition to the jurisprudence of the Australian Courts about s. 90, uniform evidence law, discussed in Chap. 5, Sect. 5.8.2.

[6]See further PACE Code C, 10.5 in England and Wales; § 752 of the Administration of Justice Act in Denmark; and for example, s. 23S(a) Crimes Act 1914 (Cth), 464 J(a) Crimes Act 1958 (Vic) and s. 89A uniform evidence law in the Evidence Act 1995 (NSW) in Australia.

For the above reasons, the broader context of the caution in Denmark is a good example of the continuity in Danish criminal proceedings as a whole.

6.2.1.2 England

In England, an interview is defined in PACE Code C (11.2) as questioning under caution. The Code C rules, at paragraph 10.5, set out the form of the caution that must be used during a police interview. A suspect must be cautioned about the following:

- You do not have to say anything;
- But it may harm your defence if your do not mention in questioning something that you later rely on in court; and
- Anything you do say may be given in evidence.

The second part of the caution is a pre-condition to the drawing of an adverse inference from the silence of the accused when questioned by the police if they then raise a defence at trial, under s. 34 of the Criminal Justice and Public Order Act 1994. In England and Wales, drawing an adverse inference from pre-trial silence in these circumstances is acknowledged as an apparent limitation of the right to silence—one that is indirectly coercive. Therefore, the English legislature has chosen to openly warn a suspect of the potential consequence of remaining silent.

In addition, what is said by a suspect during official questioning and recorded can be directly admitted as evidence at trial. Therefore, the suspect is advised about this possibility in the third part of the caution. There are stricter requirements about audio-recording of interviews in England and Wales than in Denmark.

The above is also a good illustration of the way in which a national criminal justice system is built up in a coherent manner, so that there is continuity between the investigation and the trial phases.

6.2.1.3 Australia

Like English law, Australian law allows the evidence of an interviewing officer to be led directly in evidence. This includes the admission of the transcribed record of interview as an exhibit. Therefore, in Australia, the requirements for audio recording are strict and the suspect must be cautioned in the following terms:

- You do not have to say or do anything; and
- Anything you say or do may be given in evidence.

The major difference between England and Australia in relation to the caution can be explained by the fact that the drawing of adverse inferences from the silence of the accused when questioned by the police is generally impermissible in Australia, even where a defence is introduced for the first time at trial. However, as discussed in Chap. 5, Sect. 5.5.1.2, recent legislation affecting the uniform evidence law in the state of New South Wales will allow such an inference to be drawn in cases of serious indictable offences. As pre-conditions to being able to draw such an inference at trial, the suspect must receive prior access to communicate with an Australian lawyer and must be cautioned in the same form as the English caution above. Therefore, in common law systems, where an adverse inference from pre-trial silence may be raised in a concrete manner at the trial then the suspect should be warned as to that consequence. However, in general, where such an inference is not permitted to be raised before the jury under the uniform evidence law (other than by a co-accused), the warning is not required—even though, like in Denmark, it cannot be ruled out that the jury will use this line of reasoning.

If a confession is obtained in absence of cautioning, it will be deemed unlawful under s. 139 of the uniform evidence law and may be subject to exclusion under s. 138, depending on a consideration of all the circumstances in the case. It may also fall for consideration under s. 90 of the uniform evidence law if relying on the evidence has the potential to cause a forensic disadvantage to the accused.

Also in relation to Australia, the above demonstrates continuity in the criminal proceedings as a whole, from investigation to trial.

6.2.2 Access to Legal Assistance and Other Pre-Trial Due Process Requirements

The three jurisdictions under consideration all generally apply the same basic minimum requirement that a person should be informed that they have the right to access legal advice of their own choosing prior to questioning. In all three jurisdictions, anyone charged with a criminal offence has the right to legal representation in connection with the preparation of a defence for a court hearing. What does this actually mean in the context of each of the national systems under examination?

There is some variation in the scope of the right to access legal advice and other pre-trial due process requirements, which may be significant in the transnational context. In all three jurisdictions a suspect of foreign offences may encounter significant difficulties in relation to the right. In fact, in Denmark and Australia weaknesses exist in relation to access to a lawyer for national suspects. In addition to not being frequent transit countries, this may account for why the police in these jurisdictions do not question suspects of foreign offences to any great extent. Further, Denmark does not have the facilities for audio-recording official questioning and this could also be problematic in transnational cases.

6.2.2.1 Legal Advice and Other Pre-trial Due Process Requirements in Denmark

6.2.2.1.1 Access to Legal Representation in Danish Proceedings

In Denmark, a person is *sigtet* (charged) early in an investigation, as soon as a reasonable suspicion of criminal offending arises. This status means that the *sigtet* has the right to access legal representation of own choosing or a public appointed lawyer if that is considered desirable—whether or not they are being questioned at the police station, voluntarily or in custody. Ministry of Justice guidelines say that the right to access a publicly appointed lawyer should be offered to all *sigtet* persons who are in custody or liable to punishment by imprisonment if found guilty. The general practice appears to be that the police advise all *sigtet* persons that they have a right to access a lawyer.

However, although all *sigtet* persons are offered the right to access a lawyer, in reality they are frequently interviewed without having consulted a lawyer—even those who are under arrest. This may be explained in part by the fact that the police are obliged to warn the *sigtet* prior to arranging access to a lawyer, that if he or she is found guilty they will be required to repay necessary costs of the criminal proceedings, including for the services of a defence lawyer. It may be adverse to the fairness of the proceedings if a *sigtet* is dissuaded from the protection of a lawyer prior to police interrogation by this warning.

6.2.2.1.2 Recording and Letter of Rights in Denmark

Despite the guidelines of the HR Committee (and Denmark's international obligations under the ICCPR), in accordance with § 751 of the AJA, Danish criminal investigators are only required to record what is said in a police interrogation in writing. The written record should be viewed and commented upon by the *sigtet*—although this does not always happen. As discussed above at Sect. 6.2.1, the different approach in relation to recording interviews in Denmark compared to the two common law jurisdictions can be explained by the fact that what is said during the interrogation is not automatically relied on as direct evidence of guilt. Evidence of an interview is only admissible exceptionally, if the *tiltalt* remains silent at trial or gives testimony which contradicts with what was said to the police in some respect.

The *sigtet* in Denmark does not receive a letter of rights.

6.2.2.1.3 Cross-Admissibility of Danish Confession Evidence

If a suspect is questioned by the police (for Danish offences) in accordance with the Danish national law, would that evidence be admissible, for example, in a parallel prosecution abroad?

If English prosecutors attempt to admit evidence of silence during official questioning in Denmark the trial judge would be prohibited from directing the jury in relation to drawing adverse inferences if the accused introduced a defence for the first time in his trial testimony. Furthermore, a number of factors may support a defence application to the trial judge to exclude evidence in accordance with the judicial discretion pursuant to s. 78 PACE. For example, significant silence or confessions would not be recorded or repeated in an audio-recording; the suspect might not have actually spoken with legal representative prior to questioning and might even have been dissuaded from doing so by the prospect of repaying costs; and the suspect would not have received a letter of rights. Such considerations may work against jurisdictional cross-admissibility of the suspect evidence on the ground of unfairness.

Similarly in Australia, where much emphasis is placed on audio or audio-visual recording of official questioning, confession evidence that is not recorded in this manner is likely to be excluded. Likewise, depending on the circumstances of the case, it may be considered unfair under s. 90 of the uniform evidence law to have been warned about repayment of defence costs on conviction and therefore not sought legal advice prior to custodial questioning.

The admissibility of suspect evidence at trial in the English and Australian courts depends to some extent on the fulfilment of the respective pre-trial due process requirements for investigations that are applicable in each system.

6.2.2.2 Legal Advice and Other Pre-trial Due Process Requirements in England and Wales

6.2.2.2.1 Access to Legal Representation in English Proceedings

In England and Wales, a suspect being questioning by the police outside the police station should be advised of the right to access a lawyer prior to questioning, but this is not a right as such. However, there is a great deal of focus on the importance of access to legal representation for a suspect or accused who is being questioned at a police station, whether or not they are under arrest. Overall, there is a very high level of protection of the right to access legal advice in police stations in England and Wales. A person who is at a police station can always access legal advice of their own choosing. In addition, free independent legal advice is available to all suspects at the station through a defence lawyer call centre; and in person, prior to custodial interrogation. This is probably due to the perceived coerciveness of an interview with the police in light of the warning in the caution about the consequences that may follow from the attitude of the suspect during questioning. If a person is under arrest or in police custody and requests legal advice but does not receive legal advice before being questioned, there is a restriction on drawing adverse inferences from their silence during police questioning. If, however, a person has attended the police station voluntarily, this restriction does not apply. This is on the basis that by attending voluntarily, the person is always at liberty to contact a lawyer.[7] Inferences of guilt may be drawn from the silence of the accused during pre-trial interrogation, where he then raises a defence for the first time at trial—even though inferences from silence may only be persuasive. Statements made by a suspect during police questioning can also be used to prove a prior inconsistent statement. The right to access a lawyer requires a clear waiver.[8]

Under s. 58 of the Police and Criminal Evidence Act 1984 (PACE) and the code of practice, any person who is in custody has a continuing right to communicate with a lawyer. Prior to charges being laid, legal advice is provided free to all detainees.[9] A detainee should not be interviewed without a legal representative being present.

6.2.2.2.2 Letter of Rights Prior to Questioning at the Police Station

On arrival at a police station every person should be given a written notice, which includes information about the right to access a lawyer and how to go about this. The notice will be modified according to whether the person is under arrest or attending voluntarily.[10] There is an expectation that a suspect must understand the caution (PACE Code C, 10D) and this forms part of the letter of rights.[11]

6.2.2.2.3 Recording of Police Interviews

An interview must be adequately and accurately recorded, at least in writing, and the record should be acknowledged by the person being interviewed.[12] The English Court of Appeal has found that the impropriety of not keeping a proper record of interview is significant because of the disadvantage to an accused at trial in challenging confession evidence.[13] Any interview taking place at a police station in relation to an indictable offence must be audio-recorded to ensure the accuracy

---

[7]PACE Code C, Note 10C.

[8]PACE Code C, 6.5.

[9]See also Code C, 3.1 and 6.1. After charge, the provision of publicly appointed legal representation is means tested and an accused may be required to pay some of the costs back if they are convicted.

[10]PACE Code C, 3.2 and Note 6B.

[11]See Chap. 4, Sects. 4.4.2 and 4.5.1.1.1.

[12]PACE Code C, 11.7 to 11.9.

[13]See Chap. 3, Sect. 3.7.2.

and integrity of the record.[14] The record of interview may be presented as evidence against an accused.

6.2.2.2.4 Cross-Admissibility of English Confession Evidence

If a suspect is questioned by the police under English law, it is likely that the second part of the caution, warning the suspect that remaining silent during in official questioning may harm the defence case at trial, will be viewed by Australian courts as indirectly coercive. The English caution runs contrary to the High Court decision in *Petty and Maiden*.[15] There could be good grounds for exclusion of any significant silence or confession under the unfairness discretion in s. 90 of the uniform evidence law. This may be the case despite the high level of protection provided to suspects in England in the form of access to legal representation and information on pre-trial rights prior to official questioning. In Denmark, a more lenient view may be taken given the guidelines passed down by the ECtHR on the question of drawing adverse inferences from pre-trial silence. However, the English caution generally does not entirely fit into Danish criminal proceedings. The case for jurisdictional cross-admissibility of English suspect evidence again illustrates the problems caused by taking evidence from an investigation in one legal system and using the evidence at trial in another.

6.2.2.3 Other Pre-Trial Due Process Requirements in Australia

6.2.2.3.1 Access to Legal Representation in Australian Pre-Trial Proceedings

Under federal legislation, a person who is under arrest or with an investigating official for the purposes of questioning and in respect of whom there is sufficient evidence to establish criminal offending has the right to access a lawyer of their own choosing, which includes a right for the lawyer to be present in the interview.[16] If there is exceptional reason to delay access to legal advice of own choosing then the police should arrange access to another lawyer prior to questioning. Under state legislation in Victoria, for example, a suspect who is under arrest is entitled to communicate with a lawyer.[17] However, free (means tested) publicly appointed legal representation is not available until the time charges are laid in connection

[14]PACE Code E, 3.1.

[15]*Petty and Maiden v The Queen* [1991] HCA 34; (1991) 173 CLR 95.

[16]Sections 23G and 23 L Crimes Act 1914 (Cth).

[17]Section 464C Crimes Act 1958 (Vic).

with preparation for court proceedings.[18] These types of provisions relating to access to a lawyer in pre-trial proceedings in Australia are consistent with the obligations under the ICCPR being applied in Australian law. However, the fact that all people under arrest or in custody do not have an ongoing right to access a lawyer pre-charge, including a free (means tested) publicly appointed lawyer, may mean that the level of protection of the right to access a lawyer in Australia is lower than what has been recommended by the ECtHR under art. 6 and is applied in Denmark and England.

A recent enactment in NSW does grant, upon suspicion of a serious indictable offence, the right to access an Australian lawyer prior to official questioning as well as the presence of a lawyer during questioning, even if that person is not under arrest. This is in connection with giving a special caution and openly drawing adverse inferences from the silence of the accused during official questioning if a defence is raised at trial for the first time. However, in other uniform evidence law states, comments and directions to the jury about drawing inferences of guilt from pre-trial silence are largely prohibited.

There is no written advice about procedural rights being given to suspects being questioned in Australian police stations.

6.2.2.3.2 Recording of Official Questioning in Australia

In Australia, all confession evidence of a suspect (or a person who should be regarded as a suspect) should be audio-recorded or confirmed on tape to protect its reliability and integrity.[19] This due-process requirement is given high priority and means that the large majority of police interviews will occur at the police station. Under federal legislation an interview may be recorded in writing if audio-recording is not reasonably practical.[20]

6.2.2.3.3 Cross-Admissibility of Australia Suspect Evidence

Evidence of silence during official questioning may not be used in English trials to allow the jury to draw an adverse inference about a recently invented defence at trial, unless the evidence has been gathered under s. 89A of the NSW uniform evidence law. However, it is arguable that confession evidence gathered under Australian law will otherwise be admissible in English and Danish trials, despite the fact that the Legal Aid scheme does not apply to pre-trial proceedings as perhaps it should under English and Danish law.

[18] This is in accordance with s. 25(2)(b), Charter of Human Rights and Responsibilities 2004 (Vic) and art. 14(3)(b), ICCPR.

[19] See 23 V, Crimes Act 1914; 464H, Crimes Act 1958 (Vic).

[20] 23 V, Crimes Act 1914; 464H, Crimes Act 1958 (Vic).

6.2.3 Effective Requests for Police Questioning

6.2.3.1 Danish Requests for Questioning a Suspect

The form of caution used in England and Wales and in Australia warns a suspect of the right to silence as well as the consequences or incidents of that right. As discussed above, these are not consequences that apply directly in the Danish system. Therefore, it is not necessary or rational for a suspect in a Danish prosecution to be cautioned using the same terms. Officers in England and Australia should be prepared to caution according to Danish law, simply that the suspect has the right to remain silent. However, adding the words "anything you say or do may be used against you" is not strictly contrary to § 752 and would be unlikely to prevent admissibility.

The English procedures could easily accommodate the provision of free telephone advice to a suspect of offences against Danish law attending voluntarily for official questioning—although currently there are no procedures that enable or guarantee the provision of such a service. Furthermore, as a general rule the Australian Legal Aid procedures do not currently accommodate free legal representation in pre-trial procedures at all.

6.2.3.2 English Requests for Questioning a Suspect

On the basis of the above, the four most important procedural requirements in relation to official questioning in England and Wales are cautioning; providing a written notice of right to a suspect at the police station or in custody; ensuring that the suspect has legal representation, including some form of free legal representation, according to whether or not they have attended voluntarily, and also during the interview; and audio-recording of interviews in relation to indictable offences. All of these pre-trial safeguards contribute to ensuring that fairness is maintained throughout the proceedings. However, by ensuring that a suspect has access to legal representation, there may be less need for a written notice of rights and audio-recording of police interviews. A lawyer can explain a suspect's rights, as well as the meaning and significance of the caution. The presence of a defence lawyer in a police interview provides a level of security during the interview. In addition, the defence lawyer may be able to assist the suspect in commenting on and challenging what the police write in a written record. Therefore, confession evidence obtained in Denmark or Australia, in the absence of a letter of rights or an audio-recorded interview may, nevertheless, be fair to admit against the accused in an English trial, depending on access to legal advice.

The English caution has additional requirements to the caution that is used in either Denmark or Australia. This is a reflection of the additional and more concrete ways in which evidence of the suspect's attitude during the police interview can be used in a trial. In order to meet the cautioning requirements of England and Wales, and thereby maximize the admissibility of evidence of a suspect's silence at trial, investigating officials in the other jurisdictions will have to be flexible.

However, cautioning a suspect for an English prosecution may be problematic for investigators in Australia because drawing an adverse inference at trial from the pre-trial silence of the accused is by and large prohibited by national law. In order to encourage the effectiveness of gathering confession evidence for use in an English prosecution, without missing on the suspect's 'right' to a proper caution, Australian investigators (and law-makers) will have to take a flexible approach and allow a suspect to be cautioned that it may be harmful to a trial defence to remain silent during official questioning. Otherwise, English courts will not be able to draw such an inference of guilt. The alternatives are that evidence of silence will be inadmissible or that English authorities will not seek to have a suspect questioned in Australia in relation to English offences.

Similarly, the provision of free legal advice prior to official questioning to all detainees, as well as free telephone advice to suspects attending for interview voluntarily, may cause difficulties in a request from England for assistance in Australia. In Australia, there is no right to publicly appointed legal representation until after formal charges have been laid. This is with the exception of specific enactments such as the enactment in New South Wales of s. 89A of the uniform evidence law mentioned above.

Effective international cooperation requires that where necessary publicly appointed legal advice is available—and potentially from lawyers of both the requesting and the requested state systems. It is noteworthy, however, that unless a suspect of foreign offences is already in custody in relation to local offences, or has been arrested, where necessary, on suspicion of local offences, any interview about foreign offences will only occur with the suspect's voluntary attendance. Therefore, even according to the English national rules, suspects attending voluntarily do not have the right to the attendance in person of a free defence lawyer and may stop the interview and any time and leave. Lack of access to legal advice may be an issue that would be resolved under s. 78 PACE, with reference to any other safeguards in place, such as audio-recording, and whether there are other allegations of unfairness, such as bullying.

What about an English request to Danish authorities to question a suspect about English offences? The English national provisions and procedures place emphasis on the significance of ensuring all suspects receive some form of (free) legal representation prior to being questioned, on account of the coercive nature of the English provisions, in particular the caution about the potential harmful effect of pre-trial silence to a defence at trial. On the other hand, while there is a right to access a lawyer following *sigtelse*, including a publicly appointed lawyer with respect to offences punishable by imprisonment, the Danish provisions and guidelines under § 732 of the AJA do not ensure all detainees (or a *sigtet* in relation to offences punishable by imprisonment) in fact receive legal advice prior to questioning; and the provisions may have the effect of discouraging suspects from seeking it by the advice that if they are convicted, they will have to repay the state for the lawyer's fees. In addition, there is no positive right to have a defence lawyer present during police interrogation—although, there is no rule to prohibit it.

The differences in each system in relation to publicly appointed legal advice may create hurdles in cooperation in transnational cases. In fact, it is difficult to see how a suspect of English offences can in fact enter the national systems in Denmark or Australia for publicly appointed legal representation.

6.2.3.3 Australian Requests for Questioning a Suspect

Both England and Denmark would have to be prepared to caution in accordance with the Australian legal requirements. There are no express provisions which prohibit this, although there is a practice code in England and a statutory provision in Denmark which expressly provide for other forms for cautioning to be used.

In Australia, much emphasis is placed on audio and video-recording of interviews to protect the integrity of confession evidence. This is particularly important in the national context where there is no right to access publicly appointed representation for any suspects in any jurisdiction prior to charging; and in some Australian state jurisdictions such as Victoria, there is no clear right to access a lawyer (of own choosing) for suspects who attend a police station voluntarily—although this may not always be prohibited in practice and is encouraged by the uniform evidence law.[21] Although there is a lack of facilities to audio-record official questioning in duplicate or triplicate, audio-recording could easily be accommodated in order to avoid exclusion of evidence. Furthermore, the problem may be overcome by at least ensuring that the suspect has been informed of a right to access to a lawyer of their own choosing.

6.2.3.4 General Rules to Allow for Jurisdictional Cross-Admissibility of Suspect Evidence

On the basis of the above analysis of the differences and problems associated with the procedural rules for interrogation of suspects in Denmark, England and Wales and Australia, there are a number of general observations to be made about what may be regarded as necessary, proportionate and adequate in relation to jurisdictional cross-admissibility of confession evidence in transnational cases. In order to encourage cross-admissibility of confession evidence in cases with a transnational element, pre-trial due process requirements such as cautioning and a right to access a lawyer (at least of own choosing) and audio-recording of interviews should, as a

[21]This is with the exception of official questioning in New South Wales under the new s. 89A of the NSW uniform evidence law. See also 23G Crimes Act 1914 (Cth) which states that a person accompanying an officer for official questioning may have a lawyer of their own choosing present.

general rule, be enjoyed by any person who is reasonably suspected of committing a criminal offence abroad.

On the basis of the above, there may be five key procedural guidelines that will promote the admissibility of confession evidence obtained abroad. It is envisaged that these would apply in the case of a voluntary interview as a result of a request for assistance in criminal proceedings from a foreign authority to attempt to interview a suspect; or to make enquiries of a person, during the course of which a reasonable suspicion arises:

- the suspect should be advised of the accusations against him;
- the suspect should be cautioned, as required by the law in the requesting state (or at least that he or she does not have to say anything);
- the suspect should be invited to the police station for official questioning;
- the suspect should be offered the right to legal representation (at least of own choosing) to protect the suspect's interests during the interrogation and at trial, including the presence of a legal representative during the interrogation; and
- an audio-recording should be made in preference to a written record.

If it is not reasonably practical to make an audio-recording of the interview then any written record should be viewed and commented on by the suspect. The above recommendations would also be supported by the provision of a written advice to a suspect of foreign offences about the rights that are available to them.

For effectiveness and to promote the admissibility of confession evidence, in particular, strategies must be in place in relation to cautioning in accordance with the law of the requesting state and creating effective opportunities for access to legal representation. The form of cautioning is directly related to the way is which confession evidence or evidence of silence may be relied on at trial. Providing legal representation to a suspect in transnational cases ensures equality of arms and has the potential to overcome cross-admissibility problems related to differences in or losses of procedural rights in the process of transferring evidence. For practical reasons, the legal practitioner must be a local lawyer with a local practicing certificate. However, to protect procedural fairness and the admissibility of confession evidence at trial in another jurisdiction, the lawyer should be familiar with the procedures in the requesting state.[22] Such strategies would need to apply in the case of requests to make enquires of a potential witness, or to examine a witness in court, where there is a concern that they may become a suspect; as well as in relation to suspects who are invited to attend voluntarily for official questioning.

Kaiafa-Gbandi expresses the view that the establishment of an independent EU institution, connected to (and monitored by) the bar associations of all Member States, for legal defence in transnational criminal cases ('Eurodefensor') is absolutely imperative to ensure an effective defence in the process of mutual recognition. Being an equivalent of Eurojust, this would also be of symbolic value. In

---

[22]See the discussion about legal representation and the way forward in relation to the concept of a duty lawyer call center in Chap. 8.

addition, a Eurodefence lawyer could be called on to represent individual interests in the use of certain covert measures (similar to the use of intrusion lawyers in Denmark) in transnational cases.[23]

Jurisdictions that rely on pre-trial testimonial statements, instead of statements made to the police, may request a foreign prosecutorial service to arrange and conduct a court hearing for the taking of a witness statement. It can happen that a reasonable suspicion arises in relation to a witness in the course of proceedings. In addition to advice about the accusations, in accordance with the guidelines above, questioning of a suspect should not continue unless he is cautioned about the right to remain silent in accordance with the requirements of the requesting state and provided with the opportunity to obtain legal representation before examination continues. The lawyer must be in a position to protect the suspect's interests during the pre-trial proceedings as well as the trial proceedings in another jurisdiction, as discussed above. The prosecutor would also have to be in a position to continue the conduct of the proceedings on behalf of the requesting state, as questioning of a suspect rather than a witness. It may be good practice to allow all witnesses to have a right to legal representation of their own choosing; and encourage the presence at the hearing of a prosecutor of the requesting state. In cases where a defence lawyer and a requesting state prosecutor are already present, the costs of adjourning proceedings may be minimised.

The manner in which pre-trial proceedings are conducted clearly affects the admissibility of confession evidence or evidence of silence at trial. In transnational cases this may create particular problems where evidence gathered under the procedural rules of an investigation in one system may not always be strictly admissible under the trial procedures of another. This is because of the risk that national continuity in the manner in which the balance between effectiveness and fairness is regulated is disrupted. Ignoring such interference in the national balance may create unfairness to an accused. This may occur if, for example, he does not understand information or receive the necessary advice about the consequences in another system of his attitude during official questioning. Alternatively, where official questioning is not properly recorded it may be more difficult for an accused to challenge evidence at trial.

These kinds of issues may also of relevance to the ongoing debate about the establishment of a European Public Prosecutor's Office to fight fraud against the EU budget in a cohesive manner. Member States have been debating matters such as the structure of the office and its relationship to national criminal justice systems; whether the substantive scope will involve no more than just the financial interests of the EU or will extend to other serious crimes of a cross-border nature; the *applicable procedure* and whether the procedure will focus on *harmonisation* in EPPO matters or *mutual recognition*.[24]

---

[23]Kaiafa-Gbandi (2010), pp. 396–397.

[24]See Ligeti (2013), pp. 1–4.

6.3 Cross-Admissibility of Statements Obtained Under Compulsory Powers

6.3.1 National Compulsory Information-Gathering Powers

6.3.1.1 Denmark

In Denmark, administrative compulsory information-gathering powers may not be used to obtain self-incriminating statements from a *mistænkt*. The powers may only be used to question a *mistænkt* by consent. The Danish approach may be explained by the centrality to the administration of justice of the rule contained in § 750 of the Administration of Justice Act that a person cannot be forced to make a statement in criminal proceedings. The Danish administrative powers are used in connection with regulatory investigations carried out by administrative authorities, such as the tax authority in the assessment of tax liability.

Under the provisions creating the powers, where there is a concrete suspicion of criminal offending, the *mistænkt* may be questioned by consent with the right to remain silent. Administrative powers may be used to gather information in relation to regulatory matters in parallel proceedings, even where criminal proceedings have been initiated. Questioning by written informed consent usually takes place as part of a regulatory audit procedure. However, a person being questioned by consent by the administrative authority is not entitled, as of right, to the same procedural guarantees as he would have by virtue of *sigtelse* (charging) under a criminal investigation, such as legal representation.

The provisions do not provide an absolute guarantee that self-incriminating statements obtained inadvertently or by the improper application of the provisions will be inadmissible in criminal proceedings. The question of admissibility (and weight) of such statements in criminal proceedings may be relevant to the operation of provisions such as § 752 of the AJA and is the subject of judicial power. The use of statements gathered under compulsion in these conditions would be contrary to the right against self-incrimination. Therefore, the fundamental principle against admissibility of evidence obtained contrary to the right against self-incrimination may preclude their use in criminal proceedings.

6.3.1.2 England and Australia

Although England and Australia apply different minimum standards of human rights protection, the legislative compulsory information-gathering powers function similarly. The use of compulsory information-gathering powers is closely connected to the traditional view that the prohibition on self-incrimination is a common law 'privilege', for the benefit of those giving sworn testimony, rather than a 'right'. Under both systems the common law privilege against self-incrimination is abrogated by Parliament so that self-incriminating statements may be gathered under pain of punishment in regulatory investigations. In England, the powers may be used when criminal proceedings are already on foot, even after criminal charges have been laid—although this may not be contrary to the Human Rights Act 1998 and the guidelines set by the ECtHR. In Australia, the High Court has recently indicated that compulsory questioning powers should not be used to examine a suspect who has been charged with but not yet tried for criminal offences. The Court found that to do so would be a marked departure from the accusatorial process.[25] This is a difference of approach that may even be significant in transnational cases involving England and Australia.

Of significance, under both systems, the use of compulsory questioning powers is dependent upon a use immunity, which prevents any statements from being used against a person in criminal proceedings. Without the use immunity to ensure that self-incriminating statements gathered under compulsion are inadmissible, their use against an accused in subsequent criminal proceedings will render the proceedings unfair. Nevertheless, in accordance with the legislative provisions in both systems, self-incriminating statements may be used against a person for conviction of offences against the act creating the powers, such as failure to attend an examination or giving false sworn testimony and for conviction of perjury in separate proceedings.

Various safeguards are built into each compulsory powers system. In England, the powers are used under the superintendence of the Attorney-General. Although there is no express right, in principle, a person being examined can have legal representation of their own choosing. However, it may not be of such significance to have legal representation under the English powers, where self-incriminating statements cannot be used in criminal proceedings. Nevertheless, the answers given can inform investigators and alter the course of the criminal investigation to the disadvantage of an accused.

In Australia, taking the Major Crime (Investigative Powers) Act 2004 (Vic) as an example, the compulsory powers order is applied for in the Supreme Court, with the assistance of a Public Interest Monitor; and under the additional supervision of the Victorian Inspectorate. The person being examined has the right to legal representation of their own choosing, even though self-incriminating testimony may not be used against the person being examined in future criminal or pecuniary penalty proceedings. Through the appointment of special counsel, an application can be made by an interested party for revocation of a compulsory questioning powers order. Therefore, safeguards have had a particular focus in this example of Australian compulsory powers legislation.

[25] *X7 v Australian Crime Commission* [2013] HCA 29.

6.3.2 *Cross-Admissibility of Statements Obtained by Direct Regulatory Compulsion*

In a debate in Council on 9 June 2011, the Council members agreed on the main principles in relation to the proposed EIO.[26] The types of measures that are ultimately available in the EIO include the hearing of witnesses, suspects and accused persons by videoconferencing/audio-visual transmissions (by 'competent authorities' of the issuing state) and coercive measures such as search and seizure and telecommunications interceptions.[27] At hearings, the competent authority of the executing state must be present. Witnesses may also be heard by telephone conference. In addition to use in criminal proceedings, the Council agreed that the EIO could be used by administrative authorities where there is a criminal dimension.[28] Requests for assistance by administrative authorities will depend on the ability of authorities from different jurisdictions being able to fulfil each other's requests for assistance and utilise the types of measures that are available in the cross-border perspective.

English and Australian regulatory (pre-trial) compulsory questioning powers are structurally extremely different to Danish administrative compulsory powers under the Coercive Measures Act. The English and Australian powers are based on questioning of a witness under oath (though not by a judge). By providing testimonial evidence on oath, the witness is liable to be punished for answering untruthfully. The English and Australian powers are not intended to generate criminal law evidence. They are intended to provide information to a regulatory investigation. Nevertheless, this intelligence may also be used in a criminal investigation.

Danish law does not recognise the concept of 'sworn' evidence, although a witness may be prosecuted for giving false testimony. Nevertheless, Danish law does not punish an accused person for lying while making statements in their own defence. Compelling testimony on pain of punishment is contrary to the principles enshrined in §§ 750 and 752 of the AJA. Therefore, it would be incongruous for a Danish authority to conduct a compulsory examination on behalf of England or Australia, even in the knowledge that any statements obtained could not be used as evidence in criminal proceedings. A Danish authority might question a foreign suspect by consent with the right to remain silent—but again problems arise in relation to the appropriateness of using administrative powers to question a suspect of criminal offences and to a possible lack of legal representation.

Can Denmark request England or Australian authorities to use their compulsory questioning powers to obtain evidence for a Danish criminal investigation? In the

---

[26]A debate in Council may be considered part of the *travaux preparatoires* in relation to the EIO Directive, which is now binding on most EU states (though Denmark and Ireland are not taking part) but requires implementation by the states.

[27]See Articles 24–29 of the EIO: Directive 24/14/EU of 3 April 2014.

[28]2010/0187(COD), 9/6/11, Debate in Council, European Parliament/Legislative Observatory; and see Art 4(b) of the EIO: Directive 24/14/EU of 3 April 2014.

light of the fact that such powers depend on a use immunity in relation to statements, preventing their use in criminal proceedings, again the answer would have to be ‘no’. It may arguably be possible for Denmark to request the use of the powers to facilitate the obtaining of witness evidence, which is favourable to the Danish accused (*tiltalt*) even though this is contrary to § 750. However, here the powers would be sought to be used in a way that has not been intended by the English or Australian legislature.

Nevertheless, some of the compulsory questioning powers legislation allows for intelligence sharing—and when seen in this light the transfer of information from England or Australia to a Danish law enforcement authority may be reasonable in pursuance of a legitimate aim such as law enforcement or national security—and even a Danish administrative authority could seek information believed to be held by an English or Australian witness, for use in an administrative investigation, for example to assess tax liability in a case with a criminal element.

This type of assistance is fraught with problems, however. Denmark does not use compulsory powers to question a suspect and Australian case law says that the powers do not extend to use after charges have been laid. Even if the evidence was already in existence for use in criminal proceedings, this would still be contrary to § 750 of the AJA, in addition to being contrary to the intention of the English or Australian legislature. To use statements in this way would over-ride the use immunity provided by the powers.

It is perplexing to consider the appropriateness and proportionality of a Danish administrative authority as part of administrative proceedings to send a request for mutual legal assistance in criminal matters to England or Australia, asking an equivalent administrative authority to seek to voluntarily question a suspect (who could be regarded in Denmark as *sigtet*) with the right to be advised of the accusations, the right to silence and the right to have the interview recorded in writing, but with none of the other procedural guarantees that are associated with official questioning of a *sigtet*, in particular the right to a lawyer (although this would not be precluded). Similarly, it would be unusual, if not disproportionate for English or Australian criminal investigation authorities to ask a Danish authority to question a suspect of criminal offending by consent under § 10(3) of the Danish Coercive Measures Act (referred to in Chap. 3, Sect. 3.7.2 and 3.7.5) rather than asking for a suspect to be questioned using the usual criminal evidence gathering procedures. It may be possible for the administrative authorities of any of the countries to seek information abroad from witnesses, even if the case has an element of wrongdoing by another person. Issues about the fairness of using statements obtained in the absence of legal representation and a full caution would be dealt with under s. 78 PACE or s. 90 of the uniform evidence law.

6.3.3 Using Compulsory Information Gathering to Obtain Derivative Real Evidence

Although it is generally accepted that compulsory information gathering should not be used to obtain confession evidence, the question remains whether the right against self-incrimination prevents the use of information obtained under compulsion from being used to locate real evidence that is incriminating. For example, if, contrary to Australian High Court authority, a suspect gave testimony under (English or Australian) compulsory questioning powers about committing a murder, while the answers given in testimony could not be used in a subsequent trial, the information may be used to locate the body and other real and forensic evidence. Likewise, the answers given by a fraud suspect under compulsory questioning powers may lead the police to locate a number of foreign bank accounts. Alternatively, a witness may give testimony under compulsory powers about a clandestine drug laboratory operated by two others, leading the police to find the witness's own fingerprints on the laboratory equipment. These three examples illustrate how derivative real evidence might be located by the police as a result of self-incriminating statements made under compulsion. This type of scenario is also known as the 'fruits of the poisonous tree' doctrine.

In transnational cases, the question of balancing the right against self-incrimination and reliance on derivative real evidence is really a question about information sharing. In England and Wales and Denmark, the use of derivative real evidence obtained by compulsion is unregulated—although, in Denmark, the use of compulsory questioning powers in criminal cases is contrary to § 750 AJA. Therefore, England and Denmark can share information obtained under compulsory powers with other states and use that information to obtain derivative real evidence for use in trial proceedings.

This same will also be true in most states and territories throughout Australia, subject to the condition that compulsory questioning powers are not used to question a suspect who has been charged with criminal offences. However, in Victoria, the Supreme Court has developed a 'but for' test based on Canadian jurisprudence (referred to in Chap. 5, Sect. 5.6.2), which means that on receipt of self-incriminating information obtained under compulsory questioning powers, the use of such information to obtain evidence to be relied on in a criminal trial will be subject to this test. In a Victorian case, derivative real evidence may not be used on the question of guilt in a criminal case if it would not have been located 'but for' the use of compulsion to examine a witness and obtain self-incriminating information. In Victoria, use immunity extends to derivative real evidence.

6.4 Cross-Admissibility of Confession Evidence Obtained by Unlawful Covert Surveillance

6.4.1 Using Covert Surveillance to Listen to Private Conversations

Despite the element of deception, the use of covert surveillance to listen to private conversations and capture self-incriminating statements by suspects or accused persons is a commonly accepted investigative technique. As discussed in Chaps. 3–5, in Denmark, England, and Australia covert surveillance methods are routinely used in criminal investigations of more serious offending and the power to use these coercive measures is statutorily based.

In Denmark and Australia the use of measures such as telephone interceptions, listening devices generally require judicial authority by way of a court order or warrant. In England, the use of covert surveillance measure involves political and judicial supervision. Warrants for the use of interception of telecommunications have been issued by the Secretary of State and other types of measures are generally authorised by high ranking officers under the supervision of Commissioners, who often hold or have held judicial positions. However, more recently, the use of covert surveillance has increasingly involved judicial supervision. In England there are also detailed practice codes which regulate the use of covert surveillance measures.

In all three systems the decision about whether to approve a measure for the intrusion of private conversations depends on proportionality. In general, proportionality is established by a consideration of factors such as the existence of a reasonable belief that the measure will capture the anticipated evidence, the extent of the intrusion on individual rights, the seriousness of the offending under investigation and the significance of the anticipated evidence to the criminal investigation or proceedings. In Australia and Denmark, offences generally need to carry higher maximum penalties before these types of coercive measures may be used. It is unlikely that national courts in these jurisdictions will order, for example, a listening device in a foreign case for offending carrying a lesser penalty. Doing so may support an argument that international cooperation leads to a loosening of rule of law guarantees and even encourages a 'race to the bottom'.[29]

In transnational cases, the fact that the use of covert surveillance measures in all three systems is accompanied by supervisory procedures that guard against abuse promotes the cross-admissibility of confession evidence obtained. In all three systems, the issuing of warrants or approval decisions are based on an assessment of the proportionality of using such measures in the circumstances of each case. Therefore, in general, there should be no barriers to the cross-admissibility of confession evidence obtained by covert surveillance in these jurisdictions.

[29] European Criminal Policy Initiative (2013), p. 444.

In England and Australia, it is the police who usually determine which investigative measure should be used in a particular case and to have the matter brought before the court in the proper way. In Denmark, however, the police may work in conjunction with the prosecution lawyers to decide together what measures are necessary and adequate. Further, there is an additional safeguard in place in Denmark during the court application for the use of a covert measure in the form of publicly appointed intrusion lawyers (as referred to in Chap. 3, Sect. 3.8.2). These lawyers have a right to be heard by the court to protect general individual interests and assist the court in ensuring that a measure does not become too far-reaching. Compared with the political and operational supervision of the use by the police of covert surveillance measures in England and the judicial supervision in Australia, the multi-tiered system of supervision in Denmark provides additional protection of individual rights in the investigative phase. In Denmark, the use of intrusion lawyers as an additional safeguard, along with the involvement of prosecutors and the judiciary in the use of covert surveillance measures, sits within a system which generally expects that all the relevant evidence that is gathered in a case should be presented to the court in trial proceedings. These additional safeguards in the investigative phase in Denmark may minimise the extent to which unlawfully obtained evidence is encountered in the national system.

In England, the information generated by a telecommunications interception may not be used in court proceedings in England or abroad. The information may be transferred to another state, for use in a criminal investigation. If the police have asked questions in an interview with a suspect in which the interception information is raised, then those parts of the interview must be deleted to the extent that there is no reference to the interception, while the remainder of the interview makes sense. If England has transferred intercept information to another system, there will have to be strategies in place to ensure that the prohibition on the use of intercept evidence is maintained.

6.4.2 *Exclusion of Confessions Obtained by Unlawful Covert Surveillance*

6.4.2.1 Denmark: Exclusion of Unlawfully Obtained Confession Evidence

As referred to in Chap. 3, Sect. 3.9.4, in Denmark, the general tendency is to admit unlawfully or improperly obtained evidence. This is with the exception of situations where there is a significant possibility that the evidence is unreliable and must, therefore, be excluded. In addition, there appears to a practice, as a general rule, of excluding confession evidence that has been obtained in the absence of a caution. This appears to be due to the protection of the right to silence and the right against self-incrimination—and the significance of the right of a *sigtet* in § 752(1) of the AJA to choose whether or not to speak to the police. The significance of cautioning is also illustrated by the fact that a *mistænkt* (suspect) may be questioned with the rights of a *sigtet*, namely the rights under § 752. If a *sigtet* has not been offered a right to access a defence lawyer this may perhaps be relevant to the exercise of judicial discretion to exclude a confession made in breach of the right against self-incrimination. It appears from the judgments that the decision to exclude confession evidence in such a case is not based on a balancing of competing factors, such as gravity of the offending, but rather is decided on the basis of fairness to the accused. However, it cannot be ruled out that the gravity of the offending will play a role in the exercise of judicial power in these types of cases.

Certainly where evidence has been obtained by covert methods which represent a technical breach of the AJA, the preference is to admit the evidence—and the question may not even be raised as a concrete issue in the trial. It appears, therefore, that in the case of technically unlawful confession evidence, the significance of the evidence to the case, whether it could have been obtained lawfully, the gravity of the crime and the seriousness of the breach by the police will all be relevant factors that permit the weight of the evidence to be assessed on the question of guilt.

6.4.2.2 England: Exclusion of Unlawfully Obtained Confession Evidence Under s. 78 PACE

As discussed in Chap. 4, Sect. 4.8.2, s. 78 of PACE provides for the exclusion of any kind of evidence if it was obtained in a manner that is improper, unlawful or unfair. Section 78 is a broad discretion designed to safeguard the fairness of the proceedings as a whole.

In *R v Smurthwaite*[30] (referred to in Chap. 4, Sect. 4.8.2) the English Court of Appeal found that fairness is a policy that operates in two directions: fairness to the accused and fairness to the public. Section 78 has been applied to prevent the police from obtaining a forensic advantage by breaking the rules, while ensuring that an accused is not acquitted on a mere technicality. The section therefore requires a weighing and balancing of competing interests and the weight to be attached to one circumstance depends on the existence and importance of the others. For example, the gravity of the crime, the seriousness of the breach or impropriety by the police and the probative value of the evidence may be some or all of the factors relevant to the discretion to exclude evidence under s. 78.

In *R v Khan* (Chap. 4, Sect. 4.8.2.1) the Court of Appeal found that the planting of a listening device may have constituted a breach of privacy and a civil trespass. However, in admitting the confession evidence that had been recorded, the Court highlighted the gravity of the offending, the fact that the surveillance operation was conducted in accordance with Home Office guidelines and that the operation was not specifically aimed at obtaining the evidence in question. When the case was

---

[30] *R v Smurthwaite* [1994] 1 All E.R. 898.

considered by the House of Lords the accused person's opportunity to test the evidence at the trial was emphasised.

A breach of the code of practice in relation to police questioning (Code C) may be indicative of unfairness. Any issue relating to the unfairness of a trial as a result of a lack of caution or access to a lawyer prior to questioning may be the subject of legal argument and evidence on a *voir dire* at a trial under s. 78 PACE. The significance of cautioning and the right to silence was considered by the Court of Appeal in *Smurthwaite*.[31] Whether a covert conversation between an agent and a suspect in the absence of cautioning is fair and admissible may depend on factors such as whether the undercover officer's role was active or passive, the existence of an unassailable record of any conversations, whether the covert evidence was corroborated and whether the undercover officer abused his position to ask questions which should have been asked in accordance with the codes of conduct. Therefore, in cases that have dealt with confession evidence that has been obtained in the course of an undercover operation in the absence of cautioning and access to legal advice, it is not clear to what weight is given to considerations that represent fairness to the public. The decision in *Smurthwaite* indicates that in cases involving breach of the right to silent, fairness to the accused is emphasised.

6.4.2.3 Australia: Exclusion of Unlawfully Obtained Confession Evidence Under ss. 90 and 138 of the Uniform Evidence Law

A finding of impropriety or unlawfulness in the manner in which evidence is obtained does not automatically lead to exclusion. It appears from the Australian cases discussed in Chap. 5 (Sect. 5.7), confession evidence must be considered under s. 90 and s. 138 before admissibility is finally determined.

The situation of an undercover agent eliciting admissions from a suspect in the absence of due process requirements such as cautioning is clearly a matter for exclusion of confession evidence under s. 90 of the uniform evidence law. When examining fairness under s. 90, there is no balancing of competing interests and the only consideration is fairness to the accused. Under s. 90 the gravity of the offending is not relevant to the exercise of the trial judge's discretion to exclude confession evidence. The drafters of s. 90 primarily enacted the broad provision to encompass the situations where a suspect may not have made a confession had the police investigated properly.

On the other hand, if the court must consider the admissibility of any type of improperly or unlawfully obtained evidence under s. 138, what is involved is a weighing and balancing of competing interests. The decision to exclude under s. 138 is based on the balancing of a number of competing interests which represent the cost to society of excluding the evidence, such as probative value of the evidence and gravity of the offending, and the cost to society of admitting it,

[31] *R v Smurthwaite* [1994] 1 All E.R. 898, 903.

such as the seriousness of the breach by the police and whether there was a breach of a human right under the ICCPR.

6.4.3 Maintaining Fairness: A Comparative View

What is fairness? Is the right to a fair trial only concerned with the protection of individual rights? Or is fairness defined by maintenance of balance between individual rights and the interests of the community? In relation to confessions, the answer may depend on whether the issue being examined is about pre-trial due process requirements in connection with police questioning or obtaining confessions by the use of other types of coercive measures.

In relation to the exclusion of evidence, the English courts have interpreted the broad fair trial provision in s. 78 of PACE as doubtless involving a balancing exercise between all the competing interests that represent fairness to the individual and fairness to the public. The types of factors that have been relevant to a decision about admissibility of evidence have included factors which relate to the nature of the compulsion, such as whether a breach of rights was involved, whether practice codes or legislative guidelines were followed, whether the impropriety or unlawfulness was deliberate or active. Secondly, sometimes the courts have considered the existence of safeguards in the trial phase to overcome the effects of the compulsion, for example, whether a record was made during the pre-trial phase that can be relied on at trial and whether there has been an opportunity to challenge and test the evidence. Thirdly, criteria that are relevant to the significance of the evidence at trial have also been referred to, such as whether the evidence is corroborated by other evidence. Finally, the English courts have relied on the gravity of the offending as a relevant consideration when deciding whether evidence should be excluded or admitted—the more serious the crime, the more likely that improperly or unlawfully obtained evidence will be admitted.

However, the weight placed on one interest or factor depends on the existence and importance of the others. In the cases that have dealt with a lack of pre-trial due process, such as no caution or access to a lawyer prior to questioning or questioning by an undercover officer, weight has been placed on the nature of the compulsion, whether the officer's role has been active or passive and whether codes of conduct have been followed; whether an unassailable record has been made of questioning, to enable the confession evidence and the issue of fabrication to be challenged and tested at trial; and whether there has been other evidence in support of the prosecution case. In these types of cases, it appears that little weight (or no weight in some cases) has been placed on the gravity of the offending. However, in other cases, for example unlawfully obtained covert surveillance cases, reference has been made to the gravity of the crimes involved.

Although, in Denmark, there are no express rules governing the admissibility of evidence, there is an inclination to promote seeking the truth in the free evaluation of the evidence by relying on improperly or unlawfully obtained evidence. It is then for the trial court to attach to the evidence appropriate weight according to all the circumstances of the case (where reliability is not in issue). Nevertheless, in some cases evidence is excluded (referred to, for example, in Chap. 3, Sect. 3.9.4). When considering the admissibility of evidence obtained by a technical breach of the law, for example, the court may consider the seriousness of the investigative breach and whether the evidence could have been obtained lawfully, the significance of the evidence to the case and the gravity of the offending. Some cases that deal with the admissibility of certain types of evidence, such as accomplice evidence, the court's focus is on the safeguards at trial that ensure fairness, such as disclosure and the opportunity for the accused to test the evidence, including the circumstances in which it was obtained. Similar to England, however, when a case involves the lack of caution prior to a police interrogation it appears that there is a general rule of practice in favour of exclusion, whereby the courts place weight on the fundamentality of the right to silence and right against self-incrimination and the importance of fairness to the accused. In these cases the courts do not generally refer to the gravity of the offending—although there are no definite rules.

In Australia, there is a similar distinction between a breach of pre-trial due process in obtaining confession evidence and the admissibility of improperly or unlawfully obtained evidence generally. However, the Australian rules are much clearer about what factors are relevant to the exercise of discretion to exclude evidence in the different situations. Putting aside the question of the reliability of confessions or other evidence more generally, where the probative value may be outweighed by the prejudicial effect, there are two remaining tests that are highly relevant to confessions and improperly or unlawfully obtained evidence. These tests at common law are the fairness test (in relation to confessions) and the public policy test (applying to all types of evidence). Under the uniform evidence law these two tests have been enacted in ss. 90 and 138, respectively.

Section 90 provides for the exclusion of confession evidence in circumstances where the accused may not have made a confession at all if the case had been investigated properly. The only factors that are relevant are the criteria that relate to fairness to the accused. Therefore, in cases where investigators have interfered with (and diminished) a suspect's right to choose whether to speak to the police or whether to remain silent, for example by not following pre-trial due process requirements, fairness to the accused is not offset by public policy interests.

On the other hand, s. 138 of the uniform evidence law is a provision which guides the exercise of judicial discretion to determine the admissibility of any type of improperly or unlawfully obtained evidence. The section requires the court to weigh and balance a number of factors which represent the cost to society of excluding evidence, such as the significance of the evidence in the proceedings and the nature of the offending (including the gravity), and the cost to society of admitting the evidence, such as the seriousness of the investigative impropriety and whether it brought about a breach of human rights. Section 138 does not require the court to consider the existence of safeguards at trial, which ensure that defence rights are not extinguished even if improperly or unlawfully obtained evidence is admitted. Nevertheless, the section provides that the court should consider any other relevant factors.

In all three national systems, when a case deals with the fairness of admitting a confession that has been obtained in absence of the pre-trial due process requirement of cautioning it appears that fairness to the accused is central and the gravity of the offending is not relevant. However, of all the national rules regulating the admissibility of evidence, s. 90 of the Australian uniform evidence law (and the 'fairness discretion') is the only clear expression of that position. The Australian (and Canadian) case law in relation to a lack of pre-trial due process in the context of questioning by undercover officers or agents of the police has been a major influence in the development of the scope of art. 6 by the ECtHR.

As discussed in Chap. 2, Sect. 2.3.2.2.6, in the case of *Allan v. The United Kingdom*,[32] the ECtHR expressly formulated the three pronged test to determine whether reliance on improperly or unlawfully obtained self-incriminating statements adversely affected the fairness of a criminal trial. Following this decision, the overall criteria that are relevant to a determination about the use of unlawfully obtained confession evidence are the nature and the degree of the compulsion; the existence of any relevant safeguards; and the use of the evidence at trial. The nature and degree of the compulsion may take in considerations such as the seriousness of the breach of the investigative officers, whether it was deliberate or whether there was a breach of other rights. Relevant trial safeguards may include disclosure, the opportunity to challenge and test the evidence in question, such as the making of a statement, the provision of a reasoned decision. Factors relevant to the use of the evidence include whether the evidence is the only evidence in the case, decisive or persuasive. Therefore, it can be observed that the factors relevant to a decision under art. 6 are focused on what happens at trial and the overall fairness of the proceedings, even when the issues relate to a type of unfairness which is brought about in the pre-trial proceedings. The ECtHR has expressly stated that the public interest in the gravity of the offending is not a relevant consideration in relation to reliance on improperly or unlawfully obtained confession evidence. This is despite the fact that the law and practice in England and Denmark, two jurisdictions that fall within the community of the Council of Europe, appear to provide otherwise.

The outcome of the ECtHR's jurisprudence is that all those officially accused of criminal offences have equal right to protection under art. 6 and the right to a fair trial—regardless of the seriousness of the crimes with which they are charged. This is a reasonable response to the complaint that rules about admissibility which take into account the seriousness of the offending lower the bar of protection in relation to people who are accused with serious crimes and who have the most at stake. However, beyond the national rules that relate to breaches of pre-trial due process (for example, an absence of a caution prior to questioning), this policy is not consistent with the maintenance of balance between effectiveness and fairness in the national systems.

---

[32](ECtHR) *Allan v. The United Kingdom*, Application No. 48539/99, 5 November 2002.

The approach of the English courts under s. 78 PACE is that fairness operates in two directions and that the gravity of offending is always present as a consideration in assessing the admissibility of evidence. Nevertheless, the English courts give more or less weight to the same interest depending on the circumstances of each case. It appears, therefore, that the gravity of offending might be extremely important in some cases, for example, where the unlawfulness of a covert surveillance measure involves a breach of privacy. However, in other cases the gravity of the offending appears to be of little weight, for example if the unlawfulness of the measure relates to a breach of the right to silence by a lack of caution. It is arguable that when violation of the right to silence is in issue s. 78 functions similarly to s. 90 of the uniform evidence law and even the practice to exclude confessions obtained in breach of § 752 in Denmark. In all three jurisdictions, it appears that the gravity of offending is not emphasised in cases involving a breach of the right to silence but is emphasised in cases involving a breach of other rights. On this point, therefore, despite all of the structural differences in the law about limiting and protecting the right to silence, all three systems tend to respond similarly by excluding confession evidence that has been obtained in absence of a caution. In addition, it is possible that the lack of access to a lawyer prior to police questioning will be a relevant factor to the question of admissibility of confession evidence obtained in breach of the right to silence, particularly in England and Australia. Access to legal representation may be viewed as being of less significance than cautioning, given the clearer exclusionary rule in this connection. There appears to be some consensus in the national cases in relation to exclusion of a confession obtained in absence of a caution.

6.5 Conclusion

Based on the knowledge gathered in previous chapters, this chapter has presented the essential legal rules that limit or protect the right to silence and the right against self-incrimination in the national systems of England and Wales, Denmark and Australia. Emphasis has been placed on limitation or protection of the right to silence in the gathering of confession evidence in three situations: police questioning; direct compulsion by way of administrative compulsory information gathering; and indirect limitation of the right to silence and the right against self-incrimination by gathering confession evidence under covert surveillance. This discussion has formed the basis for an examination of what happens, in theory, to the balance between effectiveness and fairness when a confession is obtained by one of the above methods in one national legal system and admitted at trial in another system.

On the transnational plane, this chapter examined whether the transfer of evidence across borders *unfairly disrupts the continuity* in the way the law limiting and protecting the right to silence functions between the investigative and the trial phases of the criminal proceedings. This question is also related to the admissibility of confession evidence in a trial in another system. The admissibility of evidence is premised on an expectation that relying on certain evidence for conviction will not adversely affect the fairness of criminal proceedings when viewed as a whole. Therefore, the use of judicial discretion to exclude evidence, particularly in common law systems, is seen as a final check and balance of fairness.

Pre-trial due process requirements, such as cautioning and providing access to legal advice prior to questioning, are mechanisms for maintaining balance in a criminal justice system as a whole. The differences in pre-trial due process requirements between the legal systems in Denmark, England and Australia may cause the transfer of confession evidence obtained during a police interrogation to adversely affect the fairness of trial proceedings. The audio-recording of interviews may also play a role in the fairness of the proceedings as it improves the ability of the accused at trial to challenge the evidence against him relating to the circumstances and substance of police questioning. This chapter has demonstrated that there is continuity between an investigative phase and a trial phase within any national system, which dictates the form that certain pre-trial due process requirements should take. In England, for example, the concrete manner of drawing adverse inferences at trial from the silence of the accused during pre-trial police questioning and the direct use at trial of what was said during official questioning is reflected in the form of the caution used prior to questioning. In addition to what may be described as the inherently coercive atmosphere of a police interrogation, particularly a custodial interrogation, the part of the caution about adverse inferences indirectly compels a suspect to provide an early account of the accusations. Therefore, in England great focus has been placed on the significance of providing all suspects at the police station with an effective right to access legal advice, including some form of free, independent legal advice. In addition, there is a requirement that police interviews in relation to indictable offences should be audio-recorded.

On the other hand, in a Danish trial there are no concrete rules about drawing adverse inferences from pre-trial silence, although this possibility is not ruled out. In addition, evidence of what was said during official questioning is only produced as an exception to the general rule that the *tiltalt*'s account of the facts should be presented directly before the court as testimony in the trial. It is only in the event that the *tiltalt* does not give evidence or gives a conflicting account in testimony that evidence of the interview is admitted into evidence. Therefore, the *sigtet* is only cautioned about the right to remain silent and not about specific consequences of that right at a contested trial. The Danish rules about the right to silence may not create the same level of compulsion to speak to the police. In turn, this may explain why there is less emphasis placed on conducting official questioning at the police station; the rules about access by a *sigtet* to a lawyer prior to interrogation are more relaxed; and there is not the same perceived need to record what is said during official questioning by way of audio-recording. Nevertheless, there are consequences that flow from answering police questions or remaining silent; and the risk that the warning to a *sigtet*, about the need to repay costs of a publicly appointed lawyer upon conviction, will most likely dissuade him or her from obtaining legal advice is concerning.

In Australia, in general, comment to the jury about drawing an inference of guilt form the accused's pre-trial silence is prohibited and considered to be entirely contrary to the right to silence. Therefore, the Australian caution does not contain the information that is considered to be coercive in the English caution. The lack of access to free legal representation prior to official questioning must be seen in this light. However, what is said during a police interview may be directly admitted into a trial and used against the accused. Therefore, Australia relies more heavily on the audio-recording of confession as a protection for accuracy and integrity of the evidence at trial. The less stringent rules in relation to legal representation also reflect the difference in the minimum standards set by the HR Committee under art. 14 of the ICCPR compared with the minimum standards of the ECtHR under art. 6 of the ECHR. Nevertheless, by delaying effective access to legal advice until after an accused has been charged, including free publicly-appointed legal representation, the Australian provisions may be inadequate in the context of transnational proceedings. This may be a particular problem in transnational cases, where the consequences of the suspect's attitude during a police interview may be more far-reaching in criminal proceedings in other systems.

If the evidence trail is followed backwards then pre-trial due process requirements are dictated by the system in which the evidence is to be admitted. By not maintaining the continuity between pre-trial and trial phases in transnational cases, without *extra care*, the suspect or accused may not benefit from the necessary safeguards that enable him or her to enjoy equality of arms and to participate in the proceedings in a deliberate and fair way. In addition, pre-trial due process leads to greater confidence in the cross-admissibility of evidence and the effectiveness of criminal proceedings as a whole.

In transnational cases, by focusing on procedural safeguards in the gathering of evidence the cross-admissibility of evidence will be promoted. This chapter suggested five rules that emphasise fairness in relation to *official questioning* of a suspect of foreign offences (a 'foreign suspect'). First, the foreign suspect must be advised of the accusations against him using the requesting country's terminology (and in a language that the foreign suspect can understand). Second, the foreign suspect should be cautioned in accordance with the law of the requesting state. In an unexpected situation the police may not be aware of the wording of a foreign caution, in which case the suspect should be told that he or she does not have to say anything. Third, the foreign suspect should be invited for official questioning at a police station; alternatively, a foreign suspect might be questioned for the first time by consent in the course of a formal hearing via video-conferencing or an audio-visual transmission.[33] A foreign suspect may even be questioned, for example, at consular offices with the consent of the local authorities and the suspect, although the legal basis is unclear. In any event, the safeguards suggested here should be utilised to maximise fairness in the transnational criminal proceedings.

[33] See, for example, Art. 24 of the European Investigation Order Directive 24/14/EU of 3 April 2014.

The fourth safeguard is extremely significant. A foreign suspect should be given the right to legal representation prior to questioning and during the interrogation/hearing. As all foreign suspects who are being questioned at a police station but are not being held in custody in relation to local offences, are questioned voluntarily and are at liberty to stop questioning at any time, then it may be reasonable that this right does not include full access to publicly appointed legal representation. However, due to the fact that significant consequences at trial can attach to pre-trial proceedings then some level of free, independent legal advice prior to the commencement of questioning is warranted—also in the context of a video-conferencing hearing. It would be most valuable if advice could be obtained from a lawyer who has knowledge of the criminal procedure in the trial country. Fifth, an audio-recording should be made of any police interview in relation to foreign offences to protect the accuracy and integrity of any evidence obtained from the suspect. This could be supported by a written record of interview.

By implication, what is being suggested here is transnational harmonisation of certain central pre-trial procedural guarantees in order to fill some of the gaps in procedural standards and strengthen the cross-admissibility of any evidence obtained and, thereby, the effectiveness of cross-border trials.[34]

This chapter also considered the cross-admissibility of confession evidence obtained by the use of administrative (or 'regulatory') compulsory questioning powers. This is potentially the most direct source of compulsion in gathering confession evidence. In important areas of administration, for example regulation of tax liability, company regulation or food safety regulation, compulsory information gathering that overrides the right to silence has been allowed in order to facilitate information-gathering investigations. Alternatively, in some systems such as England and Australia, compulsory examination powers have been used in law enforcement to obtain intelligence in relation to the types of offending that may otherwise be difficult to investigate. This would include investigations of offending by clandestine organised criminal groups. These powers generally include the power to compel a person on pain of punishment to make a statement or give testimony (and to tell the truth). Thus, the use of such powers may compel a suspect to make self-incriminating statements. Therefore, within the national systems, the legislature has responded by creating statutory mechanisms that deal with the potential problem of compelled confessions which violate the right to silence, the right against self-incrimination as well as the right to be investigated and tried in accusatorial proceedings. These legal structures are aimed at restoring the balance between effectiveness of an investigation and fairness to a suspect or accused.

The use of compulsory information gathering has evoked fundamentally different responses in the common law countries of England and Australia compared with Denmark. In general, in England and Australia the 'privilege' against self-incrimination may be abrogated by Parliament in connection with compulsory

[34]In relation to harmonisation in substantive EU criminal law and sentencing policy, see for example, Elholm (2009).

questioning powers. In this case there is no right to silence or right against self-incrimination for a person who is required to give sworn testimony against having compelled confessions used against them in criminal proceedings. Therefore, in these systems, Parliament has restored fairness to a suspect or accused by enacting a 'use immunity' so that any statements made under compulsion may not be used against the maker in future criminal proceedings.

However, in order to maintain the balance between effectiveness and fairness in transnational cases, statements made under compulsory questioning powers should not be used against an accused at trial, regardless of whether the trial happens in England or Australia or in a foreign jurisdiction. Compulsory questioning powers in England and Australia are designed to generate information to assist an investigation and are not intended to be a source of evidence in criminal proceedings inside or outside the national system. For example, it would be 'unfair' to use a statement made under compulsory powers in England or Australia in a trial in Denmark—unless, perhaps, the statement had been made by a witness who was favourable to the defence. In Danish law there is no use immunity to restore balance. In any event, the High Court of Australia has indicated that it is not appropriate to use compulsory powers to examine a suspect who is the subject of pending criminal charges. This runs contrary to the right of a suspect or accused to be tried in accusatorial proceedings where, as a general rule, the prosecution carries the burden of proof.

In Denmark, administrative authorities have been prohibited from using compulsory information gathering to question a suspect about criminal offences. Questioning of a *sigtet*, or someone who is regarded as such due to a threshold level of suspicion, in relation to criminal offences may only take place by consent (with the right to silence). However, the fairness of using these administrative powers to question a criminal suspect—in the absence of the usual procedural guarantees that a suspect may expect to benefit from under a criminal investigation, including the right to access a lawyer—is questionable, whether a future trial would take place in Denmark or abroad. There is no mechanism to avoid the Danish courts from considering confession evidence obtained in this way, even where a suspect of serious offences has not been offered access to publicly appointed legal representation prior to questioning.

The conclusion is, therefore, that using administrative compulsory questioning powers to gather confession evidence in transnational criminal cases is complex and potentially unfair.[35] Information sharing between national authorities under these powers may be a legitimate path, in which case each national system deals with the question of the use of derivative real evidence in its own way.

In general, confession evidence lawfully obtained by the use of covert surveillance measures in Denmark, England and Australia should be admissible as evidence at trial in any of these systems. All three systems use supervisory safeguards to ensure that covert surveillance is carried out in a proportionate and reasonable way.[36]

---

[35]But see Art. 4(b) of the European Investigation Order (EIO) Directive 24/14/EU of 3 April 2014.

[36]See Art. 29 of the European Investigation Order (EIO) Directive 24/14/EU of 3 April 2014.

However, the situation may not be as clear in relation to confession evidence obtained by unlawful covert surveillance. The Danish system includes additional layers of legal supervision in the investigative phase of criminal proceedings. For example, in addition to judicial supervision, Danish prosecutors work alongside the police in the investigation and provide extra regulation of the use of coercive measures. Further, 'intrusion lawyers' are appointed by the state to represent individual interests at the time an application in made by the prosecution to a judge to use a covert measure. Therefore, the problem of unlawfully obtained evidence may not exist in the Danish system to the same extent as in other systems. Additional control in the investigative phase may explain the traditional tendency of the Danish courts to consider all the relevant evidence, including evidence which is perhaps technically unlawful, in the question of guilt. By contrast, in England and Australia the question at trial of admissibility of evidence is an important in systems where there is less control in the way covert surveillance is actually carried out. There is a heavy reliance on the trial judge's role in providing a final check and balance of proportionality and fairness in the criminal proceedings as a whole.

What does this mean to the cross-admissibility of unlawfully obtained confession evidence? Other systems may perhaps be justified in being confident of the legality of covertly obtained confession evidence coming from Denmark. If confession evidence is obtained in Denmark by way of a technical breach of the legislation, the courts in England and Australia are well-equipped with rules of evidence to check proportionality and fairness. However, the Danish courts may not be inclined to question proportionality or fairness at the trial stage. Therefore, additional caution may be required on the part of prosecutors and trial courts in transnational cases. Significantly, all three systems in this work appear to distinguish between confessions that have been obtained in breach of pre-trial due process and confessions that have been obtained improperly or unlawfully by other measures. Where an investigative breach has destroyed the very essence of the right to silence, such as the absence of a caution, the general response in all the systems has been to emphasise fairness to an accused and to exclude the evidence. However, where unlawfully or improperly obtained evidence has breached rights beyond the right to silence, and which are not central to the fairness of criminal proceedings, such as privacy, then the issue is dealt with in different ways. In Denmark, the general tendency is that when technically unlawfully obtained evidence is considered on the question of guilt the fact-finders must determine the weight to be attached to the evidence in light of all the other evidence in the case. In the common law systems of England and Australia, where the essence of the right to silence in not in question, the admissibility of unlawfully obtained evidence is a matter of weighing and balancing competing interests, including the gravity of the offending.

This chapter has illustrated how the process of transferring confession evidence across borders may break the continuity of law that maintains balance between limiting and protecting the right to silence and the right against self-incrimination throughout the criminal proceedings—from investigation to trial. As a result there is potential for unfairness to be caused to the accused and/or the possibility that a court may not allow confession evidence obtained abroad to be used in a trial. Therefore, additional care must be taken in transnational cases to ensure that any potential disruption to the balance is overcome by (additional) safeguards, strategies and flexibility on either side that protect the integrity of the transfer process. Providing an effective right to access legal representation that is meaningful to the transnational context has been presented as a significant safeguard in the process.

Does the application of common minimum standards in relation to the suspect or accused person's procedural rights assist in overcoming the disruption to national balances between effectiveness and fairness? To answer this question, the following chapter will look more closely at the process of transferring confession evidence or evidence of silence between systems that fall within the same regional human rights framework under the ECHR compared with the transfer of evidence to a system that lies outside this framework, such as the ICCPR.

References

Elholm T (2009) Does EU criminal law cooperation necessarily mean increased repression? Eur J Crime Crim Law Crim Justice 17:191–226

European Criminal Policy Initiative (2013) A manifesto on European criminal procedure law. Zeitschrift für Internationale Strafrechtsdogmatik 11:430–446

Kaiafa-Gbandi M (2010) Harmonisation of criminal procedures on the basis of common principles. The EU's challenge for rule-of-law transnational crime control. In: Fijnaut C, Ouwerkerk J (eds) The future of police and judicial cooperation in the European Union. Martinus Nijhoff, Leiden

Ligeti K (2013) Introduction. In: Ligeti (ed) Towards a Prosecutor for the European Union, vol 1: A comparative analysis. Hart, Oxford

Chapter 7
Mutual Trust and the Right to Silence in International Cooperation

Abstract When considering the overarching question of whether it will be fair to use suspect evidence gathered abroad in domestic criminal proceedings, it is relevant to consider from an international perspective whether both cooperating states come under the same human rights framework; and whether that framework is influential at encouraging states to implement uniform minimum standards in the national law. Where both cooperating states come under the same human rights framework, this may be sufficient to lay the theoretical foundation for the states to build a culture of trust in relation to the minimum procedural protections surrounding the right to silence in investigations. Where there is an actual foundation for trust (because the same minimum standards are actually being applied), the fairness of using confession evidence obtained abroad can be safely assumed. On the other hand, where there is no meaningful basis for trusting the standards being applied in another criminal justice system, then states may be less inclined to cooperate and any evidence that is obtained may be inadmissible anyway. Even under frameworks such as the ECHR, with relatively well-defined minimum standards, if in reality these standards are not being consistently applied then states may be overly-trusting and, therefore, overlooking problems with fairness at the fundamental level. Therefore, this chapter explores the existence of the theoretical and actual foundations for mutual trust by investigators, prosecutors and the judiciary in relation to international cooperation in the gathering of confession evidence or evidence of silence and how the fairness of the transfer of evidence in transnational proceedings can be optimised.

7.1 Introduction

The previous chapters identified how the national laws of the three systems in Denmark, England and Wales and Australia, which deal with the gathering and use of confession evidence and evidence of a suspect or accused person's silence, function as continuum in criminal proceedings. The balance between limiting and protecting the right to silence occurs across investigations and trials so that it is achieved in the criminal proceedings as a whole. However, divergences exist in

F.M.W. Billing, *The Right to Silence in Transnational Criminal Proceedings*,
DOI 10.1007/978-3-319-42034-9_7

relation to how the national laws balancing the right to silence actually function. From a *horizontal* perspective in transnational proceedings, when suspect evidence is gathered in one system and used at trial in another, difficulties occur in maintaining balance where the continuity of national law is broken. In Chap. 6, it was argued that by ensuring the application of common minimum standards in transnational cases in relation to a small number of important pre-trial due process protections, the admissibility of evidence in transnational cases and the fairness of the proceedings as a whole will be protected. In particular, reference was made to procedural rights such as cautioning about the right to silence in the form required in the trial state, access to a lawyer who has knowledge of the trial state's criminal law and audio-recording the questioning of suspects or accused persons.

This chapter looks at the question of balancing the right to silence in transnational proceedings from a *vertical* perspective. It seeks to discover what the overall effect is on the fairness of a trial in transnational proceedings if international cooperation in the gathering of confession evidence or evidence of silence takes place within the same human rights framework, comparing the ECHR and the ICCPR. Therefore, what follows is an examination of the persuasiveness of the two human rights frameworks under consideration, the ECHR and the ICCPR, in encouraging states to apply minimum standards in relation to balancing the rules that limit and protect the right to silence—minimum standards that path the way to mutual trust.

The application of common minimum standards in relation to procedural rights lays the foundation for international cooperation in criminal proceedings based on the presence of mutual trust between cooperating states. States are more likely to cooperate with each other and will experience fewer obstacles in maintaining fairness if there is a proper basis for trust—this is a win-win situation. Therefore, mutual trust is central to the prosecutorial and judicial approaches to international cooperation in transnational cases.

The principle of 'mutual trust' acknowledges and attempts to overcome the legal differences between cooperating states rather than aiming at harmonization of laws.[1] It is based upon a presumption that cooperating states meet international standards.[2] In relation to free movement of people or the limitation of free movement, cooperation between states is built upon trust in each other's legal and political systems and confidence in the level of human rights protection provided in each state.[3] Fundamental and international human rights are, therefore, central to

[1]Ouwerkerk (2011), pp. 112 and 270; Klip (2009), pp. 23 and 31.

[2]Ouwerkerk (2011), p. 112.

[3]Klip (2009), p. 93; Suominen (2011), p. 46. The principle of mutual trust is relevant in other areas of law, such as the asylum law and the free movement of goods and services in the single EU market. For example, where goods are put on the market in one EU Member State, other Member States presume that certain standards have been complied with and may not subject the goods to further requirements for admission onto the market. See further Klip (2009), p. 93.

a number of areas that involve international cooperation, including extradition, or surrender, in addition to areas such as asylum.[4]

The free movement of goods and people, as well as the international transfer of data, also demands that investigating authorities look outside national borders in the gathering of criminal evidence. Therefore, mutual trust may also be relevant to international cooperation in the gathering of evidence. The requesting country may presume that evidence gathered lawfully in another country has been gathered in accordance with international standards of fairness. In addition, the requested country must assume that the evidence being transferred is going to be used in a *bona fide* way as part of fair criminal proceedings and with the need to protect concerns such as confidentiality in mind.

One important question is whether *absolute* mutual trust exists and whether this is a question of blind faith? Whether states can and actually do fully trust each other is difficult to assess.[5] Absolute mutual trust may be described as an unquestioning belief in the compliance of other states with fundamental and international human rights. If absolute mutual trust exists it may not always "...allow public decision makers to take into account when extraordinary circumstances apply."[6] It can never be completely ruled out that the process of international cooperation will not expose a person to violations of their fundamental and international human rights.[7] Absolute mutual trust does not allow for an examination of the actual situation—whether the situation concerns transfer of goods, asylum seekers, suspects and accused persons or criminal evidence for use at trial.[8]

In order to determine the overall effect on the international cooperation process of the international human rights frameworks, as forums for the setting of minimum standards, this chapter commences by examining how transnational investigations and prosecutions may be effected by the adherence to minimum procedural standards. Following this, judicial approaches in transnational cases to evidence obtained abroad, and the impact of the international human rights frameworks will be assessed. Finally, the principle of mutual trust will be considered in more detail in transnational cases going outside the EU and in EU law.

---

[4]Jones and Davidson (2010), p. 91; see also Peers (2011).

[5]Vernimmen-Van Tiggelen and Surano (2008), p. 20.

[6]Billing (2012), p. 88; see further Klip (2009), pp. 338–340.

[7](CJEU) *N.S. v Secretary of State for the Home Department – Opinion of Advocate General Trstenjak*, 22 September 2011, [130].

[8]See Billing (2012), p. 88.

7.2 Investigative and Prosecutorial Approaches to Procedural Rights in Transnational Cases

7.2.1 International Cooperation as Prosecutorial Tool

International cooperation in criminal matters is an investigative/prosecutorial tool to ensure the effectiveness of law enforcement even in transnational cases. Achieving the widest measure of cooperation in the gathering of evidence using mutual legal assistance in criminal matters (MLA) is really a question of determining how far a requested country is able to go. Prosecuting policy in the requesting country may in some circumstances prevent the prosecution from obtaining evidence through methods that are regarded as controversial or unlawful. Alternatively, fundamental legal principles that apply in the requested state may prevent the local authorities from meeting a foreign request. Therefore, issues at trial about admissibility of evidence will not arise at all in the absence of cooperation, where states cannot agree on the investigative methods to be used and the procedural standards that apply. The goal of effectiveness is lost.

On the other hand, it may be possible to find practical solutions to the concerns raised in the previous chapter, for example, due to the differences in pre-trial due process rules. For example, rather than contending with differences in procedure or insisting on extradition, a police officer or prosecutor may consider questioning a suspect via telephone or at an embassy, putting the other state on notice. While an embassy is not sovereign territory, it may provide a convenient location for foreign police to question a suspect by consent under their own criminal procedural rules, by consent and in the absence of the local police and, significantly, by consent of the suspect or accused. Alternatively, small concessions may be made in procedural rules by either the requesting or the requested state in order to allow for international cooperation to take place. However, these are pragmatic rather than legal solutions and may lead to a watering down of the procedural rights of a suspect or accused that enable him or her to participate meaningfully in the prosecution. The goal of fairness may be lost.

One concern in international cooperation is that asking a suspect to voluntarily attend a foreign police station for questioning alerts him (and potentially others) to the fact of an investigation and cause him to flee. Some national authorities may prefer to wait with questioning until after extradition has taken place. This ensures that differences in procedural rights, in particular in relation to pre-trial due process, will not become an issue. However, it dispenses with interrogation as a possible investigative tool prior to extradition, which is an extreme measure for the suspect or accused person involved. Not all suspects present a flight risk and some investigative agencies may be prepared to have suspects questioned abroad where there are guarantees that certain procedures and standards will be met.

Foreign suspects may be able to provide an explanation that removes suspicion. Although he or she may continue to be involved in the investigation as a witness, this is an example of the benefits (also for suspects) of requesting local police to conduct an interview of a suspect who is outside the jurisdiction. The resources and inconvenience of extradition and surrender is avoided in these types of cases and the investigation can quickly adapt to new information. Through MLA, local police may be able to interview a suspect on behalf of a foreign agency.

A structured and consistent approach to the questioning of suspects of foreign offences may assist in overcoming the problems associated with transferring confession evidence or evidence of a suspect's silence across national borders. For example, the current MLA policy in some jurisdictions, such as England, already allows as far as possible for a suspect to be cautioned according to the law of a foreign jurisdiction.[9] In addition, it is likely that foreign police will be given permission to make a list of the questions to be asked. Alternatively, MLA policy may even allow foreign police to be present to ensure consistency with trial states procedures as far as possible during an interview of a suspect. Where there is a mutual understanding of the minimum standards of due process to be maintained then both the goals of effectiveness and fairness may be realistic.

7.2.2 *Contextual Realities and Different Approaches to Utilising International Cooperation*

A careful approach to international cooperation may be required when investigative and prosecutorial authorities are cooperating with other states that may not be applying the same minimum standards of human rights protection—even when states are operating under the same human rights framework.[10]

Even within the EU and Europe, local realities are likely to influence the response to seeking and giving assistance. The authorities in each state investigate and prosecute transnational cases within the contextual realities of the individual system. Therefore, the approach in Denmark may be very different to that in England. England, in particular London, is one of the major organised crime centres in Europe. For example, in relation to trade in illicit substances and goods, it is a significant entry and exit point, as well as an important and highly populated destination. As a result, a significant number of requests go in and out of England, through the Home Office as the central authority. Denmark is primarily a country of destination. Danish authorities more often seek rather than give assistance through international cooperation. These various realities are likely to shape international cooperation goals and policies of individual countries.

In Denmark, there appears to be limited experience of having suspects of Danish investigations interviewed abroad or *vice versa*. In order to overcome the problems of differences in pre-trial due process rules, Danish police and prosecution teams

[9]Home Office (2015), pp. 18 and 20.

[10]Some of the information contained in this chapter is the result of informal interviews with police, prosecutors and legal policy officers from the three legal systems under examination.

have been known to travel abroad and conduct an interrogation of a *sigtet* at a Danish embassy, with the *sigtet*'s consent and the knowledge of the host country, under the provisions of the AJA. It is often considered necessary that the Danish police or prosecution authorities participate in interviews of witnesses and suspects outside the country.[11] Alternatively, in transnational cases, interrogation of the *sigtet* often takes place after he or she has been surrendered to Danish custody. However, the situation may arise in cross-border cases, for example in a joint investigation, where a Danish suspect will be questioned in another system or where Danish police question a person suspected of criminal offending in another country. Alternatively, cases may exist where questioning has already place abroad before the matter comes into the hands of the Danish authorities.

In Denmark, in general, it is assumed that the international cooperation process is designed to make gathering evidence across international borders more effective and is not to be used to utilise additional investigative powers other than the powers available in Denmark.[12] Therefore, the investigative steps used by foreign authorities on behalf of Danish investigators or by Danish authorities working abroad must be in accordance with Danish law and under the supervision of the Danish authority that would normally be involved in the evidence gathering process.[13] For example, the Danish police should not request a telephone intercept to be undertaken by a foreign authority in a case where there would not be the legislative and judicial authority to do so in Denmark.[14] A request to a foreign authority to carry out an investigative step, which would require the court's supervision in Denmark under the AJA, for example, must be placed before the Danish courts prior to the request being made.[15] The lawfulness of requests by Danish authorities to have a suspect/*sigtet* questioned abroad, which is an investigative step that does not normally require the approval of the court, can always be examined by the court under § 746 of the AJA.[16] Likewise, in accordance with practice or by agreement, Danish authorities will carry out an investigative step under a request from another country where there would be legislative or other authority in an equivalent national

[11]Rigsadvokaten (1996), p. 42. Due to the principle that a witness's evidence should be presented in court and tested under cross-examination, coupled with the fact that a foreign witness may only be brought voluntarily to court via mutual legal assistance, it is not uncommon for a small team of Danish police, prosecutors and defence to travel abroad and attend a hearing for a witness statement to be obtained in court and for cross-examination to take place at the same hearing.

[12]Rigsadvokaten (1996), p. 81.

[13]Rigsadvokaten (1996), p. 81.

[14]Rigsadvokaten (1996), p. 81.

[15]Justitsministeriet (2001); but see § 806 AJA whereby, in an outgoing request for production of documents abroad for use in Danish proceedings, the person or legal person against whom the request is made does not have to appear before the Danish court prior to the court making an order that the conditions for production of documents have been fulfilled. A foreign authority would then have to obtain the requisite court order for production to take place: U.2005.3108H.

[16]Rigsadvokaten (1996), p. 81; see further Chap. 3, Sect. 3.9.2.

case.[17] There appears to be, however, reticence on the part of the Danish authorities to work outside the national standards.

On account of the prominent position of English investigative authorities in transnational law enforcement, the Metropolitan Police at Scotland Yard in London has a unit which is entirely dedicated to responding to extradition requests and requests for assistance from foreign agencies. As part of their work, the officers involved question foreign suspects regularly at the request of other countries—both suspects of EU countries and suspects from countries outside the EU.[18] Interviews of suspects are always conducted on a voluntary basis. In the absence of an extradition warrant or a European arrest warrant there is no jurisdiction for English police to use English arrest powers to detain a foreign suspect, unless there are, in addition, grounds for arrest in relation to English offences. Foreign suspects are allowed the right to have a lawyer of their own choice present at the interview. However, there is no right for foreign suspects attending voluntarily to have free publicly appointed legal advice. In line with English policy, at the request of a foreign authority, the English police will usually recite a foreign caution when interviewing foreign suspects and the questions to be asked are frequently provided by the foreign investigators. Interviews are audio-recorded or recorded in writing, according to the request.

The tension between a commitment to international cooperation, on one hand, and to the application of national procedural rights, on the other, may not be so onerous among countries cooperating within the EU. In general, EU Member States are obliged, at least under international law, to apply within the national jurisdiction the minimum procedural and human rights protections that are set by EU decision makers and the ECtHR. On this basis investigators and prosecutors in EU Member States, for example, have been able to focus on finding more practical solutions to the problems of admissibility of evidence gathered via international cooperation.

The process of finding working solutions to law enforcement in transnational cases in the EU has been assisted by the operation of important networks such as the consultative EU prosecutorial network, Eurojust, and the European Judicial Network (EJN). Eurojust is based in the Hague and is made up of a group of prosecutors from different EU countries, working side by side. This makes communication between prosecuting agencies faster and more simplified and often results in practical solutions being found by agreement to most investigative and prosecutorial problems brought about by international cooperation between EU Member States. It is ultimately up to defence counsel and the courts to raise and deal with any objections relating to admissibility of evidence that has been obtained through the flexibility of cooperating authorities. Such issues may be based on a claim of unfairness and, if they arise, likewise detract from the effectiveness of transnational proceedings.

---

[17] Justitsministeriet (2001).

[18] The English police are also asked to conduct witness interviews on behalf of foreign police. Occasionally a witness may become a suspect in the course of an interview.

The need for a careful approach to international cooperation can be demonstrated by considering the Australian position in international cooperation. One example is the case of the 'Bali 9'. In April 2005, nine young Australians had been arrested by the Indonesian police in possession of a large quantity of heroin for exportation from Indonesia to Australia.[19] The arrest came after the Australian Federal Police (AFP) had contacted the Indonesian police to assist by conducting the investigation while the suspects remained on Indonesian soil. Australia's request for assistance led to six of the nine facing the death penalty.[20] Cases such as this place focus on the tension between Australia's commitment to international cooperation and law enforcement across national borders; and to the commitment to minimum fundamental human rights standards that are critical to ensuring fairness in transnational proceedings. This tension is significant in light of Australia's proximity to and relationship with neighbouring countries in South East Asia and the Pacific—even if the other state also comes under the framework of the ICCPR.

The arrest of the Bali 9 was followed closely by the decision of the Victorian Court of Appeal in the case of *The Queen v Joseph Terrence Thomas*,[21] in which the Court insisted that Australian standards under federal law in relation to legal representation prior to official questioning also applied when questioning was being conducted by the AFP outside of Australia. This appears to apply regardless of whether the requested state is a signatory to the ICCPR.[22] The *Thomas* case may be indicative of a more general policy in Australia in relation to international cooperation that unless police, prosecutors and the Attorney-General's Department (as the central authority) are assured that the same procedural standards that apply in Australia will be applied in another country, then cooperation will simply not take place. Police questioning by a foreign investigative agency of suspects or witnesses who turn out to be suspects may be problematic if the procedural rules and investigative culture are likely to be different. Likewise, the Australian government agencies may be cautious when asking neighbouring states to conduct covert surveillance or use other coercive measures. Any reticence to seek assistance is, of course, a win for those involved in organised transnational crime. Therefore, the AFP's cooperation with police forces throughout Europe and other parts of the world, including Asia, can and does result in law enforcement successes that result in arrests and extradition to Australia from other countries.[23]

---

[19]Indonesia acceded to the ICCPR on 26 February 2006, after the facts of this case had arisen.

[20]A case was brought against the AFP by the six accused in the Federal Court: see *Rush v Commissioner of Police* [2006] FCA 12.

[21]*The Queen v Joseph Terrence Thomas* [2006] VSCA 165.

[22]Pakistan ratified the ICCPR on 23 June 2010, after the hearing of the *Thomas* case.

[23]See, for example, the operation led by the Australian AFP which involved search warrants in Australia and four European countries with four arrests taking place in 2013: Australian Federal Police (2013); Annual Reports of the Commonwealth Director of Public Prosecutions frequently describe transnational cases involving the AFP.

7.3 Judicial Approaches to Confession Evidence Obtained Abroad

7.3.1 Transferring Confession Evidence Between States Within the Same Regional Human Rights Framework: The ECHR

7.3.1.1 Danish Cases

The Danish Supreme Court has found that the rules in the Administration of Justice Act (AJA) generally apply in relation to investigative steps in a criminal investigation, even when investigative steps are being carried out in another Council of Europe State such as England. Thus, in U.2005.3108H a compulsory interrogation of a witness conducted by Danish prosecutors under English compulsory questioning powers, where the witness was compelled to answer questions by way of sworn testimony under threat of punishment, was found to be contrary to § 750 of the AJA and unlawful. However, in accordance with legal tradition, the Supreme Court explicitly decided to make no ruling in relation to the admissibility of the statements of the witness in the proceedings—this was left to the discretion of the trial court. In the case, the witness evidence, which was favourable to the accused, was admitted at the trial.

In cases where a suspect has been formally interviewed by the police in a foreign jurisdiction the recording of the interview may be admissible as evidence in Danish proceedings, in a similar manner as if the interrogation had taken place in Denmark. For example, in a judgment delivered by the Glostrup City Court on 4 June 2012 the court referred to a tape-recording of an interview between an accused and the Swedish police which was played in court during the course of the accused person's evidence.[24] However, this is despite the fact that, in Sweden, witnesses and suspects may be 'summoned' to the police station for pre-trial questioning[25]—a rule which is arguably contrary to § 750 of the AJA.

As a general example of the approach in Denmark to irregularly obtained evidence, it is worthwhile considering entrapment cases, such as U.2002.340H, involving an agent of the Austrian police. In this case a majority of the Danish Supreme Court found that even though on one particular occasion evidence of importation of cocaine was the result of what would have been unlawful entrapment, contrary to § 754a of the AJA, the transaction was one of a number of 'professional' deals and therefore did not lead to the conclusion that the prosecution

[24]Glostrup Court, Udskrift af dombogen, 4 June 2012, Court no. 15-3547/2012; presumably the evidence was favourable to the accused.

[25]Wong (2013), pp. 751–752. This rule may also be in contrary to § 752 of the AJA, which requires that a suspect be advised that he or she does not have to say anything.

should be discontinued or that the accused should be acquitted.[26] Nevertheless, where the quantity of drugs was significantly greater on this occasion, a sentence reduction was justified. In the decision the High Court stated that no significance could be attached to the fact that the entrapment was carried out by a foreign agent without the involvement of the Danish authorities. Therefore, it appears from this decision that foreign evidence will usually be treated in the same way as if the Danish police had obtained the evidence.

From the above, a conclusion may be drawn that foreign confession evidence obtained in other contracting party states is generally admissible so long as the fundamental provision contained in § 752 of the AJA has been complied with and there are no pressing reasons to the contrary. In particular, this will very likely require that a suspect is informed of the right to remain silent prior to being interrogated. Due to uncertainty in the national law, it remains uncertain, however, the approach that will be taken in relation to receiving evidence of a confession obtained in a custodial setting involving a lack of a right to prior access to legal advice, including publicly appointed representation. Whether this will this result in exclusion of confession evidence in Denmark, where there is a legal tradition that favours placing all the relevant evidence before the court, despite the circumstances in which it was obtained is perhaps unlikely. It is not clear the extent to which this is considered to be a fundamental principle in Danish law, despite the findings of the ECtHR in cases such as *Salduz* and *Pischalnikov*. In the Danish context, the issue may remain a question of the weight attached to confession evidence when the court deliberates on guilt.

In addition, it is unclear precisely how a Danish court will approach the admissibility of a confession obtained unlawfully by covert surveillance evidence, if that evidence has been obtained by foreign police. It is likely that the issue of confession evidence obtained unlawfully by the use of covert surveillance will also be a question of weight, rather than 'admissibility'—although this may depend on whether the unlawfulness was deliberate or of a technical nature.

7.3.1.2 The Case-Law of England and Wales

The English judiciary has been called to adjudicate upon the fairness of using evidence obtained abroad, including confession evidence, on several occasions in a number of important decisions. A discussion in the UK of fairness and the admissibility of evidence gathered under foreign procedures in Council of Europe jurisdictions commences with the decision of the Scottish House of Lords from 1961 of *McKay*, which has had persuasive authority in the English courts. In this case, the House of Lords considered that documentary evidence seized lawfully in the Republic of Ireland[27] pursuant to a warrant against the accused was admissible,

---

[26]In a minority decision, Blok J found that the circumstances of the case were morally unacceptable and contrary to §§ 754a and 754b of the AJA.

[27]Ireland ratified the ECHR on 25 February 1953.

despite the fact that the documents were not in the accused person's name (which would have been a requirement under Scottish law). However, in obiter, Lord Wheatley took the view that the lawfulness of a particular procedure in a foreign jurisdiction does not necessarily lead to the admission of the evidence in the domestic court.[28] His Lordship referred to the possible existence of foreign procedures that are "...so offensive to our own fundamental principles of justice and fair play that the admissibility of such evidence would not be tolerated."[29] Lord Wheatley's comments, therefore, already highlighted the problem of international cooperation in the gathering of evidence—evidence should be gathered in a manner that is considered to be fair at a fundamental level in the requesting state in order to ensure cross-admissibility.

Some of the English decisions about the admissibility of foreign evidence pre-date the entry into force of the Human Rights Act 1998, which implemented the ECHR into domestic law. Although the 1990 case of *Quinn*[30] preceded the Human Rights Act, the Court of Appeal nevertheless dealt with the issue of the 'fairness' of using evidence obtained abroad under s. 78 of the Police and Criminal Evidence Act 1984 (PACE). The Court held that identification evidence obtained by foreign law enforcement agents in Dublin was admissible, despite the fact that the same evidence may not have been admissible had it been obtained in England by domestic authorities. Several weeks after the shooting of a police officer in London, the appellant stood trial in Dublin for offences committed in the Republic of Ireland. A witness to the London shooting went to Dublin and identified Quinn. The identification procedure did not conform to the provisions of PACE and the relevant codes of practice. Lord Lane CJ noted that the English courts cannot expect English procedural requirements to be followed at all times by police forces operating abroad—even in jurisdictions where the procedural rules are generally similar to the UK.[31] It was considered that s. 78 of PACE demanded the trial judge to protect the fairness of the English proceedings as a whole. In dismissing the appeal, Lord Lane found that the trial judge in the case had taken into consideration the factors that were relevant to fairness and the exercise of the discretion, such as: whether there had been sufficient opportunity to properly challenge the evidence against the appellant in cross-examination (and explore reliability); whether there had been an abuse of process, for example, because the evidence had been obtained in deliberate breach of the applicable procedural rules laid down in an official code; and the fact that the identification evidence was not the sole evidence in the case.[32]

---

[28]Obiter is of persuasive value according to the common law doctrine of precedent: see Chap. 1, Sect. 1.3.3.1.

[29]*HM Advocate v McKay* 1961 SLT 174, 179.

[30]*R v Quinn* [1990] Crim LR 581.

[31]*R v Quinn* [1990] Crim LR 581, 582.

[32]See further the Scottish case of *Holland v HM Advocate* 2005 1 SC (PC) 3 relating to whether dock identification was compatible with art. 6.

The decision in *Quinn* was followed in 1993 in the case of *Konscol*.[33] Konscol was arrested in Belgium as a passenger of a car that was found to contain a quantity of drugs. Following his arrest, a note was made by a translator of an interview conducted by a Belgian customs officer, under the authority of a Belgian examining magistrate—all in accordance with local procedures. The note went into the papers that were sent to the English authorities, and was translated. The appellant was tried in England for *inter alia* conspiracy to import drugs from Belgium. Contrary to PACE, the Belgian authorities had not offered Konscol the services of a lawyer or cautioned him of his right to remain silent. The appellant had not made any admissions in the interview, but the note recorded certain statements made by him that he conceded at trial were lies. The appellant objected to the prosecution adducing the evidence of the interview, submitting that this evidence should have been excluded in the exercise of the trial judge's discretion in s. 78 PACE. The appellant suggested that the interview would have been excluded had it been conducted in England. In applying the exclusionary discretion, the trial judge had admitted the evidence, emphasising that the trial judge was required to ensure that there was "...a just balance at all times in what is done, and that there should be an absence of dishonesty, and an absence of bullying behavior..." The Court of Appeal upheld the trial judge's refusal to exclude the evidence. In his judgment, Popplewell J stated:

> This court does not propose to lay down any guidelines as to when a court in this country should or should not admit a statement taken in a foreign country according to their rules which may not coincide with the procedure established under the Police and Criminal Evidence Act. What is fair will depend on the circumstances of each case.[34]

Despite the fact that the case touched upon the very issue of a breach of the right to silence as it is asserted in English law, the confession evidence was admitted. This result may be inconsistent with the 1994 decision of the Court of Appeal in the decision of *R v Smurthwaite*, dealing with the admissibility of evidence obtained in England.[35] However, it may indicate a more lenient approach by the English courts if denial of pre-trial due process is in accordance with foreign law.

In *Aujla*,[36] in 1998 the Court of Appeal again sanctioned the use, in an English trial, of evidence which was lawfully obtained in a Council of Europe state. The appellants were charged with conspiracy to facilitate the illegal entry of persons into England. Although English intercept data may not be used as evidence in English proceedings, in a pre-trial ruling, the trial judge admitted intercept evidence that had been obtained lawfully in the Netherlands, in accordance with Dutch procedures. Unlike the procedures in *Quinn*, the manner in which the intercept evidence had been obtained was found not to be in conflict with English law. With the proper judicial control, the Dutch police recorded conversations between the

[33] *R v Konscol* [1993] Crim LR 950.

[34] *R v Konscol* [1993] Crim LR 950.

[35] *R v Smurthwaite* [1994] 1 All E.R. 898, 903; (1994) 98 Cr.App.R. 437; see also *R v Christou and Wright* [1992] 1 Q.B. 79, 99.

[36] *R v Aujla* [1998] 2 Cr App R 16.

appellants and certain Dutch residents. Following the conviction of the Dutch residents, the trial intercept transcripts were then made available to the English police for use in the proceedings against the appellants. The authenticity and content of the recordings were never challenged and they formed part of the public record in the Netherlands. In the interlocutory appeal, the Court of Appeal found that the operation of domestic legislation did not bar the use of foreign intercept evidence. The court noted that the appellants' right to privacy in art. 8 of the Convention had not been violated and presumed that the Dutch phone tapping legislation complied with the ECHR.

The judgment in *Aujla* was affirmed by the House of Lords in *R v P and others*[37] when the court considered the admissibility of evidence of a telecommunications interception made by a foreign authority from an EU country in relation to organised drug importation activities. The procedure for conducting covert interceptions was under judicial supervision and was compliant with the jurisprudence of the ECtHR in relation to art. 8 of the ECHR. The prohibition on the use of intercept evidence generated in England by English authorities in criminal proceedings (discussed in Chap. 3, Sect. 3.6.1) did not extend to intercept evidence created by a foreign authority. It is possible that if the covert evidence had originated in a country outside the EU or the Council of Europe framework the court may not have been convinced that a scheme for obtaining covert intercept evidence would have complied with acceptable human rights standards.

The case of *Konscol* was applied by the Court of Appeal in Northern Ireland in the case of *R. A. Russell v Patrick Anthony Adams*.[38] The appellant's van was involved in an accident a town in Northern Ireland and failed to stop. That evening, on request of the Northern Ireland police (the Royal Ulster Constabulary), two members of the Irish police, the Garda Siochana, attended the appellant's girlfriend's address and found the van in a damaged condition. The two Garda spoke to the appellant at the door of the house. When they advised him of the belief that his van had been involved in the accident the appellant admitted he had been the driver. However, when he voluntarily attended the police station with the RUC, accompanied by his solicitor, he denied making the admissions. When the matter was heard in the Magistrates' Court the statements of the two Garda were admitted. The admissibility of the evidence of the Garda was examined by the Court of Appeal under the equivalent of s. 78 PACE and PACE Code C, about the requirement in Northern Ireland to caution a suspect prior to questioning upon forming a reasonable suspicion of criminal wrongdoing. Under Irish law the Garda were not obliged to caution a suspect unless they had made up their mind to charge. The Court emphasised that the local police had acted properly within the law of that jurisdiction and could not be expected to have been familiar with or implemented the requirements of the PACE Order.[39] Nor did the Court of Appeal find that admitting

[37] *R v P and Others* [2000] UKHL 72 (8 June 2000).

[38] *R. A. Russell v Patrick Anthony Adams* [2000] NICA 21 (27 November 2000).

[39] *R. A. Russell v Patrick Anthony Adams* [2000] NICA 21 (27 November 2000), 10.

evidence of the appellant's oral statement had such an adverse effect on the fairness of the proceedings that it ought to have been excluded.[40] There had been no evidence that the appellant was under pressure when he made the admission or that he would have declined to answer if he had been cautioned. Carswell LCJ concluded:

> While we appreciate and give full weight to the importance which our law attaches to the administering of cautions, in all the circumstances of the case we do not consider that it was unfair to admit the statements.

The contrast between the decisions of *Konscol* and *Adams* with the UK Supreme Court's (UKSC) judgment in 2011 of *Cadder*, dealing with the consistency of (Scottish) national provisions and art. 6, is marked. In the case of *Cadder v Her Majesty's Advocate*[41] the UKSC reconsidered the Scottish High Court of Justiciary's decision in relation to the admission of statements by an accused to the police while in pre-charge detention without access to a legal representative, in light of the decision of the Grand Chamber in *Salduz v Turkey*.[42] The UKSC rejected the HCJ's view that following *Salduz* the issue about the fairness of a trial depended on the circumstances of the case, including what arrangements existed in relation to access to legal representation and the guarantees that are otherwise in place to secure a fair trial. In particular, the HCJ had referred to the rules in Scotland relating to detainees and cautioning, audio and video-recording of interviews, rules against oppression, rules requiring a lawyer to be informed of the fact of detention and Scots law requiring corroboration of a confession.[43]

The UKSC overruled the HCJ's decision and held, following *Salduz*:

> ...contracting states [of the Council of Europe] are under a duty to organise their systems in such a way as to ensure that, unless in the particular circumstances of the case there are compelling reasons for restricting the right, a person who is detained has access to advice from a lawyer before he is subjected to police questioning.[44]

In coming to this conclusion, the UKSC emphasised the role of the ECtHR in providing "...principled solutions that are universally applicable in all the contracting states."[45] In the judgment the UKSC expressed the view that the decision in *Salduz* allowed departure from the general rule requiring access prior to the first custodial interview only if the circumstances of a case indicated that there were compelling reasons for doing so. However, the decision did not permit a systematic departure from the rule it established.[46] The UKSC noted that since the

---

[40] *R. A. Russell v Patrick Anthony Adams* [2000] NICA 21 (27 November 2000), 10.

[41] The UK Supreme Court decision in *Cadder v HM Advocate* [2010] UKSC 43 was effectively an appeal from the High Court of Justiciary in the case of *HM Advocate v Duncan McLean* [2009] HCJAC 97, 2010 SLT 73.

[42] (ECtHR GC) *Salduz v Turkey* (Application no. 36391/02) 27 November 2008.

[43] See *HM Advocate v Duncan McLean* [2009] HCJAC 97, 2010 SLT 73, [27].

[44] *Cadder v HM Advocate* [2010] UKSC 43, [48].

[45] *Cadder v HM Advocate* [2010] UKSC 43, [40].

[46] *Cadder v HM Advocate* [2010] UKSC 43, [41].

Grand Chamber's decision in *Salduz* had been handed down the majority of the contracting states that had not afforded a right to legal representation even at a custodial interview (Belgium, France, the Netherlands and Ireland) were recognising that their legal systems were inadequate in this respect.[47]

The judgment in *Cadder* provides a basis to find that *Adams* and *Konscol* would most likely be decided differently in the UK today, following the ECtHR's decisions in *Aleksandr Zaichenko v Russia*[48] and *Salduz v Turkey*.[49] In *Zaichenko* and *Salduz* the ECtHR express the clear view that cautioning prior to questioning and access to legal advice prior to questioning of a suspect who is under arrest or in custody are essential pre-trial due process requirements that ensure the fairness of the proceedings as a whole.

Historically, the English courts have taken a flexible approach to admitting foreign evidence which has been obtained lawfully within other Council of Europe contracting states but is irregular according to the rules of England and Wales. Some decisions have demonstrated that the English courts are confident in the standards of protection in other Council of Europe states, for example, in relation to the use of covert interception. In the past it has been found that even if the national procedural rules and standards of England (as requesting state) have not been followed foreign evidence is not automatically rendered inadmissible so long as the overall protective principle of fairness has not been breached. The factors that favour the admissibility of foreign evidence which has been obtained in a manner that is irregular in England are: a lack of deliberate unlawfulness, such as dishonesty, bullying or an abuse of process; the opportunity for the accused to challenge the impugned evidence and test it in cross-examination; and that the foreign evidence in question is not the sole evidence in the case. Previously, it has even been considered fair to admit confessions obtained in other contracting states without administering a caution prior to questioning or access to a lawyer prior to custodial questioning, despite the importance of these rights in the law applicable at the time under PACE and the codes of conduct.

However, following a number of key decisions in the ECtHR it is unlikely that this flexibility will still apply in relation to confession evidence that has been obtained abroad during a police interview in absence of caution or access to a legal advisor in the custodial situation—even if the interview is conducted according to national law. It is difficult to imagine that the English courts would be permitted to take the same flexible approach to issues such as lack of caution or lack of access to legal advice in light of the clear acceptance and application by the UKSC of the minimum standards of pre-trial due process laid down by the ECtHR. In accordance with the UKSC decision in *Cadder*, cautioning and access to legal advice are considered essential to protection of the right to silence and the right against self-incrimination and, therefore, are an integral part of the protective

[47] *Cadder v HM Advocate* [2010] UKSC 43, [49].

[48] (ECtHR) *Aleksandr Zaichenko v. Russia* (Application no. 39660/02) 18 February 2010.

[49] (ECtHR GC) *Salduz v Turkey* (Application no. 36391/02) 27 November 2008.

principle of fairness. Surprisingly, however, the English courts could be required to re-examine the issues that were considered in cases such as *Konscol*. Not all Council of Europe states have adjusted their criminal procedure so that the floor of rights that is applied is consistent with the guidelines laid down by the ECtHR.[50]

It is highly likely that the well-understood common law rule that unlawfully obtained evidence (obtained, for example, by covert methods such as a listening device) is not automatically inadmissible will apply equally to foreign obtained evidence as to evidence obtained within the national system.[51] All the circumstances of the case will be relevant under s. 78 PACE.

7.3.2 Transferring Confession Evidence Between States Not Operating Within the Same Regional Human Rights Framework

7.3.2.1 Danish Case Law Involving Cooperation with ICCPR Signatories

There are no grounds to suggest that confession evidence obtained lawfully abroad from a country that lies outside the framework of the ECHR will not be generally accepted and relied on in a Danish trial, provided that reliability is not in issue and the fundamental requirements of the AJA have been complied with (such as § 752). It is likely that a custodial confession made lawfully to foreign police in a country which is not a member of the Council of Europe, such as Australia, without prior access to a publicly appointed lawyer will be admitted and the weight attached to it adjusted according to all the circumstances of the case. The Danish courts tend to allow consideration of confession evidence provided that reliability and the right against self-incrimination are not in issue. Dishonesty or deliberate unlawfulness on the part of investigators may be another area where exclusion of evidence may be considered. Otherwise, the approach of the Danish courts tends to be in favour of admission.

This flexible approach to relying on relevant evidence can be demonstrated in the international context in the cases of U.2009.941H and U.2012.1604 V. In both cases the accused were convicted on the basis of accomplice evidence, where accomplices had pleaded guilty as part of a plea bargain (in the US sense).[52] In the 2009 case, the Danish Supreme Court stated that plea bargain agreements that

[50]See Chap. 1, fn 3.

[51]See *Public Prosecution Service v McKee* [2013] UKSC 32; *Re Attorney General's Reference (No. 3 of 1999)*, Re [2001] 2 W.L.R. 56; *Governor of Pentonville Prison ex Chinoy* [1992] 1 All ER 317, 330 concerning the admission in extradition proceedings conducted in England (where the applicant had been arrested) of transcripts of telephone conversations unlawfully obtained by the US in France.

[52]The United States of America ratified the ICCPR on 8 June 1992.

are lawful in the US, by which the accused is aware of the convictions and sentence to be recorded in advance of confessing, cannot be considered lawful in accordance with Danish law. Nevertheless, the Supreme Court did not find that the US plea bargain arrangements were contrary to fundamental principles of Danish law, referring to the ECtHR judgment in *Cornelis v The Netherlands*.[53] In addition, it was considered appropriate in processing cases involving international criminality to make reference to the fact that other country's procedural rules may vary to those under Danish law and that this did not necessarily preclude the use of evidence in Danish trials. Therefore, at trial, in assessing the probative value of the US accomplice evidence it was necessary to bear in mind that the evidence may not be truthful because it may be motivated by the promises of advantages flowing from the confession. In U.2012.1604 V, the Western High Court also approved of the accomplice evidence, even though it was the decisive evidence in the case, finding that it was supported by the surrounding circumstances of the case.

In July 2013, the Eastern High Court found that a person accused of weapon smuggling had not been unlawfully entrapped when an American DEA agent asked a third person for payment for drugs to be in weapons. Interestingly, in the Copenhagen City Court, the judge had found that the circumstances constituted unacceptable entrapment but allowed the evidence to be relied on for conviction anyway. Instead of acquitting, the City Court judge gave a reduction in sentence. This sentence was increased on appeal by the Eastern High Court.

7.3.2.2 The Case Law of England and Wales

7.3.2.2.1 Cooperation with ICCPR Signatories

Cases such as *R v Khyam*[54] make it clear that the courts in England and Wales will also rely on evidence obtained abroad (outside the framework of the ECHR) under procedures not recognised by English law, such as plea bargains. In *Khyam*, the most important prosecution witness had been arrested and questioned in the US. He had cooperated with the prosecution in the US with a view to a plea bargain. The trial judge had warned the jury that they should treat the witness's evidence with caution because, as the witness himself had admitted, he had cooperated with the US authorities with a view to personal gain in the form of a plea bargain.[55]

---

[53]Discussed in Chap. 2, Sect. 2.5.3.3.

[54]*R v Khyam and others* [2008] EWCA Crim 1612.

[55]See also *R v Governor of Belmarsh Prison Ex p. Francis* [1995] 3 All E.R. 634 where the Court of Appeal was asked to consider whether accomplice evidence obtained as a result of a plea bargain in the US ought to have been excluded in extradition proceedings, but found that s. 78 applied to criminal proceedings and not extradition proceedings; and see *Fernandes (Charles) v Governor of Brixton Prison* [2004] EWHC 2207 (Admin) where the court examined the admissibility of confession evidence obtained from staff members on board a ship in absence of caution or access to a lawyer in extradition proceedings and found that ss. 76 and 78 of PACE do not apply

The fact that there are more reported English cases where the admissibility of evidence obtained abroad in a 'neighbouring' Council of Europe/EU country is in issue may indicate that English police and prosecutors are less likely to cooperate with a country outside the Council of Europe/EU. Alternatively, English police and prosecutors may be more conservative in the way that MLA requests are issued and assistance is accepted, when dealing with countries out the Council or Europe/EU. Further, the discrepancy in the number of cases may indicate that English prosecutors are less likely to use and rely on evidence obtained outside the Council of Europe/EU in domestic trials, although this conclusion is inconclusive. Or the explanation could simply be that criminality involving England is at a higher level with other EU countries.

7.3.2.2.2 Cooperation with Non-ICCPR Signatories

In *Rangzieb Ahmed, Habib Ahmed v. The Queen*[56] a listening device that had been installed in a hotel room by the police in Dubai and the evidence of conversations involving the two accused was transferred to the police in Manchester and was a major plank of the prosecution case at the terrorism trial in England. It appears that the admissibility of the listening device evidence was undisputed.

7.3.2.3 Australian Case Law

7.3.2.3.1 Cooperation with ICCPR Signatories

Australian case law has dealt with the reception of confession evidence obtained extraterritorially—interstate and internationally. In general, the courts have treated the issues in the same way with a distinction being drawn between extraterritorial admissions received by an interstate or foreign investigative officer compared with the situations where admissions are received by a local police officer who is acting abroad. It appears that the Australian courts take a more flexible approach when interstate or foreign officers are the investigating officials responsible for interrogation.

In the Victorian Supreme Court trial of *R v Grainger* on 11 June 1991, Cummins J made a ruling to exclude the evidence of oral admissions allegedly made by the accused to two New South Wales police officers as they were walking from the police station cells to the underground cells of the Dubbo courthouse for the purpose of an interstate extradition application. The admissions were not subsequently recorded on tape, despite the presence at the Dubbo police station (for the

on extradition proceedings. The admissibility of confession evidence was held to be a matter for the foreign trial court to decide.

[56] *Rangzieb Ahmed, Habib Ahmed v The Queen* [2011] EWCA Crim 184, [8–9].

extradition hearing) of a Victorian police officer. The trial judge considered that the Victorian police officer was aware of the requirement under the Victorian legislation of the need to have the admissions confirmed on tape, as a significant protection of a suspect from unlawful and unfair treatment, but chose not to comply with this rule. Therefore, the confession evidence was excluded from the trial in the exercise of the fairness discretion.[57]

This ruling was subsequently followed in the case of *R v Weston* in the County Court of Victoria, where the trial judge made a ruling dated 16 March 1992 about the admissibility of evidence of a police interview conducted by officers of the Bentley Criminal Investigation Branch (CIB) in Western Australia.[58] The accused had been arrested on charges of theft and burglary and had escaped from police custody in Victoria. In doing so the accused attempted to run down a police officer with a motor vehicle and drove off. The accused was lawfully arrested and questioned by two Bentley police officers, after which they conducted a formal interview with the accused in accordance with the procedure of Western Australia. A typed record of interview was concurrently produced and at the end of the interview a Sergeant of police became involved in reading the record of interview with the accused and the confirmation by him of the interview and the lead-up questioning. The record of interview and the notes of the questioning were acknowledged (and altered in some places) and signed by the accused. Having been satisfied that the police witness testimony was truthful and that the accused's statements in the interview were voluntarily made, the trial judge considered the applicability of the Victorian legislation to the Western Australian interview. In particular, there had been no tape-recording facilities at the Bentley police station. The interview was not tape-recorded and the admissions were not confirmed on tape.[59] The legislation which controls the admissibility of evidence in Victoria was presumed not to have extraterritorial effect to bind the members of the Western Australian police force and no clear contrary intention was demonstrated. In exercising the fairness discretion, the trial judge found that the Western Australian police had conducted the interrogation lawfully and properly and that the accused had participated responsively. Although there was no tape-recording of the interview there was cogent documentary evidence of it. In reference to the probative value of the interview evidence, the trial judge found that it was not unfair to rely on it.

During the murder trial in 1995 of *R v. George Koureas* Coldrey J of the Supreme Court of Victoria was called to rule upon the admissibility of evidence of an interview between the accused and the Montreal Urban Community Police Department, which was conducted on behalf of and at the request of the Homicide

[57] At the time of the ruling the fairness discretion was a common law rule.

[58] The ruling of Cummins J was also followed in a Victorian County Court case of *R v Toubya* in December 1991.

[59] See the discussion of the Victorian rules about tape-recording in Chap. 5, Sect. 5.4.6.

Squad of the Victoria Police.[60] By facsimile and telephone the Victorian officers instructed the Montreal[61] police about using the Victorian interview 'preamble', caution and formulation of the accused's rights. The Montreal police who were unfamiliar with tape-recording of interviews were nevertheless asked to record the interview, or at least the reading of the accused's rights. Prior to the commencement of the interview at the Canadian Homicide Squad office the accused had declined the services of a lawyer. The interview was conducted in English without the services of a Greek interpreter. What resulted was a hand written record of interview, which was read back on tape by the police and the accused. However, the accused did not read back the entire contents of the written document and appeared in places to be re-answering the questions rather than reading the responses word for word. He affirmed what he said to the police in the handwritten document was the truth and agreed to sign the record. At trial it was argued on behalf of the accused that the evidence of the interview should be excluded for unfairness, generally and due to failure to comply with the provisions of the Crimes Act 1958 (Vic) in relation to recording of interviews. Coldrey J referred to the decision in *R v Weston* above and found that while the purpose and intent of protective legislative provisions were relevant to the issues of fairness and admissibility of confession evidence, non-compliance with the provisions in an extraterritorial setting will not necessarily result in exclusion of evidence. The trial judge found that the Montreal police had acted bona fide and did not deliberately breach the Victorian provisions but had misunderstood the requirements on the background of their own procedures. On the general question of fairness the trial judge did not consider that there had been unfairness on account of language difficulties and found that the accused had been informed of his rights, the questioning had been conducted appropriately and that the accused had answered responsively. However, due to the discrepancy between the hand-written document and the tape-recording, the document was excluded from evidence.

In the recent case of *The Queen v Stephen Albert Hutton*,[62] the accused was extradited from England to be prosecuted in Victoria for murder. The accused had made a series of admissions, to mental health staff (at a psychiatric hospital) and under caution at an English police station, which were decisive to the prosecution case, though supported by circumstantial evidence.[63] The accused pleaded guilty and it was acknowledged as a factor in mitigation by the sentencing court that by pleading guilty the accused had "...foregone the opportunity to make legal challenges to the admissibility of [his] confessional statements..."[64] Such legal challenges may have concerned the circumstances in which the accused's confession

---

[60] See the ruling of Coldrey J of the Victorian Supreme Court dated 1 May 1995.

[61] Canada acceded to the ICCPR on 19 May 1976.

[62] *The Queen v Stephen Albert Hutton* [2011] VSC 484. See also the English extradition hearing: *Hutton v Australia* [2009] EWHC 564 (Admin).

[63] *The Queen v Stephen Albert Hutton* [2011] VSC 484, [26].

[64] *The Queen v Stephen Albert Hutton* [2011] VSC 484, [27].

was received by the authorities in England and Australia, or were obtained by the English police in accordance with English procedures.

The approach of Australian courts to the confessional evidence of an accomplice obtained as part of an advantageous plea bargain in the US has been demonstrated by a ruling the case *DPP (Cth) v Phillip Andrew Bayly*.[65] The accused was on trial on one charge of importation of cocaine from the US to Australia and the main prosecution witness was an accomplice who gave a statement to the police about a number of instances of importation. His statement and testimony were given as part of a plea bargain in the US whereby the witness was also granted immunity in Australia, his de facto partner was spared from prosecution and he was placed on a protected witness programme. Despite the benefits of confessing and the pressure that had been placed on him in the US to 'come clean' in order to avoid lengthy gaol terms in the US and Australia, the accomplice was nevertheless considered to be a reliable witness. As well as the accused having had the opportunity to challenge and test the evidence under cross-examination, the evidence was corroborated by other evidence, such as taped telephone conversations between the accomplice and the accused, the accused's financial dealings, the accomplice's knowledge of the accused and his use of alias names and evidence of a courier to an earlier transaction. Relying on authorities, the trial judge found that there was no separate discretion for the exclusion of evidence on the ground that the accomplice was granted immunity or other benefits, which rendered his evidence unreliable. If such a discretion were to exist it would usurp the function of the jury. Instead it was found that the discretion to exclude evidence, where the prejudicial effect outweighs the probative value, applied most exceptionally and the appropriate course in such a case is to admit the evidence with appropriate warning and jury direction from the trial judge about the effect of the circumstances of the confession on the reliability of the evidence.[66]

7.3.2.3.2 Cooperation with Non-ICCPR Signatories

In the 1995 decision of the Supreme Court of Victoria of *R v Mario Perfili*,[67] the applicant appealed against his conviction for conspiracy to import prohibited wildlife into Australia. The applicant, who was in residence at a local school, and another person had been questioned by a police officer at a village airstrip in the Solomon Islands beside an aircraft, which was loaded with 191 wild parrots in 25 wire cages. After claiming to have nothing to do with the birds he then admitted that he had helped put them in the cages. The applicant and the pilot were taken to a town approximately a 2 hour drive away where they were arrested and interviewed by the police. The conversation near the aircraft was admitted in evidence against the applicant. However, the evidence of the interview was excluded and the

---

[65] *DPP v Phillip Andrew Bayly* (No. 3) [1996] SASC 5807, 13 September 1996.

[66] *DPP v Phillip Andrew Bayly* (No. 3) [1996] SASC 5807, 13 September 1996, [22–36].

[67] *R v Mario Perfili* [1995] VSC 222.

applicant remained silent at his trial.[68] The applicant appealed his conviction *inter alia* on the ground that he should have been cautioned prior to being questioned at the airstrip and the confession evidence should therefore have been excluded. Although it had been accepted that the applicant ought to have been cautioned as a matter of principle and in accordance with local practice rules of conduct, the court found that in the absence of any other investigative misconduct, the statements made by the applicant were fairly admitted. It is possible that this decision would be decided differently today. Cautioning of a person who is not under arrest or formally in custody, but who is believed to have committed an offence was introduced by the federal parliament in 2001.[69] The Australian standards would be relevant to the court's assessment of what would be fair in the circumstances. Express legislation requiring cautioning sets down what the federal parliament now considers is a fundamental principle of law. Nevertheless, the Commonwealth provisions would not be applicable to the questioning of suspects by a foreign officer so what the outcome would have been today is difficult to predict.

More recently, the Australian case from 2006 of 'Jihad Jack' has highlighted a number of problems facing law enforcement agencies in transnational cases. Firstly, the *Thomas*[70] case dealt with admissibility problems resulting from confessional statements made in the coercive environment of terrorism interrogation camps in Pakistan.[71] Secondly, and more relevantly, the case considered the impact of structural, or procedural, differences between legal systems and concluded that these differences can be prohibitive to the cross-border transfer of confessional evidence.

'Jack' Thomas was apprehended by Pakistani immigration officials at a Karachi airport on 4 January 2003 in possession of an Australian passport, which he had used when he left Australia in 2001. It had been altered to conceal his movements between these dates. He also had with him an airline ticket for travel to Australia,[72] via Indonesia, and a little over $3800 in cash.[73] Ultimately, on 26 February 2006 he was found guilty by a jury in the Supreme Court of Victoria of receiving funds from a terrorist organisation and possessing a falsified Australian passport.[74]

---

[68]Given that the evidence of the conversation at the aircraft was admitted despite there being a lack of warning, it is likely that the evidence of the interview was excluded due to reasons such as the atmosphere, the method of questioning used or a lack of recording. However, the explanation is not obvious from the written judgment.

[69]See act 136/2001.

[70]*The Queen v Joseph Terrence Thomas* [2006] VSCA 165.

[71]Pakistan ratified the ICCPR on 23 June 2010. It was, therefore, not a signatory at the relevant time.

[72]This is contrary to s 102.6(1) of the *Criminal Code Act* 1995.

[73]He was detained and remained in Pakistani custody until his release and return to Australia approximately 5 months later (6 June 2003). He was then at large for 17 months until his arrest at his Melbourne home on 18 November 2004.

[74]Contrary to s 9A(1)(e) of the *Passports Act* 1938. He was acquitted of intentionally providing resources to a terrorist organisation (Paragraph 102.7(1) of the *Criminal Code Act* 1995).

At trial, the crucial evidence against Thomas was the admissions made by him in the course of an interview with two Australian Federal Police (AFP) officers in Pakistan on 8 March 2003 ('the AFP interview'). It was not alleged that the AFP officers had conducted the AFP interview in an overbearing or offensive manner.[75] In the interview the Applicant admitted that he had altered his passport in order to conceal the amount of time he spent in Pakistan and his movements. In particular, he was minded to conceal the fact that he had been at the Al Faroq camp in Afghanistan at which al Qaeda training was conducted. He also stated that his ticket and money had been provided to him by a man named Khaled Bin Attash, who was an associate of Osama Bin Laden and a high ranking al Qaeda operative.

The truthfulness, or reliability, of these admissions was not contended at the trial. However, it was argued for the Applicant that the self-incriminating statements were made involuntarily and should not, therefore, have been adduced in the proper exercise of the trial judge's discretion. Alternatively, it was argued that the statements should have been excluded on the basis that their admission was unfair or would be contrary to public policy.

On appeal, the main crux of the case turned on the issue of the voluntariness of the applicant's admissions—whether or not the applicant had, in any real sense, a free choice to speak. The Applicant gave evidence at trial alleging the poor conditions in which he had been held in Pakistan and the coerciveness of the environment in the intelligence gathering process. The Court of Appeal found that the Applicant had been repeatedly told by his Pakistani interrogators, in whose custody he was at all times, that his fate very much depended on the extent of his cooperation, a sentiment that was impliedly endorsed by members of the AFP/ASIO joint team, one of whom was also in the AFP interview.[76] These were clearly 'persons in authority' for the purposes of inducing a confession.[77] In the circumstances, the prospect of being released by the Pakistani authorities and returning home to his family would have operated as a very powerful pressure upon the accused to speak at the time of the AFP interview on 8 March 2003.[78] Accordingly, the conviction of 'Jihad Jack' was quashed.

The second argument on appeal, made in the alternative, is highly relevant to a discussion about the transfer of confessional evidence between two systems that are not within the same human rights framework. Had the Court found that the applicant had confessed voluntarily, the remaining questions would have been whether interviewing the applicant in the absence of access to a lawyer was unfair because, if he had been given access to a lawyer, he may not have made the confession in the manner that he did or at all; or whether, in the circumstances of the case, the reception of the evidence should have been regarded as contrary to public policy.

---

75 *The Queen v Joseph Terrence Thomas* [2006] VSCA 165, [51].

76 *The Queen v Joseph Terrence Thomas* [2006] VSCA 165, [71–75].

77 *The Queen v Joseph Terrence Thomas* [2006] VSCA 165, [72].

78 *The Queen v Joseph Terrence Thomas* [2006] VSCA 165, [72, 80, 83, 84].

Of significance to the second argument, was the failure to comply with s 23G of the *Crimes Act* 1914,[79] which allows a suspect to be offered the opportunity to access a legal practitioner of his own choosing and requires an interview to be deferred for a reasonable time in order to allow a suspect the opportunity to make contact.[80] The applicant was in Pakistani custody. The Pakistanis granted the AFP access to the Applicant to conduct a formal interview. However, in accordance with Pakistani law, they refused permission for the Applicant to have access to a legal adviser.[81]

Although there was no suggestion that the AFP officers had not acted bona fide,[82] the Court of Appeal found that it would be contrary to public policy to condone a knowing non-compliance with the legal protection afforded by Australian law.[83] The court recalled that section 3A of the *Crimes Act* 1914 expressly provides for the extraterritoriality of the Act. Accordingly, the protections provided for by the Act apply to Commonwealth investigations conducted anywhere in the world.

Section 23G lays down rules that the legislature declares must be followed in order for a confession to be regarded as fairly obtained. A breach of the rule makes it prima facie unfair to use any confessional statement against an accused at trial.[84] In addition, considerations of high public policy, which favour exclusion of evidence obtained by unlawful investigative methods, will transcend any question of unfairness to an accused. It is the duty of the courts not to encourage unlawful police conduct by the appearance of acquiescence—this will require the exclusion of evidence in some circumstances to ensure the integrity of the judicial process.[85] In obiter, finding that the trial judge's discretion had miscarried on the grounds of unfairness and public policy, the Court of Appeal firmly declared that there had been only one course properly open to the AFP investigators in light of the position taken by the Pakistani authorities—to acknowledge that no formal record of interview could have been conducted so long as the applicant remained in Pakistan since any such interview would be unlawful.[86]

The decision of the Supreme Court of Victoria, Court of Appeal, in *Perfili*, highlights the more flexible approach of the Australian courts to the interrogation of

---

[79]This is an Australian federal act of parliament.

[80]This rule applies on the case of a "protected suspect" who is, for example, under arrest, is strongly suspected of wrongdoing or will not be allowed to leave during the course of questioning if he wished to do so.

[81]S 23 L creates certain exceptions where deferment is not required. It was conceded before the Court of Appeal that none of these exceptions applied [106].

[82]*The Queen v Joseph Terrence Thomas* [2006] VSCA 165, [40] and [110].

[83]*The Queen v Joseph Terrence Thomas* [2006] VSCA 165, [109].

[84]*The Queen v Joseph Terrence Thomas* [2006] VSCA 165, [113], referring to McHugh J in Pollard v R (1992) 176 CLR 177.

[85]*The Queen v Joseph Terrence Thomas* [2006] VSCA 165, [114], referring to Deane J in Pollard v R (1992) 176 CLR 177.

[86]*The Queen v Joseph Terrence Thomas* [2006] VSCA 165, [111].

suspects carried out by foreign officers. This is to be contrasted with the decision in *Thomas* and the applicability of Australian legislative standards to investigations carried out by Australian officers anywhere in the world.

7.3.2.4 Similar Approaches

Although the sample is small, the case-law outlined above in relation to admissibility of evidence obtained abroad, where issues have been raised about the fairness of the proceedings, indicate that the courts in Denmark, England (and the UK) and Australia, on the whole, take a similar approach. In the cases studied, in general, the courts in all three jurisdictions appear to take a flexible approach to admitting irregularly obtained evidence gathered by foreign investigative forces under procedures that vary from the national rules. In other words, although continuity in each national system between investigation and trial has been demonstrated, perhaps driven by the reality of globalised crime, each state demonstrated flexibility in the approach to evidence obtained by foreign forces.

This is provided that there is no evidence of other circumstances that affect the fairness of the proceedings such as coercion or deliberate abuse of process. For example, courts in all three jurisdictions have similarly allowed the prosecution to rely on confession evidence of an accomplice gathered under US plea bargain arrangements, despite the fact that this type of plea arrangement, where the benefits of confessing are precisely identifiable in advance, is not known in the three jurisdictions. However, the conclusion about judicial flexibility on the basis of the cases studied does not lead to the conclusion that a judge in England, Denmark or Australia will always admit confession evidence obtained 'lawfully' in a foreign country under local rules. For example, the Victorian Court of Appeal may not have admitted the evidence of a custodial interrogation of Thomas by the local police in Pakistan in the absence of access to a lawyer, even if there had been no clear evidence relating to the coercive atmosphere of the interrogation camp.

In England, in addition to the flexibility suggested by the cases of evidence gathered outside the EU, cases such as *Aujla* and *Konscol*, both decided in the 1990s, indicates a greater readiness on the part of the English judiciary to accept evidence from EU Member States, where the principles of the ECHR are equally applicable, even if the procedures used are contrary to English law. The fact that evidence had been obtained in another state coming under the ECHR may have encouraged judges to admit evidence on the question of guilt without considering whether the national balance between effectiveness and fairness in either state had been unfairly disturbed. However, the decision in *Konscol* might also have been the result of a lack of definition in relation to the art. 6 rights in question. The English Court of Appeal would probably decide *Konscol* differently today, following the apparent acceptance by the UK Supreme Court in *Cadder* of the application in the UK of art. 6 minimum standards of pre-trial due process laid down by the ECtHR in judgments such as *Salduz*. Today, national courts may be paying more attention to the substance of defence rights under the ECHR as determined by the ECtHR.

From the small number of Danish cases, it appears that Denmark takes a flexible approach to the admissibility of evidence gathered abroad, regardless of whether the evidence was gathered within a Convention state or not. It is likely that any concerns that surrounded the gathering of the evidence would have been dealt with by adjusting the weight to be attached. In light of the consistency of the approach in Danish case law to evidence obtained inside or outside the EU (though inconclusive), it is possible that the strongest influence on the Danish courts is the traditional tendency to favour admission. It is possible that issues about irregularly obtained evidence will not always be raised as admissibility problems but will simply be left to the court to determine weight when considering the accused's guilt.

This may explain why, in Denmark, there are fewer reported decisions about the fairness of using evidence obtained abroad under rules that are considered irregular in Denmark. However, other relevant factors may include the fact that Denmark is a smaller country which, due to its geographic location, is not a major hub of organised crime and draws less on international cooperation generally. Another explanation may be that there is not the same tradition in Denmark for reporting of cases.

In Australia, there appears to be a difference in the approach to foreign evidence that is obtained by Australian or foreign police. Australian courts have tended to be stricter in cases that have involved local investigators who are deemed to have knowledge of local rules and should, therefore, apply them. In the cases involving the gathering of evidence by interstate and oversees police investigators, the Australian courts have been generally flexible in the approach and have emphasised other factors, for example, that give the objective indications of the cogency of the evidence. In the case of *Bayley,* in relation to evidence obtained in the US under a plea bargain, the trial judge accepted the evidence and gave a jury warning about the pitfalls of relying on it. Similar to *Konscol*, it is arguable that the case of *Perfili* would have been decided differently today, where the rules about cautioning are now laid down in legislation.

The approach of the courts in the three jurisdictions appears to be similarly flexible. This is despite the differences to be found in the application of fair trial rights to investigative proceedings under art. 14 of the ICCPR and art. 6 of the ECHR, the fact that Australia does not have constitutional protection of human rights or a specific human rights enactment, and the fact that Australia does not belong to the same human rights framework as Denmark or England.[87] From a consideration of the case-law above, there is no conclusive evidence to suggest that courts within the same regional human rights framework of the ECHR are more flexible in their approach to admitting evidence obtained in other convention states. The courts of all three jurisdictions have been prepared to admit irregularly obtained evidence where there is no suggestion that suspects or accused persons

[87]In relation to criminal law cases where the High Court of Australia has referred to the decisions of the ECtHR, see, for example, *PGA v The Queen* [2012] HCA 21, [198–191]; *R v Tang* [2008] HCA 39, [30]; *Azzopardi v The Queen* [2001] HCA 25, [161]; *R v Hughes* [2000] HCA 22, [95]; *Deitrich v The Queen* [1992] HCA 57; *Antoun v The Queen* [2006] HCA 2, [37–40].

have been bullied, denied the opportunity to test the evidence or otherwise treated in such a way as to establish an abuse of process.

7.4 Mutual Trust in International Cooperation

7.4.1 Mutual Trust Between Countries That Are Not Within the Same Regional Human Rights Framework

In general, international treaties that include international cooperation provisions tend to favour cooperation, seeking the most extensive cooperation possible, and do not refer to human rights as a foundation.[88] Therefore, these treaties typically limit the grounds for refusal to provide mutual legal assistance, which do not usually refer to human rights. For example, art. 7(15) of the UN Convention Against Illicit Traffic in Narcotic Drugs and Psychotropic Substances (1988) includes a short list a discretionary grounds for refusal that do not include a general human rights ground.[89]

As an exception, however, treaties such as the UN Convention Against Torture (CAT),[90] include an obligation for the parties to promote respect for human rights in international cooperation. Indeed, human rights are central to such treaties. For example, the non-refoulement principle, usually associated with refugee law, is expressly applied in art. 3 of CAT to prohibit the extradition of a person to a State where there are substantial grounds to believe he or she is at risk of suffering torture.[91]

In addition, an important principle in extradition was established by the ECtHR in the 1989 case of *Soering v The United Kingdom*.[92] In condemning extradition of a requested person to a state that uses the death penalty, and where the requested person will spend time on death row, the Court held:

> . . .the decision by a contracting state to extradite a fugitive may give rise to an issue under Article 3, and hence engage the responsibility of that state under the Convention, where substantial grounds have been shown for believing that the person concerned, if extradited, faces a real risk of being subjected to torture or to inhuman or degrading treatment or punishment in the requesting country.[93]

[88]See, for example, the UN Security Council Resolution 1373 (2001), S/RES/1373 (28 September 2001).

[89]See also art. 46(21) of the UN Convention Against Corruption, opened for signature 9 December 2003 and entered into force 14 December 2005, UNTS v. 2349/41.

[90]Preamble, UN Convention Against Torture and Other Cruel, Inhuman or Degrading Treatment or Punishment (1984), opened for signature 10 December 1984 and entered into force 26 June 1987.

[91]See further art. 3(f) of the UN Model Treaty on Extradition (1990).

[92][ECtHR] *Soering v The United Kingdom*, Application no. 14038/88, 7 July 1989.

[93][ECtHR] *Soering v The United Kingdom*, Application no. 14038/88, 7 July 1989, [91].

This conclusion has been confirmed by the Human Rights Committee (HR Committee) in the international framework in relation to extradition in a case of the penalty of death by asphyxiation by cyanide gas.[94]

Nevertheless, occasional decisions by the HR Committee which emphasize the potential significance of human rights in international cooperation cannot lay the foundation for the operation of a fully developed principle of mutual trust between signatories of the ICCPR. Such a presumption in decision making requires objective and subjective knowledge and application of the same standards. As referred to in Chap. 2, international standards of minimum procedural rights under the ICCPR are not well-developed. It is possible that common law countries may exhibit mutual confidence in the level of procedural protection afforded to suspects and accused person based on common law principles. Today, however, there are certain significant differences in the procedure of common law countries (as discussed in the previous chapter) that would undermine the existence of mutual trust as a legal principle in cooperation between these states. The Harare Scheme,[95] which may be used as the treaty providing a basis for cooperation between common law countries, does not refer to standards of human rights protection or the need to protect fundamental rights in the cooperation process. Therefore, on the basis of the above, it is unlikely that mutual trust is a principle which operates in the area of criminal justice at the international level.[96]

In fact, in relation to cooperation within the international framework, human rights point in the opposite direction. In principle and in practice, human rights considerations may always prevent or limit cooperation. At the international level, countries are at liberty with respect to the preparedness to cooperate and the level flexibility which is deemed appropriate. Countries, such as Australia, may avoid seeking to cooperate with other countries that are signatories to the ICCPR, where there is uncertainty about the level of human rights protection in the other country. In addition, the main treaties that may be used as a basis for international cooperation generally allow a requested state to refuse a request which is contrary to the law or fundamental principles in that legal system.[97]

[94]See *Ng v Canada*, decision of the Human Rights Committee, 5 November 1993.

[95]Scheme Relating to Mutual Assistance in Criminal Matters Within The Commonwealth (1986) including amendments made by Law Ministers in April 1990, November 2002 and October 2005.

[96]See further Cryer et al. (2014), p. 78, in which it is suggested that even using notions based on *jus cogens*, such as the prohibition on torture, to justify an international standard in international cooperation is controversial.

[97]See, for example, art. 7(15)(d) of the UN Convention Against Illicit Traffic in Narcotic Drugs and Psychotropic Substances (1988).

7.4.2 *Mutual Trust Between EU Member States*

7.4.2.1 Mutual Trust in the Process of MLA

Being part of the regional human rights framework provided by the ECHR provides a stronger basis for mutual trust between EU Member States and within the wider framework of the Council of Europe. General adherence to common, well-developed minimum standards for human rights protection has facilitated a higher level of confidence in each other's systems and extensive international cooperation. As discussed in the conclusion to Chap. 2, the scope of minimum human rights standards under the ECHR creates more certainty than under the ICCPR. In addition, the system of individual complaints to the ECtHR, which produces reasoned judgments, provides a more powerful and comprehensive enforcement mechanism. Further, as referred to below, the more recent EU international cooperation criminal justice instruments place an emphasis on Member States' readiness to exercise mutual trust. Therefore, as a principle of international cooperation, the principle mutual trust is relevant in the setting of the EU, where it has no foundation in transactions that take place outside the EU.

Reinforcing fundamental human rights principles, considered to be general principles of EU law,[98] may be a necessary part of ensuring the fulfillment of Member States' obligations in international cooperation under EU law. Therefore, ensuring that minimum standards of human rights protection are met will enable Member States to assist each other in the international cooperation process. This is the underlying message of the principle of sincere cooperation, also known as the enforcement obligation, which is also relevant in criminal proceedings.[99]

7.4.2.2 Mutual Trust on a Deeper Level Through Mutual Recognition of Judicial Decisions

'Determined' to create an area of 'freedom, security and justice' within the EU and activating the conclusions of the Tampere European Council of 15 and 16 October 1999, the EU Council created two secondary EU criminal justice instruments under the former third pillar. The instruments centred upon the principle of mutual recognition of judicial decisions in order to facilitate more efficient cooperation between Member States. The principle of mutual recognition originated as a former first pillar principle used to enhance the free movement of goods and 'unblock' the single market.[100] The principle was introduced to judicial cooperation in criminal matters at the intergovernmental level under the third pillar prior to the entry into

[98]See art. 6 of the Treaty on European Union 1992.

[99]Klip (2009), pp. 16 and 428; see further art. 4(3) of the Treaty on European Union 1992.

[100]Ouwerkerk (2011), p. 54; see further Daniel et al. (2011), p. 42, fn 54.

force of the Lisbon Treaty.[101] The primary objective of mutual recognition is to build upon a high level of confidence between Member States in order to promote commitment within the EU to becoming an area of freedom that is enjoyed in conditions of security and justice and which is based on human rights, democracy and the rule of law.[102]

Extradition under the 1957 European Convention on Extradition was an important step in law enforcement within the Council of Europe states. However, a major complaint was the length of time and complexity involved in achieving surrender. Adopted in 2002,[103] the EU Council Framework Decision 2002/582[104] on the European arrest warrant (EAW) was the first criminal justice instrument that was based on the principle of mutual recognition as a cornerstone of judicial cooperation.[105] It provides for a more simplified, fast track system within the EU for the surrender of convicted persons to serve their sentence; and of persons suspected with criminal offences to stand trial.[106] A subsequent and relevant mutual recognition instrument was the Council Framework Decision 2008/978/JHA[107] on the European evidence warrant (EEW). This EU instrument was adopted to allow objects, documents and data to be gathered in one Member State and transferred quickly to another Member State under the principle of mutual recognition. Transfer of evidence is intended to occur with an emphasis on admissibility.[108] However, the types of evidence covered by the EEW are limited and it does not envisage use of coercive measures.[109] In particular, the evidence must already be in existence prior to the EEW being issued.[110] Therefore, many states may disregard the EEW and continue to cooperate in the gathering of evidence under the principles of MLA, which allows Member States to seek evidence in a manner that is tailored to a case, while seeking other forms of assistance under the one request. The downfalls of the EEW are currently being addressed by an initiative to enact a directive in relation to a new evidence gathering instrument based on mutual recognition, the European

---

[101]Billing (2012), p. 86.

[102]Tampere European Council Presidency Conclusions, 15 and 16 December 1999; see further (CJEU) Case C-396/11 *Radu* [2013] 29 January 2013, [34].

[103]See further Title VI of the Treaty on European Union (TEU) and art. 82 of the Treaty on the Functioning of the European Union.

[104]Council Framework Decision 2002/584/JHA of 13 June 2002 on the European arrest warrant and the surrender procedures between Member States, OJ 2002 L 190, 1, as amended by Council Framework Decision 2009/299/JHA of 26 February 2009, OJ 2009 L 81, 24 (EAW Framework Decision).

[105]See art. 1(2) and (3) of the EAW Framework Decision; see also Ouwerkerk (2011), p. 112.

[106]See recital 5 of the EAW Framework Decision.

[107]Council Framework Decision 2008/978/JHA of 18 December 2008 on the European evidence warrant for the purposes of obtaining objects, documents and data for use in proceedings in criminal matters, OJ L 350/72, 30 December 2008.

[108]See recital 14 of the EEW Framework Decision.

[109]See recital 14 of the EEW Framework Decision.

[110]Article 4(2) and (4) of the EEW Framework Decision.

investigation order (EIO).[111] Currently, the nature of the interrelationship between human rights and the principles of mutual trust and mutual recognition in criminal matters in the EU is only visible through the use of the EAW.

Under art. 1(2) of the EAW Framework Decision, the Member States are in principle obliged to act upon an EAW in accordance with the provisions of the Framework Decision.[112] Articles 3, 4 and 4a[113] are said to be exhaustive as to the grounds for mandatory and optional non-execution.[114] The Court of Justice in *Szymon Kozlowski* found that the executing court can only refuse to surrender on one of the mandatory or optional grounds.[115] Human rights arguments are not expressly listed as providing grounds for non-execution.

However, recital 10 of the Framework Decision emphasizes that the "…mechanism of the European arrest warrant is based on a high level of confidence between Member States." Of significance, under the heading of the definition of the EAW and the obligation to execute it, art. 1(3) provides:

> This Framework Decision shall not have the effect of modifying the obligation to respect fundamental rights and fundamental legal principles as enshrined in Article 6 of the Treaty on European Union.

Incidental to this, specific reference is made in recital 12 to the Charter of Fundamental Rights of the EU and to the principles recognized by art. 6 of the TEU, establishing the Charter as EU primary law and referring to the rights contained in the ECHR as general principles of Union law.[116] Thus, the executing judicial authority must simultaneously execute an EAW directly in accordance with 'mutual recognition', on the one hand, and in accordance with respect for fundamental rights, on the other.[117] These two obligations may, at times, be mutually exclusive. Can an EAW be refused by the executing state on the grounds of human rights or do the requirements of mutual trust and mutual recognition preclude this? Suominen has argued that it is not possible to apply mutual recognition while disregarding human rights.[118]

---

[111] See the Initiative for a Directive of the European Parliament and of the Council of the European Union regarding the European Investigation Order in criminal matters, 2010/C 165/02, OJEU C 165/22, 24 June 2010.

[112] (CJEU) Case C-396/11 *Radu* [2013] 29 January 2013, [35].

[113] Article 4a was inserted by Council Framework Decision 2009/299/JHA of 27 March 2009, OJ 2009 L 81.

[114] (CJEU) Case C-396/11 *Radu* [2013] 29 January 2013, [36].

[115] (ECJ) Case C-66/08 *Szymon Kozlowski* [2008] 17 July 2008; see further Case C-388/08 PPU *Leymann and Pustovarov* [2008] ECR I-8993, [51]; Case C-261/09 *Mantello* [2010] ECR I-11477, [37]; Case C-123/08 *Wolzenburg* [2009] ECR I-9621, [57].

[116] See further (CJEU) Case C-396/11 *Radu*, Opinion of Advocate General Sharpston, 18 October 2012 [52].

[117] This is also the foundation for using an EEW: see, for example, recitals 8, 11–13, 15, 27 and 28 and art. 1(2) and (3) of the EEW Framework Decision 2008/978/JHA.

[118] Suominen (2011), p. 225.

Judgments such as the High Court of Justice (England and Wales) decision in the 2009 case of *Andrew Symeou* suggest that EU national courts *must* take the view that the judicial systems of the countries of the EU provide minimum safeguards to guarantee a fair trial.[119] Symeou, a British citizen, was surrendered to Greece despite the English court's concerns that the issuing court had not adequately investigated allegations of violence and intimidation of witnesses, and manipulation of evidence by the police, and that the domestic arrest warrant was unlawful.

However, it is arguable that grounds for non-execution of an EAW based on fundamental rights stems from the primary law, taking precedence over the framework decision.[120] A series of non-binding opinions of Advocates General have been reported supporting the view that an executing judicial authority may refuse to execute an EAW on human rights grounds.[121] However, the question has not been clearly resolved and is particularly controversial at the national implementation level.[122] The EAW Framework Decision has indirect effect and, therefore, the purpose has been to encourage a 'conforming interpretation' by national courts of national implementing laws.[123] In its reports of 2007 and 2011, the European Commission found that the implementation of art. 1(3) and recitals 12 and 13, and Members States' obligation to respect human rights, is 'mixed'. In particular, a number of Member States have been criticized for including human rights as a mandatory ground for non-execution.[124] For example, Denmark had been criticized for requiring non-execution on the ground that there is a risk of torture, degrading treatment, and violation of due process, or if the surrender appears to be unreasonable on humanitarian grounds.[125] In the Commission's view, provisions

[119] *Andrew Symeou v Public Prosecutor's Office at the Court of Appeals, Patras, Greece* [2009] EWHC 897 (Admin), 1 May 2009, [39].

[120] Peers (2012), pp. 709–710.

[121] See Peers (2012), p. 709 f. 16; (CJEU) Case C-396/11 *Radu*, Opinion of Advocate General Sharpston, 18 October 2012, [8, 70]; and Billing (2012), pp. 85–91; see also Thunberg Schunke (2013), pp. 58–64 and 107.

[122] Commission of the European Communities, Commission Staff Working Document, *Annex to the report from the Commission on the Implementation since 2005 of the Council Framework Decision of 13 June 2002 on the European arrest warrant and the surrender procedures between Member States* [COM (2007) 407], SEC (2007) 979, 11 July 2007, 5.

[123] (ECJ) case C-103/03 *Pupino* [2005] ECR 2005 I-5285, 16 June 2005 [31–43]; see further Chalmers et al. (2014), p. 591.

[124] Commission of the European Communities, Commission Staff Working Document, *Annex to the report from the Commission on the Implementation since 2005 of the Council Framework Decision of 13 June 2002 on the European arrest warrant and the surrender procedures between Member States* [COM (2007) 407], SEC (2007) 979, 11 July 2007, 5; European Commission, *Report from the Commission to the European Parliament and the Council on the implementation since 2007 of the Council Framework Decision of 13 June 2002 on the European arrest warrant and the surrender procedures between Member States* [SEC (2011) 430], COM [2011] 175, 11 April 2011, 5. See Chap. 1, Sect. 1.3.3.2 about the value of such material, non-binding sources of law.

[125] See further recitals 12 and 13 of the EAW Framework Decision.

such as these go beyond what is required of national implementation under the Framework Decision. In England and Wales, surrender under an EAW is governed by the Extradition Act 2003. Section 21 of that Act provides that, after considering whether there is a bar to surrender on account of such matters as double jeopardy, a person's age or even specialty, the executing judge must consider whether surrender is compatible with the person's rights under the ECHR.[126] Section 21 has been interpreted as creating a rebuttable presumption, based on clear and cogent evidence, that the issuing Member State applies the minimum standards of human rights protection under the ECHR.[127] Therefore, in England, mutual trust is not absolute.

Despite its complaints about implementation of a human rights ground for refusal in some Member States, the European Commission has stated that the EAW Framework Decision does not require the executing judicial authority to surrender a person if satisfied that this would result in a breach of the requested person's fundamental rights by reason of unacceptable conditions of detention.[128] In addition, the Commission endorsed the measures referred to in the roadmap in relation to protection of procedural rights of suspected and accused persons as being relevant to implementation and application of the Framework Decision. In this way it is envisaged that Member States will reinforce mutual trust, which is imperative to the success of the EAW system.[129]

It is against this background that the referring court in the recent case of *Radu* asked the CJEU to consider whether the Framework Decision 2002/584, read in the light of Articles 47 and 48 of the Charter and of Article 6 of the ECHR, allowed the executing judicial authority to refuse to execute an EAW issued for the purposes of conducting a criminal prosecution on the ground that the issuing judicial authority did not hear the requested person before the EAW was issued contrary to the right to be heard.[130] In concluding that the EAW Framework Decision could not be

[126]See also s. 87 of the Extradition Act 2003, which creates a human rights bar in relation to extradition to a number of non-EU countries. Sections 21 and 87 are aimed at ensuring an executing judge's decision is compatible with the obligations created by s. 2 of the Human Rights Act 1998, implementing the ECHR in the UK. Double jeopardy is where a person is prosecuted for the same offending twice; specialty, also known as double-criminality, protects a person from being surrendered in relation to a non-extraditable offence.

[127]Baker et al. (2011). A rebuttable presumption is a presumption of fact, which can be reversed by the opposing party on the presentation to a court or decision maker of evidence to the contrary.

[128]European Commission, *Report from the Commission to the European Parliament and the Council on the implementation since 2007 of the Council Framework Decision of 13 June 2002 on the European arrest warrant and the surrender procedures between Member States* [SEC (2011) 430], COM [2011] 175, 11 April 2011, 7.

[129]European Commission, *Report from the Commission to the European Parliament and the Council on the implementation since 2007 of the Council Framework Decision of 13 June 2002 on the European arrest warrant and the surrender procedures between Member States* [SEC (2011) 430], COM [2011] 175, 11 April 2011, 9.

[130](CJEU) Case C-396/11 *Radu* [2013] 29 January 2013, [31–32].

interpreted in this way,[131] the CJEU emphasized the following: it was not an express ground for non-execution under the EAW Framework Decision[132]; a requested person does not have a conclusive right to be heard by the issuing court under articles 47 and 48 of the Charter[133]; allowing such a right would lead to the failure of the surrender system in so far as an arrest warrant must have an element of surprise[134]; the European legislature has ensured that the right to be heard will be observed in the executing Member State[135]; and the requested person also has the right to be assisted by counsel (and an interpreter) in the executing Member State.[136] The Court in Radu, therefore, decided the case on a narrow basis and, in reference to the right to be heard, did not give an opinion on the wider issue of the interplay between human rights and mutual recognition under the EAW Framework Decision. It appears that the case was decided in this way so as not to stymie the effectiveness of the system when balanced with individual rights in the executing member state. However, the approach appears to run contrary to the broader advisory opinion of the Advocate General in the case, which referred to the general relationship between human rights and the EAW Framework decision.[137] Although the advisory opinion of the Advocate General is not binding, it nevertheless presents an interesting view of what the law should be in relation to mutual trust.

In suggesting a resolution in the case of *Radu*, the Advocate General referred to the decision of the CJEU in the case of *N.S. v Secretary of State for the Home Department*[138] regarding the obligation of one Member State to take on the responsibility of providing asylum rather than transferring the asylum seeker to another Member State where there were substantial grounds for believing that there was a real risk that their fundamental and international human rights under the Charter and the Convention, namely the right to human dignity, freedom from

---

[131](CJEU) Case C-396/11 *Radu* [2013] 29 January 2013, [44].

[132](CJEU) Case C-396/11 *Radu* [2013] 29 January 2013, [38].

[133](CJEU) Case C-396/11 *Radu* [2013] 29 January 2013, [39].

[134](CJEU) Case C-396/11 *Radu* [2013] 29 January 2013, [40].

[135](CJEU) Case C-396/11 *Radu* [2013] 29 January 2013, [41 and 42]; see arts 14 and 19 of the EAW Framework Decision.

[136](CJEU) Case C-396/11 *Radu* [2013] 29 January 2013, [42]; see arts 11(2) and 13 of the EAW Framework Decision.

[137](CJEU) Case C-396/11 *Radu* [2013] Opinion of Advocate General Sharpston, 18 October 2012; see also the CJEU Grand Chamber decision in Case C-399/11 *Melloni*, 26 February 2013 and the opinion of Advocate General Bot of 2 October 2011 where the conclusions about the meaning of art. 4(a) of the Framework Decision 2002/584 as amended by the Framework Decision 2009/299 and refusal to execute an EAW on the ground that a conviction recorded in absentia could not be reviewed were also narrowly framed.

[138](CJEU) Case C-396/11 *Radu* [2013] Opinion of Advocate General Sharpston, 18 October 2012 referring at [76] to (CJEU) Case C-411/10 *N.S. v Secretary of State for the Home Department* (United Kingdom) [2011] 21 December 2011, joined with Case C-493/10 *M.E. v Refugee Applications Commissioner, Minister for Justice, Equality and Law Reform* (Republic of Ireland) [2011] 21 December 2011; see also [ECtHR] *M.S.S v Belgium and Greece*, Application no. 30696/09, 21 January 2011.

torture and other forms of ill-treatment, right to asylum and right to an effective remedy and a fair hearing, will be violated.[139] In the asylum decision the court found that the presumption of mutual trust must be regarded as being rebuttable rather than conclusive.[140]

On the issue of refusal of an EAW on human rights grounds, the Advocate General said:

> The competent judicial authority of the Member State executing a European arrest warrant can refuse the request for surrender without being in breach of the obligations authorised by the founding Treaties and the other provisions of European Union law, where it is shown that the human rights of the person whose surrender is requested have been infringed, or will be infringed, as part of or following the surrender process. However, such a refusal will be competent only in exceptional circumstances. In cases involving Articles 5 and 6 of the Convention and/or Articles 6, 47 and 48 of the Charter, the infringement in question must be such as *fundamentally to destroy the fairness of the process*. The person alleging infringement must persuade the decision-maker that his objections are *substantially well founded*. Past infringements that are capable of remedy will not found such an objection.[141] [emphasis added]

In line with this opinion, it makes good sense that the principle of mutual trust in international cooperation between EU Member States should not be absolute, but rather based upon a rebuttable presumption. However, as suggested by the Advocate General, it follows that not every breach or potential breach of the ECHR by the issuing state will justify refusal.[142] The interest of the issuing state, and the international community as a whole, in effective cross-border law enforcement must be recognised. This is vital to the establishment of an area of freedom that comes with security and justice. Particular attention should be paid to the rights of victims of crime. Recognising these interests means that a presumption that the standards of protection in an EU Member State are compliant with articles 5 and 6 of the ECHR should only be rebutted exceptionally.[143] These special interests in prosecution of criminal matters may distinguish EU criminal justice from EU asylum law. Therefore, in criminal cases, the test to determine whether the presumption that other EU Member States are meeting the applicable common minimum standards under articles 5 and 6 ECHR has been rebutted may require a higher standard than rebutting the presumption of mutual trust in asylum cases on the basis of a real risk of violation. The subject of an EAW may be required to show that surrender will occasion a flagrant denial of justice or, in the words of the

[139](CJEU) Case C-411/10 *N.S. v Secretary of State for the Home Department* (United Kingdom) [2011] 21 December 2011 [82–85, 92 and 94].

[140](CJEU) Case C-411/10 *N.S. v Secretary of State for the Home Department* (United Kingdom) [2011] 21 December 2011 [80 and 104].

[141](CJEU) Case C-396/11 *Radu* [2013] Opinion of Advocate General Sharpston, 18 October 2012, [108(3)].

[142](CJEU) Case C-396/11 *Radu* [2013] Opinion of Advocate General Sharpston, 18 October 2012, [74–75].

[143]See further (CJEU) Case C-396/11 *Radu* [2013] Opinion of Advocate General Sharpston, 18 October 2012, [81].

Advocate General, will fundamentally destroy the fairness of the process.[144] In addition, alternatives to immediate surrender, such as the European Supervision Order (ESO), may require further development to support the existence of the rebuttable presumption.[145]

7.4.3 Mutual Trust in Gathering Confession Evidence

Standard setting by the ECtHR has been a pre-condition to mutual recognition in EU criminal justice instruments.[146] The adoption of trust-building measures is envisaged as way to overcome the issues that may arise in executing Members States, particularly in response to the use of mutual recognition under the EAW, where Member States find themselves bound to follow unfamiliar measures that may even be construed as a violation of a constitutional right.[147] Within the EU, in 2009 the Council adopted a move to strengthen common procedural standards in relation to six priority issues: access to information, legal advice and interpreters as well as the right to communication, additional protection for vulnerable suspects and pre-trial detention.[148] This became known as the roadmap on procedural rights and has resulted in a number of Directives being adopted in these areas.

The general conclusions discussed above about mutual trust and mutual recognition in the surrender of suspects or convicted persons under an EAW should also apply to decisions in relation to evidence, including confession evidence, gathered under the EEW, the European Investigation Order (EIO)[149] or other criminal justice instruments relying on the principle. In the case of international cooperation in the gathering of evidence, it can never be completely ruled out that another country's investigative standards do not meet (in every respect) the minimum standards set by the ECtHR under art. 6 ECHR. Furthermore, it is possible that the reception of evidence gathered abroad (even lawfully) fundamentally destroys the fairness of a trial due to a disruption in the continuity of the national balancing mechanisms. The floor of rights set by the ECtHR is intended to be universally applicable in every legal system (within the Council of Europe). However, in some systems these minimum standards may not actually be applied in every respect, leading to a potential mutual trust problem; or in certain respects the international standards may not be sufficiently defined, which is a 'human rights law' problem.

---

[144] About the differences between the tests "flagrant denial" and "fundamentally destroy" see Thunberg Schunke (2013), pp. 86–89, 109–110 and 132.

[145] Billing (2012), p. 91 (Chap. 1, Sect.1.3.4).

[146] Case C-306/09 *I.B.* [2010] Opinion of Advocate General Villalón, ECR I-10341; Cryer et al. (2014), p. 77.

[147] Mitsilegas (2009), pp. 131–132.

[148] See Council of the European Union (2009).

[149] Directive 2014/21/EU of the European Parliament and of the Council of 3 April 2014.

The above concerns should be taken into account the implementation and application of the EIO. It may be wise to allow for a *rebuttable* presumption that transferring evidence will not fundamentally destroy the fairness in the circumstances of the case. In fact, mutual legal assistance in the gathering of confession evidence under any international cooperation instrument should be only take place where the transfer of evidence in transnational cases will not fundamentally destroy the fairness of the proceedings as a whole.

7.5 Conclusion

Suspect interrogation may very well be a legitimate tool for gathering evidence in transnational criminal investigations. Despite caution displayed by some state agencies, it is already a measure that is being carried out by officers of a requested state on behalf of criminal investigators of a requesting state. However, the process of gathering and transferring evidence needs to be accompanied by sufficient procedural guarantees in order to promote effectiveness and fairness as partners, rather than competitors.

This chapter has considered the approaches being taken by the various state actors, such as the police, prosecutors and the courts as well as the EU legal/judicial decision makers, to overcome the challenges involved in the use of 'irregularly' obtained evidence gathered abroad and questions of admissibility. The approach in all three jurisdictions to accepting and relying on evidence obtained abroad under foreign rules has generally been flexible, where foreign officers are the ones carrying out the investigative procedures. Local investigative authorities are expected to uphold national/state procedural standards, regardless of whether or not they are operating across national borders. However, these standards are being relaxed in transnational cases involving foreign authorities.

This chapter examined case law from the three jurisdictions of Denmark, England and Australia which has dealt with the issue of irregularly obtained foreign evidence, in particular confession evidence, obtained abroad. The sample of reported (and unreported) cases is small, particularly in relation to Denmark generally and England and Wales in respect of investigations outside Europe. Nevertheless, it may be possible to draw some general conclusions about the fairness and admissibility of using evidence obtained irregularly abroad.

In Denmark, it appears that when foreign police are using the same kinds of measures that are found in Denmark then the procedures used should also comply with Danish law. If the foreign agency uses a procedure not known to Danish law then the Danish courts have shown flexibility in relying on the evidence, while remaining mindful of the possible pitfalls of the measure used. This is same whether the requested state is a high contracting party to the ECHR or not. The above applies as long as the right to silence is not impugned, for example by not issuing a caution. The Danish courts' concern with the need for a caution is consistent with the tendency for Danish police and prosecutors to seek to question witnesses and *sigtet* persons themselves. Therefore, Denmark may be understood as preferring the application of the *locus regit actum* principle in international cooperation.

England, on the other hand, has displayed a somewhat more flexible approach to the admissibility of confession evidence obtained abroad, where the right to silence not received the same protection because, for example, no cautioning has taken place (in accordance with the law of the requested state). As well as being a European organised crime hub, this explanation for this may lie in the common law focus on rules about fairness and the admissibility of evidence. In English trials, there is an opportunity for the flexibility in connection with the receipt of foreign confession evidence to be checked and balanced at trial by the trial judge in the application of rules of evidence concerning admissibility. While the English courts have dealt with a number of transnational cases involving confession or other evidence that had been obtained in Council of Europe states, there is little evidence of the approach to the same kind of evidence obtained outside Europe. In addition, some of the English cases indicate that if the irregularly obtained evidence has originated in a country from within the ECHR framework, then it is 'assumed' that the procedure was conducted fairly. However, these judgments have not considered the requesting state's rules protecting the right to silence and pre-trial due process in any detail. The issue of fairness has frequently not been examined beyond comments that relate to factors such as a lack of bullying, a lack of abuse of process (whereby the police have not deliberately set out to avoid applying procedural rights) and the opportunity to test and challenge evidence. Cases such as these indicate that the building of trust within EU/Council of Europe states, based on the application of common minimum standards has potentially gone too far.

More recently, however, the UK Supreme Court's decision in *Cadder* has recognised that not all Council of Europe or EU states are applying the relevant standards. Furthermore, the case has put a clear focus on the need for consistently applied pre-trial due process rights, such as cautioning about the right to silence and access to a lawyer prior to custodial questioning (if not more broadly). Perhaps decisions such as Konscol would be decided differently by the English courts today.

Regardless of whether a country is a signatory of the ICCPR or not, the Australian courts have been very flexible in relation to the admissibility of confession evidence obtained across borders. Similar to the English situation, the issue of fairness where local procedural standards are not being met is 'balanced' at trial by a consideration of fairness in the context of the rules about admissibility of evidence (which are generally absent from the Danish system). However, the Australian courts have taken a strict approach when local officers are present during the gathering of relevant evidence in another jurisdiction. Like in England, some of these cases may also be decided differently today, following the passing of national legislation that provides more explicit protection of the right to silence prior to interrogation—although the provisions do not strictly apply to foreign officers.

Overall, what appears to influence these systems' broad approach to international cooperation in the gathering of confession evidence is legal tradition and trial procedure, as well as politico-geographic location. Adhering to one human rights framework or another may have influenced some of the English judgments, although this factor does not alone appear to have had a specific bearing on the cases under consideration. Of course, what is also central to all these cases is the absence of unlawful forms of coercion such as abusive, bullying behaviour on the part of investigators; and the presence of trial procedures that ensure a criminal defendant has a sufficient opportunity to test the evidence before the trial court.

However, if confession evidence comes to be gathered within the EU via a mutual recognition instrument, such as the EIO, the influence of the ECHR as a human rights framework for the setting of common minimum standards is more prominent. In this setting, the ECHR framework is the theoretical backbone to the foundation of mutual trust and the presumption of the universality of minimum standards in the EU that allows for the automatic nature of the mutual recognition of judicial decisions—at least in the absence of significant evidence to the contrary.

In light of more recent national decisions such as *Cadder*, and to some extent from the EU decisions concerning mutual trust, what can be said is that the ECHR and the jurisprudence of the ECtHR does have a significant influence on fostering awareness of the need for uniformity in relation to minimum standards. In addition, perhaps to a lesser extent, the Council of Europe framework has had some influence in encouraging actual uniformity.

Nevertheless, not all EU states are applying the same minimum standards in the same way all the time. Belgium, for example, offers the right to silence but does not provide for a caution prior to official questioning as part of that right, in the context of the national system.[150] Combining the approach of Ruth Nielsen and Tuori's legal positivist theory (referred to in Chap. 1, Sect. 1.3.3.3), the EU may be one big legal system but consisting of different levels.[151] Common minimum standards of procedural rights may lead to conformity on the surface level, while significant divergences may still exist at the medium level of legal culture, also taking into account the legal tradition that determines national principles about the right to silence and the ways in which this right should function. Furthermore, the realities of local situations in certain circumstances may be a greater force than uniformity—and hence the need for some caution in the application of the principle of mutual trust.

References

Australian Federal Police (2013) Media Release: 117 kg of drugs seized in organised crime investigation spanning five countries. Accessed 15 Jan 2014

Baker S, Perry D, Doobay A (2011) A review of the United Kingdom's extradition arrangements. Presented to the Home Secretary on 30 September 2011. Available via GOV.UK. https://www.gov.uk/government/uploads/system/uploads/attachment_data/file/117673/extradition-review.pdf. Accessed 15 Jan 2014

[150]Cape et al. (2010), pp. 77 and 86.

[151]Nielsen (2013), pp. 87, 107 and 129.

Billing F (2012) The parallel between non-removal of asylum seekers and non-execution of a European arrest warrant on human rights grounds: The CJEU Case of N.S. v. Secretary of State for the Home Department. Eur Crim Law Rev 2(1):77–91

Cape E, Namoradze Z, Smith R, Spronken T (2010) Effective criminal defence in Europe. Intersentia, Antwerpen

Chalmers D, Davies G, Monti G (2014) European Union law: text and materials, 3rd edn. CUP, Cambridge

Cryer R, Friman H, Robinson D, Wilmshurst E (2014) An introduction to international criminal law and procedure, 3rd edn. Cambridge University Press, Cambridge

Council of the European Union (2009) Press Release 14828/09: Procedural Rights in Criminal Proceedings. http://www.consilium.europa.eu/uedocs/cms_Data/docs/pressdata/en/jha/110740.pdf. Accessed 15 Jan 2014

Daniel B, Elholm T, Starup P, Steinicke M (2011) Grundlæggende EU-Ret: EU Efter Lisbontraktaten. DJØF Forlag, Copenhagen

Home Office (2015) Requests for Mutual Legal Assistance in Criminal Matters: Guidelines for authorities outside of the United Kingdom, 12th edn. Home Office, London. https://www.gov.uk/government/uploads/system/uploads/attachment_data/file/415038/MLA_Guidelines_2015.pdf. Accessed 15 Oct 2015

Jones RWD, Davidson R (eds) (2010) Extradition and mutual legal assistance handbook, 2nd edn. OUP, Oxford

Justitsministeriet (Ministry of Justice, Denmark) (2001) Vejledning om behandling af anmodninger om gensidig retshjælp i straffesager af 5 December 2001. Available via Retsinformation. https://www.retsinformation.dk/Forms/R0710.aspx?id=1879. Accessed 15 Jan 2014

Klip A (2009) European criminal law: an integrative approach. Intersentia, Antwerp

Mitsilegas V (2009) EU criminal law. Hart, Portland

Nielsen R (2013) New European legal realism – new problems, new solutions? In: Neergaard U, Nielsen R (eds) European legal method. DJØF Forlag, Copenhagen

Ouwerkerk J (2011) Quid Pro Quo? A comparative law perspective on the mutual recognition of judicial decisions in criminal matters. Intersentia, Antwerp

Peers S (2011) Court of Justice: the NS and ME opinions – the death of mutual trust? Available via Statewatch. http://www.statewatch.org/analyses/no-148-dublin-mutual-trust.pdf. Accessed 15 Dec 2011

Peers S (2012) EU justice and home affairs law. OUP, Oxford

Rigsadvokaten (1996) Anklagemyndighedens Årsberetning 1996. Ministry of Justice, Copenhagen

Suominen A (2011) The principle of mutual recognition in cooperation in criminal matters. University of Bergen

Thunberg Schunke M (2013) Whose responsibility? A study of transnational defence rights and mutual recognition of judicial decisions within the EU. Intersentia, Cambridge

Vernimmen-Van Tiggelen G, Surano L (2008) Analysis of the future of mutual recognition in criminal matters in the European Union: Final Report. Available via Europa.EU. http://ec.europa.eu/justice/criminal/files/mutual_recognition_en.pdf. Accessed 15 Jan 2014

Wong C (2013) Sweden. In: Ligeti K (ed) Towards a Prosecutor for the European Union, vol 1, A comparative analysis. Hart, Oxford

Chapter 8
Balancing the Right to Silence in Transnational Criminal Cases

Abstract The final chapter presents the main conclusions about balancing the right to silence in transnational criminal proceedings, in order to maintain fairness in the proceedings as a whole. The conclusions have been arrived at as a result of examining the question of fairness from national, transnational and international perspectives. The finding that continuity exists between the laws that limit and protect the right to silence across national investigative and trial proceedings has led to an examination of how disruption to the national balance mechanism can be minimised when transferring evidence across borders. There is a need to develop particular rules to overcome legal differences between systems that aid the criminal defendant in transnational proceedings, including rules about effective legal representation. These rules may be relevant to the future success of EU initiatives such as the European Investigation Order (EIO) and the much debated European Public Prosecutor's Office (EPPO). Considering the problem of balance from the international perspective, international minimum standards of protection that are consistently applied by member states allow for effective law enforcement based on an integrated foundation of fairness. However, a broad framework such as the ICCPR does not currently have the capacity to lay a real foundation, requiring increased efforts in the area of procedural rights.

8.1 Introduction

Issues centred on the fairness of gathering confession evidence or evidence of a suspect or accused person's silence in one legal system for use at trial in another have been examined throughout the preceding chapters. The questions raised about balancing the right to silence are central to achieving justice in transnational cases. Full protection of the right protects an individual suspect or accused person protection from unlawful compulsion and enables him or her to participate in the criminal proceedings in a knowing and willing way. In addition, investigative measures that are based on fairness promote the gathering of evidence that is not only trustworthy but evidence that can be challenged properly in a way that elevates

F.M.W. Billing, *The Right to Silence in Transnational Criminal Proceedings*,
DOI 10.1007/978-3-319-42034-9_8

the rule of law. In other words, fairness in gathering and using evidence supports effectiveness in bringing wrongdoers to justice.

The suspect or accused's right to silence was described in Chap. 1 as one circle that overlaps with another circle being the right against self-incrimination. The right to silence focuses on the suspect's choice not to make any statements at all—this is the choice whether to speak to the police or to remain silent. In order to fully exercise this right, a suspect or accused must be aware of the right; and of the consequences that flow from the right throughout the criminal proceedings. Therefore, from the *national* perspective, as with other fundamental fair trial rights, the right to silence envisages that criminal proceedings are a continuum and that there is a connection between what happens in an investigation and what happens at trial. This is a connection that has been highlighted by international bodies such as the HR Committee and the ECtHR, in their guidelines about the right to silence, especially when viewing fairness in the proceedings as a whole. Therefore, where continuity on the national level has been identified then the question is whether the balance between effective and fairness can be still be maintained even when the investigation and the trial become jurisdictionally disconnected in *transnational* proceedings. From the *international* perspective, this is more likely to be true when the cooperating states are applying the same minimum standards of fairness and there is a realistic environment of mutual trust.

In this final chapter, the overall conclusions from the work are presented, along with discussion of what the conclusions mean for international cooperation in the gathering of confession evidence or evidence of silence in the future. Specifically, with a comparison of national law as a basis, balancing the limitation and protection of the right to silence in considered in transnational and international perspectives.

8.2 The Right to Silence in Transnational Proceedings

To fully understand how the right to silence functions in transnational proceedings, it can also be useful to reflect on some broader observations about the national systems in which the right to silence has been examined.

Described briefly, Denmark can be said to have a system which has less extensive restrictions on the rights of suspects and accused persons. For example, at trial, evidence of an interrogation can only be raised in limited situations, such as if the accused fails to testify or gives testimony which is inconsistent with a prior statement; there are no rules about drawing adverse inferences from the right to silence during pre-trial interrogation (although this cannot be ruled out in accordance with the free evaluation of evidence); and administrative compulsory information gathering can only be used by consent to question a *mistænkt* (suspect). There are also significant safeguards to generally protect individual rights in an investigation, such as by the involvement of prosecutors and defence lawyers in pre-trial proceedings, in addition to more standard safeguards of the right to silence in pre-trial proceedings, including cautioning about the right to silence and a right to legal advice offered to all *sigtet* in connection with offences that are punishable by imprisonment and all detainees. *Sigtelse* is a unique process of formal accusation, equivalent to 'charging' within the autonomous meaning of the ECHR, which affords the *sigtet* pre-trial due process rights at an early stage. However, Denmark, like other civil law countries, functions with the principles of material truth and free evaluation of evidence which has influenced the traditional tendency against exclusion of evidence. Evidence may be excluded only if there are significant concerns in relation to reliability or if confession evidence has been obtained contrary to the right to silence or the right against self-incrimination, in particular as a result of a lack of caution prior to official questioning. A technical breach is not likely to result in exclusion but rather fairness and proportionality may be finally considered when the fact-finders determine the weight to be attached to specific evidence.

On the other hand, England and Wales can be characterised as a system which by law places heavier restrictions on the right to silence, such as: compelled sworn testimony in administrative compulsory questioning powers; direct use of evidence of official questioning at trial; drawing adverse inferences in certain circumstances from the pre-charge silence of an accused and the failure to give evidence. Nevertheless, there are also concrete ways in which the English law counter-balances the restrictions with significant safeguards, such as: use immunity to guarantee that self-incriminating statements obtained under administrative compulsory questioning powers cannot be used against an accused at trial; cautioning about the right to silence and the consequences of the right; and a series of procedural safeguards for suspects at police stations, even those attending voluntarily. Most significant is the right to access free legal representation for all suspects under arrest and some form of free advice for all suspects at the police station; as well as the right to have a lawyer present in the interview. The code of practice in relation to interviewing suspects clearly details other safeguards at police stations including a letter of rights, right to interpreters and the presence of independent persons for young suspects and protection of vulnerable suspects. Rules of evidence and judicial discretion to exclude evidence are important safeguards at trial, as a final check and balance on fairness throughout the proceedings.

In Australia, the picture is somewhat different again. Australian law permits certain restrictions on the right to silence and provides certain safeguards. For example, compelled sworn testimony under compulsory questioning powers in regulatory proceedings is allowed, though not if there is a real risk of interference with the administration of justice, such as by questioning a person who is already subject to criminal charges for the exact same conduct. Australian law also provides for direct use of evidence of the official questioning at trial. However, there are safeguards in place that prohibit the drawing of inferences of guilt from the pre-trial silence of the accused and stringently restrict the use that can be made of the accused's failure to testify. Other safeguards include: derivative use immunity in the form of the 'but for' test in relation to derivative real evidence; a right to a lawyer of own choosing on arrest; and audio or video-recording of interviews. Like England, Australia relies heavily on the judge's discretion to exclude unfairly or unlawfully obtained evidence, which functions as a final opportunity to restore the balance between effectiveness and fairness by checking for unfair forensic disadvantage to the accused and proportionality.

On the background of the comparison of the systems in Denmark, England and Wales and Australia this work resulted in a number of specific conclusions in relation to the maintenance of balance between law enforcement and the suspect or accused's right to silence in transnational cases. The problem of disrupting the continuity of law by transferring confession evidence or evidence of silence from the investigative phase in one system to the trial phase in another was raised in terms of jurisdictional cross-admissibility.

8.2.1 Police Questioning

Despite the national differences in pre-trial due process procedures, it should be possible for the authorities of one country to use mutual legal assistance or mutual recognition to request the police in another country to question a suspect. Chapter 7 referred to the positive experience of English police in giving this type of assistance—to countries inside and outside the EU. In fact, this may be a useful investigative tool to narrow the field of enquiry by eliminating a suspect or as a pre-curser to extradition. In the EU, complaints have been made that some countries use the EAW as a means of facilitating the opportunity to interview a suspect and can lead to months of unnecessary pre-trial detention.[1] A request for assistance to question a suspect could be used in conjunction with a European Supervision Order to avoid surrendering a suspect to be detained on remand for lengthy periods—particularly where the conditions of detention may not meet common standards. Eurojust may also be able to assist in the case of parallel transnational criminal investigations to determine where a trial is likely to take place and which procedural rules should be used in an interview or pre-trial hearing.

However, for this type of assistance to work in a way that does not significantly risk the individual's right to silence through the disruption caused to the national balance between effectiveness and fairness, requesting countries need to be satisfied that the investigators in the requested state will carry out the request with the rights specified. Requested countries need to be satisfied that the procedures suggested are consistent with relevant minimum standards of human rights protection. At the UN level, this may require a more specific approach to promoting a roadmap of procedural rights for suspects and detainees aimed at defining and promoting common standards for pre-trial due process in the states coming under the ICCPR. Harmonisation as an alternative to mutual trust can also enable human rights to facilitate rather than impede international cooperation.

---

[1]UK Parliament Human Rights Joint Committee (2011), [167–169]. Some countries outside the EU are required to provide prima facie evidence before a UK court will agree to extradite: Home Office (2014).

In relation to transnational requests for assistance in questioning suspects, based on an amelioration of the national rules, in Chap. 6 some guidelines were suggested to promote the jurisdictional cross-admissibility of confession evidence obtained during official questioning, as follows:

- the suspect should be advised as soon as possible of the accusations against him;
- the suspect should be cautioned (upon the basis that the police/authorities in the requesting state have established a suspicion of criminal offending on reasonable grounds) prior to being questioned in the form required by the law in the requesting state or at least be told that he has does not have to say anything;
- the suspect should be invited to the police station for official police questioning;
- the suspect should be offered the right to legal representation (at least of own choosing) to protect the suspect's interests during the interrogation/hearing and at trial, including the presence of a legal representative during questioning; and
- an audio-recording should be made whenever possible or a written record, which is preferably commented on or acknowledged.

Whether as 'minimum standards' or 'harmonised standards' these types of guarantees could overcome the rights gaps between national jurisdictions that can create unfairness in the cooperation process. For example, understanding the context of the English caution and the requirement for access to a lawyer, if a requested country was prepared to caution a suspect of English offences in accordance with the English caution in the presence of a relevantly qualified lawyer, then the rules relating to drawing adverse inferences in the English trial could be maintained. Alternatively, states would have to agree that no such inferences may be drawn in transnational cases. Similarly, by making an audio-recording of a suspect interview or hearing, this defence right in an Australian trial could be upheld instead of being given up.

Certainly within the EU, through the operation of mutual trust and by expanding on the understanding of pre-trial due process in other systems, countries can offer each other assistance where that requires the police to use specific rules from other systems in order to conduct voluntary interviews. Alternatively, the above minimum rights relating to pre-trial questioning could be agreed as applying in transnational proceedings. Such an agreement could also assist police operating in joint investigative teams, for example.

In addition, using police interrogation as an effective tool in transnational investigations must be supported by guaranteeing an effective right to access to legal representation prior to questioning. This does, of course, present a dilemma of striking an appropriate balance between competing interests. On the one hand, there is clearly a public interest in the police, for example, being able to obtain unguarded responses to questioning and moreover to encourage a suspect to give an early account. This is evidently in the interests of victims of crime and can avoid the trauma and public expense of ongoing criminal investigations and prosecutions. However, there are also important individual interests at stake because the attitude of an accused during the pre-trial phase, and in official questioning in particular, can have significant affects during the criminal trial in various legal systems, in the way confession evidence and evidence of silence may be used. Further, interviews conducted at a police station (and maybe even questioning in-court) inherently involve a power imbalance and may appear coercive to the individuals involved, without any impropriety on the part of the interviewer. This is particularly true in the custodial setting, where the suspect is not at liberty to leave; but may also apply where an individual is being questioned by 'the authority' voluntarily. Therefore, the ECtHR has found that suspects should be provided an effective right to prior access to a lawyer, particularly those who are questioned in a custodial setting. This includes access to legal advice from a free publicly appointed defence lawyer when necessary in the interests of justice. Criminal laws are complex and confusing, particularly in a transnational context.[2] There is, therefore, arguably a special need for the legal assistance of a lawyer for a suspect who is being questioned voluntarily in relation to foreign offences. This is an issue that needs to be considered in detail (see Sect. 8.4.1 below).

Rules about admissibility of confession evidence or evidence of the suspect's pre-trial silence are one way that the courts can openly deal with any issues about unfairness in the evidence gathering process—also in transnational cases. Courts in England and Australia, with lay jury trials for serious offences, rely more heavily on judicial exclusion of evidence as a check and balance in relation to fairness in the criminal proceedings. Danish courts, where the principle of material truth and the free evaluation of evidence are significant, tend not to exclude evidence and adjust the weight attached to unfairly or unlawfully/irregularly obtained evidence (in conjunction with an assessment of the other evidence in the case). However, even in Denmark there is some evidence of a practice to exclude evidence from a trial if it has been obtained contrary to the right to silence, meaning that no 'caution' about the right was given.

Using the current national rules, courts in the three jurisdictions have been able make assessments about the fairness or appropriateness of using confessions and other types of evidence obtained in other legal systems. From a small sample of cross-border cases, on the whole, the courts in all three countries appeared to take a flexible approach to cross-admissibility of irregularly obtained foreign evidence. However, in England and Wales, in the context of evidence coming from other ECHR countries, there appears to have been a shift from being more relaxed about the types of due process procedures followed in other ECHR countries, to being overall more conscious of the need to apply common standards of fairness in particular in relation to police questioning.

[2] See (ECtHR) *Stojkovic v France and Belgium*, Application no. 25303/08, 27 October 2011.

8.2.2 The Use of Silence as Evidence of Guilt

It is generally accepted that using evidence of a suspect or accused's persons silence—either when being questioned by the police or at trial—as evidence of their guilt is a clear limitation on the right to silence. On the other hand, ambushing the prosecution with a defence at trial is frustrating and arguably provides a reasonable basis for allowing adverse inferences.

The three national systems examined in this work have divergent rules in relation to whether adverse inferences of guilt are permissible at trial and the conditions that allow for this. In Denmark, there are no express rules allowing or prohibiting the drawing of adverse inferences. This type of evidence may simply form part of the assessment of the evidence by the court. In England and Wales, a judge may direct the jury on the circumstances in which it may infer guilt from silence. The conditions for drawing adverse inferences from pre-trial silence are related to the giving of an extended caution to explain the consequences of silence and the receipt of legal advice to back this up. In Australia, it is generally considered to be unacceptable to draw adverse inferences from trial and legislation is in place to prevent such inferences in most circumstances. However, for example in NSW in relation to certain serious offences, legislation has been introduced to allow for adverse inferences from silence in similar circumstances to the UK legislation. Here, the gravity of offending has been relevant to the limitation of the right to silence.

The form of caution used in the national setting is clearly linked to the approach being taken in the national courts as to whether pre-trial silence, when a person is being questioned by the police under caution, may be used to infer guilt at trial. This explains, at least in part, that differences exist in pre-trial procedure and the reasons why it is important to consider the form of caution used in transnational proceedings.

8.2.3 Administrative Questioning Powers

Requesting via international cooperation administrative compulsory information gathering be used to gather information from a potential suspect of criminal offences in another jurisdiction is fraught with difficulties. Of itself, the dual role of seeking information under compulsion for administrative purposes at the same time as questioning suspects voluntarily (while respecting the right not to incriminate oneself) walks a faint, fine and rapidly shifting line. Transnational cases involve additional crevasses. In particular, the structure of administrative compulsory powers may be unrecognisable between systems which leads to problems with continuity and balance. For example, compulsory questioning powers in England and Australia frequently rely on use immunity to overcome the significant compulsion of witnesses to answer questions during sworn testimony on pain of punishment. Such use immunity does not exist in Danish law, where administrative questioning powers are used by consent (with a caution about the right to silence) in relation to suspects of criminal offences. However, the Danish administrative questioning powers do not allow a right to access a lawyer or require an audio or audio-visual recording of the interrogation, two rights that arguably ensure an 'effective defence' according to international standards.

In addition, if criminal charges are pending, or 'anticipated', the ECtHR and the High Court of Australia have condemned the use of administrative proceedings to question a criminal suspect in relation to the same conduct, contrary to the right against self-incrimination and the accusatorial principle that the prosecution bears the burden of proof throughout the proceedings.[3] While this prohibition may not apply to the use of Danish administrative questioning powers by consent, it clearly applies to English and Australian compulsory questioning powers—a factor which highlights the need for cooperating states to understand the precise ways in which administrative powers work in other jurisdictions.

Even where charges are not pending or imminent, the propriety of using administrative powers to question a witness who may—in a very short space of time—turn out to be a suspect must be carefully considered. This is especially true in the transnational context, where administrative proceedings do not generally give the same procedural rights as a suspect would receive in a criminal investigation. It is possible for administrative authorities to seek assistance of foreign authorities to question witnesses in administrative investigations with a criminal element. However, in some states, administrative compulsory information-gathering powers are not, as a general rule, intended to generate evidence for use in criminal proceedings. It is always open for states to use administrative compulsory powers for information gathering and information sharing in transnational cases.

8.2.4 Covert Surveillance Measures to Listen to Private Conversations

In general, the three systems under consideration all provide for statutory powers in relation to intrusive covert surveillance measures to be used by the police when investigating serious crime within systems of supervision that provide safeguards against abuse. Therefore, despite differences in the details of how that supervision is actually provided, it is generally acceptable to use international cooperation to gather evidence via these types of measures. However, issues may arise given that the definition of serious crime in England, being crime which would generally lead to a term of imprisonment of at least 3 years, appears to set the threshold somewhat lower than the usual threshold for serious offending in Denmark (6 years) and Australia (7 years). It may be considered against fundamental principles and

[3]This applies apart from in a very small number of situations, such proving self-defence.

proportionality for an Australian or Danish court to authorise the use of a listening device warrant for less serious types of offending than that contemplated by the national legislature. National courts would probably not be prepared to do this. This may be a case where the court in the requested state should hear arguments on behalf of an individual about whether this would constitute a flagrant breach of rights before approving of a measure. Turning the scenario around, it is arguably inappropriate for Australian or Danish police to seek evidence via international cooperation to have a listening device installed in England in relation to offences that are less serious than the Australian or Danish legislation provides for. This may be considered to be an abuse of process. Again, perhaps the situation is different when the discussion is not about evidence-gathering but about information gathering and sharing in parallel investigations in transnational cases.

8.3 The Relationship Between Human Rights and International Cooperation

Looking at the right to silence from the international perspective, it may be concluded that the relationship between human rights and international cooperation works in two directions. If states are applying the same procedural standards in evidence gathering then the effectiveness of international cooperation is enhanced; on the other hand, a lack of mutual trust in the procedural protections being applied in another state may hinder the international process altogether. Investigators, prosecutors and governments may not be prepared to seek or accept confession evidence that comes from a country where procedures and practices are unpredictable or fundamentally at odds with national law. Although national courts are generally flexible in accepting irregularly obtained evidence from another state, there is a risk that irregularly obtained evidence will be deemed unfair or inadmissible. Ironically, perhaps it should be. On the one hand, where the national courts are being flexible in relation to evidence gathered by foreign investigators who 'cannot be expected' to be familiar with local procedures and standards, indicates a bias towards effectiveness of prosecution of transnational crime. On the other hand, the stakes are high when the issue is the successful prosecution of crimes such as drug trafficking, terrorism or child pornography. Therefore, establishing well-defined common minimum standards that close the gaps in the international cooperation process is critical. Narrowing of the gaps explains why the principle of mutual trust is able to exist at all in the EU, under the framework of the ECHR, and the common standards stipulated by the ECtHR, and under the influence of EU trust-building initiatives such as the roadmap of procedural rights. Nevertheless, mutual trust is not always enough.[4]

[4]See (ECtHR) *Stojkovic v France and Belgium*, Application no. 25303/08, 27 October 2011.

The presumption of mutual trust based on common human rights standards under the framework of the ECHR can facilitate far-reaching cooperation in the EU. However, it should not be allowed to operate conclusively. It can never be ruled out that other states within the EU, for example, are not applying common minimum standards of human rights. In additional, from time to time guidelines about common minimum standards may be too imprecise to guarantee that all states will apply them in the same way. International cooperation that is based on mutual trust or a mutual confidence in one another's standards of human rights protection should always take place on the *rebuttable* assumption that common minimum standards are being applied.

Due to the particular complexity of transnational cases, certain procedural safeguards should always be in place in relation to suspects and accused persons of this category of cases. For example, in addition to the provision of an interpreter as necessary, in relation to pre-trial questioning, a suspect should always be cautioned about the right to silence, preferably in the form required by the national law of the trial state; suspects attending voluntarily to be questioned about foreign offences should always have an effective right to access independent legal advice, including some form of free legal representation (perhaps even from both the requesting and requested systems); and third, formal foreign suspect interviews conducted at a police station or pre-trial hearings should always be audio-recorded to ensure the accuracy and the integrity of any evidence obtained. Harmonization rather cooperation based on mutual trust could achieve the result of clearly defining the most fundamental procedural rights in transnational cases. This path is being followed in the EU with the instruments being passed under the roadmap of procedural rights initiative.

8.4 The Way Forward: Confession Evidence Across Borders

8.4.1 Right of Access to Legal Assistance in Transnational Cases

In Chap. 6, the need for a right of access to effective legal representation for suspects being questioned in relation to foreign offences out of the jurisdiction was raised. Currently, there are no clear standards about legal representation for suspects and persons accused of foreign offences and these individuals are being left in a vulnerable position. While it is linked to the right not to say anything that these suspects are able to access a lawyer of their own choosing, this will not assist those suspects who cannot afford or do not have the capacity to employ the services of a lawyer. Police questioning of foreign suspects will generally occur on a voluntary basis and a right of access to legal representation, beyond a lawyer of own choosing, will not be necessary in order to apply the guidelines of the ECtHR or the HR Committee.

The position of the suspect of foreign offences is somewhat unique. If they are a foreign national they should receive a right to the services of an interpreter. If they are a young or otherwise vulnerable person, they should receive a right to the presence of an independent third person. Significantly, they should receive an entitlement to effective legal representation. But, in this respect, a person being questioned about foreign offences is clearly at a disadvantage, where effective legal representation would require receiving legal advice from a lawyer who is familiar with the legal system where the charges originated. The establishment of a Eurodefence body that can be called upon to give legal advice in transnational cases would be a significant step forwards.

A primary concern may be means tested (or free) legal representation at surrender hearings, which are custodial in nature. In England, the rules that previously allowed independent means tested legal advice in relation to all extradition hearings were abolished. However, there is a recommendation for the reintroduction of such rules. In addition, there is clearly an argument for access to some form of free legal representation for suspects who participate in their own investigation—particularly where foreign rules may be confusing to themselves and even to a local defence lawyer of own choosing, if there is one.

Foreign legal advice does not necessarily have to be given in person. By the formation of a defence lawyer's network such as 'Eurodefence' (or even 'Interdefence'), foreign suspects at police stations, or their local legal representatives, could access a free duty lawyer call centre to obtain advice from a lawyer who has in depth knowledge of the foreign system in question. Other options for a call centre could include video-link or face-time facilities.

There is an argument that suspects being interviewed voluntarily do not have the same basis to claim a right to access free legal advice as persons being interviewed in a custodial setting. Any right to access legal advice must be balanced against the community interest in the police being able to obtain unguarded statements from suspects and even economic concerns. However, in transnational cases it is arguable that some form of effective legal representation is necessary to ensure that the suspect understands the nature of the right to silence and the right against self-incrimination in the context of the system producing the charges and running the prosecution.

The significance of an effective right to legal advice prior to the first official interrogation of a suspect or accused person, in addition to the importance of making audio or audio-visual recordings of questioning, attaches to all criminal proceedings, whether national, transnational or international.

8.4.2 The European Investigation Order

The European Investigation Order (EIO) was adopted on 3 April 2014[5]; and participating Member States have until 22 May 2017 to implement the directive.[6] Taking precedence over mutual legal assistance,[7] the EIO Directive goes part of the way to determining some of the issues discussed in the previous chapters that arise in relation to forum and form of requests for international cooperation in the gathering of evidence—many of which can lead to challenges in relation to the jurisdictional cross-admissibility of evidence in criminal trials. The United Kingdom will be taking part in the measure; though Denmark and Ireland are not.[8]

As a starting point, the EIO does not modify Member States' obligations to respect either fundamental rights or fundamental legal principles. Article 1(2) provides that the EIO shall be executed in accordance with the principle of mutual recognition; though under art. 4 the obligation to execute does not modify the executing state's obligation to respect fundamental rights. Likewise, art. 9 provides that the executing state shall recognise the requested measure and ensure its execution; unless this would be contrary to fundamental principles.

In relation to fundamental rights, in recitals 12 and 39, states are reminded of the obligation to respect fundamental rights, referring specifically to the Title VI Charter, Art 6 TEU, international law and international agreements. In particular, in recital 12, Member States should ensure full respect for defence rights in art. 48 Charter, while acknowledging that these rights may be limited in accordance with proportionality. Under art. 11, it is a clear ground for non-recognition or non-execution of an EIO that there are *substantial grounds* to believe that the investigative measure in question would be incompatible with art. 6 TEU and the Charter; though not if an alternative measure is possible in accordance with art. 10 (2) of the Directive.

Further, the EIO Directive has enacted a rebuttable presumption of mutual trust in recital 19; though implementation of the rebuttable presumption is not strictly obligatory. Under recital 19, a presumption of compliance by other Member States with EU law and fundamental rights (referring to 'mutual confidence') is rebuttable. Therefore, if the executing authority has *substantial grounds* for believing that execution would result in a breach of a fundamental right and to execute the EIO would be to "*disregard its obligations concerning the protection of fundamental rights*" then execution *should* be refused. This wording goes some way to ending a conclusive presumption of mutual trust in relation to gathering evidence in criminal

[5]Directive of the European Parliament and of the Council of the European Union regarding the European Investigation Order in criminal matters, 2014/41/EU, OJEU L 130/1, 1 May 2014 (hereinafter EIO Directive 2014/41/EU).

[6]Art. 36, EIO Directive 2014/41/EU.

[7]Recital 35, EIO Directive 2014/41/EU.

[8]Recitals 43–45, EIO Directive 2014/41/EU.

matters, but does not fully ensure that the rebuttable presumption is effective for the individuals who are the subject of the EIO.

In relation to fundamental legal principles, additionally under art. 11, it is a ground for non-recognition or non-execution of an EIO if the same measure would not be authorised under the domestic law of the executing state (unless an alternative measure is available under art. 10(2)). Thus, while under art. 9 the preference and significance of *forum regit actum* is established, the *locus regit actum* principle is maintained as a 'default' position. In fact, in relation to a covert investigation measure, the law of the executing state always applies. While the EIO Directive clearly makes the issuing state law a priority to the greatest extent possible, the effect of art. 9 is that the position according to mutual legal assistance is not greatly changed and cooperating states will still meet the same challenges when their procedural rules are at odds with each other. Many of the same differences encountered today in relation to the right to silence and the gathering of confession evidence will remain.

The EIO includes specific measures that may allow for the questioning of suspects and accused persons. Under recitals 25 and 26, as a potential effective alternative to an EAW, the EIO may/should be used for obtaining evidence from a suspect or accused by a temporary transfer to the issuing state or by conducting a video conference hearing. However, if the suspect or accused is to be prosecuted, including standing trial, the EAW should be used. Under recital 24, extra conditions may need to be stipulated and/or agreed on in relation to various measures, including temporary transfer, video-conference hearings and covert surveillance.

Article 22 provides for the temporary transfer of persons already in custody with their consent, if their presence is required for the gathering of certain evidence. Article 24 provides that an EIO may be used to allow for the 'hearing' of a suspect or accused person by video conference or audio-visual transmission by the 'competent authority' of the issuing state, with the consent of that person. Under the provisions of sub-paragraph 5, the hearing is to be conducted according to the issuing state's laws and the suspect or accused should be notified in advance of the procedural rights that shall accrue to him or her. However, a 'competent authority' of the executing state must be present to ensure the identity of the person being heard and to ensure that the hearing is conducted in accordance with fundamental legal principles in the executing state. Therefore, there appears to be a potential conflict of laws that has not been resolved by the Directive. Thus, states are being left to resolve potential or actual problems between themselves, much as they would under mutual legal assistance. Of significance, while it is certainly a step in the right direction that the individual being heard will be notified in advance of their procedural rights, this measure needs to be supported by additional measures securing effective legal representation for suspects or accused persons. The same may be said in relation to witnesses who might become or become suspects or accused persons in the course of questioning. In reality, states will have to consider whether all potential suspects are questioned with the rights of a suspect or accused person (and with effective legal representation), so that the resources that are invested in arranging these types of hearings are not lost in the event that a reasonable suspicion may be formed during the questioning.

Other articles worthy of mention include art. 4(b), which allows administrative authorities to use the EIO in relation to an infringement where criminal proceedings may result. The exact wording and scope of the provision in unclear. However, if this provision is to be read consistently with the jurisprudence of the ECtHR it means that where criminal proceedings are anticipated a violation of art. 6 ECHR will occur if compulsory administrative powers are used to obtain self-incriminating information.[9] Further, if criminal proceedings are not on foot and self-incriminating statements are obtained under administrative compulsory powers, the use of the statements as evidence in subsequent criminal proceedings against the maker will also amount to a violation of art. 6.[10] This demonstrates the difficulty of using administrative powers in relation to criminal proceedings.

In accordance with art. 6 of the Directive, an EIO may be issued in cases where the measure sought would also be available in like domestic cases. Member States, therefore, are prevented from forum shopping.

Based on the above, in conclusion, the EIO Directive settles certain debates and develops broad principles in relation to international cooperation in gathering evidence in the EU, though many of the same challenges that are present under mutual legal assistance still exist. The EIO needs to be supported by further measures such as developing the right of a suspect or accused person the subject of an EIO to access effective legal representation, for example, by establishing a Eurodefence office. Given the complexities that still exist, the need for such an office is highlighted.

8.4.3 The European Public Prosecutor's Office

Article 86 of the Treaty on the Functioning of the EU (TFEU) provides for the establishment of the European Public Prosecutor's Office (EPPO)—envisaged as a supranational body aimed at prosecuting financial crime affecting EU interests. Article 86 stipulates that the EPPO will be established out of Eurojust. However, before the EPPO becomes a reality there are still many questions that require answering. It appears as though a decentralised model is the preferable one, whereby each state has delegated prosecutors who act as national prosecutors in national cases, as well as prosecuting proceedings that are related to EPPO prosecutions. Although the EPPO will be a supranational body, appeals will presumably be heard in the national appellate courts.

---

[9]See [ECtHR] *Weh v Austria*, Application no. 38544/97, 8 April 2004 [43–45 and 50]; and see Chap. 1, Sect. 1.2.2 and Chap. 2, Sect. 2.3.2.4.

[10]See [ECtHR] *Weh v Austria*, Application no. 38544/97, 8 April 2004 [43–45 and 50]; and see Chap. 1, Sect. 1.2.2 and Chap. 2, Sect. 2.3.2.4.

Along with Chap. 6, this chapter has highlighted some areas of differences and potential difficulties between legal systems that will also have to be addressed in the course of creating the EPPO. In relation to the right to silence and the right against self-incrimination alone, real divergences exist that may be problematic in transnational EPPO cases and also from the point of view of consistency in and between EPPO prosecutions being conducted in different states.

As discussed at paragraph 2 above, administrative compulsory information-gathering powers and the legitimacy of using evidence in a criminal trial that was gathered from a suspect in an administrative investigation; the way in which the accused person's pre-trial silence may be used to establish guilt; the manner in which the evidence of the interview between the police and the accused may be raised at trial; and rules about the admissibility of improperly and unlawfully gathered evidence are all areas which separate an English from a Danish criminal trial. In addition, these differences have influenced the form of pre-trial due process that is considered appropriate and fair in the respective systems. Further, the accruing of pre-trial procedural rights in the two systems has been evidently influenced by the structure of pre-trial proceedings and differences in the understanding of when a suspect must be 'charged' and what that actually means (see Chap. 1 Appendix, Table 1.1). These differences exist despite the fact that the systems in both countries operate within the framework of the ECHR and by and large apply the fair trial rights under art. 6. The meaning of 'fairness' and the significance of the gravity of an accused's criminal offending to the fairness of relying on improperly or unfairly obtained evidence is also a matter that needs to be clarified to ensure that any harmonisation of procedural rules is consistent with the ECtHR's case law.

Article 30(1) of the proposal for an EPPO presented by the European Commission on 13 July 2013[11] suggests that admissibility of evidence shall be determined in every EPPO proceedings in reference to the accused person's fundamental fair trial rights, regardless of whether an 'admissibility' procedure exists in the national procedure or not. As discussed in the previous chapters of this book, by reference to the criminal procedure in Denmark and England and Wales, the traditional lack of a procedure questioning the admissibility of evidence (other than in relation to a lack of caution about the right to silence) in Denmark compared to the relative significance of tests of admissibility in English law, reflect the way that various limitations and protections of procedural rights in criminal proceedings demonstrate a continuity in balancing rights throughout the investigation and the trial. By requiring an admissibility proceeding to take place in all EPPO prosecutions, regardless of the legal tradition of the trial state, may be to impose an additional check in those systems that throws the criminal procedure out of balance. Nevertheless, a breach of fundamental rights may be something that would devalue evidence in the process of free evaluation in those states, so that the result is not so different. This potential weighting in favour of the accused would have less significance if all the EU

[11]COM (2013) 534 final.

member states involved in the EPPO were applying the same minimum standards for a fair trial.

8.4.4 Reinforcing Procedural Rights in International Cooperation Outside the EU

Among the lessons learned from the European experience of international cooperation is that a closer level of assistance in transnational criminal cases is optimal if it is established upon the trust-building foundation of common minimum standards of procedural fairness. In other words, effectiveness is best achieved by engaging with fairness.

It may be over-ambitious to model mutual legal assistance in criminal matters at the international level directly on the European design. However, in addition to the current focus on highly controversial issues such as the death penalty and anti-torture measures, as a matter of secondary prioritisation the HR Committee may consider launching a project that focuses on developing an international roadmap of procedural rights in pre-trial proceedings.[12] This would focus harmonisation efforts on key issues such as translation and interpretation; being informed of the accusations and of the right to silence; effective legal advice, including some form of free legal advice if necessary, for all suspects being questioned in relation to foreign offences; and a right for persons in detention to communicate with a relative, employer and consular advisors. The UN Human Rights Council (HRC) has a working group on arbitrary detention, which has published statements about the procedural rights mentioned here.[13] In addition, the UN HR Committee's Special Rapporteur on the promotion and protection of human rights and fundamental freedoms while countering terrorism also deals with aspects of a fair trial in a report presented to the UN General Assembly on 6 August 2008.[14] However, a more specific approach to particular key rights may assist the international cooperation process outside Europe.

[12]See Resolution of the Council of 30 November 2009 on a roadmap for strengthening procedural rights of suspected or accused persons in criminal proceedings, OJEU C 295/1, 4 December 2009.

[13]See in particular the Body of principles for the protection of all persons under any form of detention or imprisonment, adopted by the UN General Assembly on 9 December 1988.

[14]Scheinin (2008).

8.5 Concluding Remarks

A trial may be rendered unfair if confession evidence or evidence of the accused's pre-trial silence has been gathered in another system, due to differences in the way procedural rights such as the right to silence operate in cooperating national systems. Nevertheless, judicial practice in Denmark, England and Wales and Australia indicates that domestic courts will generally be flexible in relation to using evidence that has been obtained abroad in a manner that is considered irregular according to the national law. This is on the proviso that the evidence has not been gathered abroad by members of their own national authorities and that there was no unlawful or abusive coercion involved. The courts tend to be flexible regardless of whether the foreign state comes under the ICCPR or the ECHR; and regardless of the significance of the irregularity on the operation of the right to silence. However, Danish courts will not accept confession evidence obtained in absence of cautioning about the right.

Due to a greater level of trust and obligation to cooperate within the EU, the authorities in Denmark and England and Wales may be more willing to cooperate with authorities from other European/EU states; and certainly the obligations involved in mutual recognition (including the EIO) will encourage EU states' willingness to cooperate with each other. Therefore, it is arguable that investigators and prosecutors within the EU (and to some extent outside the EU) can achieve the greatest amount of effectiveness in transnational proceedings by simply taking a practical approach and using international cooperation, with the help of networks such as Eurojust, to obtain evidence gathered by foreign agencies. Due to the flexible approach of the national courts, often this may be successful, without paying attention to the detail of how fundamental rights, such as the right to silence, might be infringed. However, decisions that are more recent indicate that the national courts are becoming aware of the need to apply the minimum standards set by the ECtHR, especially judgments of the Grand Chamber. In addition, the minimum standards developed by the ECtHR are arguably applicable in the law of some states, such as in Denmark, as part of the incorporation of the ECHR.

Despite the tendency towards flexibility, it is nevertheless always preferable to base effectiveness in gathering evidence on a sound theoretical foundation. This can be achieved by ensuring that the pathways to international cooperation remain open because they operate on a *well-founded* presumption of mutual trust, or on the acknowledgement of the challenges that are involved in relation to fundamental rights. Thus, effectiveness is achieved through fairness. This will add to the legitimacy of the criminal justice systems involved.

References

Home Office (UK) (2014) Extradition: processes and review. https://www.gov.uk/guidance/extradition-processes-and-review. Accessed 15 Jan 2014

Sheinin M (2008) Report of the Special Rapporteur on the promotion and protection of human rights and fundamental freedoms while countering terrorism, A/63/223. Available via the Office of the High Commissioner for Human Rights. http://www.ohchr.org/EN/Issues/Terrorism/Pages/Annual.aspx. Accessed 15 Jan 2014

UK Parliament Human Rights Joint Committee (2011) Fifteenth report: the human rights implications of UK extradition policy, 22 June 2011. http://www.publications.parliament.uk/pa/jt201012/jtselect/jtrights/156/15602.htm. Accessed 15 Jan 2014

Table of Cases

Australia

Adam v The Queen [2001] HCA 57
Antoun v The Queen [2006] HCA 2
Azzopardi v The Queen [2001] HCA 25
Bunning v Cross [1978] HCA 22
Clarke v The Queen [2007] HCA 39
Coates v R [2005] HCA 1
Coco v The Queen [1994] HCA 15, (1994) 179 CLR 427
Collins v The Queen (1980) 31 ALR 257
Controlled Consultants Pty Ltd v Commissioner for Corporate Affairs [1985] HCA 6
Cornwall v The Queen [2007] HCA 12
Das v Victorian Human Rights and Equal Opportunity Commission [2009] VSC 381
Davis v The Queen [2001] HCA 25
Dawson v The Queen [1961] HCA 74
Dietrich v The Queen [1992] HCA 57
Director of Public Prosecutions (NSW) v Alderman (1998) 45 NSWLR 52627
Director of Public Prosecutions (NSW) v Attallah [2001] NSWCA 171
Director of the Fair Work Building Industry Inspectorate v Construction, Forestry, Mining and Energy Union [2015] FCA 47
DPP v Cook [2006] TASSC 75
DPP v Phillip Andrew Bayley (No. 3) [1996] SASC 5807
Driscoll v The Queen [1977] HCA 43
Dupas v The Queen [2012] VSCA 328
Edwards v The Queen [1993] HCA 6
Em v The Queen [2007] HCA 46
EPA v Caltex [1993] HCA 74

F.M.W. Billing, *The Right to Silence in Transnational Criminal Proceedings*,
DOI 10.1007/978-3-319-42034-9

Esso Australia Resources Ltd v Federal Commissioner of Taxation [1999] HCA 67
Foster v The Queen (1993) 67 ALJR 550
George v Rockett [1990] HCA 26
Hamberger (Employment Advocate) v Williamson and CFMEU [2000] FCA 1644
Hamilton v Oades [1989] HCA 21
Hammond v The Commonwealth of Australia (1982) 152 CLR 188
Heatherington v The Queen [1994] HCA 19
Hill v The Queen [2007] HCA 39
JB v Regina [2012] NSWCCA 12
Jones v R [2005] NSWCCA 443
Kelly v The Queen [2004] HCA 12
King v McLellan (1974) VR 773
King v The Queen (1986) 15 FCR 427
KMJ v. Tasmania [2011] TASCCA 7
Lee v NSW Crime Commission [2013] HCA 39
Lee v The Queen [1998] HCA 60
Lee v The Queen [2014] HCA 20
Mann v Carnell [1999] HCA 66
The Queen [2005] HCA 68
Marks v The Queen [2007] HCA 39
McDermott v The Queen [1948] HCA 23; (1948) 76 CLR 501
McKinney v The Queen [1991] HCA 6
Michael George Valentine v Technical and Further Education Commission & Anor [2007] NSWCA 208
Nicholls v The Queen [2005] HCA 1
Palmer v The Queen [1998] HCA 2
Papakosmas v The Queen [1999] HCA 37
Petty and Maiden v The Queen [1991] HCA 34
PGA v The Queen [2012] HCA 21
Pollard v The Queen [1992] HCA 69
Pyneboard Pty Ltd v Trade Practices Commission [1983] HCA 9
Rees v. Kratzmann (1965) 114 CLR 63
Ridgeway v The Queen [1995] HCA 66
R v Amad [1962] VR 545
R v Anderson [2002] NSWCCA 141
R v Barrett [2007] VSCA 95; (2007) 16 VR 240
R v Burt [2000] 1 Qd R 28
R v Carr [1972] 1 NSWLR 608
R v Clarke (1997) 97 A Crim R 414
R v Dalley [2002] NSWCCA 284
R v Esposito (1998) 105 A Crim R
R v Frangulis [2006] NSWCCA 363
R v GH [2000] FCA 1618; (2000) 105 FCR 419
R v Horton (1998) 45 NSWLR 426
R v Hughes [2000] HCA 22

R v Ireland (1970) 126 CLR 321
R v Kempley (1944) 18 ALJR 118
R v L [1998] NSWSC 225
R v Lee (1950) 82 CLR 133
R v Lowe [1997] 2 VR 465
R v M [2002] QCA 486; (2002) A Crim R 324
R v Mario Perfili [1995] VSC 222
R v Medcalfe [2002] ACTSC 83
R v Phan (2001) 123 A Crim R 30
R v Phung and Huynh [2001] NSWSC 115
R v Pimental [1995] NSWCCA 401
R v Quach (2002) 137 A Crim R 345; [2002] NSWCCA 519
R v Schaeffer [2005] VSCA 306
R v Sean Sonnet (Ruling No. 2) [2011] VSC 551
R v Sharp (2003) 143 A Crim R 344
R v Simmons; R v Moore (No. 3) [2015] NSWSC 189
R v Smith [1984] 1 NSWLR 462
R v Suckling [1999] NSWCCA 36
R v Swaffield; Pavic v R [1998] HCA 1; (1998) 192 CLR 15
R v Tang [2008] HCA 39
R v Taylor [1999] ACTSC 47
R v Thomas [2006] VSCA 165
R v Ul-Haque [2007] NSWSC 1251
R v Villar; R v Zugecic [2004] NSWCCA 302
R v Walker [2000] NSWCCA 130
R v XY [2013] NSWCCA 121
R v Yammine and Chami [2002] NSWCCA 289
RPS v The Queen (2000) 199 CLR 620
Rush v Commissioner of Police [2006] FCA 12
Sinclair v The King [1946] HCA 55; (1946) 73 CLR 316
Sonnet v The Queen [2010] VSCA 315
Sorby v Commenwealth [1983] HCA 10
The Queen v Joseph Terrence Thomas [2006] VSCA 165
The Queen v Stephen Albert Hutton [2011] VSC 484
The Queen v Willis (Ruling No. 1) [2015] VSC 261
Thorson v Pine (2004) 139 FCR 5
Tofilau v The Queen [2007] HCA 39
Van Der Meer v The Queen (1988) 35 A Crim R 232
Weininger v The Queen [2003] HCA 14
Weissensteiner v The Queen [1993] HCA 65
X7 v Australian Crime Commission [2013] HCA 29
Zhao v Commissioner of the Australian Federal Police [2014] VSCA 137

Denmark

Ugeskrift for Retsvæsen

U.1965.410H
U.1968.68H
U.1970.901V
U.1972.164V
U.1976.236.Ø
U.1979.912V
U.1990.866Ø
U.1991.255H
U.1991.881Ø
U.1992.300H
U.1993.512V
U.1995.426H
U.1995.752H
U.1995.856H
U.1996.6V
U.1996.699V
U.1996.1384V
U.1996.1434/2Ø
U.1996.1488V
U.1997.82H
U.1997.336V
U.1997.1422/1V
U.1999.1744Ø
U.2000.1309H
U.2001B.337
U.2001.661V
U.2001.945Ø
U.2002.15H
U.2002.2729V
U.2003.903V
U.2003.1029H
U.2003.1896V
U.2005.3007V
U.2005.3287V
U.2005.753H
U.2005.3108H
U.2006.1967Ø
U.2006.2895H
U.2009.2287Ø
U.2009.2640V

U.2010.1250Ø
U.2012.272H

Tidsskrift for Kriminalvidenskab

TfK 2003.435/1Ø
TfK 2004.760E
TfK 2008 209 V
TfK 2008.710Ø

Tidsskrift for Strafferet

TfS 1996, 313
TfS 2007, 199

Rulings

Glostrup Court, Udskrift af dombogen, 4 June 2012, Court no. 15-3547/2012

England and Wales

AF v Secretary of State for the Home Department [2009] UKHL 29
Andrew Symeou v Public Prosecutor's Office at the Court of Appeals, Patras, Greece [2009] EWHC 897 (Admin), 1 May 2009
Ambrose v Harris, Procurator Fiscal, Oban (Scotland) [2011] UKSC 43
A v Secretary of State [2005] UKHL 71; [2005] 3 W.L.R. 1249
Brown v Stott (Procurator Fiscal, Dunfermline) [2001] 2 All E.R. 97; [2003] 1 A.C. 681
Cadder v HM Advocate [2010] UKSC 43
DPP v Lawrence [2007] EWHC 2154 (Admin); [2008] 1 Cr. App. R. 10
DPP v Kavanagh [2005] EWHC 820
Fernandes (Charles) v Governor of Brixton Prison [2004] EWHC 2207 (Admin)
Governor of Pentonville Prison ex Chinoy [1992] 1 All ER 317
Hamilton and Others v Naviede [1995] 2 A.C. 75
Hutton v Australia [2009] EWHC 564
Ibrahim v R [1914] A.C. 599
Lam Chi-Ming v The Queen [1991] 2 A.C. 212

Loosely v R [2001] UKHL 53
Morgans v DPP [2001] 1 A.C. 315
Omar v Chief Constable of Bedfordshire [2002] EWHC 3060
Public Prosecution Service v McKee [2013] UKSC 32
Rangzieb Ahmed, Habib Ahmed v The Queen [2011] EWCA Crim 184
Re Attorney General's Reference (No. 3 of 1999), Re [2001] 2 W.L.R. 56
Roche Registration Limited, R (on the Application of) v Secretary of State for Health (Rev 2) [2014] EWHC 2256 (Admin) (9 July 2014)
R v Armstrong [2009] EWCA Crim 643
R v Aspinall [1999] 2 Cr. App. R. 115
R v Aujla [1998] 2 Cr App R 16
R v Bailey (1993) 897 Cr. App. R. 365
R v Baker [1895] 1 Q.B. 797
R v Becouarn [2993] EWCA Crim 1154
R v Benn [2005] EWCA Crim 2100
R v Betts; R v Hall [2001] EWCA Crim 224
R v Birchall [1999] Crim. L.R. 310
R v Bowden (Brian Thomas) [1999] 2 Crim. App. R. 176
R v Brophy [1982] A.C. 476
R v Bryce [1992] 4 All E.R. 567; (1992) 95 Cr. App. R. 320
R v Cadette [1995] Crim. L.R. 229
R v Christie [1914] AC 545
R v Compton [2002] EWCA Crim 2835
R v Cowan [1996] Q.B. 373
R v Christou and Wright [1992] 1 Q.B. 79
R v Dervish [2001] EWCA Crim 2789; [2002] 2 Cr. App. R. 6
R v Director of the Serious Fraud Office, Ex p. Smith [1993] A.C. 1
R v Dougall [2010] EWCA Crim 1048
R v Drake [1996] Crim L.R. 109
R v Dunn (1990) 91 Cr. App. R. 237
R v Essa [2009] EWCA Crim 43
R v Friend [1997] 2 All E.R. 1011
R v Goodway (1994) 98 Cr. App. R. 11
R v Governor of Belmarsh Prison Ex p. Francis [1995] 3 All E.R. 634
R v Hasan (Aytach) [2005] 2 A.C. 467
R v Horncastle [2009] UKSC 14
R v Howell [2003] EWCA Crim 1; [2005] 1 Cr. App. R. 1
R v J [2003] EWCA Crim 3309
R v Johnstone [2003] UKHL 28; [2003] 1 W.L.R. 1736
R v Keenan [1989] 3 All E.R. 599
R v Khan [1994] 4 All E.R. 426
R v Khan [1997] A.C. 558
R v. Khyam and others [2008] EWCA Crim 1612
R v Kirk [2000] 1 W.L.R. 567
R v Konscol [1993] Crim LR 950

R v Latif and Shahzad [1994] Crim. L.R. 750
R v Lesley [1996] 1 Cr. App. R. 39
R v Lucas [1981] Q.B. 720
R v Makanjuola [1995] 1 W.L.R. 1348 CA
R v Mason [1987] 3 All E.R. 481; (1988) 86 Cr. App. R. 349
R v P and Others [2000] UKHL 72
R v P (Telephone Intercepts: Admissibility of Evidence) [2002] 1 AC 146
R v Petkar [2004] 1 Cr. App. R. 270
R v Pointer [1997] Crim. L.R. 676
R v Quinn [1990] Crim LR 581
R v Samuel [1988] Q.B. 615
R v Sang [1980] A.C. 402
R v Sargent [2001] UKHL 54; [2003] 1 A.C. 347
R v Scott [1991] Crim. L.R. 56
R v Seelig [1992] 1 W.L.R. 148
R v S(F) [2008] EWCA Crim 2177
R v Smurthwaite [1994] 1 All E.R. 898
R v Stavrinos (2003) 140 A Crim R 594
R v Van Gelderen [2008] EWCA Crim 422
R v Walsh (1990) 91 Cr. App. R. 161
R v Warickshall (1783) 1 Leach 263; 168 ER 234
R v Webber [2004] UKHL 1

Human Rights Committee

Ng v Canada, decision of the Human Rights Committee, 5 November 1993
Paul Kelly v. Jamaica, Communication No. 253/1987, U.N. Doc. CCPR/C/41/D/253/1987 (1991)
Sánchez López v. Spain, Communication No. 777/1997, U.N. Doc. CCPR/C/67/D/777/1997 (25 November 1999)

Court of Justice of the European Union

Case C-41/74 *Van Duyn v Home Office*[1974] 4 December 1974
Case C-66/08 *Szymon Kozlowski* [2008] 17 July 2008
Case C-105/03 *Pupino* [2005] ECR I-5285, 16 June 2005
Case C-123/08 *Wolzenburg* [2009]
Case C-261/09 Mantello [2010]
Case C-306/09 *I.B.* Opinion of Advocate General Villalón, ECR I - 10341 [2010]
Case C-388/08 *PPU Leymann and Pustovarov* [2008]
Case C-396/11 *Radu* [2013] 29 January 2013

Case C-399/11 *Melloni* [2013] 26 February 2013
Case C-411/10 *N.S. v Secretary of State for the Home Department* [2011] 12 December 2011
Case C-493/10 *M.E. v Refugee Applications Commissioner, Minister for Justice, Equality and Law Reform* (Republic of Ireland) [2011] 21 December 2011.

European Court of Human Rights

Abas v The Netherlands, Application no. 27943/95, Decision as to Admissibility, 26 February 1997.
Adamkiewicz v. Poland, Application no. 54729/00, 2 March 2010.
Aleksandr Zaichenko v. Russia, Application no. 39660/02, 18 February 2010.
Allan v. The United Kingdom, Application No. 48539/99, 5 November 2002.
Artico v. Italy , Application no. 6694/74, 13 May 1980.
Averill v The United Kingdom, Application no. 36408/97, 6 June 2000.
Beckles v. The United Kingdom, Application no. 44652/98, 8 October 2002.
Brennan v. The United Kingdom, Application no. 39846/98, 16 October 2001.
Brusco v France, Application no. 1466/07, 14 October 2010.
Bykov v. Russia, Application no. 4378/02, 10 March 2009 (GC)
Condron v. The United Kingdom, Application no. 35718/97, 2 May 2000.
Cornelis v. The Netherlands, Application no. 994/03, Decision as to Admissibility, 25 May 2004.
Croissant v. Germany, Application no. 13611/88, 25 September 1992.
Dayanan v. Turkey, Application no. 7377/03, 13 October 2009.
Deweer v. Belgium, Application no. 6903/75, 27 February 1980.
Edwards v The United Kingdom, Application no. 13071/87, 16 December 1992.
Engel and Others v The Netherlands, Application no. 5100/71 and others, 8 June 1976.
Fox, Campbell and Hartley v The United Kingdom, Application no. 12244/86; 12245/86; 12383/86, 30 August 1990.
Funke v. France, Application no. 10828/84, 25 February 1993.
Goodwin v The United Kingdom, Application no. 17488/90, 11 July 2002 (GC)
Gäfgen v Germany, Application no. 22978/05, 1 June 2010 (GC)
H and J v The Netherlands, Application nos 978/09 and 992/09, 13 November 2014.
Heaney and McGuinness v Ireland Application no. 34720/97, 21 March 2001.
Heglas v. the Czech Republic, Application no. 5935/02, 1 March 2007.
Ibrahim and Others v the United Kingdom, Application no. 50541/08, 16 December 2014.
IJL, GMR and AKP v The United Kingdom, Application no. 29522/95, 30056/96 and 30574/96, 19 September 2000.
Imbrioscia v Switzerland Application no. 13972/88, 24 November 1993.
Jalloh v Germany Application no. 54810/00, 11 July 2006 (GC)
J.B. v Switzerland, Application no. 31827/96, 3 May 2001.

Khan v. The United Kingdom, Application no. 35394/97, 12 May 2000.
King v The United Kingdom, Application no. 13881/02, Decision as to Admissibility, 8 April 2003.
Klass and Others v. Germany, Application no. 5029/71, 6 September 1978.
Lazarenko v The Ukraine, Application no. 22313/04, 28 October 2010.
Magee v. The United Kingdom, Application no. 28135/95, 6 June 2000.
Marttinen v Finland, Application no. 19235/03, 21 April 2009.
Miailhe v. France (No. 2), Application no. 18978/91, 26 September 1996.
Mortensen v Denmark, Application no. 24967/94, Decision of the Commission, 15 May 1996.
M.S.S v Belgium and Greece, Application no. 30696/09, 21 January 2011.
Murray v. The United Kingdom Application no. 18731/91, 8 February 1996 (GC)
O'Halloran and Francis v The United Kingdom, Application nos. 15809/02 and 25624/02, 29 June 2007 (GC)
Öcalan v Turkey, Application no. 46221/99, 12 August 2005 (GC)
Panovits v Cyprus, Application no. 4268/04, 11 December 2008.
Perry v. The United Kingdom, Application no. 63737/00, Decision as to Admissibility, 26 September 2002.
PG and JA v The United Kingdom, Application no. 44787/98, 25 September 2001.
Pishchalnikov v. Russia Application no. 7025/04, 24 September 2009.
Poitrimol v France, Application no. 14032/88, 23 November 1993.
Salduz v Turkey, Application no. 36391/02, 27 November 2008.
Saunders v. The United Kingdom Application no. 19187/91, 17 December 1996 (GC)
Schenk v. Switzerland, Application no. 10862/84, 12 July 1988.
Sejdovic v. Italy, Application no. 56581/00, 1 March 2006 (GC)
Shabelnik v Ukraine, Application no. 16404/03, 19 February 2009.
Soering v The United Kingdom, Application no. 14038/88, 7 July 1989.
Stewart v. The United Kingdom, Application no. 10044/82, Decision of the Commission, 10 July 1984.
Stojkovic v France and Belgium, Application no. 25303/08, 27 October 2011.
Tony Michael Chalkley v. The United Kingdom, Application no. 63831/00, 12 June 2003.
Trymbach v The Ukraine, Application no. 44385/02, 12 January 2012.
Tsonyo Tvsonev v. Bulgaria (no. 2), Application no. 2376/03, 14 January 2010.
Tsonyo Tvsonev v Bulgaria (no. 3) Application no. 21124/04, 16 October 2013.
Unterpertinger v Austria, 9120/80, 24 November 1986.
Weh v. Austria, Application no. 38544/97, 8 April 2004.

Other

HM Advocate v Duncan McLean [2009] HCJAC 97, 2010 SLT 73 (Scotland)
HM Advocate v McKay, 1961 SLT 174 (Scotland)
Holland v HM Advocate, 2005 1 SC (PC) 3(Scotland)
Murphy v Waterfront Commission 378 US 52 (1964) (US)
Nix v. Williams, 11 June 1984, 467 US 431 (1984) (US)
R. A. Russell v Patrick Anthony Adams [2000] NICA 21 (27 November 2000) (Northern Ireland)
R. v. Broyles [1991] 3 Supreme Court Reports 595 (Canada)
R. v. Herbert [1990] 2 Supreme Court Reports 151 (Canada)
R. v. Liew [1999] 3 Supreme Court Reports 227 (Canada)

Table of Statutes and National Legislation

Australia

Australian Uniform Evidence Law

Evidence Act 1995 (Commonwealth)
Evidence Act 1995 (New South Wales)
Evidence Act 2001 (Tasmania)
Evidence Act 2004 (Norfolk Island)
Evidence Act 2008 (Victoria)
Evidence Act 2011 (Australian Capital Territory)
Evidence (National Uniform Legislation) Act 2011 (Northern Territory)

Commonwealth

Building and Construction Industry Improvement Act 2005.
Crimes Act 1914.
Criminal Code Act 1995.
Fair Work Act 2009.
Human Rights (Parliamentary Scrutiny) Act 2011.
Passports Act 1938.

Victoria

Charter of Human Rights and Responsibilities Act 2006.
Crimes Act 1958.

F.M.W. Billing, *The Right to Silence in Transnational Criminal Proceedings*
DOI 10.1007/978-3-319-42034-9

Criminal Procedure Act 2009.
Major Crime (Investigative Powers) Act 2004.

Australian Capital Territory

Human Rights Act 2004.

New South Wales

Crimes Act 1900.
Criminal Procedure Act 1986.
New South Wales Police Commissioner's Code of Practice for Custody, Rights, Investigation, Management and Evidence

Denmark

Administration of Justice Act
Coercive Measures Act
Ministry of Justice, *Bekendtgørelse om vejledning af sigtede om adgangen til at begære en forsvarer beskikket*, BEK nr. 467 af 26/09/1978
Ministry of Justice, *Bekendtgørelse af lov om adgang til forsvarerbistand under en administrative skatte- eller afgiftsstraffesag (forsvarerbistandsloven)*, LBK nr. 827 af 25/08/2005
Ministry of Justice, *Cirkulæreskrivelse* nr. 9155, 18 March 2010.
Tax Control Act

England and Wales

Banking Act 1987.
Bankrupt Law Consolidation Act 1849.
Companies Act 1985.
Criminal Attempts Act 1981.
Criminal Evidence Act 1898.
Criminal Justice Act 1987.
Criminal Justice Act 2003.
Criminal Justice and Public Order Act 1994.
Criminal Law Act 1977.

Criminal Procedure and Investigations Act 1996.
European Communities Act 1972.
Excise Management Act 1979.
Extradition Act 2003.
Financial Services Act 1986.
Human Rights Act 1998.
Insolvency Act 1986.
Intelligence Services Act 1994.
Interception of Communications Act 1985.
Interpretation Act 1978.
Magistrates' Court Act 1980.
Magna Carta 1215.
Part III of the Police Act 1997.
Perjury Act 1911.
Police and Criminal Evidence Act 1984 (and Codes of Practice)
Proceeds of Crime Act 2002.
Regulation of Investigatory Powers Act 2000.
Road Traffic Act 1988.
Serious Organised Crime and Police Act 2005.
Terrorism Act 2000.
Theft Act 1968.
Value Added Tax Act 1994.
Youth Justice and Criminal Evidence Act 1999.
Judges' Rules and Administrative Directions to the Police, Home Office Circular 89/1978

Table of International Treaties and Legislation

United Nations

Convention Against Corruption, opened for signature 9 December 2003.
Convention Against Illicit Traffic in Narcotic Drugs and Psychotropic Substances (1988)
Convention Against Torture and Other Cruel, Inhuman or Degrading Treatment or Punishment (1984)
Convention on the Rights of the Child (1989)
International Covenant on Civil and Political Rights and the First Optional Protocol (1966)
International Covenant on Economic, Social and Cultural Rights (1966)
Rome Statute of the International Criminal Court (1998)
UN Model Treaty on Extradition (1990)
Universal Declaration of Human Rights (1948)
Vienna Convention on the Law of Treaties (1969)

Council of Europe

Convention for the Protection of Human Rights and Fundamental Freedoms 1950.
Council of Europe Prison Rules 2006.

European Union

Charter of Fundamental Rights of the European Union 2007, OJEU C 326/391, 26 October 2012.

F.M.W. Billing, *The Right to Silence in Transnational Criminal Proceedings*,
DOI 10.1007/978-3-319-42034-9

Directive 2010/64/EU of the European Parliament and of the Council of 20 October 2010 on the right to interpretation and translation in criminal proceedings, OJEU L 280/1, 26 October 2010.

Directive 2012/13/ EU of the European Parliament and of the Council of 22 May 2012 on the right to information in criminal proceedings, OJEU L 142/1, 1 June 2012.

Directive 2013/48/EU of the European Parliament and of the Council of 22 October 2013 on the right of access to a lawyer in criminal proceedings and in European arrest warrant proceedings, and on the right to have a third party informed upon deprivation of liberty and to communicate with third persons and with consular authorities while deprived of liberty, OJEU L 294/1, 6 November 2013.

Directive 2014/41/EU of the European Parliament and of the Council of 3 April 2014 regarding the European Investigation Order in criminal matters, OJEU L 130/1, 1 May 2014.

Treaty on European Union (1992) in the Lisbon Treaty OJEU C 306, 17 December 2007.

Treaty on the Functioning of the EU (1997) in the Lisbon Treaty OJEU C 306, 17 December 2007.

Other Treaties

African Charter on Human and People's Rights (1981)

American Convention of Human Rights (1969)

Arab Charter on Human Rights 1994 (revised 2004).

Commonwealth of Independent States Convention on Human Rights and Fundamental Freedoms (1995)

Scheme Relating to Mutual Assistance in Criminal Matters Within The Commonwealth (1986) including amendments made by Law Ministers in April 1990, November 2002 and October 2005.

Other Official Documents

Council of Europe

Committee of Ministers, Resolution (73) 5, Standard Minimum Rules for the Treatment of Prisoners, adopted 19 January 1973.

European Union

Commission of the European Communities, Commission Staff Working Document, Annex to the report from the Commission on the Implementation since 2005 of the Council Framework Decision of 13 June 2002 on the European arrest warrant and the surrender procedures between Member States [COM (2007) 407], SEC (2007) 979, 11 July 2007, 5

Court of Justice of the European Union, Opinion 2/13 of the Court (Full Court) on Accession to the European Convention for the Protection of Human Rights and Fundamental Freedoms, 18 December 2014, ECLI:EU:C:2014:2454

European Commission, Report from the Commission to the European Parliament and the Council on the implementation since 2007 of the Council Framework Decision of 13 June 2002 on the European arrest warrant and the surrender procedures between Member States [SEC (2011) 430], COM [2011] 175, 11 April 2011, 5

Tampere European Council Presidency Conclusions, 15 and 16 December 1999.

United Nations

Basic Principles on the Role of Lawyers agreed at the 8th UN Congress on the Prevention of Crime and the Treatment of Offenders, Havana, Cuba, A/CONF.144/28/Rev.1, 118 (1990)

Committee Against Torture, *General Comment No. 2*, CAT/C/GC/2, 24 January 2008 (United Nations, Geneva, 2008).

Human Rights Committee, *General Comment No. 29: States of Emergency (Article 4)*, CCPR/C/21/Rev.1/Add.11, 31 August 2001 (United Nations, Geneva, 2001).

Human Rights Committee, *General Comment No. 32, Article 14: Right to equality before courts and tribunals and to a fair trial*, CCPR/C/GC/32, 23 August 2007 (United Nations, Geneva, 2007).

Security Council Resolution 1373 (2001), S/RES/1373 (28 September 2001).

Index

F.M.W. Billing, *The Right to Silence in Transnational Criminal Proceedings*,
DOI 10.1007/978-3-319-42034-9

The manufacturer's authorised representative in the EU is Springer Nature Customer Service Centre GmbH, Europaplatz 3, 69115 Heidelberg, Germany. If you have any concerns regarding our products, please contact ProductSafety@springernature.com

Printed and bound by CPI Group (UK) Ltd, Croydon, CR0 4YY
30/09/2026
02225653-0020